# Dynamic Programming and Stochastic Control

This is Volume 125 in
MATHEMATICS IN SCIENCE AND ENGINEERING
A Series of Monographs and Textbooks
Edited by RICHARD BELLMAN, *University of Southern California*

The complete listing of books in this series is available from the Publisher upon request.

# Dynamic Programming and Stochastic Control

*DIMITRI P. BERTSEKAS*

MASSACHUSETTS INSTITUTE OF TECHNOLOGY
CAMBRIDGE, MASSACHUSETTS

ACADEMIC PRESS
*A Subsidiary of Harcourt Brace Jovanovich, Publishers*
New York London
Paris San Diego San Francisco São Paulo Sydney Tokyo Toronto 1976

ACADEMIC PRESS, INC.
111 Fifth Avenue, New York, New York 10003

*United Kingdom Edition published by*
ACADEMIC PRESS, INC. (LONDON) LTD.
24/28 Oval Road, London NW1

**Library of Congress Cataloging in Publication Data**

Bertsekas, Dimitri P
Dynamic programming and stochastic control.

(Mathematics in science and engineering)
Bibliography: p.
1. Dynamic programming. 2. Stochastic processes.
I. Title. II. Series.
T57.83.B48 519.7'03 76-16143
ISBN 0-12-093250-4

PRINTED IN THE UNITED STATES OF AMERICA

83 9 8 7 6 5 4

*To my mother*
*and the memory of my father*

# Contents

## *Chapter 4* Problems with Imperfect State Information



## *Chapter 5* Computational Aspects of Dynamic Programming —Suboptimal Control



# Part II CONTROL OF UNCERTAIN SYSTEMS OVER AN INFINITE HORIZON

## *Chapter 6* Minimization of Total Expected Value—Discounted Cost



## *Chapter 7* Minimization of Total Expected Value—Undiscounted Cost

# Preface

This text evolved from an introductory course on optimization under uncertainty that I taught at Stanford University in the spring of 1973 and at the University of Illinois in the fall of 1974. It is aimed at graduate students and practicing analysts in engineering, operations research, economics, statistics, and business administration. As a textbook it could be used, for example, in a one-semester first-year graduate course, which could cover primarily the first five chapters, the first half of Chapter 6, and parts of Chapter 8. It could also be used in a two-quarter graduate course, which would probably cover the whole text. Depending on the students' backgrounds and interests, some material could be omitted or added by the instructor.

The basic objective of the book is to provide a unified framework for sequential decision making under uncertainty and to stress the few fundamental concepts underlying the treatment of uncertainty and the technique of dynamic programming. These concepts (risk, feedback, sufficient statistics, adaptivity, contraction mappings, and the principle of optimality) are emphasized and developed in a framework that is devoid, to the extent possible, of structural assumptions on the problem considered. This is accomplished by considering general dynamic systems defined over arbitrary state and control spaces. Thus our formulation allows the simultaneous treatment of several important classes of problems, such as stochastic control problems (popular in modern control theory) as well as problems of control of finite state Markov chains (popular in operations research and statistics). However, rigor is claimed only in the case where the underlying probability space is a finite or countable set, while other cases are treated in a formal manner.

While most of the theoretical developments are carried out in a general framework, a large portion of the text is devoted to applications from specific problem areas. This serves the dual purpose of illustrating the theoretical developments and of presenting material that is important in its own right. My objective has not been to develop any particular application area in great depth but rather to emphasize those aspects which are strongly related with the dynamic programming technique and associated concepts.

The mathematical prerequisite for the text is a good knowledge of introductory probability theory and undergraduate mathematics. This includes the equivalent of a one-semester first course in probability theory together with the usual calculus, real analysis, vector–matrix algebra, and elementary optimization theory most undergraduate students are exposed to by their fourth year of studies. A summary of this material together with appropriate references is provided in the appendixes. An effort has been made to keep the mathematics at the lowest possible level consistent with rigor. However, it is inevitable that some mathematical maturity is required on the part of the reader so that he is able to think in relatively abstract terms. In addition, the last part of the text, which deals with infinite-horizon problems, is mathematically more sophisticated (particularly after Section 6.3) than the first part and requires a firm grasp of some of the basic convergence concepts of analysis. Readers with a somewhat weak analysis background may find the developments of Sections 6.4–6.7 and Chapter 7 difficult to follow. Familiarity with the notions associated with finite state Markov chain theory is not required, except in Chapter 8 and the last section of Chapter 7. Even there, however, Markov chain theory is used in a peripheral manner, and the reader should be able to follow the development of the material after a reading of Appendix D perhaps supplemented by an hour or two of instruction. While prior courses or background on dynamic system theory, optimization, or control will undoubtedly be helpful to the reader, it is felt that the material in the text is reasonably self-contained.

The nature of the subject of this book makes the choice of level of presentation rather difficult. The reason is that dynamic programming is a very simple and general technique, which nonetheless requires the extremely complicated machinery of measure-theoretic probability theory if it is to be presented in a mathematically rigorous way and within a general setting (i.e., in general spaces and in the presence of uncertainty). My choice has been to adopt a somewhat freewheeling style of mathematical presentation in the first five chapters, which deal with finite-horizon problems, and to raise substantially the level of mathematical precision in the last three chapters, which deal with infinite-horizon problems. The developments in Chapters 1–5 are carried out in a very general setting. However, the mathematical framework is not entirely rigorous, as explained in Section 2.1. In Chapters 6–8 the class of problems under consideration is restricted. Limitations are placed on the probability space in the interest of keeping the mathematics simple. This approach, which may seem somewhat unorthodox, was dictated by two basic considerations. First, since the validity of many of the concepts to be presented (principle of optimality, sufficient statistics, feedback, adaptivity, etc.) depends very little on structure, I felt that the reader should be given the opportunity to appreciate the power and generality of these concepts without being impeded by irrelevant structural assumptions or complicated mathematical details. Second, I felt that mathematical precision is essential for the treatment of infinite-horizon problems, much more so than for finite-horizon problems. In addition, I wanted to provide a systematic and up-to-date treatment of infinite-horizon problems in the hope that this treatment will be of some value to the research community as well as to practicing analysts.

The text starts with an introductory chapter on formulation of decision problems under uncertainty. The aim here is to provide a broad framework within which sequential decision problems under uncertainty can be appropriately placed. The material in this chapter is of fundamental importance. However, it is not used in an essential manner in the remainder of the text, and the reader may proceed directly to Chapter 2 if he so wishes.

Part I deals with finite-horizon problems. A single-model problem of broad applicability is employed throughout this part. The model is based on a state variable representation of a dynamic system that is standard in modern control theory. Transition probability models can be embedded within the framework of the model utilized by means of a simple reformulation. The dynamic programming algorithm is developed and illustrated in several applications of independent interest. Linear quadratic stochastic control problems and inventory control problems are treated in considerable detail. Several additional topics from operations research, economics, and statistics are also considered. Problems with imperfect state information are treated as special cases of the basic model problem. Since implementation of optimal and suboptimal controllers for such problems often requires the use of estimators, I have added a sizable appendix covering least squares estimation and Kalman filtering. The emphasis throughout the text is on optimization of dynamic stochastic systems. However, in view of the limitations of the dynamic programming technique, the subject of suboptimal control is of undeniable importance. For this reason I have included material on suboptimal control with emphasis placed on those techniques that are of broad applicability.

While Part I emphasizes conceptual aspects of sequential decision problems, Part II concentrates on the mathematical aspects of infinite-horizon problems. The coverage is quite thorough and includes discounted, undiscounted, and average cost problems. The basic model problems are set up in such a way that the underlying probability spaces are countable, thus eliminating all mathematical difficulties of a measure-theoretic nature.

However, the class of dynamic systems considered is much broader than the class of Markov chains with countable state space. For example, it includes all deterministic systems defined over arbitrary state and control spaces. This part also contains several new results developed either in the text or in the problem sections.

The problems at the end of each chapter are of four basic varieties: drill problems, examples or counterexamples, problems illustrating additional results of specific application nature, and theoretical problems that extend and supplement the developments of the text. Some of the problems in the last category (particularly in Chapters 6 and 7) are quite difficult to solve, and hints as well as references have been supplied where appropriate. The serious reader will benefit a great deal by going over the theoretical problems, which constitute a significant component of the text.

# Acknowledgments

I feel indebted to several individuals and institutions who helped make the book possible. I am thankful to my teachers at the Massachusetts Institute of Technology, where I was introduced to the subject of stochastic control, particularly Mike Athans, Sanjoy Mitter, and Ian Rhodes. My perspective of the subject was broadened considerably during the time I taught at Stanford University; the stimulating atmosphere of the Engineering–Economic Systems Department and my interaction with David Luenberger in particular played an important role in shaping the framework of the book. The Coordinated Science Laboratory at the University of Illinois provided research support and an appropriate environment for completing the project. As with every book that develops through classroom instruction, the comments and suggestions of my students led to several clarifications and improvements. My greatest debt of gratitude to a single individual goes to Steve Shreve, who read with exceptional care the entire preliminary version of the manuscript and made a plethora of penetrating comments and suggestions. As a result of my interaction with Steve, the quality of the manuscript improved substantially. Finally, I wish to give my thanks to my family for their support during the course of the work that led to this book.

*Chapter 1*

# Introduction

This chapter sets the stage for the remainder of this text. Rather than dealing with methods for analysis or solution of decision problems under uncertainty, it examines various approaches for formulating such problems. Since the subject is of fundamental importance and at the same time far from trivial, it is worth examining even in the context of an introductory text. On the other hand, the material in this chapter is not used in a direct way later and it is not essential for the reader to have a firm grasp of it in order to proceed to subsequent chapters.

Optimization problems under uncertainty possess several important characteristics that are not present in the absence of uncertainty, i.e., in deterministic optimization problems. The two most important such characteristics are the need to take into account *risk* in the problem formulation and the possibility of *information gathering* (feedback) during the decision process.

To illustrate the first characteristic consider a problem of dividing an amount of capital $x$ between two different investment opportunities $A$ and $B$. Assume that $A$ offers with certainty \$1.499 per dollar invested. If the total profit is to be maximized and $B$ offers with certainty \$1.5 per dollar invested, then the capital $x$ should be invested in its totality in $B$. Assume now instead that $B$ offers \$1.5 per dollar *on the average* but not with certainty. For example, assume that $B$ offers \$0 per dollar with probability 0.8 and \$7.5 per dollar with probability 0.2. Then if the average (expected) profit is to be maximized, again the whole capital should be invested in $B$. In fact, the same is true for any probability distribution on the return of $B$ with expected value \$1.5 per dollar invested. However, most investors would object to such an allocation. Not wanting to take any risk of losing part of their capital, some would invest exclusively in the sure opportunity $A$, while others would invest at least a positive fraction of $x$ in $A$. This indicates that a problem formulation whereby expected profit is maximized may be inappropriate since in this formulation the risk associated with each allocation is not reflected at all in the cost functional and hence does not influence the optimal decision. At the same time, the question is posed as to what is an appropriate formulation of the problem.

As a more dramatic example of the need to take into account risk in the problem formulation consider the following situation (the so-called St. Petersburg paradox). An individual is offered the opportunity of paying $x$ dollars in exchange for participation in the following game. A fair coin is flipped sequentially and the individual is paid $2^k$ dollars, where $k$ is the number of times heads have come up before tails come up for the first time. The decision that the individual must make is whether to accept or reject participation in the game. Now if he accepts, his expected (average) profit from the game is

$$\sum_{k=0}^{\infty} \frac{1}{2^{k+1}} \cdot 2^k - x = \infty,$$

so that if his acceptance criterion is based on maximization of his expected profit, he is willing to pay any amount $x$ to enter the game. This, however, is in strong disagreement with the behavior of individuals due to the risk element involved in entering the game and shows again that a different formulation of the problem is necessary. Some of the aspects of formulating decision problems under uncertainty so that risk is properly taken into account are dealt with in the next two sections.

A second important feature of many decision problems under uncertainty is the possibility of carrying out the decisions in stages while gathering information between stages about some of the uncertain parameters involved in the problem. This information may be used with advantage when making future decisions. The fundamental role of information gathering in problems of decision under uncertainty will become evident as we progress through this text. For the moment let us consider the following simple example:

EXAMPLE A two-stage scalar dynamic system evolves according to the equations

$$x_1 = u_0 + w_0, \qquad x_2 = x_1 + u_1.$$

The scalar $w_0$ is an uncertain parameter taking values 1 or $-1$ with probability $\frac{1}{2}$. The problem is to choose $u_0$ and $u_1$ so as to minimize the expected absolute value of $x_2$:

$$E\{|x_2|\} = E\{|u_0 + u_1 + w_0|\} = 0.5\{|u_0 + u_1 + 1| + |u_0 + u_1 - 1|\}.$$

It is easy to see that the optimal value is

$$\min_{u_0, u_1} E\{|x_2|\} = 1$$

and the optimal values of $u_0$, $u_1$ are those for which $-1 \leqslant u_0 + u_1 \leqslant 1$. Consider now the situation where $u_0$ and $u_1$ are chosen sequentially in a way that $x_1$ *is known to the decision maker when selecting* $u_1$. Then clearly the optimal selection policy for $u_0$ and $u_1$ is to take $u_0$ to be any scalar and to take $u_1$ equal to $-x_1$. With this selection policy the optimal cost is reduced to zero. The reduction was made possible by means of proper use of the information received (i.e., the value of $x_1$). Note that the optimal value of $u_1$ depends on the value of $x_1$, i.e., it is a function of the information received. Note also that if there is no uncertainty, for example, if $w_0 = 0$ with probability one, then the optimal cost is zero whether the value of $x_1$ becomes known prior to selecting $u_1$ or not. This is a manifestation of the intuitively obvious fact that information gathering can be of no help when there are no uncertain parameters in the problem.

## 1.1 The Problem of Decision under Uncertainty†

A decision problem in one of its simplest and most abstract forms consists of three nonempty sets $\mathscr{D}$, $\mathscr{N}$, and $\mathscr{O}$, a function $f\colon \mathscr{D} \times \mathscr{N} \to \mathscr{O}$, and a complete and transitive relation $\preccurlyeq$ on $\mathscr{O}$:

- $\mathscr{D}$ the set of possible decisions
- $\mathscr{N}$ indexes the uncertainty in the problem and may be called the set of "states of nature"
- $\mathscr{O}$ the set of outcomes of the decision problem
- $f$ the function that determines which outcome will result from a given decision and state of nature, i.e., if decision $d \in \mathscr{D}$ is selected and state of nature $n \in \mathscr{N}$ prevails, then the outcome $f(d, n) \in \mathscr{O}$ occurs

† As mentioned earlier, the concepts in this and subsequent sections in this chapter are not essential for the understanding of the remainder of the text. The reader may proceed directly to Chapter 2 if he so wishes.

$\preccurlyeq$ a relation determining our preference among the outcomes.† Thus for $O_1, O_2 \in \mathcal{O}$, by $O_1 \preccurlyeq O_2$ we mean that outcome $O_2$ is at least as preferable as outcome $O_1$. By completeness of the relation we mean that every two elements of $\mathcal{O}$ are related in the sense that given any $O_1, O_2 \in \mathcal{O}$ either $O_1 \preccurlyeq O_2$, but not $O_2 \preccurlyeq O_1$, or $O_2 \preccurlyeq O_1$, but not $O_1 \preccurlyeq O_2$, or both $O_2 \preccurlyeq O_1$ and $O_1 \preccurlyeq O_2$. By transitivity we mean that $O_1 \preccurlyeq O_2$ and $O_2 \preccurlyeq O_3$ implies $O_1 \preccurlyeq O_3$ for any three elements $O_1, O_2, O_3 \in \mathcal{O}$.

EXAMPLE Consider the following situation. An individual may bet \$1 on the toss of a coin or not bet at all. If he bets and guesses correctly, he wins \$1 and if he does not guess correctly, he loses \$1. Here $\mathcal{D}$ consists of three elements {bet on heads, bet on tails, not bet}, $\mathcal{N}$ consists of two elements {heads, tails}, and $\mathcal{O}$ consists of three elements, the three possible final fortunes of the player {\$0, \$1, \$2}. The preference relation on $\mathcal{O}$ is the natural one, i.e., $0 \preccurlyeq 1$, $0 \preccurlyeq 2$, $1 \preccurlyeq 2$, and the values of the function $f$ are given in Table 1.1 for each value of $d$ and $n$.

TABLE 1.1

| | | $\mathcal{D}$ | | |
|---|---|---|---|---|
| | | $H$ | $T$ | Not bet |
| $\mathcal{N}$ | $H$ | \$2 | \$0 | \$1 |
| | $T$ | \$0 | \$2 | \$1 |

Now the relative order by which we rank outcomes is usually clear in any given situation. On the other hand, in order for the decision problem to be completely formulated we need a *ranking among decisions* that is consistent in a well-defined sense with our ranking of outcomes. Furthermore, in order to be able to apply mathematical methods for the analysis of the decision problem we would like to have this ranking determined by a numerical function $F : \mathcal{D} \to R$ ($R$ is the set of real numbers) such that

$$d_1 \preccurlyeq d_2 \Leftrightarrow F(d_1) \leqslant F(d_2) \qquad \forall d_1, d_2 \in \mathcal{D}, \tag{1}$$

where the notation $d_1 \preccurlyeq d_2$ implies that the decision $d_2$ is at least as preferable as the decision $d_1$. It is by no means clear what this ranking among decisions

† The symbol $\preccurlyeq$ in this chapter will be (somewhat loosely) used to denote a preference relation within either the set of outcomes or the set of decisions. The precise meaning should be clear from the context and hopefully the use of the same symbol to denote different preference relations will create no confusion to the reader.

should be or how one should go about determining and characterizing such a ranking. For example, in the gambling problem above different people will have different preferences as to accepting or refusing the gamble. There are a number of approaches and viewpoints for determing a ranking among decisions and this section deals with some of these.

*Payoff Functions, Dominant and Noninferior Decisions*

Let us consider the case where it is possible to assign to each element of $\mathcal{O}$ a real number in such a way that the order between elements of $\mathcal{O}$ agrees with the usual order of the corresponding numbers. That is, there exists a real-valued function $G: \mathcal{O} \to R$ with the property

$$G(O_1) \leqslant G(O_2) \Leftrightarrows O_1 \preccurlyeq O_2 \qquad \forall O_1, O_2 \in \mathcal{O}. \tag{2}$$

Such a function does not always exist (see Problem 2). However, its existence can be guaranteed under quite general assumptions. In particular, one may show that it exists if $\mathcal{O}$ is a countable set. Also if such a function $G$ exists, it is far from unique, since if $\Phi$ is any monotonically increasing function $\Phi: R \to R$, the composite function $\Phi \cdot G$ (defined by $(\Phi \cdot G)(O) = \Phi[G(O)]$) has the same property (2) as $G$. For instance, in the example given earlier a function $G: \{0, 1, 2\} \to R$ satisfies (2) if and only if $G(0) < G(1) < G(2)$ and there is an infinity of such functions.

Now for any choice of $G$ as in (2) we define the function $J: \mathcal{D} \times \mathcal{N} \to R$ by means of

$$J(d, n) = G[f(d, n)]$$

and call it a *payoff function*.

Given a payoff function $J$ it is possible to obtain a complete ranking of decisions by means of a numerical function in the special *case of certainty* (the case where the set $\mathcal{N}$ of states of nature consists of a single element $\bar{n}$). By defining

$$F(d) = J(d, \bar{n})$$

we have

$$d_1 \preccurlyeq d_2 \Leftrightarrows F(d_1) \leqslant F(d_2) \Leftrightarrows f(d_1, \bar{n}) \preccurlyeq f(d_2, \bar{n})$$

and the numerical function $F$ defines a complete ranking of decisions.

When $\mathcal{N}$ contains more than one element, the order on $\mathcal{O}$ induces only a *partial order* on $\mathcal{D}$ by means of the relations

$$\begin{aligned} d_1 \preccurlyeq d_2 &\Leftrightarrows J(d_1, n) \leqslant J(d_2, n) \qquad \forall n \in \mathcal{N}, \\ &\Leftrightarrows f(d_1, n) \preccurlyeq f(d_2, n) \qquad \forall n \in \mathcal{N}. \end{aligned} \tag{3}$$

In this partial order it is not necessary that every two elements of $\mathscr{D}$ be related, i.e., for some $d, d' \in \mathscr{D}$ we may have neither $d \preccurlyeq d'$ nor $d' \preccurlyeq d$. If, however, for two decisions $d_1, d_2 \in \mathscr{D}$ we have $d_1 \preccurlyeq d_2$ in the sense of (3), then we can conclude that $d_2$ is at least as preferable as $d_1$ since the resulting outcome $f(d_2, n)$ is at least as preferable as $f(d_1, n)$ *regardless of the state of nature n that will occur*.

A decision $d^* \in \mathscr{D}$ is called a *dominant decision* if

$$d \preccurlyeq d^* \qquad \forall d \in \mathscr{D},$$

where $\preccurlyeq$ is understood in the sense of the partial order defined by (3). Naturally such a decision need not exist. If, however, such a decision does exist, then it may be viewed as optimal. In most problems of interest to an analyst, however, there exists no dominant decision. For instance, in the gambling example considered earlier a dominant decision does not exist. In fact, no two decisions are related in the sense of (3) for this example.

In the absence of a dominant decision one can consider the set $\mathscr{D}_m \subset \mathscr{D}$ of all *noninferior decisions*, where $d_m \in \mathscr{D}_m$ if for every $d \in \mathscr{D}$ the relation $d_m \preccurlyeq d$ implies $d \preccurlyeq d_m$ in the sense of the partial order defined by (3). In terms of a payoff function $J$, noninferior decisions may be characterized by

$$\begin{aligned} d_m \in \mathscr{D}_m \Leftrightarrow \text{there does not exist any } d \in \mathscr{D} \text{ such that}& \\ J(d_m, n) \leqslant J(d, n) \quad \forall n \in \mathscr{N} \quad \text{and}& \\ J(d_m, n) < J(d, n) \quad \text{for some} \quad n \in \mathscr{N}.& \end{aligned}$$

Clearly it makes sense to consider only the decisions in $\mathscr{D}_m$ as candidates for optimality since any decision that is not in $\mathscr{D}_m$ is dominated by one that belongs to $\mathscr{D}_m$. Furthermore, it may be proved that the set $\mathscr{D}_m$ is nonempty when the set $\mathscr{D}$ is a finite set, so that at least for this case there exists at least one noninferior decision. However, in practice the set $\mathscr{D}_m$ of noninferior decisions often is either difficult to determine or contains too many elements. For instance, in the gambling example given earlier, the reader may easily verify that every decision is noninferior.

Whenever the partial order (3) fails to produce a satisfactory ranking among decisions, one must turn to other approaches to formulate the decision problem. Approaches that we shall examine assume a notion of a *generalized outcome* of a decision and introduce a complete order on the set of these generalized outcomes consistent with the original order on the set of outcomes $\mathscr{O}$. The complete order on the set of generalized outcomes in turn induces a complete order on the set of decisions.

### *The Min–Max Approach*

The min–max (or max–min) approach takes the point of view that the generalized outcome of a decision $d$ is the set of all possible outcomes resulting from $d$:

$$f(d, \mathscr{N}) = \{O \in \mathscr{O} \mid \text{there exists } n \in \mathscr{N} \text{ with } f(d, n) = O\}.$$

In addition, it adopts a pessimistic attitude and ranks the sets $f(d, \mathscr{N})$ on the basis of their worst possible element. This is accomplished by introducing a complete order on the set of all subsets of $\mathscr{O}$ by means of the relation

$$\mathscr{O}_1 \preccurlyeq \mathscr{O}_2 \Leftrightarrow \inf_{O \in \mathscr{O}_1} G(O) \leqslant \inf_{O \in \mathscr{O}_2} G(O) \qquad \forall \mathscr{O}_1, \mathscr{O}_2 \subset \mathscr{O}, \tag{4}$$

where $\mathscr{O}_1, \mathscr{O}_2$ is any pair of subsets of $\mathscr{O}$ and $G$ a numerical function consistent with the order on $\mathscr{O}$ in accordance with (2). From (4) we have a complete order on the set $\mathscr{D}$ by means of

$$d_1 \preccurlyeq d_2 \Leftrightarrow f(d_1, \mathscr{N}) \preccurlyeq f(d_2, \mathscr{N}) \Leftrightarrow \inf_{n \in \mathscr{N}} G[f(d_1, n)] \leqslant \inf_{n \in \mathscr{N}} G[f(d_2, n)]$$

or in terms of a payoff function $J$,

$$d_1 \preccurlyeq d_2 \Leftrightarrow \inf_{n \in \mathscr{N}} J(d_1, n) \leqslant \inf_{n \in \mathscr{N}} J(d_2, n).$$

Thus by using the min–max approach the decision problem is formulated concretely in that it reduces to maximizing over $\mathscr{D}$ the numerical function

$$F(d) = \inf_{n \in \mathscr{N}} J(d, n).$$

Furthermore, it can easily be shown that the elements of $\mathscr{D}$ that maximize $F(d)$ above will not change if $J$ is replaced by $\Phi \cdot J$, where $\Phi: R \to R$ is any monotonically increasing function. Nonetheless, the min–max approach is pessimistic in nature and will often produce an unduly conservative decision. Characteristically in the earlier gambling example the optimal decision according to the min–max approach is to refuse the gamble.

The second approach for formulating decision problems that we shall examine is considered in the next section.

## 1.2 Expected Utility Theory and Risk

It is often the case that in a given decision problem under uncertainty we have additional information concerning the mechanism by which states of nature occur. In particular we are often in a position to know that the states of nature occur in accordance with a given probabilistic mechanism, which may depend on the decision $d$ adopted. To be specific, assume for convenience that the set of states of nature $\mathcal{N}$ is either a finite or a countable set† and that for every decision $d \in \mathcal{D}$ we know that states of nature occur according to a given probability distribution $P(\cdot \mid d)$ defined on $\mathcal{N}$. Now each decision $d \in \mathcal{D}$ specifies the probability of each outcome via the function $f(d, \cdot)$ and the relation

$$P_d(O) = P(\{n \mid f(d, n) = O\} \mid d) \qquad \forall O \in \mathcal{O}.$$

In this relation $P_d(O)$ denotes the probability that the outcome $O$ will occur when the decision $d$ is adopted. One may view the probability measure $P_d$ associated with each $d \in \mathcal{D}$ as a "probabilistic outcome" (or "generalized outcome" to use the term of the previous section) corresponding to $d$, since $P_d$ specifies the probabilistic mechanism by which outcomes occur once $d$ is selected. We shall also use the term *lottery*‡ for a probability measure on the set of outcomes. In the gambling example given earlier, the decision "bet on heads" has as a generalized outcome the probability distribution (or lottery) $(\frac{1}{2}, 0, \frac{1}{2})$ on the set of outcomes $\mathcal{O} = \{\$0, \$1, \$2\}$. The decision "bet on tails" has the same generalized outcome while the decision "not bet" has as a generalized outcome the probability distribution (0, 1, 0).

The basic idea of the expected utility approach is the following: We already have a complete ranking of the outcomes, i.e., the elements of $\mathcal{O}$. If we had a complete ranking of *all lotteries* on the set of outcomes (presumably consistent with the original ranking on $\mathcal{O}$ in the sense that if the outcome $O_1$ is preferable to the outcome $O_2$, then the lottery assigning probability one to $O_1$ is preferable to the lottery assigning probability one to $O_2$), then we could in turn obtain a complete ranking of all decisions in $\mathcal{D}$. This is true simply because we could rank any two decisions $d_1, d_2 \in \mathcal{D}$ according to the relative order of their corresponding lotteries $P_{d_1}$, $P_{d_2}$, i.e., by means of the relation

$$d_1 \preccurlyeq d_2 \leftrightharpoons P_{d_1} \preccurlyeq P_{d_2}.$$

The fundamental premise of the expected utility approach is to assume at the outset that *the decision maker has a complete ranking of all lotteries on the set of outcomes*, i.e., the decision maker is in a position to express his preference between any two probability distributions on the set of outcomes. This in turn settles the question of ranking decisions in view of the preceding relation.

† For the benefit of the advanced reader we mention that when $\mathcal{N}$ is not countable it is necessary that a probability space structure be introduced on $\mathcal{N}$ and $\mathcal{O}$ as in Appendix C. Furthermore, it is necessary that the function $f(d, \cdot)$ satisfy certain (measurability) assumptions in order that the probability measure $P_d$ be well defined.

‡ The term "lottery" is associated with the conceptually convenient device of viewing outcomes as prizes of some sort and viewing a fixed probabilistic mechanism for winning a prize as a lottery.

Furthermore, if there exists a numerical function $G$ by means of which preferences on the set of lotteries can be expressed,

$$P_{d_1} \precsim P_{d_2} \Leftrightarrow G(P_{d_1}) \leqslant G(P_{d_2}),$$

then decisions can be ranked by means of a numerical function $F$,

$$d_1 \precsim d_2 \Leftrightarrow F(d_1) \leqslant F(d_2),$$

where $F(d) = G(P_d)$ for all $d \in \mathscr{D}$.

The aspect of this formulation that is extremely appealing, however, from an analytical point of view is that the ordering of decisions can be expressed not only by a function $G$ as above, but also by means of an essentially unique numerical function called the *utility function*. This function, denoted $U$, maps the space of outcomes into the set of real numbers and satisfies

$$d_1 \precsim d_2 \Leftrightarrow P_{d_1} \precsim P_{d_2} \Leftrightarrow E\{U[f(d_1, n)] \mid d_1\} \leqslant E\{U[f(d_2, n)] \mid d_2\}, \qquad (5)$$

where the expectations are taken with respect to the corresponding probability distribution $P(\cdot \mid d)$ on $\mathscr{N}$. The problem of selecting an optimal decision is thus reduced to the problem of maximizing over $\mathscr{D}$ the expected value of the numerical function $U$.

In order to clarify the problem formulation based on the approach of this section and to illustrate the advantages resulting from the introduction of a utility function let us consider the following example:

Investment Example Consider a problem of allocating one unit of capital between two investment opportunities $A$ and $B$. Opportunity $A$ yields \$1.5 per dollar invested with certainty, while opportunity $B$ yields \$1 per dollar invested with probability $\frac{1}{2}$ and \$3 per dollar invested with probability $\frac{1}{2}$. The problem is to decide on the fractions $d$ and $(1 - d)$ of the capital to be invested in opportunities $A$ and $B$, respectively, where $0 \leqslant d \leqslant 1$.

In terms of the framework of the decision problem of Section 1.1, the set of decisions $\mathscr{D}$ consists of the interval $[0, 1]$, i.e., the set of values that the fraction $d$ invested in $A$ can take. The set of states of nature $\mathscr{N}$ consists of two elements $n_1, n_2$, where $n_1$: $B$ yields \$1 per dollar invested, and $n_2$: $B$ yields \$3 per dollar invested. The set of outcomes $\mathscr{O}$ may be taken to be the interval $[1, 3]$, which is the set of possible final fortunes of the investor resulting from all possible decisions and states of nature. The function $f$ that determines the outcome corresponding to any decision $d$ and state of nature $n$ is given by

$$f(d, n) = \begin{cases} 1.5d + \phantom{3}(1 - d) & \text{if} \quad n = n_1 \\ 1.5d + 3(1 - d) & \text{if} \quad n = n_2. \end{cases}$$

The preference relation on the set of outcomes is the natural one, i.e., a final fortune $O_1$ is at least as preferable as a final fortune $O_2$ if $O_1$ is numerically greater than or equal to $O_2$ (i.e., $O_2 \precsim O_1$ if $O_2 \leqslant O_1$).

Let us first note that, since $B$ has a higher expected rate of return, the decision that maximizes expected value of profit is to invest exclusively in opportunity $B$ ($d^* = 0$). On the other hand the optimal decision based on the max–min approach is to invest exclusively in $A$ ($d^* = 1$) since in this approach one maximizes profit based on the assumption that the most unfavorable state of nature will occur. Mathematically this can be verified by noting that $d^* = 1$ maximizes over $[0, 1]$ the function $F(d)$ given by

$$F(d) = \min\{1.5d + (1 - d), 1.5d + 3(1 - d)\}.$$

Note that the approach of maximizing expected profit and the max–min approach lead to very different decisions. Yet it is safe to assume that many decision makers would settle on a decision that differs from both decisions mentioned above and that invests a positive fraction of the capital in both opportunities $A$ and $B$.

Now in the expected utility approach the fundamental assumption is that the decision maker has a complete ranking of all lotteries on the set of outcomes. In other words, given any two probability distributions on the interval of final fortunes $[0, 3]$ the decision maker can express his preference between the two, in the sense that he can point out the distribution in accordance with which he would rather have his final fortune selected. Now the probability distribution on the set of final fortunes corresponding to a decision $d$ is the one that assigns probability $\frac{1}{2}$ to $[1.5d + (1 - d)]$ and probability $\frac{1}{2}$ to $[1.5d + 3(1 - d)]$. According to the expected utility approach a decision $d$ is optimal if its corresponding probability distribution is at least as preferable as all other distributions of the type described above. It should be clear, however, that a mathematical formulation of the corresponding optimization problem is very cumbersome since it is difficult to visualize or conjecture the form of a numerical function by means of which these probability distributions can be ranked. On the other hand, let us assume that a utility function $U$ satisfying (5) exists (and it does exist under mild assumptions). Then an optimal decision is one that solves the problem

$$\begin{aligned} &\text{maximize} \quad \underset{n}{E}\{U[f(d, n)]\} \\ &\text{subject to} \quad 0 \leqslant d \leqslant 1. \end{aligned}$$

Substituting the data of the problem we have

$$\underset{n}{E}\{U[f(d, n)]\} = \tfrac{1}{2}\{U[1.5d + (1 - d)] + U[1.5d + 3(1 - d)]\}$$

and the maximization problem above is formulated in a rather convenient form.

As an example let us assume that the decision maker's utility function is quadratic of the form

$$U(O) = \alpha O - O^2,$$

where $\alpha$ is some scalar. We require that $6 < \alpha$ so that $U(O)$ is increasing in the interval $[0, 3]$. This is necessary for the preference relation on the set of outcomes specified by the utility function to be consistent with the original preference relation. Solution of the maximization problem above yields the optimal decision $d^*$, where

$$d^* = \begin{cases} 0 & \text{if} \quad 8 \leqslant \alpha \\ (8 - \alpha)/5 & \text{if} \quad 6 < \alpha < 8. \end{cases}$$

Note that for $6 < \alpha < 8$ a positive fraction of the capital is invested in opportunity $A$ even though it offers a return that is less than the average return of $B$.

It is to be noted, of course, that different decision makers faced with the same decision problem may have different utility functions, so that before the problem can be numerically solved the form of the utility function must be specified. This can be done experimentally if necessary (see Problem 3). However, the importance of the notion of a utility function satisfying (5) lies primarily with the fact that under relatively mild assumptions it exists and can serve as the starting point for analysis of the decision problem. The reason is that important conclusions about optimal decisions can often be obtained based on either incomplete knowledge of the utility function or fairly general assumptions on its form. Several examples of such situations will be given subsequently.

We provide below the theorem of existence of a utility function for the case where the set of outcomes $\mathcal{O}$ is a finite set. For more general cases see the book by Fishburn [F3].

Consider the set $\mathcal{O}$ of outcomes and assume that it is a finite set, $\mathcal{O} = \{O_1, O_2, \ldots, O_N\}$. Let $\mathcal{P}$ be the set of all probability distributions $P = (p_1, p_2, \ldots, p_N)$ on $\mathcal{O}$, where $p_i$ is the probability of outcome $O_i$, $i = 1, \ldots, N$. For any $P_1, P_2 \in \mathcal{P}$, $P_1 = (p_1^1, \ldots, p_N^1)$, $P_2 = (p_1^2, \ldots, p_N^2)$, and any $\alpha \in [0, 1]$, we use the notation

$$\alpha P_1 + (1 - \alpha)P_2 = (\alpha p_1^1 + (1 - \alpha)p_1^2, \ldots, \alpha p_N^1 + (1 - \alpha)p_N^2) \in \mathcal{P}.$$

Let us make the following assumptions:

**A.1** There exists a complete and transitive relation $\preccurlyeq$ on $\mathcal{P}$. (For any $P_1, P_2 \in \mathcal{P}$ we write $P_1 \sim P_2$ if $P_1 \preccurlyeq P_2$ and $P_2 \preccurlyeq P_1$, and we write $P_1 \prec P_2$ if $P_1 \preccurlyeq P_2$ but not $P_1 \sim P_2$).

**A.2** If $P_1 \sim P_2$, then for all $\alpha \in [0, 1]$ and all $P \in \mathscr{P}$

$$\alpha P_1 + (1 - \alpha)P \sim \alpha P_2 + (1 - \alpha)P.$$

**A.3** If $P_1 \prec P_2$, then for all $\alpha \in (0, 1]$ and all $P \in \mathscr{P}$

$$\alpha P_1 + (1 - \alpha)P \prec \alpha P_2 + (1 - \alpha)P.$$

**A.4** If $P_1 \prec P_2 \prec P_3$, there exists an $\alpha \in (0, 1)$ such that

$$\alpha P_1 + (1 - \alpha)P_3 \sim P_2.$$

Before proving the expected utility theorem let us provide a brief discussion of the above assumptions. It is convenient for interpretation purposes to view each of the outcomes $O_1, O_2, \ldots, O_N$ as a monetary prize. Consider any probability distribution $P = (p_1, p_2, \ldots, p_N)$ on the set of outcomes. Imagine a pointer that spins in the center of a circle divided into $N$ regions. We shall assume that the pointer is spun in such a way that when it stops it is equally likely to be pointing in any given direction. The region associated with each prize $O_i$, $i = 1, \ldots, N$, occupies a fraction $p_i$ of the circumference of the circle. Then we associate with $P$ the game (or lottery) whereby we spin the wheel and win the prize corresponding to the region within which the pointer stops. Now given any two probability distributions $P_1$ and $P_2$ and a scalar $\alpha \in [0, 1]$ we can associate with the probability distribution

$$\alpha P_1 + (1 - \alpha)P_2$$

the following game. A pointer is spun in the center of a circle divided in two regions, say 1 and 2, occupying respective fractions $\alpha$ and $(1 - \alpha)$ of its circumference. Depending on whether the pointer stops in region 1 or region 2 the game corresponding to $P_1$ or $P_2$ is played and a prize is won accordingly.

Assumption A.1 requires that we are able to state our preference between games such as the above, which correspond to any two probability distributions $P_1$ and $P_2$. Furthermore, our preference relation must be transitive, i.e., if $P_1 \preccurlyeq P_2$ and $P_2 \preccurlyeq P_3$, then $P_1 \preccurlyeq P_3$. This assumption forms the core of the expected utility approach. Assumptions A.2 and A.3 have obvious interpretations and both seem reasonable. Assumption A.4 is a continuity assumption requiring that if $P_1 \prec P_2 \prec P_3$, one is indifferent to the game associated with $P_2$ and a game the outcome of which decides with respective probabilities $\alpha$ and $(1 - \alpha)$ whether the game associated with $P_1$ or $P_3$ will be played. This assumption is inconsistent with a worst-case viewpoint whereby one ranks lotteries according to the worst outcome that can occur with positive probability and has been the subject of some controversy. For example, consider the extreme situation where there are three outcomes $O_1$ = death, $O_2$ = receive nothing, and $O_3$ = receive \$1. Then it appears reasonable that any probability distribution that assigns a positive probability to $O_1$ (death) cannot be preferable or equivalent to any probability distribution that assigns a zero probability to $O_1$. Yet Assumption A.4 requires that for some $\alpha$ with $0 < \alpha < 1$ we are indifferent between the status quo and a game whereby we receive \$1 with probability $(1 - \alpha)$ and die with probability $\alpha$. On the other hand it is possible to argue that if the probability of death $\alpha$ is extremely close to zero, then this might actually be the case.

The following theorem is the central result of the expected utility theory. It states that a preference relation on the set of all lotteries satisfying Assumptions A.1–A.4 can be characterized numerically by means of an essentially unique function, the utility function. It is worth pointing out that *this result concerns an arbitrary preference relation on lotteries on the set of outcomes and is thus completely decoupled from any decision problem that one may be considering*.

**Theorem** Under Assumptions A.1–A.4 there exists a real-valued function $U: \mathcal{O} \to R$ called *utility function*, such that for all $P_1, P_2 \in \mathcal{P}$

$$P_1 \preccurlyeq P_2 \Leftrightarrow \mathop{E}_{P_1}\{U(O)\} = \sum_{i=1}^{N} p_i^1 U(O_i) \leqslant \sum_{i=1}^{N} p_i^2 U(O_i) = \mathop{E}_{P_2}\{U(O)\}, \tag{6}$$

where we denote by $E_{P_i}\{\cdot\}$ the expectation with respect to the probability distribution $P_i$. Furthermore, $U$ is unique up to a positive linear transformation, i.e., if $U^*$ is another function with the above property, there exist scalars $s_1 > 0$, $s_2$, such that

$$U^*(O) = s_1 U(O) + s_2 \qquad \forall O \in \mathcal{O}.$$

*Proof* We first show the following statement:

S If $P_1 \prec P_3$ and $P_2$ is such that $P_1 \preccurlyeq P_2 \preccurlyeq P_3$, then there exists a *unique* scalar $\alpha \in [0, 1]$ such that

$$\alpha P_1 + (1 - \alpha) P_3 \sim P_2. \tag{7}$$

Furthermore, if $P_2'$ is such that $P_1 \preccurlyeq P_2 \preccurlyeq P_2' \preccurlyeq P_3$ and $\alpha'$ corresponds to $P_2'$ as in (7), then $\alpha \geqslant \alpha'$.

Indeed if $P_1 \sim P_2 \prec P_3$, then $\alpha = 1$ is the unique scalar satisfying (7) since if for some $0 \leqslant \alpha < 1$ we had

$$\alpha P_1 + (1 - \alpha) P_3 \sim P_2 \sim \alpha P_1 + (1 - \alpha) P_2,$$

then Assumption A.3 would be contradicted. Similarly if $P_1 \prec P_2 \sim P_3$, then $\alpha = 0$ is the unique scalar satisfying (7). Assume now that $P_1 \prec P_2 \prec P_3$. Then by Assumption A.4 there exists an $\alpha_1 \in (0, 1)$ satisfying (7). Assume that

$\alpha_1$ is not unique and there exists another scalar $\alpha_2 \in (0, 1)$ such that (7) is satisfied, i.e.,

$$\alpha_1 P_1 + (1 - \alpha_1)P_3 \sim P_2 \sim \alpha_2 P_1 + (1 - \alpha_2)P_3. \tag{8}$$

Let us assume that $0 < \alpha_1 < \alpha_2 < 1$. Then we have

$$P_3 = \frac{\alpha_2 - \alpha_1}{1 - \alpha_1} P_3 + \frac{1 - \alpha_2}{1 - \alpha_1} P_3, \tag{9}$$

$$\alpha_2 P_1 + (1 - \alpha_2)P_3 = \alpha_1 P_1 + (1 - \alpha_1)\left\{\frac{\alpha_2 - \alpha_1}{1 - \alpha_1} P_1 + \frac{1 - \alpha_2}{1 - \alpha_1} P_3\right\}. \tag{10}$$

Since $P_1 \prec P_3$ we have by Assumption A.3 and (9)

$$\frac{\alpha_2 - \alpha_1}{1 - \alpha_1} P_1 + \frac{1 - \alpha_2}{1 - \alpha_1} P_3 \prec \frac{\alpha_2 - \alpha_1}{1 - \alpha_1} P_3 + \frac{1 - \alpha_2}{1 - \alpha_1} P_3 = P_3.$$

Again using A.3 and (10) we have

$$\alpha_2 P_1 + (1 - \alpha_2)P_3 \prec \alpha_1 P_1 + (1 - \alpha_1)P_3.$$

However, this contradicts (8) and hence the uniqueness of the scalar $\alpha$ in (7) is proved.

To show that if $P_1 \preccurlyeq P_2 \preccurlyeq P_2' \preccurlyeq P_3$, then $\alpha \geqslant \alpha'$ assume the contrary, i.e., $\alpha < \alpha'$. Then we have, using A.3,

$$\begin{aligned}
P_2' &\sim \alpha' P_1 + (1 - \alpha')P_3 \\
&= (1 - \alpha + \alpha')\left\{\frac{\alpha}{1 - \alpha + \alpha'} P_1 + \frac{1 - \alpha'}{1 - \alpha + \alpha'} P_3\right\} + (\alpha' - \alpha)P_1 \\
&\prec (1 - \alpha + \alpha')\left\{\frac{\alpha}{1 - \alpha + \alpha'} P_1 + \frac{1 - \alpha'}{1 - \alpha + \alpha'} P_3\right\} + (\alpha' - \alpha)P_3 \\
&= \alpha P_1 + (1 - \alpha)P_3 \sim P_2.
\end{aligned}$$

Hence $P_2' \prec P_2$, which contradicts the assumption $P_2 \preccurlyeq P_2'$. It follows that $\alpha \geqslant \alpha'$ and Statement S is proved.

Now consider the probability distributions

$$\bar{P}_1 = (1, 0, \ldots, 0), \qquad \bar{P}_2 = (0, 1, \ldots, 0), \quad \ldots, \quad \bar{P}_N = (0, 0, \ldots, 0, 1).$$

Assume without loss of generality that $\bar{P}_1 \preccurlyeq \bar{P}_2 \preccurlyeq \cdots \preccurlyeq \bar{P}_N$ and assume further that $\bar{P}_1 \prec \bar{P}_N$ (if $\bar{P}_1 \sim \bar{P}_2 \sim \cdots \sim \bar{P}_N$, the proof of the proposition is trivial). Let $A_1$, $A_N$ be any scalars with $A_1 < A_N$ and define

$$U(O_1) = A_1, \qquad U(O_N) = A_N.$$

Let $\alpha_i$, $i = 1, \ldots, N$, be the unique scalar $\alpha_i \in [0, 1]$ such that

$$\alpha_i \bar{P}_1 + (1 - \alpha_i)\bar{P}_N \sim \bar{P}_i, \qquad i = 1, \ldots, N, \tag{11}$$

and define

$$U(O_i) = A_i = \alpha_i A_1 + (1 - \alpha_i)A_N, \qquad i = 1, \ldots, N. \tag{12}$$

We shall prove that the function $U : \mathcal{O} \to R$ as defined above has the desired property (6). Indeed for any probability distribution $P = (p_1, \ldots, p_N)$ it is easily shown that $\bar{P}_1 \preccurlyeq P \preccurlyeq \bar{P}_N$ and thus we can define $\alpha(P)$ to be the unique scalar in [0, 1] such that

$$\alpha(P)\bar{P}_1 + [1 - \alpha(P)]\bar{P}_N \sim P. \tag{13}$$

From Statement S we obtain for all $P$, $P'$

$$P \preccurlyeq P' \Leftrightarrow \alpha(P) \geqslant \alpha(P'). \tag{14}$$

Now from (11) we have

$$P = \sum_{i=1}^{N} p_i \bar{P}_i \sim \sum_{i=1}^{N} p_i[\alpha_i \bar{P}_1 + (1 - \alpha_i)\bar{P}_N] \sim \sum_{i=1}^{N} p_i \alpha_i \bar{P}_1 + \left(1 - \sum_{i=1}^{N} p_i \alpha_i\right)\bar{P}_N. \tag{15}$$

Comparing (13) and (15) we obtain

$$\alpha(P) = \sum_{i=1}^{N} p_i \alpha_i,$$

and from (14)

$$P_1 \preccurlyeq P_2 \Leftrightarrow \sum_{i=1}^{N} p_i^1 \alpha_i \geqslant \sum_{i=1}^{N} p_i^2 \alpha_i. \tag{16}$$

From (12) we have $\alpha_i = (A_N - A_i)/(A_N - A_1)$, and substituting in (16) we obtain

$$P_1 \preccurlyeq P_2 \Leftrightarrow \sum_{i=1}^{N} p_i^1 A_i \leqslant \sum_{i=1}^{N} p_i^2 A_i,$$

which is equivalent to (6), which was to be proved.

It remains to show that the function $U$ defined by (12) is unique up to a positive linear transformation. Indeed if $U^*$ were another utility function satisfying (6), then, by denoting $U^*(O_i) = A_i^*$, $i = 1, \ldots, N$, by (11) and (6) we would have

$$U^*(O_i) = \alpha_i U^*(O_1) + (1 - \alpha_i)U^*(O_N).$$

This implies

$$\alpha_i = \frac{A_N - A_i}{A_N - A_1} = \frac{A_N^* - A_i^*}{A_N^* - A_1^*}$$

from which

$$A_i^* = \frac{A_N^* - A_1^*}{A_N - A_1} A_i + A_N^* - \frac{A_N(A_N^* - A_1^*)}{A_N - A_1}.$$

This proves the theorem. Q.E.D.

Returning now to the decision problem, once we assume the existence of a preference relation on the set of lotteries characterized by a utility function, we can rank decisions as follows. Given the probability distribution $P(\cdot\,|d)$ on the set of states of nature $\mathcal{N}$, every decision $d \in \mathcal{D}$ induces a probability distribution (or lottery) $P_d$ on the set of outcomes $\mathcal{O}$. Under the assumptions of the expected utility theorem there exists a utility function $U: \mathcal{O} \to R$ such that for any $d_1, d_2 \in \mathcal{D}$

$$P_{d_1} \preccurlyeq P_{d_2} \Leftrightarrow \underset{P_{d_1}}{E}\{U(O)\} \leqslant \underset{P_{d_2}}{E}\{U(O)\}.$$

We have, however,

$$\underset{P_d}{E}\{U(O)\} = E\{U[f(d, n)]\,|\,d\} \qquad \forall d \in \mathcal{D},$$

where the expectation on the left is taken with respect to $P_d$ and the expectation on the right is taken with respect to the probability distribution $P(\cdot\,|d)$ on $\mathcal{N}$. Hence

$$P_{d_1} \preccurlyeq P_{d_2} \Leftrightarrow E\{U[f(d_1, n)]\,|\,d_1\} \leqslant E\{U[f(d_2, n)]\,|\,d_1\}.$$

By ranking decisions $d \in \mathcal{D}$ in accordance with the ranking of the corresponding probability measures $P_d$,

$$d_1 \preccurlyeq d_2 \Leftrightarrow P_{d_1} \preccurlyeq P_{d_2} \Leftrightarrow E\{U[f(d_1, n)]\,|\,d_1\} \leqslant E\{U[f(d_2, n)]\,|\,d_2\},$$

we obtain a complete order on the set $\mathcal{D}$ induced by the utility function $U$. The optimal decision is found by maximization of the numerical function $F: \mathcal{D} \to R$, where

$$F(d) = E\{U[f(d, n)]\,|\,d\}$$

and the decision problem is formulated in a way that is amenable to mathematical analysis.

*The Notion of Risk*

Consider a decision maker possessing a utility function $U$ defined over an interval $X$ of real numbers. We say that the decision maker is *risk averse* if

$$\underset{P}{E}\{U(x)\} \leqslant U[\underset{P}{E}\{x\}] \tag{17}$$

for every probability distribution $P$ on $X$ for which the expectations above are finite. In other words, a decision maker is risk averse if he always prefers the expected value of the lottery over the lottery itself. Such behavior characterizes most decision makers. One may show that risk aversion is equivalent to concavity of the utility function (see Appendix A for the definition and properties of concave and convex functions). On the other hand, we say that the decision maker is *risk preferring* if the opposite inequality holds in (17), which is the case of a convex utility function. A gambler playing an unbiased roulette and receiving no reward or pleasure from gambling per se is a typical example of a risk preferring decision maker. Finally, a decision maker having a linear utility function is said to be *risk neutral.*

The notion of risk is important since it captures a basic attribute of the attitudes of the decision maker and often characterizes important aspects of his behavior. An important and widely accepted measure of risk has been proposed by Pratt [P7]. In his formulation the function

$$r(x) = -U''(x)/U'(x) \tag{18}$$

[where $U''$, $U'$ denote the second and first derivative of $U$, respectively, and it is assumed that $U'(x) \neq 0$] measures locally (at the point $x$) the risk aversion of the decision maker and is called the *index of absolute risk aversion*. It can be interpreted as follows.

Let $x$ be a gamble over the set of real numbers (i.e., a random variable) with given probability distribution and expected value $\bar{x} = E\{x\}$. Let us denote by $y$ the amount of insurance the decision maker is willing to pay in order to avoid the gamble $x$, and instead receive the expected value $\bar{x}$ of the gamble. In other words, $y$ is such that

$$U(\bar{x} - y) = E\{U(x)\}. \tag{19}$$

Intuitively, $y$ provides a natural measure of risk aversion. Proceeding formally we have by a Taylor series expansion around $\bar{x}$,

$$U(\bar{x} - y) = U(\bar{x}) - yU'(\bar{x}) + o(y), \tag{20}$$

where by $o(\alpha)$ we denote a quantity that is negligible compared with the scalar $\alpha$ provided $\alpha$ is close to zero, i.e., $\lim_{\alpha \to 0} [o(\alpha)/\alpha] = 0$. Also we have

$$\begin{aligned} E\{U(x)\} &= E\{U(\bar{x}) + (x - \bar{x})U'(\bar{x}) + \tfrac{1}{2}(x - \bar{x})^2 U''(\bar{x}) + o[(x - \bar{x})^2]\} \\ &= U(\bar{x}) + \tfrac{1}{2}\sigma^2 U''(\bar{x}) + E\{o[(x - \bar{x})^2]\}, \end{aligned} \tag{21}$$

where $\sigma^2$ is the variance of $x$. From (19)–(21), we have

$$yU'(\bar{x}) = -\tfrac{1}{2}\sigma^2 U''(\bar{x}) + o(y) + E\{o[(x - \bar{x})^2]\}.$$

From this equation and (18) it follows that the amount of insurance or risk premium $y$ that the decision maker is willing to pay is proportional (to first order) to the index of absolute risk aversion $r(\bar{x})$, thus justifying the use of $r(x)$ as a measure of local risk aversion. Notice that in the preceding investment example the index $r(y)$ is equal to $2/(\alpha - 2y)$ and tends to decrease as $\alpha$ increases. This fact is reflected in the optimal investment, where an increasing fraction of the capital is invested in the risky asset as $\alpha$ increases.

The index $r(x)$ often plays an important role in the analysis of behavior of decision makers. It is generally accepted that for most decision makers $r(x)$ is a decreasing or at least nonincreasing function of $x$, i.e., the decision maker more readily accepts risk as his wealth is increased. On the other hand, for the quadratic utility function $U(x) = -\tfrac{1}{2}x^2 + bx + c$, the index $r(x)$ is equal to $(b - x)^{-1}$ and is an increasing function of $x$ (for $x < b$). For this reason the quadratic utility function is often considered inappropriate or at least accepted with reservation in economics applications, despite the analytical simplifications often resulting from its use.

## 1.3 Some Nonsequential Decision Problems

In this section we provide some examples of decision problems under uncertainty. In addition to illustrating the general approach of the previous section, some of these examples constitute applications that are important in their own right and will be useful later on. Their common characteristic is that the decision problems are formulated as *single-stage problems* as opposed to sequential decision problems, which are the subject of the remainder of this text.

### 1.3.1 *Quadratic Cost Functionals*†

Many decision problems under uncertainty that are of interest involve the minimization of the expected value of a function that is quadratic in the decision variable. Not only do such functionals arise naturally in many problems of interest but also there is an incentive in using a quadratic cost functional whenever this is reasonable. This is due to the great analytical simplification associated with such cost functionals as will be seen shortly. We consider below the simplest case. The results obtained have analogs in

† We shall often be considering minimization of $E\{-U\}$, where $U$ is a utility function, instead of maximization of $E\{U\}$ and we shall refer to $\{-U\}$ as a *cost functional*.

more complicated cases involving sequential decision making. These cases will be considered in Chapters 3, 4, 6, and 8.

Consider the problem of finding a vector $u \in R^m$ ($R^m$ denotes $m$-dimensional Euclidean space) that minimizes the cost functional

$$J(u) = \underset{R,x}{E} \{\tfrac{1}{2}u'Ru + x'u\}.$$

In this relation $R$ is an $m \times m$ symmetric random matrix, $x$ an $m$-dimensional random vector, and prime denotes transposition. The joint probability distribution of $R$ and $x$ is given and we assume that the expected values

$$\bar{R} = E\{R\}, \qquad \bar{x} = E\{x\},$$

are finite. Furthermore $\bar{R}$ is assumed to be a positive definite matrix.

By taking the expectation with respect to $R$, $x$, we have

$$J(u) = \tfrac{1}{2}u'\bar{R}u + \bar{x}'u,$$

and the minimizing $u$ is given by

$$u^* = -\bar{R}^{-1}\bar{x}.$$

Thus the solution for this case is obtained analytically. Furthermore, the optimal vector $u^*$ depends linearly on the mean $\bar{x}$. This linear dependence is characteristic of quadratic cost functionals. Note also that the solution $u^*$ is the same as the one that would be obtained if the problem were formulated as a deterministic problem (no uncertain parameters present) with all random quantities in the cost functional replaced by their expected values. This property is sometimes called the *certainty equivalence principle*. For single-stage problems it holds not only when the cost functional is quadratic but also for many other cost functionals. For example, it holds if the cost functional is linear with respect to the random quantities, such as in the cost functional

$$J(u) = \underset{r_1,\ldots,r_n}{E} \left\{ \sum_{i=1}^{n} r_i f_i(u) \right\},$$

where $r_i$ are random variables and $f_i$ given functions. On the other hand, as will be seen later, for quadratic cost functionals the certainty equivalence principle holds in satisfying form even in multistage decision problems, a fact that does not hold true for a wide class of cost functionals.

Finally, consider the situation where the choice of $u$ is made after observing a random vector $z$ that is probabilistically related to $R$, $x$, and furthermore the conditional probability distribution $P(R, x \mid z)$ is known for every $z$. Then clearly the optimal decision is given by

$$u^*(z) = -E\{R \mid z\}^{-1} E\{x \mid z\}$$

and depends on $z$. Whenever $R$ and $z$ are independent and $x$, $z$ are Gaussian random vectors, then $E\{x|z\}$ is a linear function of $z$ (a fact to be shown later) and the decision $u^*$ depends in turn linearly on $z$.

### 1.3.2 *Least-Squares Estimation of Parameters*

Assume that we wish to determine an estimate of the value of a random vector $w \in R^n$ given a known measurement $z \in R^m$. The joint probability distribution function $F(w, z)$ is known. We formulate this problem as follows.

Given $z$, find a vector $\hat{w}(z)$ that minimizes over all vectors $\beta \in R^m$ the quadratic cost functional

$$J(\beta) = E\{\|w - \beta\|^2 | z\},$$

where $\|x\|$ denotes the usual Euclidean norm of a vector $x$ ($\|x\| = (x'x)^{1/2}$). This cost functional is a reasonable one and at the same time it results in a conveniently simple solution.

Assuming that all expected values appearing below are finite, we have

$$J(\beta) = E\{\|w\|^2 - 2w'\beta + \|\beta\|^2 | z\} = E\{\|w\|^2 | z\} - 2E\{w|z\}'\beta + \|\beta\|^2$$

and minimization with respect to $\beta$ yields the minimizing vector

$$\hat{w}(z) = E\{w|z\}.$$

The corresponding optimal value of the cost functional is given by

$$J[\hat{w}(z)] = E\{\|w - E\{w|z\}\|^2 | z\}.$$

The fact that the conditional mean $E\{w|z\}$ is the optimal (least-squares) estimate of $w$ given $z$ is a fundamental result of parameter estimation theory. In general it may be difficult to obtain a convenient analytic expression for $E\{w|z\}$. If, however, the joint probability distribution of $(w, z)$ is a Gaussian distribution, then it is possible to show that the conditional mean $E\{w|z\}$ is a linear function of $z$. This is a well-known result and will be proved in the appendix to Chapter 4.

### 1.3.3 *Inventory Control*

Consider a problem of ordering a quantity of a certain item to meet a stochastic demand for the item. The cost of purchasing or producing $u$ units of the item is

$$C(u) = \begin{cases} K + cu & \text{if } u > 0, \\ 0 & \text{if } u = 0, \end{cases} \tag{22}$$

where $K \geqslant 0$ is a fixed cost and $c > 0$ is cost per unit. The demand $w$ is a random variable taking values in a bounded interval $[0, b]$, where $b > 0$, with given probability distribution. There is a holding cost $h \geqslant 0$ per unit inventory left over at the end of the period under consideration and a depletion cost $p \geqslant 0$ per unit demand that is unmet. The holding and depletion costs may have a variety of more specific interpretations depending on the particular problem at hand.

Given an initial inventory $x$ the problem is to determine the additional inventory $u$ to be ordered so as to minimize the total expected cost:

$$J(u) = C(u) + E\{p \max(0, w - x - u) + h \max(0, x + u - w)\}. \quad (23)$$

By denoting

$$L(y) = pE\{\max(0, w - y)\} + hE\{\max(0, y - w)\}, \quad (24)$$

the expected cost is written

$$J(u) = \begin{cases} K + cu + L(x + u) & \text{if} \quad u > 0 \\ L(x) & \text{if} \quad u = 0. \end{cases} \quad (25)$$

We first note that $L(y)$ is a convex function of $y$ as can be easily seen from its definition. Hence the function $G$ defined by

$$G(y) = cy + L(y)$$

is also a convex function. Now let $S$ be a value of $y$ that minimizes $cy + L(y)$ and $s$ be the smallest value of $y$ for which $cy + L(y) = K + cS + L(S)$ (see Fig. 1.1). If $c < p$, one may show that such values exist. This can be seen from the fact $\lim_{y\to\infty} L'(y) = h$, $\lim_{y\to-\infty} L'(y) = -p$, where $L'$ denotes the first derivative of $L$. Hence if $c < p$, we have $\lim_{|y|\to\infty} G(y) = \infty$ and the existence of a minimizing scalar $S$ is guaranteed.

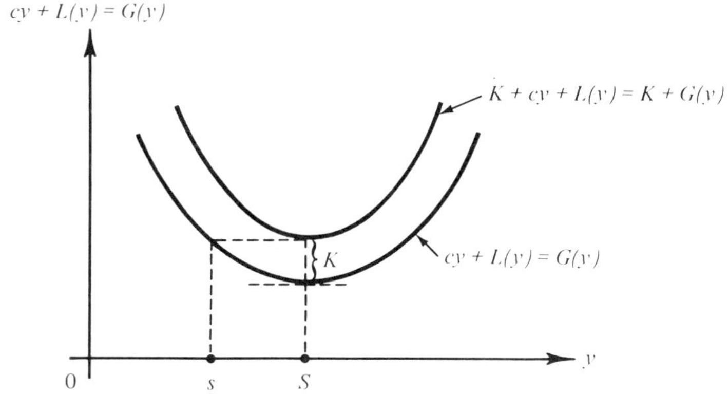


FIGURE 1.1

We will show that a value of $u$ that, for a fixed value of $x$, minimizes the expected cost is given by the equation

$$u^* = \begin{cases} S - x & \text{if} \quad x < s, \\ 0 & \text{if} \quad x \geqslant s. \end{cases} \tag{26}$$

In other words, the *optimal ordering policy is characterized by a "critical" level s below which a positive quantity must be ordered and by a "target" level S to which the total inventory* $(x + u^*)$ *should be raised when a positive order* $u^*$ *is placed.*

Indeed if $x \geqslant s$, then from Fig. 1.1 it follows that

$$cx + L(x) \leqslant K + c(x + u) + L(x + u) \qquad \forall u > 0,$$

or equivalently

$$L(x) \leqslant K + cu + L(x + u).$$

In view of the form of the cost functional [cf. Eqs. (23) and (25)] the above relation shows that for $x \geqslant s$ no ordering ($u = 0$) is optimal.

If $x < s$, then from Fig. 1.1 again it follows that

$$K + cS + L(S) < L(x) + cx$$
$$K + cS + L(S) \leqslant K + c(x + u) + L(x + u) \qquad \forall u > 0,$$

or equivalently

$$J(S - x) = K + c(S - x) + L(S) < L(x)$$
$$J(S - x) = K + c(S - x) + L(S) \leqslant K + cu + L(x + u) \qquad \forall u > 0.$$

These relations in view of (25) show that for $x < s$, ordering $S - x$ is optimal. Thus the optimality of $u^*$ as given by (26) has been proved.

An inventory ordering policy of form (26) is usually called an $(s, S)$ *policy*. Note that if $K = 0$ and if $S$ is the smallest minimizing value of $cy + L(y)$ (for example, if it is the unique minimizing value), then one has $s = S$ and the optimal policy is characterized by a single critical level. Such a policy may be called a "degenerate" $(s, S)$ policy. Inventory control will be reexamined in a sequential framework in Section 3.2.

### 1.3.4 *Choice between Risky and Secure Assets*

An individual with given initial wealth $\alpha$ wishes to invest part of it in a risky asset offering a rate of return $e$, and the rest in a secure asset offering rate of return $s > 0$. We assume that $s$ is known with certainty while $e$ is a random variable with known probability distribution $P$. If $x$ is the amount invested in the risky asset, then the final wealth of the decision maker is given by

$$y = s(\alpha - x) + ex = s\alpha + (e - s)x.$$

The decision to be made by the individual is to choose $x$ so as to maximize

$$J(x) = E\{U(y)\} = E\{U[s\alpha + (e - s)x]\}$$

subject to the constraint $x \geqslant 0$. We shall assume that $U$ is a concave, monotonically increasing, twice continuously differentiable function with negative second derivative, and with index of absolute risk aversion denoted

$$r(y) = -U''(y)/U'(y). \tag{27}$$

We also assume that the probability distribution of $e$ is such that all expected values appearing below are finite, and furthermore we assume that the utility function $U$ is such that the maximization problem has a solution (necessary and sufficient conditions for this to occur are given by Bertsekas [B12]).

Now given $\alpha$, the amount $x^*$ to be invested in the risky asset is determined from the necessary condition

$$dJ(x^*)/dx = E\{(e - s)U'[s\alpha + (e - s)x^*]\} = 0 \qquad \text{if} \quad x^* > 0, \tag{28}$$

$$dJ(x^*)/dx \leqslant 0 \qquad \text{if} \quad x^* = 0.$$

Now since $U'$ is everywhere positive it follows that if $E\{(e - s)\} > 0$, then we cannot have $x^* = 0$ since

$$dJ(0)/dx = E\{(e - s)\}U'(s\alpha) > 0.$$

Hence

$$E\{(e - s)\} > 0 \rightarrow x^* > 0,$$

or in words, a positive amount will be invested in the risky asset if its expected rate of return is greater than the rate of return of the secure asset.

Assume now that $E\{(e - s)\} > 0$ and denote by $x^*(\alpha)$ the amount invested in the risky asset when the initial wealth is $\alpha$. We would like to investigate the effect of changes in initial wealth $\alpha$ on the amount invested $x^*(\alpha)$. By differentiating (28) with respect to $\alpha$ (using the implicit function theorem) we obtain

$$E\{(e - s)U''[s\alpha + (e - s)x^*(\alpha)][s + (e - s)\,dx^*(\alpha)/d\alpha]\} = 0,$$

from which

$$\frac{dx^*(\alpha)}{d\alpha} = -\frac{E\{s(e - s)U''[s\alpha + (e - s)x^*(\alpha)]\}}{E\{(e - s)^2U''[s\alpha + (e - s)x^*(\alpha)]\}}.$$

Since the denominator is always negative and the constant $s$ is positive, the sign of $dx^*(\alpha)/d\alpha$ is the same as the sign of $E\{(e - s)U''[s\alpha + (e - s)x^*(\alpha)]\}$, which by (27) is equal to

$$f(\alpha) = -E\{(e - s)U'[s\alpha + (e - s)x^*(\alpha)]r[s\alpha + (e - s)x^*(\alpha)]\}.$$

Now assume that the risk aversion index $r(y)$ is monotonically decreasing, i.e., $r(y_1) > r(y_2)$ if $y_1 < y_2$. Then

$$\begin{aligned} f(\alpha) &= -\int_s^{\infty} (e - s)U'[s\alpha + (e - s)x^*(\alpha)]r[s\alpha + (e - s)x^*(\alpha)]\, dP(e) \\ &\quad - \int_{-\infty}^{s} (e - s)U'[s\alpha + (e - s)x^*(\alpha)]r[s\alpha + (e - s)x^*(\alpha)]\, dP(e) \\ &> -r(s\alpha)\left\{\int_s^{\infty} (e - s)U'[s\alpha + (e - s)x^*(\alpha)]\, dP(e)\right. \\ &\quad \left. + \int_{-\infty}^{s} (e - s)U'[s\alpha + (e - s)x^*(\alpha)]\, dP(e)\right\} \\ &= -r(s\alpha)\, dJ[x^*(\alpha)]/dx = 0, \end{aligned}$$

where we have utilized the fact that

$$(e - s)r[s\alpha + (e - s)x^*(\alpha)] \leqslant (e - s)r(s\alpha)$$

with strict inequality if $e \neq s$.

Thus we have $f(\alpha) > 0$ and hence $dx^*(\alpha)/d\alpha > 0$ if $r(x)$ is monotonically decreasing. Similarly, we obtain $dx^*(\alpha)/d\alpha < 0$ if $r(x)$ is monotonically increasing. In words, the individual, given more wealth, will invest more (less) in the risky asset if his utility function has decreasing (increasing) index of absolute risk aversion. Aside from its intrinsic value, this result reaffirms the important role of the index of risk aversion in shaping significant aspects of a decision maker's behavior.

### 1.3.5 *Investment Problems—Mean-Variance Analysis*

Consider an investor who wishes to allocate a certain amount of wealth $A$ among $n$ different risky investment opportunities offering corresponding rates of return $e_1, \ldots, e_n$. We assume that $e_1, \ldots, e_n$ are random variables with given joint probability distribution. If amount $x_i$, $i = 1, \ldots, n$, is allocated to investment opportunity $i$, the final wealth of the decision maker is a random variable $y$ given by

$$y = \sum_{i=1}^{n} e_i x_i.$$

Let us consider the problem of selecting $x_1, \ldots, x_n$ so as to maximize

$$E\{U(y)\} = E\left\{U\left(\sum_{i=1}^{n} e_i x_i\right)\right\},$$

where $U$ is a utility function. The maximization is subject to $\sum_{i=1}^{n} x_i = A$ and other constraints on $x_1, \ldots, x_n$, which we denote by $(x_1, \ldots, x_n) \in X \subset R^n$.

This is a very significant problem and arises often in a practical setting. It has been studied extensively for many years and various approaches have been suggested for its formulation and solution. First of all, it should be mentioned that it is essential to consider the problem in terms of a nonlinear utility function, for if one were to formulate the problem as one of maximization of the expected revenue $E\{\sum_{i=1}^{n} e_i x_i\}$ subject, for example, to the constraints $\sum_{i=1}^{n} x_i = A$, $x_i \geqslant 0$, $i = 1, \ldots, n$, the resulting optimal allocation would be to invest exclusively in the asset that gives the greatest expected return regardless of the risk that such an allocation would entail. This behavior is, however, contrary to real-life observations, which clearly indicate that most people find it advantageous to diversify their investments in the presence of uncertainty.

Now the solution process of a problem of the type described may be divided into three phases, which may partially overlap. The first phase is to collect and analyze data that will be used to determine the probability distribution of the rates of return $e_1, \ldots, e_n$. The second phase is to examine the probabilistic properties of the total return $y = \sum_{i=1}^{n} e_i x_i$ for various feasible allocations $(x_1, \ldots, x_n)$, while the third phase is actually to determine an optimal allocation $(x_1^*, \ldots, x_n^*)$ that maximizes the expected utility $E\{U(y)\}$. A major difficulty with the actual solution is that while the first two phases will ordinarily be carried out by specialists the last phase will by necessity involve the (usually nonspecialist) investor since he is the only one who, through his attitude toward risk (or equivalently through his utility function), will determine the character of the allocation that he prefers .It is of course conceivable that the investor could reveal to the specialist his utility function via experimentation and the specialist could in turn derive the optimal allocation by solving the corresponding maximization problem. However, aside from the fact that such a procedure is time consuming and meaningless to the nonspecialist investor, it has the further disadvantage that a separate problem must be solved for each investor and for each level of investment. A very popular alternative approach to the problem is the so-called *mean-variance approach*. In this approach each allocation $(x_1, \ldots, x_n)$ is characterized by the mean and the variance of the corresponding total return $y = \sum_{i=1}^{n} e_i x_i$. The investor is called upon to decide the combination of mean and variance that he prefers—a much more meaningful process to him.

The approach is also often very economical from the computational point of view as we shall explain shortly.

The basic assumption of the mean-variance approach is that the objective $E\{U(y)\}$ can be expressed in terms of the mean

$$E\{y\} = \bar{y}(x_1, \ldots, x_n)$$

and the variance

$$E\{(y - E\{y\})^2\} = \sigma^2(x_1, \ldots, x_n)$$

of the random variable $y$. This fact holds true if $U$ is a quadratic utility function or if the distribution of $y$ is characterized completely by its mean and variance for every $x_1, \ldots, x_n$ (for example, if $y$ has a Gaussian distribution or more generally a two-parameter distribution). Under these circumstances the problem becomes one of maximizing

$$E\{U(y)\} = G[\bar{y}(x_1, \ldots, x_n), \sigma^2(x_1, \ldots, x_n)]$$

subject to the constraints on $x_1, \ldots, x_n$, where $G$ is the function expressing $E\{U(y)\}$ in terms of $\bar{y}$, $\sigma^2$.

Now for most decision makers it seems reasonable that $G$ should be increasing as $\bar{y}$ is increasing and decreasing as $\sigma^2$ is increasing. In other words, the decision maker prefers a large average return but wishes to avoid a large variance, which is associated with large risk—an attitude consistent with risk aversion. We assume that this is indeed so. Then if $x_1^*, \ldots, x_n^*$ is a solution of the problem, it is clear that $x_1^*, \ldots, x_n^*$ must also solve the problem

$$\begin{aligned} &\text{minimize} \quad \sigma^2(x_1, \ldots, x_n) \\ &\text{subject to} \quad \sum_{i=1}^n x_i = A, \quad (x_1, \ldots, x_n) \in X, \quad \bar{y}(x_1, \ldots, x_n) \geqslant m^*, \end{aligned}$$

where $m^* = \bar{y}(x_1^*, \ldots, x_n^*)$.

Thus we need only look for possible solutions of problems of the form

$$\begin{aligned} &\text{minimize} \quad \sigma^2(x_1, \ldots, x_n) \\ &\text{subject to} \quad \sum_{i=1}^n x_i = A, \quad (x_1, \ldots, x_n) \in X, \quad \bar{y}(x, \ldots, x_n) \geqslant m, \end{aligned}$$

where $m$ is a scalar. This problem can be written

$$\begin{aligned} &\text{minimize} \quad x'Qx \\ &\text{subject to} \quad \sum_{i=1}^n x_i = A, \quad x \in X, \quad \sum_{i=1}^n \bar{e}_i x_i \geqslant m, \end{aligned} \tag{29}$$

where $x = (x_1, \ldots, x_n)$, $\bar{e}_i = E\{e_i\}$, and $Q$ is the covariance matrix of the vector $(e_1, \ldots, e_n)$:

$$Q = E\left\{\begin{bmatrix} (e_1 - \bar{e}_1)^2 \cdots (e_1 - \bar{e}_1)(e_n - \bar{e}_n) \\ (e_n - \bar{e}_n)(e_1 - \bar{e}_1) \cdots (e_n - \bar{e}_n)^2 \end{bmatrix}\right\}.$$

It is a problem that can be solved by standard nonlinear programming techniques.

By solving the problem above repeatedly for various values of $m$ we obtain a locus of pairs $\{|\sigma|(m), \bar{y}(m)\}$, where $|\sigma|$ is the standard deviation (square root of $\sigma^2$) as shown in Fig. 1.2. The pairs $\{|\sigma|(m), \bar{y}(m)\}$ correspond to $m$ via the optimization problem above. Usually a higher mean $\bar{y}$ is also associated with a higher $|\sigma|$ (higher risk), in which case the curve has a positive slope as shown in Fig. 1.2.

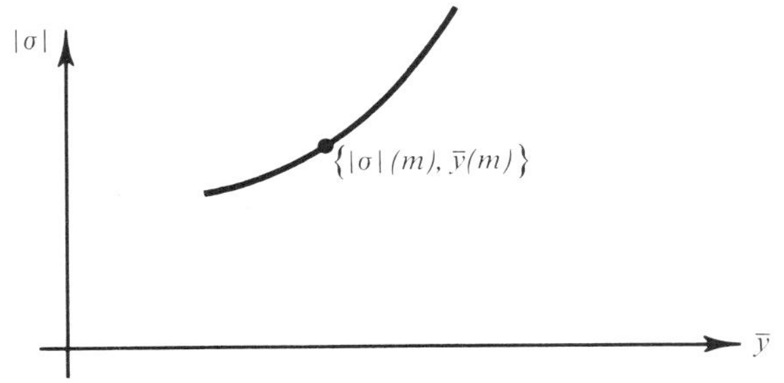


FIGURE 1.2

As a result of the preceding analysis the investment problem has been reduced to choosing a point on the mean standard deviation locus, a task that is relatively easy and meaningful to an investor. In this way the various tasks involved in the solution process can be separated. The probability distribution of the returns of the risky assets will be determined by professionals, and the mean standard deviation locus will be determined via computational solution of the related optimization problems. Then the individual investor will be called upon to decide on a point on the mean standard deviation locus that will determine the amount he will invest in each opportunity.

Important simplifications and computational economies result in the frequent case where the constraint set $X$ is a cone with vertex at the origin, i.e., when $X$ has the property that $(\lambda x_1, \ldots, \lambda x_n) \in X$ for every $\lambda \geqslant 0$ and $(x_1, \ldots, x_n) \in X$. This is the case, for example, when $X = R^n$ or when $X$ is the positive orthant

$$X = \{(x_1, \ldots, x_n) | x_i \geqslant 0, i = 1, \ldots, n\}.$$

Under these circumstances one may formulate the minimization problem (29) in terms of the fractions of total investment

$$\alpha_i = x_i/A, \qquad i = 1, \ldots, n,$$

rather than the actual amounts $x_1, \ldots, x_n$ to be invested. Thus when $X$ is a cone with vertex at the origin, the minimization problem (29) may be written

$$\begin{aligned} &\text{minimize} \quad A^2(\alpha' Q \alpha) \\ &\text{subject to} \quad \sum_{i=1}^{n} \alpha_i = 1, \quad \alpha \in X, \qquad \sum_{i=1}^{n} \bar{e}_i \alpha_i \geqslant m/A. \end{aligned}$$

This problem may be solved for $A = 1$ and various values of $m$. The corresponding mean standard deviation locus may be viewed as the *set of mean standard deviation pairs "per unit of total investment."* Given a pair $\{|\sigma|, \bar{y}\}$ on this locus it is easy to see that the corresponding standard deviation and mean associated with a total investment of $A$ monetary units will be $A|\sigma|$ and $A\bar{y}$, respectively. In this way a mean standard deviation locus constructed for a single level of total investment may be used for any level of investment. Thus the approach allows a once and for all computation, valid for every level of investment and hence for every investor.

## 1.4 A Model for Sequential Decision Making

The class of decision problems considered in the first two sections of this chapter is very broad. In this text we are primarily interested only in a subclass of such decision problems. These problems involve a dynamic system. Such systems have an input–output description and furthermore in such systems inputs are selected sequentially after observing past outputs. This allows the possibility of feedback. Let us first give an abstract description of the type of problems with which we shall be dealing.

Let us consider a system characterized by three sets $U$, $W$, and $Y$ and a function $S: U \times W \to Y$. We shall call $U$ the *input set*, $W$ the *uncertainty set*, $Y$ the *output set*, and $S$ the *system function*. Thus an input $u \in U$ and an uncertain quantity $w \in W$ produce an output $y = S(u, w)$ through the system function $S$ (Fig. 1.3). Implicit here is the assumption that the choice of the input $u$ is somehow controlled by a decision maker or device to be designed while $w$ is chosen, say, by Nature according to some mechanism, probabilistic or not.

In many problems that evolve in time, the input is a time function or sequence and there may be a possibility of observing the output $y$ as it evolves in time. Naturally this output may provide some information about the

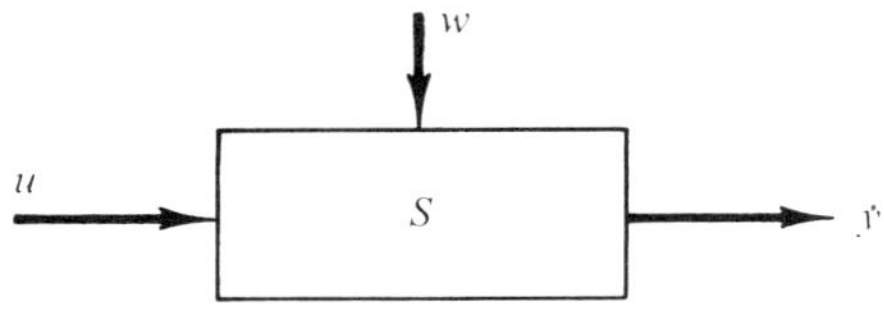


FIGURE 1.3

uncertain quantity $w$, which can be fruitfully taken into account in choosing the input $u$ by means of a feedback mechanism.

**Definition** We say that a function $\pi: Y \to U$ is a *feedback controller* (otherwise called *policy* or *decision function*) for the system if for each $w \in W$ the equation

$$u = \pi[S(u, w)]$$

has a unique solution (dependent on $w$) for $u$.

Thus for any fixed $w$, a feedback controller $\pi$ generates a unique input $u$ and hence a unique output $y$ (Fig. 1.4). In any practical situation the class of

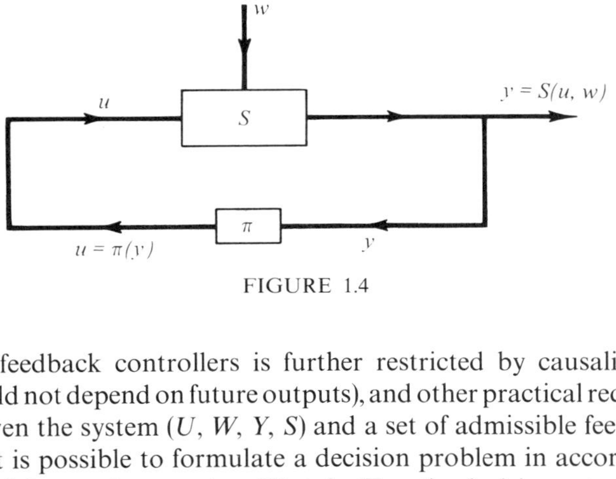


FIGURE 1.4

admissible feedback controllers is further restricted by causality (present inputs should not depend on future outputs), and other practical requirements.

Now given the system $(U, W, Y, S)$ and a set of admissible feedback controllers $\Pi$ it is possible to formulate a decision problem in accordance with the theory of the previous section. We take $\Pi$ as the decision set and $W$ as the set of states of nature. We take as the set of outcomes the Cartesian product of $U$, $W$, and $Y$, i.e., $\mathcal{O} = (U \times W \times Y)$. Now a feedback controller $\pi \in \Pi$ and a state of nature $w \in W$ generate a unique outcome $(u, w, y)$, where $u$ is the unique solution of the equation $u = \pi[S(u, w)]$ and $y = S(u, w)$. Thus we may write $(u, w, y) = f(\pi, w)$, where $f$ is the function determined by the system function $S$.

If $G$ is a numerical function ordering our preferences on $\mathcal{O}$, $J$ the corresponding payoff function for the decision problem above, and a min–max viewpoint adopted, then the problem takes the form of finding $\pi \in \Pi$ that maximizes

$$F(\pi) = \inf_{w \in W} J(\pi, w) = \inf_{w \in W} G(u, w, y)$$

where $u$ and $y$ are expressed in terms of $\pi$ and $w$ by means of $u = \pi[S(u, w)]$ and $y = S(u, w)$.

If $w$ is selected in accordance with a known probabilistic mechanism, i.e., a given probability measure that may depend on $\pi$, and the function $S$ and the elements of $\Pi$ satisfy suitable (measurability) assumptions, then it is possible to formulate a decision problem by means of a utility function. Find $\pi \in \Pi$ that maximizes

$$F(\pi) = \underset{w}{E}\{U(u, w, y)\}$$

where $u$ and $y$ are expressed in terms of $\pi$ and $w$ by means of $u = \pi[S(u, w)]$ and $y = S(u, w)$.

We shall be mostly dealing with problems of this second type.

Of course, on the basis of the formulation given one can say that problems of decision under uncertainty can be reduced to problems of decision under certainty—the problem of maximizing over $\Pi$ the numerical function $F(\pi)$. However, it is important to realize that due to the feedback possibility *the set* $\Pi$ *is a set of functions* (of the system output). This fact renders deterministic optimization techniques such as mathematical programming or Pontryagin's maximum principle inapplicable, and the only formulation that offers some possibility of analysis is the so-called method of *dynamic programming* (DP). In DP the problem of minimizing $F(\pi)$ is decomposed into a sequence of much simpler optimization problems that are solved backwards in time.

*A Discrete-Time Sequential Decision Model*

In this text we are primarily interested in the analysis of DP techniques that are applicable to a specific but quite broad class of optimization problems under uncertainty. This class of problems involves a dynamic system evolving in discrete time according to the equation

$$x_{k+1} = f_k(x_k, u_k, w_k), \qquad k = 0, 1, \ldots, N-1, \tag{30}$$

where $x_k$ denotes the state of system, $u_k$ a control input, and $w_k$ an uncertain parameter or disturbance. The functions $f_k$ are given and $x_k$, $u_k$, $w_k$ are elements of appropriate sets (or spaces). The system operates over a finite number of stages $N$ (a *finite horizon*). We shall mostly assume that $w_k$ is selected according to a probabilistic mechanism that depends on the current state $x_k$ and control input $u_k$, but does not depend on the values of the prior uncertain parameters $w_0, w_1, \ldots, w_{k-1}$. It will be assumed that the control inputs $u_k$ are selected by a decision maker or controller that has knowledge of the current state $x_k$ (*perfect state information*). Thus a feedback controller of the form

$$\{\mu_0(x_0), \mu_1(x_1), \ldots, \mu_{N-1}(x_{N-1})\}$$

is sought. This feedback controller is a sequence of functions of the current state and can be viewed as a plan that tells us that when at time $k$ the state is $x_k$, then the control $u_k = \mu_k(x_k)$ should be applied. This controller must satisfy various constraints [for example, $\mu_k(x_k) \in U_k$ for all $x_k$, where $U_k$ is a given constraint set] that depend on the problem at hand.

In terms of our earlier model the system input is $u = \{u_0, u_1, \ldots, u_{N-1}\}$, the uncertain quantity is $w = \{w_0, w_1, \ldots, w_{N-1}\}$ (perhaps together with the initial state $x_0$, if $x_0$ is uncertain), the output is $y = \{x_0, x_1, \ldots, x_N\}$, and the system function $S$ is determined in an obvious manner from the system equation (30). The class $\Pi$ of admissible feedback controllers is the set of sequences of functions $\pi = \{\mu_0, \ldots, \mu_{N-1}\}$, where $\mu_k$ is a function of the output $y$ of the form $\mu_k(y) = \mu_k(x_k)$. Furthermore, $\mu_k$ must satisfy constraints depending on the problem.

We shall assume that the utility function has an additive structure of the form

$$U(u, w, y) = U_N(x_N) + \sum_{k=0}^{N-1} U_k(x_k, u_k, w_k).$$

Thus the problem becomes one of finding an admissible $\pi^* = \{\mu_0^*, \ldots, \mu_{N-1}^*\}$ that maximizes

$$J_\pi = J(\mu_0, \ldots, \mu_{N-1}) = E\left\{U_N(x_N) + \sum_{k=0}^{N-1} U_k[x_k, \mu_k(x_k), w_k]\right\}$$

subject to the system equation constraints

$$x_{k+1} = f_k[x_k, \mu_k(x_k), w_k], \qquad k = 0, 1, \ldots, N-1,$$

and any other existing constraints on the feedback controller. We note that the description of the problem given above is intended to be informal. A precise definition together with examples will be provided in the next chapter.

Dynamic programming is directly applicable to problems of this form. Furthermore, as will be shown later, various other dynamic optimization problems involving, for example, correlated disturbances, imperfect state information, and nonadditive cost function, can be directly reformulated into the problem described above (perhaps by introducing a more complex structure). Finally it is possible to let the final time $N$ go to infinity, which is the case of an *infinite time horizon*.

We finally should point out that while we refer to $k$ as the time index and to system (30) as a discrete-time system, it is not necessary that the index $k$ have a time interpretation. For example, consider the problem of finding quantities $u_0, \ldots, u_{N-1}$ of $N$ products to be purchased so as to satisfy the constraint

$$\sum_{k=0}^{N-1} \beta_k u_k = A$$

and minimize the cost

$$\sum_{k=0}^{N-1} g_k(u_k),$$

where $A, \beta_0, \ldots, \beta_{N-1}$ are given scalars and $g_0, \ldots, g_{N-1}$ are given functions. By defining the system equation to be

$$x_{k+1} = x_k + \beta_k u_k, \qquad k = 0, 1, \ldots, N-1, \tag{31}$$

with an initial state $x_0 = 0$, the problem becomes one of minimizing

$$\sum_{k=0}^{N-1} g_k(u_k)$$

subject to the system equation (31) and the terminal state constraint $x_N = A$. Clearly this problem can be put into the framework considered in this section.

*Relation to Transition Probability Models*

It is to be noted that often the system equation is not given in form (30); rather one is given the set of transition probabilities $P(x_{k+1} | x_k, u_k)$ of the system moving from state $x_k$ to state $x_{k+1}$ when the control input is $u_k$. This is typically the case when the controlled system is a finite state Markov chain, or when the problem involves analysis of a decision tree [R1]. In this case it is still possible to define a system equation of form (30) by letting the space of uncertain parameters $w_k$ be the same as the space of $x_{k+1}$ and by defining the system functions $f_k$ to be of the form

$$f_k(x_k, u_k, w_k) = w_k$$

and letting the probability distribution of $w_k$ be equal to the given transition probability distribution $P(x_{k+1} | x_k, u_k)$. The system equation is simply $x_{k+1} = w_k$ and the occurrence of a particular value of $w_k$ is equivalent to transition to the corresponding state $x_{k+1}$. The transition probabilities depend on the current state $x_k$ and the control $u_k$ that is applied.

EXAMPLE Consider a controlled process that can be in one of two possible states 1 and 2. In other words $S = \{1, 2\}$, where $S$ is the state space. Let there be two possible controls $u^1$ and $u^2$ at each state, i.e., the control space is $C = \{u^1, u^2\}$. Let the transition probabilities $p_{ij}(u^n)$ of moving from state $i$ to state $j$ when the control $u^n$ is applied, $i, j, n = 1, 2$, be given by

$$p_{11}(u^1) = \alpha_1, \quad p_{12}(u^1) = 1 - \alpha_1, \quad p_{11}(u^2) = \alpha_2, \quad p_{12}(u^2) = 1 - \alpha_2,$$
$$p_{21}(u^1) = \beta_1, \quad p_{22}(u^1) = 1 - \beta_1, \quad p_{21}(u^2) = \beta_2, \quad p_{22}(u^2) = 1 - \beta_2,$$

where $\alpha_1, \alpha_2, \beta_1, \beta_2 \in [0, 1]$. Then this controlled process may be represented by the system equation

$$x_{k+1} = w_k,$$

where $w_k$ is a random variable taking the values 1 and 2 with probability distribution depending on $x_k$ and $u_k$ as follows:

$$p(w_k \mid x_k = 1, u_k = u^1) = \begin{cases} \alpha_1 & \text{if } w_k = 1, \\ 1 - \alpha_1 & \text{if } w_k = 2, \end{cases}$$

$$p(w_k \mid x_k = 2, u_k = u^1) = \begin{cases} \beta_1 & \text{if } w_k = 1, \\ 1 - \beta_1 & \text{if } w_k = 2, \end{cases}$$

$$p(w_k \mid x_k = 1, u_k = u^2) = \begin{cases} \alpha_2 & \text{if } w_k = 1, \\ 1 - \alpha_2 & \text{if } w_k = 2, \end{cases}$$

$$p(w_k \mid x_k = 2, u_k = u^2) = \begin{cases} \beta_2 & \text{if } w_k = 1, \\ 1 - \beta_2 & \text{if } w_k = 2. \end{cases}$$

It is to be noted that we could also consider the reverse transformation whereby we start from the system equation $x_{k+1} = f_k(x_k, u_k, w_k)$ and construct an equivalent model specified by means of transition probabilities $P(x_{k+1} \mid x_k, u_k)$. The choice of model is clearly a matter of taste and does not affect in any way the results to be obtained.

## 1.5 Notes

Utility theory and its relation to decision making were first clarified by von Neumann and Morgenstern [N3]. For an extensive account of related results see the text by Fishburn [F3] and for a bibliography see the survey paper by the same author [F2]. The text [F3] also contains Savage's expected utility theory, which is based on the notion of subjective probability [S4]. Other sources describing the expected utility theory are Luce and Raiffa [L7], Blackwell and Girshick [B23], Owen [O3], and DeGroot [D1]. Blackwell and Girshick [B23] also describe Wald's statistical decision theory, which formulates the decision problem as a game against Nature. The expected utility approach is by far the most popular approach for formulating problems of decision under uncertainty. The min–max approach receives serious consideration occasionally. Another approach, which has not met with particular favor, is based on the min–max regret criterion of Savage [S3].

The material on quadratic cost functionals and least-squares estimation is standard. For detailed expositions see, for example, the work of Aoki [A1], Meditch [M6], the *IEEE* special issue [I1], and the references listed therein. Inventory policies of the $(s, S)$ form were considered and analyzed in a pioneering paper by Arrow *et al.* [A6], which stimulated a great deal of work on the subject. The material on the investment problem is taken from Arrow [A4] and Mossin [M8]. For detailed expositions of the mean-variance approach see Markovitz [M4] and Sharpe [S9].

For an excellent treatment of decision problems under uncertainty that involve dynamic systems see the work of Witsenhausen [W4].

## Problems

**1.** Show that there exists a function $G: \mathcal{O} \to R$ satisfying relation (2) of Section 1.1 provided the set $\mathcal{O}$ is countable. Show also that if the set of decisions is finite, there exists at least one noninferior decision.

**2.** Let $\mathcal{O} = [-1, 1]$. Define an order on $\mathcal{O}$ by means of

$$O_1 \prec O_2 \Leftrightarrow |O_1| < |O_2| \qquad \text{or} \qquad O_1 < O_2 = |O_1|.$$

Show that there exists no real-valued function $G$ on $\mathcal{O}$ such that

$$O_1 \prec O_2 \Leftrightarrow G(O_1) < G(O_2) \qquad \forall O_1, O_2 \in \mathcal{O}.$$

*Hint* Assume the contrary and associate with every $O_1 \in (0, 1)$ a rational number $r(O_1)$ such that

$$G(-O_1) < r(O_1) < G(O_1).$$

Show that if $O_1 \neq O_2$, then $r(O_1) \neq r(O_2)$.

**3.** *Experimental Measurement of Utility* Consider an individual faced with a decision problem with a finite collection of outcomes $O_1, O_2, \ldots, O_N$. It is assumed that the individual has a preference relation over the set of lotteries on the set of outcomes satisfying Assumptions A.1–A.4 of the expected utility theorem, and hence a utility function over the set of outcomes exists. Suppose that $O_1 \preccurlyeq O_2 \preccurlyeq \cdots \preccurlyeq O_N$ and furthermore that $O_1 \prec O_N$.

(a) Show that the following method will determine a utility function. Define $U(O_1) = 0$, $U(O_N) = 1$. Let $p_i$ with $0 \leqslant p_i \leqslant 1$ be the probability for which one is indifferent between the lottery $\{(1 - p_i), 0, \ldots, 0, p_i\}$ and $O_i$

occurring with certainty. Then let $U(O_i) = p_i$. Try the procedure on yourself for $O_i = \$50i$ with $i = 0, 1, \ldots, 10$.

(b) Show that the following procedure will also yield a utility function. Determine $U(O_{N-1})$ as in (a) but set

$$U(O_{N-2}) = \tilde{p}_{N-2}U(O_{N-1}),$$

where $\tilde{p}_{N-2}$ is the probability for which one is indifferent between the lottery $\{(1-\tilde{p}_{N-2}), 0, \ldots, \tilde{p}_{N-2}, 0\}$ and $O_{N-2}$ occurring with certainty. Similarly set $U(O_i) = \tilde{p}_i U(O_{i+1})$, where $\tilde{p}_i$ is the appropriate probability. Again try this procedure on yourself for $O_i = \$50i$ with $i = 0, 1, \ldots, 10$ and compare the results with the ones obtained in (a).

(c) Devise a third procedure whereby the utilities $U(O_1)$, $U(O_2)$ are specified initially and $U(O_i)$, $i = 3, \ldots, N$, is obtained from $U(O_{i-2})$, $U(O_{i-1})$ through a comparison of the type considered above. Try the procedure for $O_i = \$50i$, $i = 0, 1, \ldots, 10$.

**4.** Suppose that two people $A$ and $B$ desire to make a bet. Person $A$ will pay \$1 to person $B$ if a specific event occurs and person $B$ will pay $x$ dollars to person $A$ if the event does not occur. Person $A$ believes that the probability of the event occurring is $p_A$ with $0 < p_A < 1$, while person $B$ believes that the corresponding probability is $p_B$ with $0 < p_B < 1$. Suppose that the utility functions $U_A$ and $U_B$ of persons $A$ and $B$ are strictly increasing functions of monetary gain. Let $\alpha$, $\beta$ be such that

$$U_A(\alpha) = \frac{U_A(0) - p_A U_A(-1)}{1 - p_A}, \qquad U_B(-\beta) = \frac{U_B(0) - p_B U_B(1)}{1 - p_B}.$$

Show that if $\alpha < \beta$, then any value of $x$ between $\alpha$ and $\beta$ is a mutually satisfactory bet.

**5.** In the problems of Sections 1.3.1–1.3.5 identify a set of decisions, a set of states of nature, a set of outcomes, and a utility function (or cost function).

**6.** In the inventory problem assume that the holding and depletion costs do not have the linear form considered but are such that the sum of the expected holding and depletion costs is a convex function of $(x + u)$. Show under appropriate assumptions that the optimal ordering policy is of the $(s, S)$ type.

**7.** Consider a roulette wheel that is partitioned into $k$ disjoint events $A_1, \ldots, A_k$ such that the probability of occurrence of $A_i$ is $p_i$, $i = 1, \ldots, k$, and $\sum_{i=1}^{k} p_i = 1$. Suppose a person is given an amount of $x$ dollars that he can allocate among the $k$ events $A_1, \ldots, A_k$ and that he receives as his reward the amount $x_i$ that he has allocated to the event $A_i$ that actually occurs. The constraints are $x_i \geqslant 0$, $\sum_{i=1}^{k} x_i = x$. Suppose that his utility function over positive monetary rewards $r$ is $U(r) = \ln r$. Show that his preferred allocation is given by $x_i = p_i x$, $i = 1, \ldots, k$.

**8.** In an investment problem such as the one considered in Section 1.3.5 assume that there are two investment opportunties with $\bar{e}_1 = 1.5$, $\bar{e}_2 = 1.8$, and $E\{(e_1 - \bar{e}_1)^2\} = 0.5$, $E\{(e_2 - \bar{e}_2)^2\} = 1$, and $E\{(e_1 - \bar{e}_1)(e_2 - \bar{e}_2)\} = 0$. Compute the mean standard deviation locus for a unit of capital invested and for the two cases where $X = R^2$ and $X = \{(x_1, x_2) | x_1 \geqslant 0, x_2 \geqslant 0\}$.

*Part I*

# Control of Uncertain Systems over a Finite Horizon

*Chapter 2*

# The Dynamic Programming Algorithm

## 2.1 The Basic Problem

In this section we formulate the basic problem of optimal control of a dynamic system over a finite horizon with which we shall be dealing. The formulation is very general since the state space, control space, and uncertainty space are arbitrary and may vary from one state to the next. In particular, the system may be defined over a finite or infinite state space. The problem is characterized by the fact that the number of stages of evolution of the system is finite and fixed (finite horizon), and by the fact that the control law is a function of the current state (perfect state information). However, problems where the termination time is not fixed or where the controller may decide to terminate the process prior to the final time can be reduced to the case of fixed termination time (see Problem 8). The situation in which the controller has imperfect state information can also be reduced to a problem with perfect state information as will be seen in Chapter 4. A variety of related problems can also be reduced into the form of the basic problem (see Section 2.3 and the problem section).

*Basic Problem*

Given is the discrete-time dynamic system

$$x_{k+1} = f_k(x_k, u_k, w_k), \qquad k = 0, 1, \ldots, N-1, \tag{1}$$

where the state $x_k$ is an element of a space $S_k$, $k = 0, 1, \ldots, N$, the control $u_k$ is an element of a space $C_k$, $k = 0, 1, \ldots, N-1$, and the random "disturbance" $w_k$ is an element of a space $D_k$, $k = 0, 1, \ldots, N-1$. The control $u_k$ is constrained to take values from a given nonempty subset $U_k(x_k)$ of $C_k$, which depends on the current state $x_k$ [$u_k \in U_k(x_k)$ for all $x_k \in S_k$ and $k = 0, 1, \ldots, N-1$]. The random disturbance $w_k$ is characterized by a probability measure $P_k(\cdot \mid x_k, u_k)$ defined on a collection of events in $D_k$ (in the sense of Appendix C). This probability measure may depend explicitly on $x_k$ and $u_k$ but not on values of prior disturbances $w_{k-1}, \ldots, w_0$. We consider the class of control laws (also called "policies") that consist of a finite sequence of functions $\pi = \{\mu_0, \mu_1, \ldots, \mu_{N-1}\}$, where $\mu_k: S_k \to C_k$ and such that $\mu_k(x_k) \in U_k(x_k)$ for all $x_k \in S_k$. Such control laws will be termed *admissible*.

Given an initial state $x_0$, the problem is to find an admissible control law $\pi = \{\mu_0, \mu_1, \ldots, \mu_{N-1}\}$ that minimizes the cost functional

$$J_\pi(x_0) = \underset{\substack{w_k \\ k=0,\ldots,N-1}}{E} \left\{ g_N(x_N) + \sum_{k=0}^{N-1} g_k[x_k, \mu_k(x_k), w_k] \right\} \tag{2}$$

subject to the system equation constraint

$$x_{k+1} = f_k[x_k, \mu_k(x_k), w_k], \qquad k = 0, 1, \ldots, N-1. \tag{3}$$

The real-valued functions

$$g_N: S_N \to R, \qquad g_k: S_k \times C_k \times D_k \to R, \qquad k = 0, 1, \ldots, N-1,$$

are given.

Let us provide an example that illustrates the nature of the basic problem above. Many other examples will be given subsequently.

Inventory Control Example Consider an $N$-period version of the inventory problem of Section 1.3. We assume that inventory can be ordered at the beginning of each of $N$ time periods rather than just once. Naturally each inventory order is made with knowledge of the current stock so that instead of seeking optimal numerical values for the inventory orders we are now interested in an optimal rule that tells us how much to order for every level of current stock and for each time period. Let us denote

- $x_k$ stock available at the beginning of the $k$th period
- $u_k$ stock ordered (and immediately delivered) at the beginning of the $k$th period

$w_k$ demand during $k$th period with given probability distribution
$h$ holding cost per unit item remaining unsold at the end of the $k$th period
$c$ cost per unit stock ordered
$p$ shortage cost per unit demand unfilled

We assume that $w_0, \ldots, w_{N-1}$ are independent random variables and that excess demand is backlogged and filled as soon as additional inventory becomes available. This is represented by negative stock in the system equation, which is given by

$$x_{k+1} = x_k + u_k - w_k.$$

The cost functional (representing expected loss) to be minimized is given by

$$E\left\{\sum_{k=0}^{N-1} [cu_k + p \max(0, w_k - x_k - u_k) + h \max(0, x_k + u_k - w_k)]\right\}.$$

Here we assume that unfilled demand at the beginning of the $N$th period is lost and the stock $x_N$ that is left over has no value (i.e., the terminal cost is zero).

The objective is to find an ordering policy $\{\mu_0, \ldots, \mu_{N-1}\}$, $u_k = \mu_k(x_k)$, $k = 0, \ldots, N-1$, that minimizes the expected cost. Clearly this problem is within the framework of the basic problem of the previous section with the identifications

$$g_N(x_N) = 0,$$

$$g_k(x_k, u_k, w_k) = cu_k + p \max(0, w_k - x_k - u_k) + h \max(0, x_k + u_k - w_k),$$

$$f_k(x_k, u_k, w_k) = x_k + u_k - w_k,$$

$$U_k(x_k) = [0, \infty).$$

*Optimal Value of the Basic Problem*

It is to be noted that while the objective in the basic problem is to find an admissible control law that minimizes the cost functional $J_\pi(x_0)$, it is not guaranteed a priori that such a minimizing control law exists. It is possible that $J_\pi(x_0)$ can attain arbitrarily small values by suitable choice of the control law $\pi$, in which case we say that the optimal value is $-\infty$ and no control law attains this value.† It is also possible that the set of values that $J_\pi(x_0)$ attains as $\pi$ ranges over the set of admissible policies $\Pi$ is bounded below but no optimal control law exists (in the same way that, for example, the scalar function $e^t$ is bounded below by zero but it is not minimized by any scalar $t$).

† Such a situation will almost never occur in practice if our problem is well formulated.

The greatest lower bound of the set of real numbers $\{J_\pi(x_0) | \pi \in \Pi\}$, i.e., the set of all values of the cost function $J_\pi(x_0)$ that can be attained by using admissible policies $\pi \in \Pi$, is denoted

$$J^*(x_0) = \inf_{\pi \in \Pi} J_\pi(x_0)$$

and is called the *optimal value* of the problem. When an optimal policy $\pi^*$ exists, we write

$$J^*(x_0) = J_{\pi^*}(x_0) = \min_{\pi \in \Pi} J_\pi(x_0).$$

When there does not exist an optimal policy we may be interested in finding an $\varepsilon$-optimal policy, i.e., a policy $\bar{\pi}$ such that

$$J^*(x_0) \leqslant J_{\bar{\pi}}(x_0) \leqslant J^*(x_0) + \varepsilon,$$

where $\varepsilon$ is some small number implying an acceptable deviation from optimality.

Note that the optimal value of the problem depends on the initial state $x_0$. Naturally, different initial states may have different optimal values associated with them. We shall refer to the function $J^*$ that assigns to each initial state $x_0$ the corresponding optimal value $J^*(x_0)$ as the *optimal value function.*

*The Role of Information Gathering in the Basic Problem*

As indicated in the basic problem every control law $\{\mu_0, \mu_1, \ldots, \mu_{N-1}\}$ consists of functions of the current state. The physical interpretation of the associated process is that at each time $k$ the controller (which may be a device or a decision maker, depending on the situation) observes the exact value of the current state $x_k$ and applies a control $\mu_k(x_k)$ as shown in Fig. 2.1. Subsequently the disturbance $w_k$ occurs and the next state $x_{k+1}$ is generated according to system equation (3). This state is observed by the controller, which subsequently applies the control $\mu_{k+1}(x_{k+1})$ specified by the function

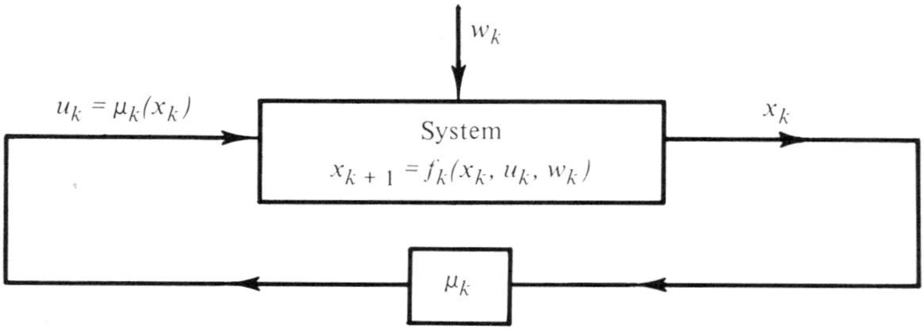


FIGURE 2.1

$\mu_{k+1}$, and the process is repeated. Thus a control law $\{\mu_0, \mu_1, \ldots, \mu_{N-1}\}$ may be viewed as a plan that specifies the control to be applied at each time for every state that may occur at that time.

It is important to realize that this mode of operation of the control law implies *information gathering* during the control process. The information received by the controller is the value of the current state at each time. Furthermore, this information is utilized directly during the control process since the control at time $k$ depends on the current state $x_k$ via the function $\mu_k$. The effects of the availability of this information by the controller may be significant indeed. The reason is that if this information were not available, then the controls applied would not depend on the current state, i.e., the control $u_k$ would have to be the same for every value of the state $x_k$ that may occur. As a result the cost functional would have to be optimized over all sequences of controls $\{u_0, u_1, \ldots, u_{N-1}\}$ satisfying the constraints of the problem and the resulting optimal value would be higher than the optimal value of the basic problem. This fact is intuitively clear and is illustrated well by the example given immediately before Section 1.1. It may be proved mathematically by observing that the set of control laws that consist of constant functions (i.e., functions independent of the current state) is only a subset of the set of admissible control laws. As another example, the reader may consider the inventory control problem considered earlier in this section where the information possessed by the inventory manager at the beginning of each period $k$ is the current inventory stock $x_k$. Clearly this information can be very important in deciding on the appropriate amount $u_k$ to be ordered at the $k$th period.

It is to be noted, however, that it is not necessary that knowledge and utilization of the information provided by the value of the current state lead to a net reduction of the value of the cost functional. For instance, in deterministic control problems (where no uncertain disturbances are present), optimization of the cost functional over all sequences $\{u_0, u_1, \ldots, u_{N-1}\}$ of controls leads to the same optimal value as optimization over all admissible control laws. This fact is illustrated in the example given prior to Section 1.1 and will be further discussed in the following. The same fact may be true even in some stochastic control problems. An important class of such problems is described in Problem 13.

### *Theoretical Limitations of the Formulation of the Basic Problem*

Before proceeding with the development of the dynamic programming algorithm, it is necessary to clarify certain aspects of our problem and to delineate certain probabilistic elements in the formulation that *do not lie on firm mathematical ground.*

First of all, once an admissible control law $\{\mu_0, \mu_1, \ldots, \mu_{N-1}\}$ is adopted, the following underlying sequence of events is envisioned:

*Stage* 0 (1) The controller observes $x_0$ and applies $u_0 = \mu_0(x_0)$.

(2) The disturbance $w_0$ is generated according to the given probability measure $P_0(\cdot \mid x_0, \mu_0(x_0))$.

(3) The cost $g_0[x_0, \mu_0(x_0), w_0]$ is incurred.

(4) The next state $x_1$ is generated according to the system equation

$$x_1 = f_0[x_0, \mu_0(x_0), w_0].$$

*Stage k* (1) The controller observes $x_k$ and applies $u_k = \mu_k(x_k)$.

(2) The disturbance $w_k$ is generated according to the given probability measure $P_k(\cdot \mid x_k, \mu_k(x_k))$.

(3) The cost $g_k[x_k, \mu_k(x_k), w_k]$ is incurred and added to previous costs.

(4) The next state $x_{k+1}$ is generated according to the system equation

$$x_{k+1} = f_k[x_k, \mu_k(x_k), w_k].$$

*Last Stage* $(N - 1)$ (1) The controller observes $x_{N-1}$ and applies

$$u_{N-1} = \mu_{N-1}(x_{N-1}).$$

(2) The disturbance $w_{N-1}$ is generated according to the given probability measure $P_{N-1}(\cdot \mid x_{N-1}, \mu_{N-1}(x_{N-1}))$.

(3) The cost $g_{N-1}[x_{N-1}, \mu_{N-1}(x_{N-1}), w_{N-1}]$ is incurred and added to previous costs.

(4) The final state $x_N$ is generated according to

$$x_N = f_{N-1}[x_{N-1}, \mu_{N-1}(x_{N-1}), w_{N-1}].$$

(5) The terminal cost $g_N(x_N)$ is incurred and added to previous costs.

The above process is well defined and couched in precise probabilistic terms. Formidable complications, however, are introduced by the need to view the cost functional

$$g_N(x_N) + \sum_{k=0}^{N-1} g_k[x_k, \mu_k(x_k), w_k]$$

as a *well-defined random variable* with well-defined expected value. The framework of probability theory requires that for each $\{\mu_0, \mu_1, \ldots, \mu_{N-1}\}$ we define an underlying probability space, i.e., a set $\Omega$, a collection of events in $\Omega$, and a probability measure on these events. Furthermore, the cost functional above must be a well-defined random variable on this space in the sense of Appendix C (a measurable function from the probability space into the real line in the terminology of measure-theoretic probability theory). In order for this to be true, additional (measurability) assumptions on the functions $f_k$, $g_k$, and $\mu_k$ may be required and it may be necessary to introduce additional structure on the spaces $S_k$, $C_k$, and $D_k$. Furthermore, these assumptions may restrict the class of admissible control laws since the functions $\mu_k$ may be constrained to satisfy additional (measurability) requirements.

Thus unless these additional assumptions and structure are specified, the problem we consider is formulated inadequately and the reader should view subsequent developments in the light of this fact. On the other hand, a rigorous formulation of the basic problem for general classes of state, control, and disturbance spaces is well beyond the mathematical and probabilistic framework of this introductory text and will not be undertaken here. Nonetheless we feel that these mathematical difficulties (at least for finite horizon problems) are mainly of a technical nature and do not affect in a substantial manner the basic results to be obtained and the basic structural properties of the problem with which we are concerned. For this reason we find it convenient to proceed with formal derivations and arguments in much the same way as in most introductory texts and journal literature on the subject.

We would like to stress, however, that under the assumption that *the disturbance spaces $D_k$, $k = 0, 1, \ldots, N - 1$, are countable sets* all the mathematical difficulties mentioned above disappear since for this case, *with the only additional assumption that the expected values of all terms in the expression of the cost functional* (2) *exist and are finite for every admissible policy* $\pi$, one can provide a sound framework for the problem.

One easy way to dispense with all the difficulties of a probabilistic nature when the disturbance spaces $D_k$ are countable sets is to rewrite all expected values in the cost functional as infinite sums in terms of the probabilities of the elements of $D_k$. Thus if $w_k^1, w_k^2, \ldots$ are the elements of $D_k$,

$$D_k = \{w_k^1, w_k^2, \ldots\},$$

and $p_k^i(x_k, u_k)$ denotes the probability of $w_k^i$, $k = 0, 1, \ldots, N - 1, i = 1, 2, \ldots,$ the value of the cost functional (2) corresponding to an admissible policy $\pi$ could be written alternatively as

$$J_\pi(x_0) = J_\pi^0(x_0),$$

where $J_\pi^0(x_0)$ is given by the last step of the recursive algorithm, which proceeds backwards from stage $N - 1$ to stage 0:

$$\begin{aligned} J_\pi^{N-1}(x_{N-1}) = \sum_{i=1}^{\infty} & p_{N-1}^i[x_{N-1}, \mu_{N-1}(x_{N-1})] \\ & \times [g_N[f_{N-1}(x_{N-1}, \mu_{N-1}(x_{N-1}), w_{N-1}^i)] \\ & + g_{N-1}[x_{N-1}, \mu_{N-1}(x_{N-1}), w_{N-1}^i]], \end{aligned}$$

$$\begin{aligned} J_\pi^k(x_k) = & \sum_{i=1}^{\infty} p_k^i[x_k, \mu_k(x_k)] [g_k[x_k, \mu_k(x_k), w_k^i] \\ & + J_\pi^{k+1}[f_k(x_k, \mu_k(x_k), w_k^i)]], \\ & \qquad k = 0, 1, \ldots, N - 2. \end{aligned}$$

In these equations the form of $J_\pi^{N-1}(x_{N-1})$ is obtained once we substitute the system equation in the expression

$$\underset{w_{N-1}}{E}\{g_N(x_N) + g_{N-1}[x_{N-1}, \mu_{N-1}(x_{N-1}), w_{N-1}]\}$$

which represents expected cost for stage $N - 1$ plus the expected terminal cost when the state at stage $N - 1$ is equal to $x_{N-1}$ and the policy $\pi$ is used. Similarly $J_\pi^k(x_k)$ may be viewed as expected cost for the last $N - k$ stages when the state at stage $k$ is $x_k$ and the policy $\pi$ is used. In the above expressions we have written out the expected values explicitly in terms of the probabilities $p_k^i[x_k, \mu_k(x_k)]$. When the sets $D_k$ are known to be finite then all infinite sums may be replaced by finite sums. If one makes the additional assumption that all the infinite sums in the above expression are well defined and finite for every admissible policy, then one can work directly with the expression of the cost functional above and bypass any need for posing the problem in a precise probabilistic manner. In this formulation the use of expectations becomes merely a convenient shorthand notation.

There is another way, more probabilistic in nature, to ensure that the cost

$$E\left\{g_N(x_N) + \sum_{k=0}^{N-1} g_k[x_k, \mu_k(x_k), w_k]\right\}$$

corresponding to a control law $\pi = \{\mu_0, \mu_1, \ldots, \mu_{N-1}\}$ is well defined when $D_k$ are countable sets. One may rewrite the value of the cost functional $J_\pi(x_0)$ as

$$J_\pi(x_0) = \underset{x_1, \ldots, x_N}{E}\left\{g_N(x_N) + \sum_{k=0}^{N-1} \tilde{g}_k[x_k, \mu_k(x_k)]\right\}, \tag{4}$$

where

$$\tilde{g}_k[x_k, \mu_k(x_k)] = \underset{w_k}{E}\{g_k[x_k, \mu_k(x_k), w_k] \mid x_k, \mu_k(x_k)\},$$

with the expectation above taken with respect to the probability distribution $P_k(\cdot \mid x_k, \mu_k(x_k))$ defined on the countable set $D_k$. Then one may take as the basic probability space the Cartesian product of $\tilde{S}_1, \tilde{S}_2, \ldots, \tilde{S}_N$, where

$$\tilde{S}_1 = \{x_1 \in S_1 \mid x_1 = f_0[x_0, \mu_0(x_0), w_0], w_0 \in D_0\},$$

$$\tilde{S}_{k+1} = \{x_{k+1} \in S_{k+1} \mid x_{k+1} = f_k[x_k, \mu_k(x_k), w_k], x_k \in \tilde{S}_k, w_k \in D_k\},$$
$$k = 1, 2, \ldots, N - 1.$$

The set $\tilde{S}_k$ is the subset of $S_k$ of all states that can be reached at time $k$ when the control law $\{\mu_0, \mu_1, \ldots, \mu_{N-1}\}$ is employed. The fact that $D_0, D_1, \ldots, D_{N-1}$ are countable sets ensures that sets $\tilde{S}_1, \ldots, \tilde{S}_N$ are also countable (this is true in view of the fact that the union of any countable collection of countable sets is a countable set). Now the system equation (3), the probability distributions $P_k(\cdot \mid x_k, \mu_k(x_k))$, the initial state $x_0$, and the control law $\{\mu_0, \mu_1, \ldots, \mu_{N-1}\}$ define a probability distribution on the countable set $\tilde{S}_1 \times \tilde{S}_2 \times \cdots \times \tilde{S}_N$ and the expectation in (4) is defined with respect to this latter distribution. In fact, it is easy to see that in this formulation the states $\{x_0, x_1, x_2, \ldots, x_N\}$ form a finite Markov sequence [P2] that can be described completely by the corresponding conditional probabilities $P(x_{k+1} \mid x_k)$, which depend on the control law $\{\mu_0, \mu_1, \ldots, \mu_{N-1}\}$ employed and the data of the problem. In light of this fact, the cost functional (4) may be written

$$
\begin{aligned}
J_\pi(x_0) = \tilde{g}_0[x_0, \mu_0(x_0)] + \underset{x_1}{E}\Big\{\tilde{g}_1[x_1, \mu_1(x_1)] \\
+ \underset{x_2}{E}\Big\{\cdots + \underset{x_{N-1}}{E}\Big\{\tilde{g}_{N-1}[x_{N-1}, \mu_{N-1}(x_{N-1})] \\
+ \underset{x_N}{E}\{g_N(x_N) \mid x_{N-1}\} \mid x_{N-2}\Big\} \cdots \mid x_1\Big\} \mid x_0\Big\}
\end{aligned}
$$

where the conditional probability distributions $P(x_{k+1} \mid x_k)$, depend on the control law $\pi = \{\mu_0, \mu_1, \ldots, \mu_{N-1}\}$. This Markovian property lies at the heart of the dynamic programming algorithm.

In conclusion the basic problem has been formulated in a mathematically rigorous way only for the case where the disturbance spaces $D_0, \ldots, D_{N-1}$, are countable sets. In the absence of countability of $D_k$ the reader should interpret subsequent results and conclusions only as plausible (imprecise) statements or conjectures. In fact, when discussing infinite horizon problems (where the need for precision is much greater) we shall make the countability assumption explicit. It is to be noted, however, that *the advanced reader will have little difficulty in establishing rigorously most of our subsequent results concerning specific applications in Chapters 3 and 4. This can be done in a manner explained in the Notes to this chapter and in Problem 12.*

## 2.2 The Dynamic Programming Algorithm

The problem of the previous section appears quite formidable since the cost functional (2) must be minimized over a class of functions of the current state. This fact together with the complexity of the cost functional make the use of variational optimization techniques impossible in almost every case. The dynamic programming (DP) technique decomposes the problem into a sequence of simpler minimization problems that are carried out over the control space rather than over a space of functions of the current state.

The DP technique rests on a very simple idea, the so-called *principle of optimality*. The name is due to Bellman, who contributed a great deal to the popularization of DP and to its transformation into a systematic tool. Roughly the principle of optimality says the following rather obvious fact.

Suppose that $\{\mu_0^*, \mu_1^*, \ldots, \mu_{N-1}^*\}$ is an optimal control law for the basic problem. Consider the subproblem whereby we are at state $x_i$ at time $i$ and wish to minimize the "cost-to-go" from time $i$ to time $N$

$$E\left\{g_N(x_N) + \sum_{k=i}^{N-1} g_k[x_k, \mu_k(x_k), w_k]\right\}.$$

Then the (truncated) control law $\{\mu_i^*, \mu_{i+1}^*, \ldots, \mu_{N-1}^*\}$ is also optimal for this subproblem.

It is perhaps helpful to introduce the dynamic programming algorithm by means of an example:

Inventory Control Example (continued) Consider the inventory control example of the previous section and let us utilize the following procedure for determining the optimal inventory ordering policy starting with the last time period and proceeding backward in time.

*$N - 1$ Period* Assume that at the beginning of period $N - 1$ the stock available is $x_{N-1}$. Clearly no matter what happened in the past the inventory manager should order inventory $u_{N-1}^* = \mu_{N-1}^*(x_{N-1})$, which minimizes over $u_{N-1}$ the sum of the ordering, holding, and depletion costs for the last time period, which is equal to

$$\underset{w_{N-1}}{E}\{cu_{N-1} + h\max(0, x_{N-1} + u_{N-1} - w_{N-1}) + p\max(0, w_{N-1} - x_{N-1} - u_{N-1})\}.$$

Let us denote the optimal cost for the last period by $J_{N-1}(x_{N-1})$:

$$J_{N-1}(x_{N-1}) = \min_{u_{N-1}\geqslant 0}\ \underset{w_{N-1}}{E}\{cu_{N-1} + h\max(0, x_{N-1} + u_{N-1} - w_{N-1}) + p\max(0, w_{N-1} - x_{N-1} - u_{N-1})\}.$$

Naturally $J_{N-1}$ is a function of the stock $x_{N-1}$. It is calculated for each $x_{N-1}$ either analytically or numerically (in which case a table is used for computer storage of the function $J_{N-1}$). In the process of calculating $J_{N-1}$ we obtain the optimal inventory ordering policy $\mu_{N-1}^*(x_{N-1})$ for the last period, where $\mu_{N-1}^*(x_{N-1}) \geqslant 0$ minimizes the right-hand side of the above equation for each value of $x_{N-1}$.

*$N - 2$ Period* Assume that at the beginning of period $N - 2$ the inventory is $x_{N-2}$. Now it is clear that the inventory manager should order

inventory $u_{N-2} = \mu_{N-2}^*(x_{N-2})$, which minimizes not only the expected cost of period $N - 2$ but rather the

(expected cost of period $N - 2$) + (expected cost of period $N - 1$, given that an optimal policy will be used at period $N - 1$).

This however is equal to

$$\underset{w_{N-2}}{E} \{cu_{N-2} + h \max(0, x_{N-2} + u_{N-2} - w_{N-2}) + p \max(0, w_{N-2} - x_{N-2} - u_{N-2})\} + \underset{w_{N-2}}{E} \{J_{N-1}(x_{N-1})\}.$$

Using the system equation $x_{N-1} = x_{N-2} + u_{N-2} - w_{N-2}$ the last term in the above sum is also written $E_{w_{N-2}} \{J_{N-1}(x_{N-2} + u_{N-2} - w_{N-2})\}$.

Thus the optimal cost $J_{N-2}(x_{N-2})$ for the last two periods, given that we are at state $x_{N-2}$, is written

$$\begin{aligned} J_{N-2}(x_{N-2}) = \min_{u_{N-2} \geq 0} \underset{w_{N-2}}{E} \{&cu_{N-2} + h \max(0, x_{N-2} + u_{N-2} - w_{N-2}) \\ &+ p \max(0, w_{N-2} - x_{N-2} - u_{N-2}) \\ &+ J_{N-1}(x_{N-2} + u_{N-2} - w_{N-2})\}. \end{aligned}$$

Again $J_{N-2}(x_{N-2})$ is calculated for every $x_{N-2}$. At the same time the optimal ordering policy $\mu_{N-2}^*(x_{N-2})$ is also computed.

*k Period* Similarly we have that at period $k$ and for initial inventory $x_k$ the inventory manager should order $u_k$ to minimize

(expected cost of period $k$) + (expected cost of periods $k + 1, \ldots, N - 1$, given that an optimal policy will be used for these periods).

By denoting by $J_k(x_k)$ the optimal value, we have

$$\begin{aligned} J_k(x_k) = \min_{u_k \geq 0} \underset{w_k}{E} \{&cu_k + h \max(0, x_k + u_k - w_k) + p \max(0, w_k - x_k - u_k) \\ &+ J_{k+1}(x_k + u_k - w_k)\} \end{aligned} \tag{5}$$

which is actually the dynamic programming equation for this problem.

The functions $J_k(x_k)$ denote the optimal expected cost for the remaining periods when starting at period $k$ and with initial inventory $x_k$. These functions are computed recursively backward in time starting at period $N - 1$ and ending at period 0. The value $J_0(x_0)$ is the optimal expected cost for the process when the initial inventory at time 0 is $x_0$. During the calculations the optimal inventory policy $\{\mu_0^*(x_0), \mu^*(x_1), \ldots, \mu_{N-1}^*(x_{N-1})\}$ is simultaneously computed from minimization of the right-hand side of (5) for every $x_k$ and $k$.

We now state the dynamic programming algorithm for the basic problem and show its optimality.

**Proposition** Let $J^*(x_0)$ be the optimal value of the cost functional (2) in the problem of Section 2.1. Then

$$J^*(x_0) = J_0(x_0),$$

where the function $J_0$ is given by the last step of the following dynamic programming algorithm, which proceeds backward in time from period $N - 1$ to period 0:

$$J_N(x_N) = g_N(x_N) \tag{6}$$

$$J_k(x_k) = \inf_{u_k \in U_k(x_k)} \underset{w_k}{E} \{g_k(x_k, u_k, w_k) + J_{k+1}[f_k(x_k, u_k, w_k)]\} \qquad (7)†$$

$$k = 0, 1, \ldots, N - 1.$$

Furthermore, if $u_k^* = \mu_k^*(x_k)$ minimizes the right-hand side of (7) for each $x_k$ and $k$ the control law $\pi^* = \{\mu_0^*, \ldots, \mu_{N-1}^*\}$ is optimal.

*Proof* The fact that the probability measure characterizing $w_k$ depends only on $x_k$ and $u_k$ and not on prior values of disturbances $w_0, \ldots, w_{k-1}$ allows us to write the optimal value of the cost $J^*(x_0)$ in the form

$$\begin{aligned} J^*(x_0) &= \inf_{\mu_0, \ldots, \mu_{N-1}} J(x_0, \mu_0, \ldots, \mu_{N-1}) \\ &= \inf_{\mu_0, \ldots, \mu_{N-1}} \Bigg[ \underset{w_0}{E} \Bigg\{ g_0[x_0, \mu_0(x_0), w_0] + \underset{w_1}{E} \Bigg\{ g_1[x_1, \mu_1(x_1), w_1] + \cdots \\ &\quad + \underset{w_{N-1}}{E} \{ g_{N-1}[x_{N-1}, \mu_{N-1}(x_{N-1}), w_{N-1}] + g_N(x_N) \} \cdots \Bigg\}\Bigg\}\Bigg], \end{aligned}$$

where the expectation over $w_k$, $k = 0, 1, \ldots, N - 1$, is conditional on $x_k$ and $\mu_k(x_k)$. The above expression may also be written

$$\begin{aligned} J^*(x_0) = \inf_{\mu_0} \Bigg[ \underset{w_0}{E} \Bigg\{ g_0[x_0, \mu_0(x_0), w_0] + \inf_{\mu_1} \Bigg[ \underset{w_1}{E} \Bigg\{ g_1[x_1, \mu_1(x_1), w_1] + \cdots \\ + \inf_{\mu_{N-1}} \Bigg[ \underset{w_{N-1}}{E} \{ g_{N-1}[x_{N-1}, \mu_{N-1}(x_{N-1}), w_{N-1}] + g_N(x_N) \} \Bigg]\Bigg\}\Bigg] \cdots \Bigg\}\Bigg]. \end{aligned} \tag{8}$$

† Both the DP algorithm and its proof are, of course, rigorous only if the basic problem is rigorously formulated. As explained in the previous section, this is the case when the disturbance spaces $D_k$, $k = 0, 1, \ldots, N - 1$, are countable sets and the expected values of all terms in the expression of the cost functional (2) are well defined and finite for every admissible policy $\pi$. In addition, it is assumed that the expected value in (7) exists and is finite for all $u_k \in U_k(x_k)$ and all $x_k \in S_k$. We further note that, although not explicitly denoted, the expectation in (7) is taken with respect to the probability measure characterizing $w_k$, which depends on both $x_k$ and $u_k$. We also write inf instead of min since the minimum of the expected value may not be attained. If the minimum is known to be attained by some $u_k \in U_k(x_k)$ then inf can be replaced by min.

In the above equations the minimizations indicated are over all functions $\mu_k$ such that $\mu_k(x_k) \in U_k(x_k)$ for all $x_k$ and $k$. In addition, the minimization is subject to the ever-present system equation constraint

$$x_{k+1} = f_k[x_k, \mu_k(x_k), w_k].$$

Now we use the fact that for any function $F$ of $x$, $u$, we have

$$\inf_{\mu \in M} F[x, \mu(x)] = \inf_{u \in U(x)} F(x, u),$$

where $M$ is the set of all functions $\mu(x)$ such that $\mu(x) \in U(x)$ for all $x$.

By applying this fact in (8), using the substitution $x_{k+1} = f_k(x_k, u_k, w_k)$, and introducing the functions $J_k$ of (7) in (8) we obtain the desired result:

$$J^*(x_0) = J_0(x_0).$$

It is also clear that $\{\mu_0^*, \ldots, \mu_{N-1}^*\}$ is an optimal control law if $\mu_k^*(x_k)$ minimizes the right-hand side of (7) for each $x_k$ and $k$. Q.E.D.

In the DP algorithm of the above proposition, ideally, we would like to be able to determine closed-form expressions for the "cost-to-go" functions $J_k$. This is possible in a number of important special cases. In any case even if a closed-form expression for $J_k$ or the optimal control law $\mu_k^*$ cannot be obtained, one hopes to obtain characterizations of $J_k$ or $\mu_k^*$ that are of interest. We shall examine such cases in the next chapter.

In the absence of an analytical solution one has to resort to a numerical solution of the DP equations. This may be quite difficult and expensive since the minimization in (7) must be carried out for each value of $x_k$. Typically the state space is discretized and the minimization is carried out for a finite number of states $x_k$. The computational requirements are proportional to the number of discretization points. Thus for complex multidimensional problems the computational burden may be overwhelming even for the most potent of presently existing computers. More about the computational aspects of DP will be presented in Chapter 5. However, at this point it must be recognized that DP cannot provide a complete solution to every problem of the type we are considering. Nonetheless, it is the only general approach for attacking sequential optimization problems under uncertainty.

We now provide an example that demonstrates the computational aspects of the DP algorithm.

EXAMPLE 1 Consider an inventory control problem similar to the one of Section 2.1 but different in that *inventory and demand are nonnegative integer variables*. Furthermore assume that *there is an upper bound on the stock* $(x_k + u_k)$ that can be stored and also assume *that the excess demand* $(w_k - x_k - u_k)$ *is lost*. As a result the stock equation takes the form

$$x_{k+1} = \max(0, x_k + u_k - w_k).$$

Assume that the maximum capacity $(x_k + u_k)$ for stock is 2 units, that the planning horizon $N$ is 3 periods, and that the holding cost $h$ and the ordering cost $c$ are both 1 unit. The shortage cost $p$ is assumed to be 3 units, and the demand $w_k$ has the same probability distribution for all periods, given by

$$p(w_k = 0) = 0.1, \qquad p(w_k = 1) = 0.7, \qquad p(w_k = 2) = 0.2.$$

There is no fixed cost (i.e., $K = 0$) and the initial stock $x_0$ is zero.

Let us solve this problem by applying the DP algorithm (6), (7). For our case this algorithm takes the form

$$J_3(x_3) = 0,$$

$$J_k(x_k) = \min_{\substack{0 \leq u_k \leq 2 - x_k \\ u_k = 0, 1, 2}} \mathop{E}_{w_k} \{u_k + \max(0, x_k + u_k - w_k) + 3 \max(0, w_k - x_k - u_k)$$

$$+ J_{k+1}[\max(0, x_k + u_k - w_k)]\}, \qquad k = 0, 1, 2,$$

where $x_k$, $u_k$, $w_k$ can take the values 0, 1, and 2.

*Stage* 2 We compute $J_2(x_2)$ for each of the three possible states:

$$J_2(0) = \min_{u_2 = 0, 1, 2} \mathop{E}_{w_2} \{u_2 + \max(0, u_2 - w_2) + 3 \max(0, w_2 - u_2)\}$$

$$= \min_{u_2 = 0, 1, 2} \{u_2 + 0.1[\max(0, u_2) + 3 \max(0, -u_2)]$$

$$+ 0.7[\max(0, u_2 - 1) + 3 \max(0, 1 - u_2)] + 0.2[\max(0, u_2 - 2)$$

$$+ 3 \max(0, 2 - u_2)]\}.$$

We calculate the expectation of the right-hand side for each of the three possible values of $u_2$:

$$u_2 = 0: \quad E\{\cdot\} = 0.7 \times 3 \times 1 + 0.2 \times 3 \times 2 = 3.3,$$

$$u_2 = 1: \quad E\{\cdot\} = 1 + 0.1 \times 1 + 0.2 \times 3 \times 1 = 1.7,$$

$$u_2 = 2: \quad E\{\cdot\} = 2 + 0.1 \times 2 + 0.7 \times 1 = 2.9,$$

Hence we have by selecting the minimizing $u_2$,

$$\triangleright \qquad J_2(0) = 1.7, \qquad \mu_2^*(0) = 1. \qquad \triangleleft$$

For $x_2 = 1$ we have

$$J_2(1) = \min_{u_2=0,1} \mathop{E}_{w_2} \{u_2 + \max(0, 1 + u_2 - w_2) + 3\max(0, w_2 - 1 - u_2)\}$$

$$= \min_{u_2=0,1} \{u_2 + 0.1[\max(0, 1 + u_2) + 3\max(0, -1 - u_2)]$$

$$+ 0.7[\max(0, u_2) + 3\max(0, -u_2)]$$

$$+ 0.2[\max(0, u_2 - 1) + 3\max(0, 1 - u_2)]\},$$

$$u_2 = 0: \quad E\{\cdot\} = 0.1 \times 1 + 0.2 \times 3 \times 1 = 0.7,$$

$$u_2 = 1: \quad E\{\cdot\} = 1 + 0.1 \times 2 + 0.7 \times 1 = 1.9.$$

Hence

▷ $$J_2(1) = 0.7, \qquad \mu_2^*(1) = 0.$$ ◁

For $x_2 = 2$ we have

$$J_2(2) = \mathop{E}_{w_2} \{\max(0, 2 - w_2) + 3\max(0, w_2 - 2)\}$$

$$= 0.1 \times 2 + 0.7 \times 1 = 0.9,$$

▷ $$J_2(2) = 0.9, \qquad \mu_2^*(2) = 0.$$ ◁

*Stage* 1 Again we compute $J_1(x_1)$ for each of the three possible states $x_2 = 0, 1, 2$ using the values $J_2(0)$, $J_2(1)$, $J_2(2)$ obtained in the previous stage:

$$J_1(0) = \min_{u_1=0,1,2} \mathop{E}_{w_1} \{u_1 + \max(0, u_1 - w_1) + 3\max(0, w_1 - u_1)$$

$$+ J_2[\max(0, u_1 - w_1)]\},$$

$$u_1 = 0: \quad E\{\cdot\} = 0.1 \times J_2(0) + 0.7[3 \times 1 + J_2(0)]$$
$$+ 0.2[3 \times 2 + J_2(0)] = 5.0,$$

$$u_1 = 1: \quad E\{\cdot\} = 1 + 0.1[1 + J_2(1)] + 0.7 \times J_2(0)$$
$$+ 0.2[3 \times 1 + J_2(0)] = 3.3,$$

$$u_1 = 2: \quad E\{\cdot\} = 2 + 0.1[2 + J_2(2)] + 0.7[1 + J_2(1)]$$
$$+ 0.2 \times J_2(0) = 3.82,$$

▷ $$J_1(0) = 3.3, \qquad \mu_1^*(0) = 1,$$ ◁

$$J_1(1) = \min_{u_1 = 0, 1} \mathop{E}_{w_1} \{u_1 + \max(0, 1 + u_1 - w_1) + 3 \max(0, w_1 - 1 - u_1)$$
$$+ J_2[\max(0, 1 + u_1 - w_1)]\}$$

$$u_1 = 0: \quad E\{\cdot\} = 0.1[1 + J_2(1)] + 0.7 \times J_2(0) + 0.2[3 \times 1 + J_2(0)] = 2.3,$$

$$u_1 = 1: \quad E\{\cdot\} = 1 + 0.1[2 + J_2(2)] + 0.7[1 + J_2(1)] + 0.2 \times J_2(0) = 2.82,$$

$$\triangleright \qquad J_1(1) = 2.3, \qquad \mu_1^*(1) = 0, \qquad \triangleleft$$

$$J_1(2) = E\{\max(0, 2 - w_1) + 3 \max(0, w_1 - 2) + J_2[\max(0, 2 - w_1)]$$
$$= 0.1[2 + J_2(2)] + 0.7[1 + J_2(1)] + 0.2 \times J_2(0) = 1.82,$$

$$\triangleright \qquad J_1(2) = 1.82, \qquad \mu_1^*(2) = 0. \qquad \triangleleft$$

*Stage* 0 Here we need only compute $J_0(0)$ since the initial state is known to be zero. We have

$$J_0(0) = \min_{u_0 = 0, 1, 2} \mathop{E}_{w_0} \{u_0 + \max(0, u_0 - w_0) + 3 \max(0, w_0 - u_0)$$
$$+ J_1[\max(0, u_0 - w_0)]\},$$

$$u_0 = 0: \quad E\{\cdot\} = 0.1 \times J_1(0) + 0.7[3 \times 1 + J_1(0)] + 0.2[3 \times 2 + J_1(0)] = 6.6,$$

$$u_0 = 1: \quad E\{\cdot\} = 1 + 0.1[1 + J_1(1)] + 0.7 \times J_1(0) + 0.2[3 \times 1 + J_1(0)] = 4.9,$$

$$u_0 = 2: \quad E\{\cdot\} = 2 + 0.1[2 + J_1(2)] + 0.7[1 + J_1(1)] + 0.2 \times J_1(0) = 5.352,$$

$$\triangleright \qquad J_0(0) = 4.9, \qquad \mu_0^*(0) = 1. \qquad \triangleleft$$

If the initial state were not known a priori we would have to compute in a similar manner $J_0(1)$ and $J_0(2)$ as well as the minimizing $u_0$. These calculations yield

$$\triangleright \qquad J_0(1) = 3.9, \qquad \mu_0^*(1) = 0, \qquad \triangleleft$$

$$\triangleright \qquad J_0(2) = 3.352, \qquad \mu_0^*(2) = 0. \qquad \triangleleft$$

Thus the optimal ordering policy for each period is to order one unit if the current stock is zero, and order nothing otherwise.

It is worth noting that DP can be applied (in a form that is of comparable simplicity to the one of the previous proposition) to problems where the cost functional does not have the additive structure of the one of the basic problem but rather the form

$$E\left\{\beta \exp\left(\alpha\left[g_N(x_N) + \sum_{k=0}^{N-1} g_k(x_k, u_k, w_k)\right]\right)\right\},$$

where $\alpha$ and $\beta$ are given scalars. This type of cost functional can be called a *risk-sensitive cost functional* since it corresponds to an exponential utility function expressing risk aversion or risk preference (depending on the sign of $\beta$) on the part of the decision maker. Such a cost functional is particularly interesting when the expression

$$g_N(x_N) + \sum_{k=0}^{N-1} g_k(x_k, u_k, w_k)$$

has a monetary interpretation. For the form of the DP algorithm corresponding to this case as well as the case of a discounted cost functional see Problems 6, 7, and 9 at the end of this chapter.

We finally point out that DP is fully applicable to deterministic problems involving no uncertain parameters. Such problems can be embedded within the framework of the basic problem simply by considering disturbance spaces $D_k$ having a single element. Thus the algorithm of this section may be used for the solution of such problems. In addition, other algorithms of the DP type, which proceed forward in time (rather than backward), may be used for deterministic problems (see Kaufmann and Cruon [K6]). It is important to note that, in contrast to stochastic problems, *using feedback in deterministic problems results in no advantage in terms of reduction of the value of the cost functional.* In other words, minimizing the cost functional over the class of admissible control laws $\{\mu_0, \ldots, \mu_{N-1}\}$ results in the same optimal value of the cost functional as minimizing over the class of *sequences of control vectors* $\{u_0, \ldots, u_{N-1}\}$ with $u_k \in U_k(x_k)$ for all $k$. This is true simply because if $\{\mu_0^*, \ldots, \mu_{N-1}^*\}$ is an optimal control law for a deterministic problem, then the control sequence $\{u_0^*, \ldots, u_{N-1}^*\}$, where

$$u_k^* = \mu_k^*(x_k^*), \qquad k = 0, \ldots, N-1,$$

and the states $x_0^*, \ldots, x_{N-1}^*$ are defined by

$$x_{k+1}^* = f_k(x_k^*, u_k^*), \qquad x_0^* = x_0, \qquad k = 0, 1, \ldots, N-1,$$

also achieves the optimal value of the problem. For this reason we may minimize the cost functional over sequences of controls, a task that may be achieved by variational deterministic optimal control algorithms such as steepest descent, conjugate gradient, and Newton's method. These algorithms, when applicable, are usually more efficient than DP. On the other hand *DP has a wider scope of applicability since it can handle difficult constraint sets such as integer or discrete sets.* Furthermore, *DP leads to a globally optimal solution* as opposed to variational techniques, for which this cannot be guaranteed in general.

Our main objective in this text is the analysis of stochastic optimization problems and the ramifications of the presence of uncertainty. For this reason we will pay little attention to deterministic problems. However, we offer here an example of a deterministic problem, the so-called optimal path problem, for which DP is a principal method of solution and which demonstrates some of the computational aspects of DP .

EXAMPLE 2 Let $\{1, 2, \ldots, N + 1\}$ be a set of points and let $C_{ij}$, $i, j = 1, \ldots, N + 1$, represent the cost of moving directly from point $i$ to point $j$ in one move. We assume that $0 \leqslant C_{ij} < \infty$ and $C_{ii} = 0$ for all $i, j$. We would like to find the optimal path from point $i$ to point $N + 1$, i.e., the sequence of moves that minimizes total cost to get to the point $N + 1$ starting from each of the points $1, 2, \ldots, N$.

Now it is clear that an optimal path need not take more than $N$ moves, so that we may limit the number of moves to $N$. We formulate the problem as one where *we require exactly N moves but allow degenerate moves from a point i into itself.* We denote for $i = 1, \ldots, N$, $k = 0, 1, \ldots, N - 1$,

$J_{N-1}(i)$ optimal cost for getting to $(N + 1)$ from $i$ in one move,

$J_k(i)$ optimal cost for getting to $(N + 1)$ from $i$ in $(N - k)$ moves.

Then the cost of the optimal path from $i$ to $(N + 1)$ is $J_0(i)$. While it is possible to formulate this problem within the framework of the basic problem and subsequently apply the DP algorithm, we bypass this formulation and write directly the DP equation, which for our problem takes the intuitively clear form:

$$\text{optimal cost from } i \text{ to } (N+1) \text{ in } (N-k) \text{ moves} = \min_{j=1,\ldots,N} \{C_{ij} + \text{optimal cost from } j \text{ to } (N+1) \text{ in } (N-k-1) \text{ moves}\},$$

or

$$J_k(i) = \min_{j=1,\ldots,N} \{C_{ij} + J_{k+1}(j)\}, \qquad k = 0, 1, \ldots, N-2,$$

with

$$J_{N-1}(i) = C_{i(N+1)}, \qquad i = 1, 2, \ldots, N.$$

The optimal policy when at point $i$ after $k$ moves is to move to point $j^*$, where $j^*$ minimizes over all $j = 1, \ldots, N$ the expression in braces above. Note that a degenerate move from $i$ to $i$ is not excluded. If the optimal path obtained from the algorithm contains such degenerate moves this simply means that its duration is less than $N$ moves.

To demonstrate the algorithm consider the problem shown in Fig. 2.2a where the costs $C_{ij}$, $i \neq j$ (we assume $C_{ij} = C_{ji}$), are shown along the connecting line segments. The DP algorithm is shown in Fig. 2.2b with the "cost-to-go" $J_k(i)$ shown at each point $i$ and index $k$ together with the optimal path. The optimal paths are

$$1 \to 5, \quad 2 \to 3 \to 4 \to 5, \quad 3 \to 4 \to 5, \quad 4 \to 5.$$

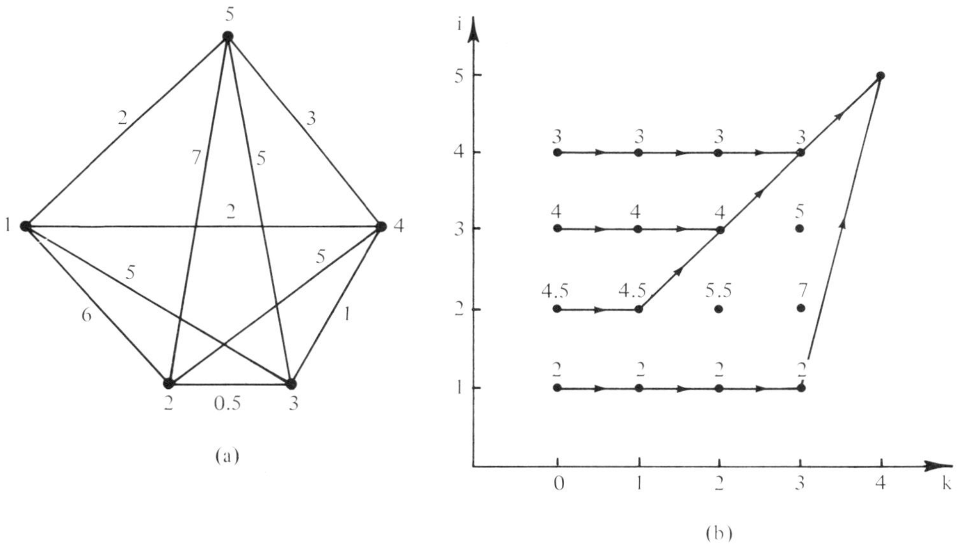


FIGURE 2.2

Note from the figure that the costs-to-go $J_k(i)$ are monotonically nondecreasing with $k$, which is easy to prove in general for optimal path problems of the type considered here.

## 2.3 Time Lags, Correlated Disturbances, and Forecasts

This section deals with various situations where some of the assumptions inherent in the basic problem formulation are not satisfied. We shall consider the case where there are time lags in the state and the control appearing in the system equation, i.e., the next state $x_{k+1}$, depends explicitly not only on the present state and control $x_k$, $u_k$ but also on past states $x_{k-1}, x_{k-2}, \ldots, x_{k-n}$

and controls $u_{k-1}, u_{k-2}, \ldots, u_{k-m}$. We will consider the case where the disturbances $w_k$ are correlated, and the case where at time $k$ a forecast on the future uncertainties $w_k, w_{k+1}, \ldots$ becomes available, thus updating the corresponding probability distributions. The situation where the system evolution may terminate prior to the final time either due to a random event or due to an action of the decision maker is covered in the problems. Generally speaking, in all these cases it is possible to reformulate the problem into the framework of the basic problem by using the device of *state augmentation*. The (unavoidable) price paid, however, is an increase in complexity of the reformulated problem.

*Time Lags*

First let us consider the case of time lags. For simplicity assume that there is at most a single period time lag in the state and control, i.e., assume a system equation of the form

$$\begin{aligned} x_{k+1} &= f_k(x_k, x_{k-1}, u_k, u_{k-1}, w_k), \qquad k = 1, 2, \ldots, N-1, \\ x_1 &= f_0(x_0, u_0, w_0). \end{aligned} \tag{9}$$

The case where there are time lags of more than one period is a straightforward extension of the single-period case.

Now if we introduce additional state variables $y_k$ and $s_k$ and make the identifications $y_k = x_{k-1}$, $s_k = u_{k-1}$, the system equation (9) yields for $k = 1, 2, \ldots, N-1$,

$$\begin{bmatrix} x_{k+1} \\ y_{k+1} \\ s_{k+1} \end{bmatrix} = \begin{bmatrix} f_k(x_k, y_k, u_k, s_k, w_k) \\ x_k \\ u_k \end{bmatrix} \tag{10}$$

By defining $\tilde{x}_k = (x_k, y_k, s_k)$ as the new state we have

$$\tilde{x}_{k+1} = \tilde{f}_k(\tilde{x}_k, u_k, w_k), \tag{11}$$

where the system function $\tilde{f}_k$ is defined in an obvious manner from (10). By using (11) as the system equation and by making a suitable reformulation of the cost functional the problem is reduced to the basic problem without time lags. Naturally the control law $\{\mu_0, \ldots, \mu_{N-1}\}$ that is sought will consist of functions $\mu_k$ of the new state $\tilde{x}_k$, or equivalently *$\mu_k$ will be a function of the present state $x_k$ as well as past state $x_{k-1}$ and control $u_{k-1}$*. The DP algorithm (in terms of the variables of the original problem) is

$$J_N(x_N) = g_N(x_N),$$

$$\begin{aligned} J_{N-1}(x_{N-1}, x_{N-2}, u_{N-2}) = \inf_{u_{N-1} \in U_{N-1}(x_{N-1})} \; \underset{w_{N-1}}{E} \{ & g_{N-1}(x_{N-1}, u_{N-1}, w_{N-1}) \\ & + J_N[f_{N-1}(x_{N-1}, x_{N-2}, u_{N-1}, u_{N-2}, w_{N-1})]\}, \end{aligned}$$

$$\begin{aligned} J_k(x_k, x_{k-1}, u_{k-1}) = \inf_{u_k \in U_k(x_k)} \; \underset{w_k}{E} \{ & g_k(x_k, u_k, w_k) \\ & + J_{k+1}[f_k(x_k, x_{k-1}, u_k, u_{k-1}, w_k), x_k, u_k]\}, \\ & \qquad k = 1, \ldots, N-2, \end{aligned}$$

$$\begin{aligned} J_0(x_0) = \inf_{u_0 \in U_0(x_0)} \; \underset{w_0}{E} \{ & g_0(x_0, u_0, w_0) \\ & + J_1[f_0(x_0, u_0, w_0), x_0, u_0]\}. \end{aligned}$$

We note that similar reformulations are possible when time lags appear in the cost functional, for example, in the case where the expression to be minimized is of the form

$$E\left\{ g_N(x_N, x_{N-1}) + \sum_{k=0}^{N-1} g_k(x_k, x_{k-1}, u_k, w_k) \right\}.$$

The extreme case of time lags in the cost functional is when it has the non-additive form

$$E\{g_N(x_N, x_{N-1}, \ldots, x_0, u_{N-1}, \ldots, u_0, w_{N-1}, \ldots, w_0)\}.$$

Then in order to reduce the problem to the form of the basic problem the augmented state $\tilde{x}_k$ at time $k$ must include

$$(x_k, x_{k-1}, \ldots, x_0, u_{k-1}, \ldots, u_0, w_{k-1}, \ldots, w_0)$$

and the reformulated cost functional takes the form

$$E\{g_N(\tilde{x}_N)\}, \qquad \text{where} \quad \tilde{x}_N = (x_0, \ldots, x_N, u_0, \ldots, u_{N-1}, w_0, \ldots, w_{N-1}).$$

The control law sought consists of functions $\mu_k$ of the present and past states $x_k, \ldots, x_0$, the past controls $u_{k-1}, \ldots, u_0$, and the past disturbances $w_{k-1}, \ldots, w_0$. Naturally we must assume that past disturbances are known to the controller for otherwise we are faced with a problem with imperfect state information. The DP algorithm takes the form

$$J_{N-1}(x_0, \ldots, x_{N-1}, u_0, \ldots, u_{N-2}, w_0, \ldots, w_{N-2})$$
$$= \inf_{u_{N-1} \in U_{N-1}(x_{N-1})} \underset{w_{N-1}}{E} \{g_N(x_0, \ldots, x_{N-1}, f_{N-1}(x_{N-1}, u_{N-1}, w_{N-1}),$$
$$u_0, \ldots, u_{N-1}, w_0, \ldots, w_{N-1})\},$$

$$J_k(x_0, \ldots, x_k, u_0, \ldots, u_{k-1}, w_0, \ldots, w_{k-1})$$
$$= \inf_{u_k \in U_k(x_k)} \underset{w_k}{E} \{J_{k+1}(x_0, \ldots, x_k, f_k(x_k, u_k, w_k),$$
$$u_0, \ldots, u_k, w_0, \ldots, w_k)\}, \qquad k = 0, \ldots, N-2.$$

Similar algorithms may be written for the case where the control constraint set depends on past states or controls, etc.

*Correlated Disturbances*

We turn now to the case where the disturbances $w_k$ are correlated. Here we shall assume that the $w_k$ are elements of a Euclidean space and that the probability distribution of $w_k$ does not depend explicitly on the current state $x_k$ and control $u_k$ but rather it depends explicitly on the prior values of the disturbances $w_0, \ldots, w_{k-1}$. By using statistical methods it is often possible to represent the process $w_0, w_1, \ldots, w_{N-1}$ by means of a linear system

$$y_{k+1} = A_k y_k + \xi_k, \quad k = 0, 1, \ldots, N-1, \quad y_0 = 0,$$
$$w_k = C_k y_{k+1},$$

where $A_k$, $C_k$ are matrices of appropriate dimension and $\xi_k$ are independent random vectors with given statistics. In other words, the correlated process $w_0, \ldots, w_{N-1}$ is represented as the output of a linear system perturbed by a white process, i.e., a process consisting of independent random vectors as shown below:

$$\xrightarrow{\xi_k} y_{k+1} = A_k y_k + \xi_k \xrightarrow{y_{k+1}} C_k \rightarrow w_k.$$

By considering now $y_k$ as additional state variables we have a new system equation

$$\begin{bmatrix} x_{k+1} \\ y_{k+1} \end{bmatrix} = \begin{bmatrix} f_k[x_k, u_k, C_k(A_k y_k + \xi_k)] \\ A_k y_k + \xi_k \end{bmatrix} \tag{12}$$

By taking as the new state the pair $\tilde{x}_k = (x_k, y_k)$ and as new disturbance the vector $\xi_k$, we can write (12) as

$$\tilde{x}_{k+1} = \tilde{f}_k(\tilde{x}_k, u_k, \xi_k).$$

By suitable reformulation of the cost functional the problem is reduced to the form of the basic problem. Note that it is necessary that $y_k, k = 1, \ldots, N - 1$, can be observed by the controller in order for the problem to be one of perfect state information. This is, for example, the case when the matrix $C_{k-1}$ is the identity matrix and $w_{k-1}$ is observable. The DP algorithm takes the form

$$J_N(x_N, y_N) = g_N(x_N),$$

$$J_k(x_k, y_k) = \inf_{u_k \in U_k(x_k)} \underset{\xi_k}{E} \{g_k[x_k, u_k, C_k(A_k y_k + \xi_k)] + J_{k+1}[f_k[x_k, u_k, C_k(A_k y_k + \xi_k)], A_k y_k + \xi_k]\}.$$

When $C_k$ is the identity matrix, the optimal controller is of the form

$$\{\mu_0^*(x_0), \mu_1^*(x_1, w_0), \ldots, \mu_{N-1}^*(x_{N-1}, w_{N-2})\}.$$

*Forecasts*

Finally let us consider the case where at time $k$ the decision maker has access to a forecast $y_k$ that results in a reassessment of the probability distribution of $w_k$ and possibly of future disturbances. For example, $y_k$ may be an exact prediction of $w_k$ or perhaps an exact prediction that the probability distribution of $w_k$ is a specific one out of a finite collection of distributions. Forecasts that can be of interest in practice are, for example, probabilistic predictions on the state of the weather, the interest rate for money, demand for inventory, etc., depending on the nature of the problem at hand.

Generally, forecasts can be handled by state augmentation although the reformulation into the form of the basic problem may be quite complex. We shall treat here only a simple situation.

Consider the case where the probability distribution of $w_k$ does not depend on $x_k, u_k, w_{k-1}, \ldots, w_0$. Assume that at the beginning of each period $k$ the decision maker receives an accurate prediction that the next disturbance $w_k$ will be selected in accordance with a particular probability distribution out of a finite collection of given distributions $\{P_{k|1}, \ldots, P_{k|n}\}$, i.e., if the forecast is $i$, then $w_k$ is selected according to $P_{k|i}$. The a priori probability that the forecast at time $k$ will be $i$, $i = 1, \ldots, n$, is $p_i^k$ and is given. Thus the forecasting process can be represented by means of the equation

$$y_{k+1} = \xi_k, \tag{13}$$

where $y_{k+1}$ can take the values $1, 2, \ldots, n$ and $\xi_k$ is a random variable taking the values $1, 2, \ldots, n$ with probabilities $p_1^{k+1}, \ldots, p_n^{k+1}$. The interpretation here is that when $\xi_k$ takes the value $i$, then $w_{k+1}$ will occur in accordance with the probability distribution $P_{k+1|i}$.

Now by combining the system equation and (13) we obtain an augmented system given by

$$\begin{bmatrix} x_{k+1} \\ y_{k+1} \end{bmatrix} = \begin{bmatrix} f_k(x_k, u_k, w_k) \\ \xi_k \end{bmatrix} = \tilde{f}_k(\tilde{x}_k, u_k, \tilde{w}_k).$$

The new state is $\tilde{x}_k = (x_k, y_k)$ and the new "disturbance" is $\tilde{w}_k = (w_k, \xi_k)$. The probability distribution of $\tilde{w}_k$ is given in terms of the distributions $P_{k|i}$ and the probabilities $p_i^k$, and depends explicitly on $\tilde{x}_k$ (via $y_k$) but not on the prior disturbances $\tilde{w}_{k-1}, \ldots, \tilde{w}_0$. Thus by suitable reformulation of the cost functional the problem can be cast into the framework of the basic problem. It is to be noted that the *control applied at each time is a function of both the current state and the current forecast*. The DP algorithm takes the form

$$J_N(x_N, y_N) = g_N(x_N),$$

$$J_k(x_k, y_k) = \inf_{u_k \in U_k(x_k)} \underset{w_k}{E} \left\{ g_k(x_k, u_k, w_k) + \sum_{i=1}^{n} p_i^{k+1} J_{k+1}[f_k(x_k, u_k, w_k), i] \,|\, y_k \right\},$$

$$k = 0, 1, \ldots, N-1,$$

where the expectation over $w_k$ is taken with respect to the probability distribution $P_{k|y_k}$, where $y_k$ may take the values $1, 2, \ldots, n$. Extension to forecasts covering several periods can be handled similarly, albeit at the expense of increased complexity. Problems where forecasts can be affected by the control action also admit a similar treatment.

It should be clear from the preceding discussion that state augmentation is a very general and potent device for reformulating problems of decision under uncertainty into the basic problem form. One should also realize that there are many ways to reformulate a problem by augmenting the state in different ways. The basic guideline to follow is to *select as the augmented state at time k only those variables the knowledge of which can be of benefit to the decision maker when making the kth decision.* For example, in the case of single period time lags it appears intuitively obvious that the controller can benefit from knowing at time $k$ the values of $x_k, x_{k-1}, u_{k-1}$, since these variables affect the value of the next state $x_{k+1}$ through the system equation. It is clear though that the controller has nothing to gain from knowing at time $k$ the values of $x_{k-2}, x_{k-3}, \ldots, u_{k-2}, \ldots,$ and for this reason these past states and controls need not be included in the augmented state although their inclusion is technically possible. The theme of considering as state variables in the reformulated problem only those variables the knowledge of which would be beneficial to the decision making process will be predominant in the discussion of problems with imperfect state information (Chapter 4).

Finally, we note that while state augmentation is a convenient device, it tends to introduce both analytical and computational complexities, which in many cases are insurmountable.

## 2.4 Notes

Dynamic programming is a simple mathematical technique that has been used for many years by engineers, mathematicians, and social scientists in a variety of contexts. It was Bellman, however, who realized in the early 1950s that DP could be developed (in conjunction with the then appearing digital computer) into a systematic tool for optimization. Bellman contributed a great deal to focusing attention on the broad scope of DP. In addition, many mathematical results related to DP (particularly for the infinite horizon case) are due to him. His early books [B3, B4] are still popular reading. Other texts related to DP are those by Howard [H16], Kaufmann and Cruon [K6], Kushner [K10], Nemhauser [N2], and White [W3]. For a rigorous treatment of sequential decision problems in general spaces see Striebel [S18a], Hinderer [H9], Blackwell *et al.* [B23a], and Freedman [F3a]. Risk-sensitive criteria (Problem 7) have been considered by Howard and Matheson [H17] in the context of finite-state Markov decision processes and by Jacobson [J1] in the context of automatic control of a linear system. However, the fact that sequential decision problems with multiplicative cost functionals can be treated by DP has been noted by Bellman [B3] (see Problems 9 and 11).

As mentioned in Section 2.1 the formulation of the basic problem and the subsequent developments are rigorous only for the case where the disturbance spaces are countable sets. *Nonetheless, the DP algorithm can often be utilized in a simple way when the countability assumption is not satisfied and there are further restrictions (such as measurability) on the class of admissible control laws.* The advanced reader will understand how this can be done by solving Problem 12, which shows that if one can find within a subset of control laws (such as those satisfying certain measurability restrictions) a control law that attains the minimum in the DP algorithm, then this control law is optimal. This fact may be used to establish rigorously many of our subsequent results concerning specific applications in Chapters 3 and 4. For example, in linear-quadratic problems (Section 3.1) one determines from the DP equations a control law that is a linear function of the current state. When $w_k$ can take uncountably many values it is necessary that admissible control laws consist only of functions $\mu_k$, which are Borel measurable. Since the linear control law belongs to this class, the result of Problem 12 guarantees that this control law is optimal.

## Problems

**1.** Use the DP algorithm to solve the following two problems:

(a) minimize $\sum_{i=0}^{3} x_i^2 + u_i^2$
subject to $x_0 = 0,\ x_4 = 8,\ u_i =$ nonnegative integer,
$x_{i+1} = x_i + u_i,\ i = 0, 1, 2, 3;$

(b) minimize $\sum_{i=0}^{3} x_i^2 + 2u_i^2$
subject to $x_0 = 5,\ u_i \in \{0, 1, 2\},$
$x_{i+1} = x_i - u_i,\ i = 0, 1, 2, 3,$

**2.** Air transportation is available between $n$ cities, in some cases directly and in others through intermediate stops and change of carrier. The air fare between cities $i$ and $j$ is denoted $C_{ij}$ ($C_{ij} = C_{ji}$) and for notational convenience we write $C_{ij} = \infty$ if there is no direct flight between $i$ and $j$. The problem is to find the cheapest possible air fare for going from any city $i$ to any other city $j$ perhaps through intermediate stops. Formulate a DP algorithm for solving this problem. Solve the problem for $n = 6$ and $C_{12} = 30$, $C_{13} = 60$, $C_{14} = 25$, $C_{15} = C_{16} = \infty$, $C_{23} = C_{24} = C_{25} = \infty$, $C_{26} = 50$, $C_{34} = 35$, $C_{35} = C_{36} = \infty$, $C_{45} = 15$, $C_{46} = \infty$, $C_{56} = 15$.

**3.** Suppose we have a machine that is either running or broken down. If it runs throughout one week, it makes a gross profit of \$100. If it fails during the week, gross profit is zero. If it is running at the start of the week and we perform preventive maintenance, the probability that it will fail during the week is 0.4. If we do not perform such maintenance, the probability of failure is 0.7. However, maintenance will cost \$20. When the machine is broken down at the start of the week it may either be repaired at a cost of \$40 in which case it will fail during the week with a probability of 0.4 or it may be replaced at a cost of \$150 by a new machine that is guaranteed to run through its first week of operation. Find the optimal repair, replacement, and maintenance policy that maximizes total profit over four weeks, assuming a new machine at the start of the first week.

**4.** A game of the blackjack variety is played by two players as follows: Both players throw a die. The first player, knowing his opponent's result, may stop or may throw the die again and add the result to the result of his previous throw. He then may stop or throw again and add the result of the new throw to the sum of his previous throws. He may repeat this process as many times as he wishes. If his sum exceeds seven (i.e., he busts), he loses the game. If he stops before exceeding seven, the second player takes over and throws the die successively until the sum of his throws is four or higher. If the sum of the second player is over seven, he loses the game. Otherwise the player with the larger sum wins, and in case of a tie the second player wins. The problem is to determine a stopping strategy for the first player that maximizes his probability of winning for each possible initial throw of the second player. Formulate the problem in terms of DP and find an optimal stopping strategy for the case where the second player's initial throw is three.

*Hint* Take $N = 6$ and a state space consisting of the following 15 states:

$x^1$: busted,

$x^{1+i}$: already stopped at sum $i$ $(1 \leqslant i \leqslant 7)$,

$x^{8+i}$: current sum is $i$ but the player has not yet stopped $(1 \leqslant i \leqslant 7)$.

The optimal strategy is to throw until the sum is four or higher.

**5.** *Min–Max Problems* In the framework of the basic problem consider the case where the disturbances $w_0, w_1, \ldots, w_{N-1}$ do not have a probabilistic description but rather are known to belong to corresponding given sets $W_k(x_k, u_k) \subset D_k$, $k = 0, 1, \ldots, N-1$, which may depend on the current value of the state $x_k$ and control $u_k$. Consider the problem of finding a control law $\pi = \{\mu_0, \ldots, \mu_{N-1}\}$ with $\mu_k(x_k) \in U_k(x_k)\ \forall x_k, k$, which minimizes the cost functional

$$J_\pi(x_0) = \sup_{\substack{w_k \in W_k[x_k, \mu_k(x_k)] \\ k=0,1,\ldots,N-1}} \left\{ g_N(x_N) + \sum_{k=0}^{N-1} g_k[x_k, \mu_k(x_k), w_k] \right\}.$$

Obtain a DP algorithm for the solution of this problem.

**6.** *Discounted Criteria* In the framework of the basic problem consider the case where the cost functional is of the form

$$E\left\{ \alpha^N g_N(x_N) + \sum_{k=0}^{N-1} \alpha^k g_k(x_k, u_k, w_k) \right\},$$

where $\alpha$ is a discount factor with $0 < \alpha < 1$. Show that an alternate form of the DP algorithm is given by

$$V_N(x_N) = g_N(x_N),$$
$$V_k(x_k) = \inf_{u_k \in U_k(x_k)} E\{g_k(x_k, u_k, w_k) + \alpha V_{k+1}[f_k(x_k, u_k, w_k)]\}.$$

**7.** *Risk-Sensitive Criteria* In the framework of the basic problem consider the case where the cost functional is of the form

$$\underset{\substack{w_k \\ k=0,1,\ldots,N-1}}{E} \left\{ \exp\left[ g_N(x_N) + \sum_{k=0}^{N-1} g_k(x_k, u_k, w_k) \right] \right\}.$$

(a) Show that the optimal value $J^*(x_0)$ of the problem is equal to $J_0(x_0)$ where the function $J_0$ is obtained from the last step of the DP algorithm

$$J_N(x_N) = \exp[g_N(x_N)],$$

$$J_k(x_k) = \inf_{u_k \in U_k(x_k)} \underset{w_k}{E} \{J_{k+1}[f_k(x_k, u_k, w_k)] \exp[g_k(x_k, u_k, w_k)]\}.$$

Show that the algorithm above yields an optimal control law if one exists.

(b) Define the functions $V_k(x_k) = \ln J_k(x_k)$. Assume also that $g_k$ is a function of $x_k$ and $u_k$ only (and not of $w_k$). Show that the above DP algorithm can be rewritten

$$V_N(x_N) = g_N(x_N),$$

$$V_k(x_k) = \inf_{u_k \in U_k(x_k)} \left[ g_k(x_k, u_k) + \ln \underset{w_k}{E} \{\exp V_{k+1}[f_k(x_k, u_k, w_k)]\} \right].$$

**8.** *Terminating Process* Consider the case in the basic problem where the system evolution terminates at time $i$ when a certain given value $\bar{w}_i$ of the disturbance at time $i$ occurs, or when a termination decision $u_i$ is made by the controller. If termination occurs at time $i$, the resulting cost is

$$T + \sum_{k=0}^{i} g_k(x_k, u_k, w_k),$$

where $T$ is a termination cost. If the process has not terminated up to the final time $N$, the resulting cost is $g_N(x_N) + \sum_{k=0}^{N-1} g_k(x_k, u_k, w_k)$. Reformulate the problem into the framework of the basic problem.

*Hint* Augment the state space by introducing a "termination" state.

**9.** In the framework of the basic problem consider the case where the cost functional is of the multiplicative form

$$\underset{\substack{w_k \\ k=0,\ldots,N-1}}{E} \{g_N(x_N) \cdot g_{N-1}(x_{N-1}, u_{N-1}, w_{N-1}) \cdots g_0(x_0, u_0, w_0)\}$$

Devise an algorithm of the DP type that is applicable to this problem under the assumption $g_k(x_k, u_k, w_k) \geqslant 0$ for all $x_k, u_k, w_k$, and $k$.

**10.** Assume that we have a vessel whose maximum weight capacity is $z$ and whose cargo is to consist of different quantities of $N$ different items. Let $v_i$ denote the value of the $i$th type of item, $w_i$ the weight of $i$th type of item, and $x_i$ the number of items of type $i$ that are loaded in the vessel. The problem of determining the most valuable cargo is that of maximizing $\sum_{i=1}^{N} x_i v_i$ subject to the constraints $\sum_{i=1}^{N} x_i w_i \leqslant z$ and $x_i = 0, 1, 2, \ldots$. Formulate this problem in terms of DP.

**11.** Consider a device consisting of $N$ stages connected in series, where each stage consists of a particular component. The components are subject to failure and in order to increase the reliability of the device duplicate components are provided. For $j = 1, 2, \ldots, N$ let $(1 + m_j)$ be the number of components for the $j$th stage, let $p_j(m_j)$ be the probability of successful operation of the $j$th stage when $(1 + m_j)$ components are used, and let $c_j$ denote the cost of a single component at the $j$th stage. Consider the problem of finding the number of components at each stage that maximize the reliability of the device expressed by

$$p_1(m_1) \cdot p_2(m_2) \cdots p_N(m_N)$$

subject to the cost constraint $\sum_{j=1}^{N} c_j m_j \leqslant A$, where $A > 0$ is given. Formulate the problem in terms of DP.

**12.** Consider a variation of the basic problem whereby we seek

$$\inf_{\pi \in \tilde{\Pi}} J_\pi(x_0),$$

where $\tilde{\Pi}$ is some given subset of the set of sequences $\{\mu_0, \mu_1, \ldots, \mu_{N-1}\}$ of functions $\mu_k: S_k \to C_k$ with $\mu_k(x_k) \in U_k(x_k)$ for all $x_k \in S_k$. Assume that

$$\pi^* = \{\mu_0^*, \mu_1^*, \ldots, \mu_{N-1}^*\}$$

belongs to $\tilde{\Pi}$ and satisfies for all $k = 0, 1, \ldots, N-1$ and $x_k \in S_k$

$$\begin{aligned} \tilde{J}_k(x_k) &= \underset{w_k}{E}\,\{g_k[x_k, \mu_k^*(x_k), w_k] + \tilde{J}_{k+1}[f_k(x_k, \mu_k^*(x_k), w_k)]\} \\ &= \inf_{u_k \in U_k(x_k)} \underset{w_k}{E}\,\{g_k(x_k, u_k, w_k) + \tilde{J}_{k+1}[f_k(x_k, u_k, w_k)]\}, \end{aligned}$$

with $\tilde{J}_N(x_N) = g_N(x_N)$ and furthermore the functions $\tilde{J}_k$ are real valued and the expectations above are well defined and finite. Show that

$$\tilde{J}_0(x_0) = \inf_{\pi \in \tilde{\Pi}} J_\pi(x_0) = J_{\pi^*}(x_0).$$

**13.** *Semilinear Systems* Consider a problem involving the system

$$x_{k+1} = A_k x_k + f_k(u_k) + w_k,$$

where $x_k \in R^n$, $f_k$ are given functions, and $A_k$ and $w_k$ are random $n \times n$ matrices and $n$-vectors, respectively, with given probability distributions that do not depend on $x_k$, $u_k$ or prior values of $A_k$ and $w_k$. Assume that the cost functional is of the form

$$\underset{\substack{A_k, w_k \\ k=0,1,\ldots,N-1}}{E} \left\{ c_N' x_N + \sum_{k=0}^{N-1} [c_k' x_k + g_k[\mu_k(x_k)]] \right\},$$

where $c_k$ are given vectors and $g_k$ given functions. Show that if the optimal value for this problem is finite and the control constraint sets $U_k(x_k)$ are independent of $x_k$, then the "cost-to-go" functions of the DP algorithm are affine (linear plus constant). Assuming that there is at least one optimal policy, show that there exists an optimal policy that consists of constant functions $\mu_k^*$, i.e., $\mu_k^*(x_k) =$ const for all $x_k \in R^n$.

**14.** A farmer annually producing $x_k$ units of a certain crop stores $(1 - u_k)x_k$ units of his production, where $0 \leqslant u_k \leqslant 1$, and invests the remaining $u_k x_k$ units, thus increasing the next year's production to a level $x_{k+1}$ given by

$$x_{k+1} = x_k + w_k u_k x_k, \qquad k = 0, 1, \ldots, N-1.$$

The scalars $w_k$ are independent random variables with identical probability distributions which do not depend either on $x_k$ or $u_k$. Furthermore $E\{w_k\} = \bar{w} > 0$. The problem is to find the optimal investment policy that maximizes the total expected product stored over $N$ years

$$\underset{\substack{w_k \\ k=0,1,\ldots,N-1}}{E} \left\{ x_N + \sum_{k=0}^{N-1} (1 - u_k)x_k \right\}.$$

Show that one optimal control law is given by:

(a) If $\bar{w} > 1$, $\mu_0^*(x_0) = \cdots = \mu_{N-1}^*(x_{N-1}) = 1$.
(b) If $0 < \bar{w} < 1/N$, $\mu_0^*(x_0) = \cdots = \mu_{N-1}^*(x_{N-1}) = 0$.
(c) If $1/N \leqslant \bar{w} \leqslant 1$,

$$\begin{aligned} \mu_0^*(x_0) &= \cdots = \mu_{N-\bar{k}-1}^*(x_{N-\bar{k}-1}) = 1, \\ &\;\;\vdots \\ \mu_{N-\bar{k}}^*(x_{N-\bar{k}}) &= \cdots = \mu_{N-1}^*(x_{N-1}) = 0, \end{aligned}$$

where $\bar{k}$ is such that $1/(\bar{k}+1) < \bar{w} \leqslant 1/\bar{k}$.

Note that this control law consists of constant functions.

**15.** Let $x_k$ denote the number of educators in a certain country at time $k$ and let $y_k$ denote the number of research scientists at time $k$. New scientists (potential educators or research scientists) are produced during the $k$th period by educators at a rate $\gamma_k$ per educator, while educators and research scientists leave the field due to death, retirement, and transfer at a rate $\delta_k$. The scalars $\gamma_k$, $k = 0, 1, \ldots, N-1$, are independent identically distributed random variables taking values within a closed and bounded interval of positive numbers. Similarly $\delta_k$, $k = 0, 1, \ldots, N-1$, are independent identically distributed and take values in an interval $[\delta, \delta']$ with $0 < \delta \leqslant \delta' < 1$. By means of incentives a science policy maker can determine the proportion

$u_k$ of new scientists produced at time $k$ who become educators. Thus the number of research scientists and educators evolves according to the equations

$$\begin{aligned} x_{k+1} &= (1 - \delta_k)x_k + u_k \gamma_k x_k, \\ y_{k+1} &= (1 - \delta_k)y_k + (1 - u_k)\gamma_k x_k, \end{aligned} \qquad k = 0, 1, \ldots, N - 1.$$

The initial numbers $x_0, y_0$ are known and it is required to find a policy $\{\mu_0^*(x_0, y_0), \ldots, \mu_{N-1}^*(x_{N-1}, y_{N-1})\}$ with

$$0 < \alpha \leqslant \mu_k^*(x_k, y_k) \leqslant \beta < 1, \qquad \forall x_k, y_k, \quad k = 0, 1, \ldots, N - 1,$$

which maximizes $E_{\gamma_k, \delta_k}\{y_N\}$, i.e., the expected final number of research scientists after $N$ periods. The scalars $\alpha$ and $\beta$ are given.

(a) Show that the "cost-to-go" functions $J_k(x_k, y_k)$ are linear, i.e., for some scalars $\lambda_k, \mu_k$

$$J_k(x_k, y_k) = \lambda_k x_k + \mu_k y_k.$$

(b) Derive an optimal policy $\{\mu_0^*, \ldots, \mu_{N-1}^*\}$ under the assumption $E\{\gamma_k\} > E\{\delta_k\}$, and show that this optimal policy can consist of constant functions.

(c) Assume that the actual proportion of new scientists who become educators at time $k$ is $u_k + \varepsilon_k$ (rather than $u_k$), where $\varepsilon_k$ are identically distributed independent random variables that are also independent of $\gamma_k, \delta_k$ and take values in the interval $[-\alpha, 1 - \beta]$. Derive the form of the "cost-to-go" functions and the optimal policy.

**16.** *Optimal Path Algorithm (Dijkstra)* Consider the problem of Example 2 in Section 2.3 with the difference that $C_{ij}$ can take the value $\infty$, and $C_{ii} = \infty$ for all $i$. For $i = 1, 2, \ldots, N$ define $V_{N-1}(i) = C_{i,(N+1)}$ and let $s_{N-1}$ minimize $V_{N-1}(i)$ over all $i$. For $k = 0, 1, \ldots, N - 2$ define

$$V_k(i) = \begin{cases} V_{k+1}(i) \text{ if } i \in \{s_{k+1}, s_{k+2}, \ldots, s_{N-1}\} \\ \min\{V_{k+1}(i), C_{is_{k+1}} + V_{k+1}(s_{k+1})\} \text{ otherwise,} \end{cases}$$

and let $s_k$ minimize $V_k(i)$ over all $i \notin \{s_{k+1}, s_{k+2}, \ldots, s_{N-1}\}$. Show that, for $k = 0, 1, \ldots, N - 1$, the minimal cost to get from $s_k$ to $N + 1$ equals $V_k(s_k)$.

**17.** *Deterministic Finite State Problems* Consider the deterministic case of the basic problem ($D_k$ has a single element) where the spaces $S_k$ and $C_k$ are finite sets.

(a) Show that such a problem can be reduced to an optimal path problem such as the one of Problem 16.

(b) Use the algorithm of Problem 16 to solve Problems 1(a) and 1(b). *Hint*: Associate a node with each state in $S_k$, $k = 0, 1, \ldots, N$. Add a constant if necessary to $g_k$ to ensure that the cost per stage is nonnegative.

*Chapter 3*

# Applications in Specific Areas

## 3.1 Linear Systems with Quadratic Cost Functional —The Certainty Equivalence Principle

In this section we consider the special case of a linear system

$$x_{k+1} = A_k x_k + B_k u_k + w_k, \qquad k = 0, 1, \ldots, N-1,$$

where the objective is to find a control law $\{\mu_0^*(x_0), \ldots, \mu_{N-1}^*(x_{N-1})\}$ minimizing the quadratic cost functional

$$\underset{\substack{w_k \\ k=0,\ldots,N-1}}{E} \left\{ x_N' Q_N x_N + \sum_{k=0}^{N-1} (x_k' Q_k x_k + u_k' R_k u_k) \right\}.$$

In the above expressions $x_k \in R^n$, $u_k \in R^m$, and the matrices $A_k$, $B_k$, $Q_k$, $R_k$ are given and have appropriate dimension. We assume that $Q_k$ are symmetric positive semidefinite matrices and $R_k$ are symmetric and positive definite. The disturbances $w_k$ are independent random vectors with given probability distributions that do not depend on $x_k$, $u_k$. Furthermore, the vectors $w_k$ have zero mean and finite second moments. The control $u_k$ is unconstrained.

The problem above represents a popular formulation of a regulation problem whereby we desire to keep the state of the system close to the origin. Such problems are common in the theory of automatic control of a motion or a process. The quadratic cost functional is often a reasonable one since it induces a high penalty for large deviations of the state from the origin but a relatively small penalty for small deviations. However, even in cases where the quadratic cost functional is not entirely justified, it is still used since it leads to an elegant analytical solution that can often be implemented with relative ease in engineering applications. A number of variations and generalizations of this problem also have similar solutions. For example, the disturbances $w_k$ could have nonzero means and the quadratic cost functional could have the form

$$E\left\{(x_N - \bar{x}_N)'Q_N(x_N - \bar{x}_N) + \sum_{k=0}^{N-1} [(x_k - \bar{x}_k)'Q_k(x_k - \bar{x}_k) + u_k' R_k u_k]\right\}.$$

This cost functional expresses a desire to keep the state of the system close to a certain given trajectory $(\bar{x}_0, \bar{x}_1, \ldots, \bar{x}_N)$ rather than close to the origin. The analysis of the corresponding problem is very similar to the one of the present section and is left to the reader. Another generalization is the case where $A_k$, $B_k$ are independent random matrices, rather than being known. This case will be considered subsequently in this section.

Applying now the DP algorithm we have

$$J_N(x_N) = x_N' Q_N x_N, \tag{1}$$

$$J_k(x_k) = \min_{u_k} E\{x_k' Q_k x_k + u_k' R_k u_k + J_{k+1}(A_k x_k + B_k u_k + w_k)\}. \tag{2}$$

It turns out that the "cost-to-go" functions $J_k$ are quadratic and as a result the control law is a linear function of the state. These facts can be verified by straightforward calculation. By expansion of the quadratic form (1) in (2) for $k = N - 1$, and by using the fact that $E\{w_{N-1}\} = 0$ to eliminate the term $E\{w_{N-1}' Q_N(A_{N-1}x_{N-1} + B_{N-1}u_{N-1})\}$, we have

$$\begin{aligned} J_{N-1}(x_{N-1}) = x_{N-1}' Q_{N-1} x_{N-1} + \min_{u_{N-1}} [&u_{N-1}' R_{N-1} u_{N-1} \\ &+ u_{N-1}' B_{N-1}' Q_N B_{N-1} u_{N-1} + x_{N-1}' A_{N-1}' Q_N A_{N-1} x_{N-1} \\ &+ 2x_{N-1}' A_{N-1}' Q_N B_{N-1} u_{N-1}] + E\{w_{N-1}' Q_N w_{N-1}\}. \end{aligned}$$

By differentiating with respect to $u_{N-1}$ and setting the derivative equal to zero, we obtain

$$(R_{N-1} + B_{N-1}' Q_N B_{N-1}) u_{N-1} = -B_{N-1}' Q_N A_{N-1} x_{N-1}.$$

The matrix multiplying $u_{N-1}$ on the left is positive definite (and hence invertible), since $R_{N-1}$ is positive definite and $B_{N-1}' Q_N B_{N-1}$ is positive semidefinite. As a result the minimizing control vector is given by

$$u_{N-1}^* = -(R_{N-1} + B_{N-1}' Q_N B_{N-1})^{-1} B_{N-1}' Q_N A_{N-1} x_{N-1}.$$

By substitution into the expression for $J_{N-1}$, we have

$$J_{N-1}(x_{N-1}) = x'_{N-1}K_{N-1}x_{N-1} + E\{w'_{N-1}Q_N w_{N-1}\},$$

where the matrix $K_{N-1}$ is obtained by straightforward calculation and is given by

$$K_{N-1} = A'_{N-1}[Q_N - Q_N B_{N-1}(B'_{N-1}Q_N B_{N-1} + R_{N-1})^{-1}B'_{N-1}Q_N]A_{N-1} + Q_{N-1}.$$

The matrix $K_{N-1}$ is clearly symmetric. It is also positive semidefinite. This can be seen from the fact that from the calculation given above we have for every $x \in R^n$

$$x'K_{N-1}x = \min_u[x'Q_{N-1}x + u'R_{N-1}u + (A_{N-1}x + B_{N-1}u)'Q_N(A_{N-1}x + B_{N-1}u)].$$

Since $Q_{N-1}$, $R_{N-1}$, and $Q_N$ are positive semidefinite, the expression within brackets is nonnegative. Since minimization over $u$ preserves nonnegativity if follows that $x'K_{N-1}x \geqslant 0$ for all $x \in R^n$. Hence $K_{N-1}$ is positive semidefinite.

Now in view of the fact that $J_{N-1}$ above is a positive semidefinite quadratic function (plus an inconsequential constant term) we may proceed in an entirely similar manner and obtain from the DP equation (2) the optimal control law for stage $N-2$. Similarly we show that $J_{N-2}$ is a positive semidefinite quadratic function, and proceeding sequentially we obtain the optimal control law for every $k$. This control law has the form

$$\mu_k^*(x_k) = L_k x_k, \tag{3}$$

where the gain matrices $L_k$ are given by the equation

$$L_k = -(B'_k K_{k+1} B_k + R_k)^{-1} B'_k K_{k+1} A_k, \tag{4}$$

and where the symmetric positive semidefinite matrices $K_k$ are given recursively by the algorithm

$$K_N = Q_N, \tag{5}$$

$$K_k = A'_k[K_{k+1} - K_{k+1}B_k(B'_k K_{k+1} B_k + R_k)^{-1}B'_k K_{k+1}]A_k + Q_k. \tag{6}$$

The optimal value of the cost functional is given by

$$J_0(x_0) = x'_0 K_0 x_0 + \sum_{k=0}^{N-1} E\{w'_k K_{k+1} w_k\}.$$

The attractive aspect of the solution of this problem is the relative ease with which the control law (3) can be computed and implemented in engineering applications. The current state $x_k$ is being fed back as input through the

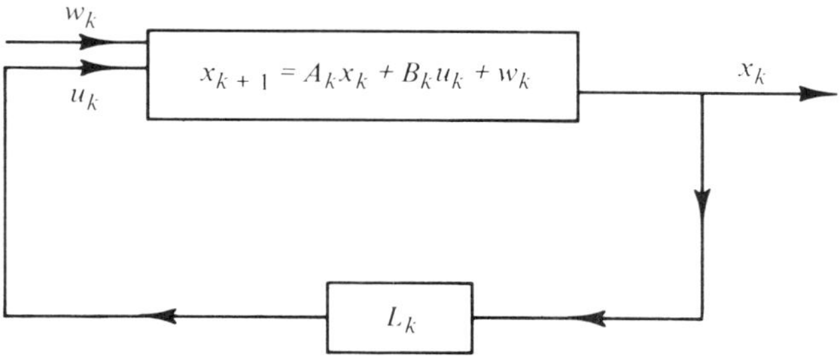


FIGURE 3.1

linear feedback gain matrix $L_k$ as shown in Fig. 3.1. This fact accounts for the great popularity of the linear–quadratic formulation. As we shall see in Chapter 4 the linearity of the control law is still maintained even for problems where the state $x_k$ of the system is not completely observable (imperfect state information).

*The Riccati Equation and Its Asymptotic Behavior*

Equation (6) is called the *discrete matrix Riccati equation* since it is the discrete-time analog of (matrix) Riccati differential equations. It plays an important role in modern control theory. Its properties have been studied extensively and exhaustively. One interesting property of the Riccati equation is that whenever the matrices $A_k$, $B_k$, $Q_k$, $R_k$ are constant and equal to $A$, $B$, $Q$, $R$, respectively, then as $k \to -\infty$ the solution $K_k$ converges (under mild assumptions) to a steady-state solution $K$ satisfying the so-called *algebraic matrix Riccati equation*

$$K = A'[K - KB(B'KB + R)^{-1}B'K]A + Q. \tag{7}$$

We shall shortly provide a proof of this important property, which will also be useful in Chapter 6. Based on this property, when one is faced with a problem involving the stationary linear system

$$x_{k+1} = Ax_k + Bu_k + w_k, \qquad k = 0, 1, \ldots, N-1, \tag{8}$$

and the number of stages $N$ is large, one can reasonably approximate the control law (3) by a linear *stationary* control law of the form $\{\mu^*\mu^*, \ldots, \mu^*\}$ where

$$\mu^*(x) = Lx, \tag{9}$$

$$L = -(B'KB + R)^{-1}B'KA, \tag{10}$$

and $K$ is the steady-state solution of the Riccati equation (6) satisfying (7). This control law is even more attractive for implementation purposes in many engineering applications.

We now turn to proving convergence of the sequence of matrices $\{K_k\}$ generated by the Riccati equation (5) and (6). We first introduce the notions of controllability and observability, which are of major importance in modern control theory.

**Definition** A pair $(A, B)$, where $A$ is an $n \times n$ matrix and $B$ an $n \times m$ matrix is said to be *controllable* if the $n \times nm$ matrix

$$[B, AB, A^2B, \ldots, A^{n-1}B]$$

has full rank (i.e., has linearly independent rows). A pair $(A, C)$, where $A$ is an $n \times n$ matrix and $C$ an $m \times n$ matrix, is said to be *observable* if the pair $(A', C')$ is controllable, where $A'$ and $C'$ denote the transposes of $A$ and $C$, respectively.

One may easily prove that if the pair $(A, B)$ is controllable, then for any initial state $x_0$ there exists a sequence of control vectors $u_0, u_1, \ldots, u_{n-1}$ that force the state $x_n$ of the system

$$x_{k+1} = Ax_k + Bu_k \tag{11}$$

to be equal to zero at time $n$. This is true simply because from the system equation we obtain

$$x_n = A^n x_0 + Bu_{n-1} + ABu_{n-2} + \cdots + A^{n-1}Bu_0$$

or equivalently

$$x_n - A^n x_0 = [B, AB, \ldots, A^{n-1}B]\begin{bmatrix} u_{n-1} \\ \vdots \\ u_0 \end{bmatrix}. \tag{12}$$

Now if $(A, B)$ is controllable, the matrix $[B, AB, \ldots, A^{n-1}B]$ has full rank and as a result the right-hand side of (12) can be made equal to any vector in $R^n$ by appropriate selection of $(u_0, u_1, \ldots, u_{n-1})$. In particular, one can choose $(u_0, u_1, \ldots, u_{n-1})$ so that the right-hand side of (12) is equal to $-A^n x_0$, which implies $x_n = 0$. This property explains the name "controllable pair" and in fact is often used to define controllability. The notion of observability has an analogous interpretation in the context of estimation problems (see e.g., Meditch [M6]).

**Definition** We say that an $n \times n$ matrix $D$ is *stable* if $\lim_{k\to\infty} D^k x = 0$ for every vector $x \in R^n$ (or equivalently $\lim_{k\to\infty} D^k = 0$).

This definition is motivated by the fact that if $D$ is a stable matrix, then the state $x_k$ of the system $x_{k+1} = Dx_k$ tends to zero as $k \to \infty$ for an arbitrary state $x_0$. The notion of stability is, of course, of paramount importance in control theory. In the context of our problem it is important to be assured that the stationary control law, Eqs. (9) and (10), results in a stable system, i.e., the matrix $(A + BL)$ is a stable matrix, and hence, in the absence of input disturbance, the state $x_k$ of the closed-loop system resulting upon substitution of the control law (9),

$$x_k = (A + BL)x_{k-1} = (A + BL)^k x_0, \qquad k = 0, 1, \ldots,$$

tends to zero as $k \to \infty$.

We now have the following proposition, which shows that, for a stationary system and constant matrices $Q_k$, $R_k$, under controllability and observability conditions the solution of the Riccati equation (5) and (6) converges to a positive definite matrix $K$ for an arbitrary positive semidefinite initial matrix. By matrix convergence we mean that every element of the matrices of the sequence converges to the corresponding element of the limit matrix. In addition, the proposition shows that the corresponding control law, Eqs. (9) and (10), results in a stable system.

**Proposition** Let $A$ be an $n \times n$ matrix, $B$ an $n \times m$ matrix, $Q$ an $n \times n$ symmetric positive semidefinite matrix, and $R$ an $m \times m$ symmetric positive definite matrix. Consider the discrete-time Riccati equation

$$P_{k+1} = A'[P_k - P_k B(B'P_k B + R)^{-1}B'P_k]A + Q, \qquad k = 0, 1, \ldots, \tag{13}$$

where the initial matrix $P_0$ is an arbitrary positive semidefinite symmetric matrix. Assume that the pair $(A, B)$ is controllable. Assume also that $Q$ may be written as $C'C$, where the pair $(A, C)$ is observable.† Then:

(a) There exists a positive definite symmetric matrix $P$ such that for every positive semidefinite symmetric initial matrix $P_0$ we have

$$\lim_{k \to \infty} P_k = P.$$

Furthermore, $P$ is the unique solution of the algebraic matrix equation

$$P = A'[P - PB(B'PB + R)^{-1}B'P]A + Q \tag{14}$$

within the class of positive semidefinite symmetric matrices.

(b) The matrix

$$D = A + BL, \tag{15}$$

where

$$L = -(B'PB + R)^{-1}B'PA, \tag{16}$$

is a stable matrix.

† Notice that if $r$ is the rank of $Q$, there exists an $r \times n$ matrix $C$ of rank $r$ such that $Q = C'C$ (see Appendix A).

*Proof* The proof proceeds in several steps. First we show convergence of the sequence generated by (13) when the initial matrix $P_0$ is equal to zero. Next we show that the corresponding matrix $D$ of (15) is stable. Subsequently we show convergence of the sequence generated by (13) when $P_0$ is any positive semidefinite symmetric matrix, and finally we show uniqueness of the solution of (14).

*Initial Matrix* $P_0 = 0$ Consider the optimal control problem of finding a sequence $u_0, u_1, \ldots, u_{k-1}$ that minimizes

$$\sum_{i=0}^{k-1} (x_i' Q x_i + u_i' R u_i) \tag{17}$$

subject to

$$x_{i+1} = A x_i + B u_i, \qquad i = 0, 1, \ldots, k-1, \tag{18}$$

where $x_0$ is given. The optimal value of this problem, according to the theory of this section, is

$$x_0' P_k(0) x_0,$$

where $P_k(0)$ is given by the Riccati equation (13) with $P_0 = 0$. We have

$$x_0' P_k(0) x_0 \leqslant x_0' P_{k+1}(0) x_0 \qquad \forall x_0 \in R^n, \quad k = 0, 1, \ldots,$$

since for any control sequence $(u_0, u_1, \ldots, u_k)$ we have

$$\sum_{i=0}^{k-1} (x_i' Q x_i + u_i' R u_i) \leqslant \sum_{i=0}^{k} (x_i' Q x_i + u_i' R u_i)$$

and hence

$$\begin{aligned} x_0' P_k(0) x_0 &= \min_{u_i,\, i=0,\ldots,k-1} \sum_{i=0}^{k-1} (x_i' Q x_i + u_i' R u_i) \\ &\leqslant \min_{u_i,\, i=0,\ldots,k} \sum_{i=0}^{k} (x_i' Q x_i + u_i' R u_i) = x_0' P_{k+1}(0) x_0, \end{aligned}$$

where both minimizations are subject to the system equation constraint $x_{i+1} = A x_i + B u_i$. Furthermore, for a fixed $x_0$ and for every $k$, $x_0' P_k(0) x_0$ is bounded above by the cost corresponding to a control sequence that forces $x_0$ to the origin in $n$ steps and applies zero control after that. Such a sequence exists by the controllability assumption. Thus the sequence $\{x_0' P_k(0) x_0\}$ is increasing and bounded above and therefore converges to some real number for every $x_0 \in R^n$. It follows that the sequence $\{P_k(0)\}$ converges to some matrix $P$ in the sense that each of the sequences of the elements of $P_k(0)$ converge to the corresponding elements of $P$. To see this take $x_0 = (1, 0, \ldots, 0)$.

It follows that the sequence of first diagonal elements of $P_k(0)$ converges to the first diagonal element of $P$. Similarly by taking $x_0 = (0, \ldots, 0, 1, 0, \ldots, 0)$ with the one in the $i$th coordinate, for $i = 2, \ldots, n$, it follows that all the diagonal elements of $P_k(0)$ converge to the corresponding diagonal elements of $P$. Next take $x_0 = (1, 1, 0, \ldots, 0)$ to show that the second elements of the first row converge. Similarly proceeding we obtain

$$\lim_{k \to \infty} P_k(0) = P,$$

where $P_k(0)$ are generated by (13) with $P_0 = 0$. Furthermore, the limit matrix $P$ is positive semidefinite and symmetric. Now by taking the limit in (13) it follows that $P$ satisfies

$$P = A'[P - PB(B'PB + R)^{-1}B'P]A + Q. \tag{19}$$

Furthermore, if we define

$$L = -(B'PB + R)^{-1}B'PA, \qquad D = A + BL \tag{20}$$

by direct calculation we can verify the following equality, which will be useful subsequently in the proof:

$$P = D'PD + Q + L'RL. \tag{21}$$

*Stability of* $D = A + BL$ Consider the system

$$x_{k+1} = (A + BL)x_k = Dx_k \tag{22}$$

for an arbitrary initial state $x_0$. Since

$$x_k = D^k x_0,$$

it will be sufficient to show that $x_k \to 0$ as $k \to \infty$. Now we have for all $k$ by using (21)

$$x'_{k+1}Px_{k+1} - x'_k Px_k = x'_k(D'PD - P)x_k = -x'_k(Q + L'RL)x_k.$$

Hence

$$x'_{k+1}Px_{k+1} = x'_0 Px_0 - \sum_{i=0}^{k} x'_i(Q + L'RL)x_i. \tag{23}$$

Since the left-hand side of the equation is bounded below by zero it follows that

$$x'_k(Q + L'RL)x_k \to 0.$$

Using the fact that $R$ is positive definite and $Q$ may be written as $C'C$, we obtain

$$\lim_{k \to \infty} Cx_k = 0, \qquad \lim_{k \to \infty} Lx_k = 0. \tag{24}$$

From (22) we have

$$\begin{bmatrix} C\left(x_{k+n-1} - \sum_{i=1}^{n-1} A^{i-1}BLx_{k+n-i-1}\right) \\ C\left(x_{k+n-2} - \sum_{i=1}^{n-2} A^{i-1}BLx_{k+n-i-2}\right) \\ \vdots \\ C(x_{k+1} - BLx_k) \\ Cx_k \end{bmatrix} = \begin{bmatrix} CA^{n-1} \\ CA^{n-2} \\ \vdots \\ CA \\ C \end{bmatrix} x_k. \tag{25}$$

By (24) the left-hand side tends to zero and hence the right-hand side tends to zero also. By the observability assumption, however, the matrix multiplying $x_k$ on the right side of (25) has full rank. It follows that $x_k \to 0$ and hence the matrix $D$ of (21) is stable.

*Positive Definiteness of P* Assume the contrary, i.e., that there exists some $x_0 \neq 0$ such that $x_0' P x_0 = 0$. Then from (23) we obtain

$$x_k'(Q + L'RL)x_k = 0 \qquad \forall k = 0, 1, \ldots,$$

where $x_k = D^k x_0$. This in turn implies [cf. Eq. (24)]

$$Cx_k = 0, \qquad Lx_k = 0 \qquad \forall k = 0, 1, \ldots.$$

Consider now (25) for $k = 0$. By the above equalities the left-hand side is zero and hence

$$0 = \begin{bmatrix} CA^{n-1} \\ \vdots \\ CA \\ C \end{bmatrix} x_0.$$

Since the matrix multiplying $x_0$ above has full rank by the observability assumption we obtain $x_0 = 0$, which contradicts the hypothesis $x_0 \neq 0$. Hence $P$ is positive definite.

*Arbitrary Initial Matrix $P_0$* Next we show that the sequence of matrices $\{P_k(P_0)\}$ defined by (13) when the starting matrix is an arbitrary positive semidefinite matrix $P_0$ converges to $P = \lim_{k\to\infty} P_k(0)$. Indeed, the optimal value of the optimal control problem of minimizing

$$x_k' P_0 x_k + \sum_{i=0}^{k-1} (x_i' Q x_i + u_i' R u_i) \tag{26}$$

subject to (18) equals $x_0' P_k(P_0) x_0$. Hence we have for every $x_0 \in R^n$

$$x_0' P_k(0) x_0 \leqslant x_0' P_k(P_0) x_0.$$

Consider now the cost (26) corresponding to the controller $\mu(x_k) = u_k = L x_k$, where $L$ is defined by (20). This cost is given by

$$x_0' \left[ D'^k P_0 D^k + \sum_{i=0}^{k-1} [D'^i (Q + L'RL) D^i] \right] x_0$$

and is greater than $x_0' P_k(P_0) x_0$, which is, of course, the optimal value of (26). Hence we have for all $k$ and $x \in R^n$

$$x' P_k(0) x \leqslant x' P_k(P_0) x \leqslant x' \left[ D'^k P_0 D^k + \sum_{i=0}^{k-1} [D'^i (Q + L'RL) D^i] \right] x. \quad (27)$$

Now we have proved

$$\lim_{k \to \infty} P_k(0) = P, \quad (28)$$

and we also have (using the fact that $\lim_{k \to \infty} D'^k P_0 D^k = 0$)

$$\lim_{k \to \infty} \left\{ D'^k P_0 D^k + \sum_{i=0}^{k-1} [D'^i (Q + L'RL) D^i] \right\}$$

$$= \lim_{k \to \infty} \left\{ \sum_{i=0}^{k-1} [D'^i (Q + L'RL) D^i] \right\} = P, \quad (29)$$

where the last equality may be verified easily using (21). Combining (27)–(29) we obtain

$$\lim_{k \to \infty} P_k(P_0) = P,$$

for an arbitrary positive semidefinite symmetric $P_0$.

*Uniqueness of Solution* If $\tilde{P}$ were another positive semidefinite solution of (14), we would have

$$\lim_{k \to \infty} P_k(\tilde{P}) = \mathrm{P}.$$

Since $\tilde{P}$ is a solution of (14), however, it follows that

$$P_k(\tilde{P}) = \tilde{P} \qquad \forall k = 0, 1, \ldots,$$

and hence $\tilde{P} = P$. Q.E.D.

We note that this proposition may be sharpened by substituting the controllability and observability assumptions by weaker assumptions of stabilizability and detectability. We refer the reader to the work of Kucera [K9], Payne and Silverman [P4], and Wonham [W12] for an exposition of this refinement.

*Random System Matrices*

We consider now the more general case where the matrices $A_k$ and $B_k$ are not known but rather are independent random matrices over time and independent of $w_k$. Their probability distributions are given and they are assumed to have finite second moments. This problem falls again within the framework of the basic problem by considering as "disturbance" at each time $k$ the triplet $(A_k, B_k, w_k)$. The DP algorithm is written

$$J_N(x_N) = x'_N Q_N x_N,$$

$$J_k(x_k) = \min_{u_k} \underset{w_k, A_k, B_k}{E} \{x'_k Q_k x_k + u'_k R_k u_k + J_{k+1}(A_k x_k + B_k u_k + w_k)\}.$$

Calculations very similar to those for the case where $A_k$, $B_k$ are not random show that the optimal control law is of the form

$$\mu_k^*(x_k) = L_k x_x, \tag{30}$$

where the gain matrices $L_k$ are given by

$$L_k = -[R_k + E\{B'_k K_{k+1} B_k\}]^{-1} E\{B'_k K_{k+1} A_k\}, \tag{31}$$

and where the matrices $K_k$ are given by the recursive equation

$$K_N = Q_N, \tag{32}$$

$$K_k = E\{A' K_{k+1} A_k\} - E\{A'_k K_{k+1} B_k\}[R_k + E\{B'_k K_{k+1} B_k\}]^{-1} \times E\{B'_k K_{k+1} A_k\} + Q_k. \tag{33}$$

Finally we further pursue an observation made in Chapter 1, which is related to the nature of the quadratic criterion. Consider the minimization over $u$ of the quadratic form

$$\underset{w}{E}\{(ax + bu + w)^2\},$$

where $a$, $b$ are given scalars and $w$ is a random variable. The optimum is attained for

$$u^* = -(a/b)x - (1/b)E\{w\}.$$

Now $u^*$ is independent of the particular probability distribution of the random vector $w$ and depends only on the mean $E\{w\}$. In particular, the result of the optimization is the same as for the corresponding deterministic problem where $w$ is known with certainty and equal to $E\{w\}$. As mentioned in Section

1.3, this property is called the *certainty equivalence principle* and appears in various forms in many (but not all) stochastic control problems involving linear systems and quadratic criteria. For the first problem of this section ($A_k$, $B_k$ known), the certainty equivalence principle is expressed by the fact that the control law (3) is the same as the one that would be obtained from the corresponding deterministic problem where $w_k$ is not random but rather is known and equal to zero (its expected value). However, for the problem where $A_k$, $B_k$ are random the certainty equivalence principle does not hold since if one replaces $A_k$, $B_k$ with their expected values in Eq. (33), the resulting control law need not be optimal.

## 3.2 Inventory Control

We consider now the $N$-period version of the inventory control problem considered in Sections 2.1 and 2.2. For simplicity we shall assume initially that the *fixed cost* $K$ *is zero*. Furthermore, we will assume that excess demand at each period is backlogged and is filled when additional inventory becomes available. This is represented by negative inventory in the system equation

$$x_{k+1} = x_k + u_k - w_k, \qquad k = 0, 1, \ldots, N-1.$$

We assume that the successive demands $w_k$ are bounded and independent, the unfilled demand at the end of the $N$th period is lost, and the inventory leftover at the end of the $N$th period has zero value. We shall also assume for convenience that the demands $w_0, \ldots, w_{N-1}$ are characterized by identical probability measures. The results of this section can also be proved without this restriction by trivial modifications of the proofs given here. Under these circumstances the total expected cost to be minimized is given by the expression

$$\underset{\substack{w_k \\ k=0,\ldots,N-1}}{E} \left\{ \sum_{k=0}^{N-1} [cu_k + p \max(0, w_k - x_k - u_k) + h \max(0, x_k + u_k - w_k)] \right\}.$$

The assumptions made in Section 1.3.3 ($c > 0$, $h \geqslant 0$, $p > c$) will also be in effect here.

By applying the DP algorithm for the basic problem we have

$$J_N(x_N) = 0, \tag{34}$$

$$J_k(x_k) = \min_{u_k \geqslant 0} [cu_k + L(x_k + u_k) + E\{J_{k+1}(x_k + u_k - w_k)\}], \tag{35}$$

where the function $L$ is defined by

$$L(y) = p\, E\{\max(0, w_k - y)\} + h\, E\{\max(0, y - w_k)\}.$$

We now show that an optimal control law is determined by a sequence of scalars $\{S_0, S_1, \ldots, S_{N-1}\}$ and has the form

$$\mu_k^*(x_k) = \begin{cases} S_k - x_k & \text{if} \quad x_k < S_k, \\ 0 & \text{if} \quad x_k \geqslant S_k. \end{cases} \tag{36}$$

For each $k$, the scalar $S_k$ minimizes the function

$$G_k(y) = cy + L(y) + E\{J_{k+1}(y - w)\}. \tag{37}$$

It is evident from the form of the DP algorithm and the discussion of the single-period problem of Section 1.3.3 that the optimal control law is indeed of form (36) provided the cost-to-go functions $J_k$ [and hence also the functions $G_k$ of (37)] are convex, and furthermore $\lim_{|y| \to \infty} G_k(y) = \infty$, so that the minimizing scalars $S_k$ exist.

Indeed, proceeding inductively we have that $J_N$ is convex [cf. Eq. (34)]. By the single-period problem solution, an optimal policy at time $N - 1$ is given by

$$\mu_{N-1}^*(x_{N-1}) = \begin{cases} S_{N-1} - x_{N-1} & \text{if} \quad x_{N-1} < S_{N-1}, \\ 0 & \text{if} \quad x_{N-1} \geqslant S_{N-1}. \end{cases}$$

Furthermore from the DP equation (35) we have

$$J_{N-1}(x_{N-1}) = \begin{cases} c(S_{N-1} - x_{N-1}) + L(S_{N-1}) & \text{if} \quad x_{N-1} < S_{N-1}, \\ L_{N-1}(x_{N-1}) & \text{if} \quad x_{N-1} \geqslant S_{N-1}, \end{cases}$$

which is a convex function by the convexity of $L$ and the fact that $S_{N-1}$ minimizes $cy + L(y)$ (see Fig. 3.2). Thus given the convexity of $J_N$ we were able to prove the convexity of $J_{N-1}$.

Similarly one can see that $\lim_{|y| \to \infty} G_k(y) = \infty$, since $c < p$, and we have

$$J_k(x_k) = \begin{cases} c(S_k - x_k) + L(S_k) + E\{J_{k+1}(S_k - w_k)\} & \text{if} \quad x_k < S_k, \\ L(x_k) + E\{J_{k+1}(x_k - w_k)\} & \text{if} \quad x_k \geqslant S_k, \end{cases}$$

where $S_k$ minimizes $cy + L(y) + E\{J_{k+1}(y - w)\}$. Again the convexity of $J_{k+1}$ [which implies convexity of $L(x) + E\{J_{k+1}(x - w)\}$] is sufficient to show the convexity of $J_k$ and thus the optimality of the control law (36) is demonstrated.

*Positive Fixed Cost*

We now turn to the somewhat more complicated case where there is a nonzero fixed cost $K > 0$ associated with a positive inventory order. In other words, we consider the case where the cost for ordering inventory $u \geqslant 0$ is given by

$$C(u) = \begin{cases} K + cu & \text{if} \quad u > 0, \\ 0 & \text{if} \quad u = 0. \end{cases}$$

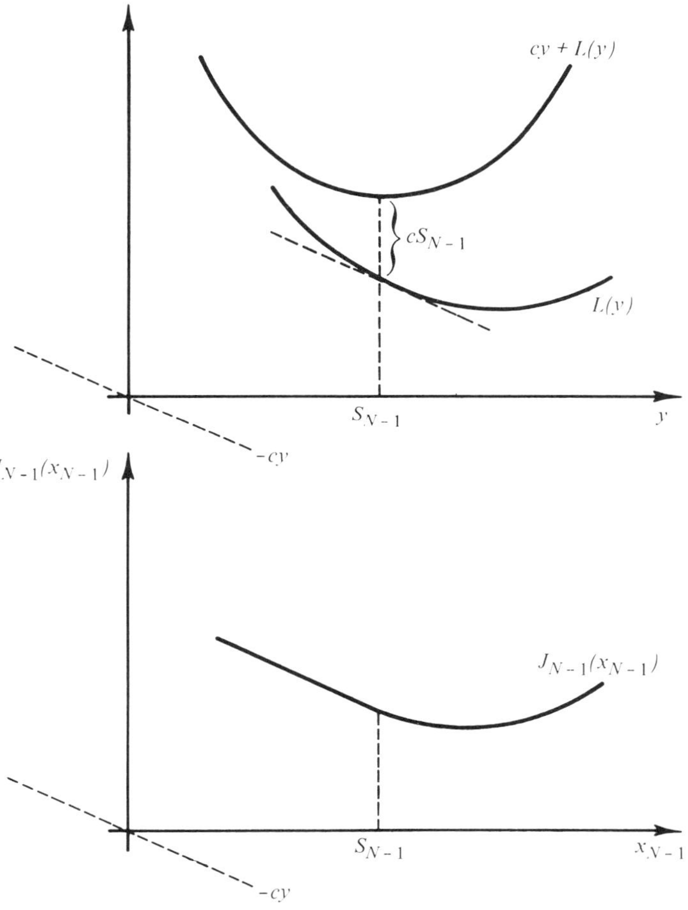


FIGURE 3.2

The DP algorithm takes the form

$$J_N(x_N) = 0,$$

$$J_k(x_k) = \min_{u_k \geq 0} [C(u_k) + L(x_k + u_k) + E\{J_{k+1}(x_k + u_k - w_k)\}],$$

with $L$ defined as earlier by

$$L(y) = p\ E\{\max(0, w - y)\} + h\ E\{\max(0, y - w)\}.$$

Consider again the functions $G_k$

$$G_k(y) = cy + L(y) + E\{J_{k+1}(y - w)\}. \qquad (3\text{8}$$

If we could prove that the functions $G_k$ were convex functions, then based on the analysis of Section 1.3.3 it would follow that a policy of the $(s, S)$ type

$$\mu_k^*(x_k) = \begin{cases} S_k - x_k & \text{if } x < s_k, \\ 0 & \text{if } x \geqslant s_k, \end{cases} \tag{39}$$

would be optimal, where $S_k$ is a value of $y$ that minimizes $G_k(y)$ and $s_k$ is the smallest value of $y$ for which $G_k(y) = K + G_k(S_k)$. Unfortunately when $K > 0$ it is not necessarily true that $J_k$ or $G_k$ are convex functions. This opens the possibility of functions $G_k$ having the form shown in Fig. 3.3. For this case the

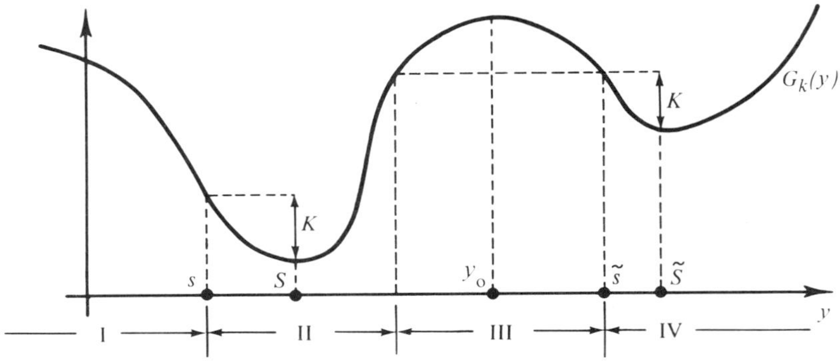


FIGURE 3.3

optimal policy is to order $(S - x)$ in interval I, zero in intervals II and IV, and $(\tilde{S} - x)$ in interval III. However, we will show that even though the functions $G_k$ may not be convex they have the property

$$K + G_k(z + y) \geqslant G_k(y) + z\left[\frac{G_k(y) - G_k(y - b)}{b}\right] \qquad \forall z \geqslant 0, \quad b > 0, \quad y. \tag{40}$$

This property is called *K-convexity* and was first utilized by Scarf [S6] to show the optimality of multiperiod $(s, S)$ policies. Now if (40) holds, then the situation shown in Fig. 3.3 is impossible; for if $y_0$ is the local maximum in the interval III, then we must have, for sufficiently small $b > 0$,

$$[G_k(y_0) - G_k(y_0 - b)]/b \geqslant 0,$$

and from (40) it follows that

$$K + G_k(\tilde{S}) \geqslant G_k(y_0),$$

which contradicts the construction shown in Fig. 3.3. More generally it is easy to show by essentially repeating the analysis of Section 1.3.3 [using part (d) of the lemma below] that if (40) holds, then an optimal policy takes the form (39).

**Definition** We say that a function $g\colon R \to R$ is *K-convex*, where $K \geqslant 0$, if

$$K + g(z + y) \geqslant g(y) + z\left[\frac{g(y) - g(y - b)}{b}\right] \qquad \forall z \geqslant 0, \quad b > 0, \quad y.$$

Some properties of $K$-convex functions are provided in the following lemma. The last part of the lemma essentially proves the optimality of the $(s, S)$ policy (39) when $G_k$ satisfies (40).

**Lemma** (a) A convex function $g\colon R \to R$ is also 0-convex and hence also $K$-convex for all $K \geqslant 0$.

(b) If $g_1(y)$ and $g_2(y)$ are $K$-convex and $L$-convex ($K \geqslant 0$, $L \geqslant 0$), respectively, then $\alpha g_1(y) + \beta g_2(y)$ is $(\alpha K + \beta L)$-convex for all positive $\alpha$ and $\beta$.

(c) If $g(y)$ is $K$-convex, then $E_w\{g(y - w)\}$ is also $K$-convex provided $E_w\{|g(y - w)|\} < \infty$ for all $y$.

(d) If $g\colon R \to R$ is a continuous $K$-convex function and $g(y) \to \infty$ as $|y| \to \infty$, then there exist scalars $s$ and $S$ with $s \leqslant S$ such that

(i) $g(S) \leqslant g(y)$, $\forall y \in R$;

(ii) $g(S) + K = g(s) < g(y)$, $\forall y < s$;

(iii) $g(y)$ is a decreasing function on $(-\infty, s)$;

(iv) $g(y) \leqslant g(z) + K$ for all $y, z$ with $s \leqslant y \leqslant z$.

*Proof* Part (a) follows from elementary properties of convex functions and parts (b) and (c) follow directly from the definition of a $K$-convex function. We shall thus concentrate on proving part (d).

Since $g$ is continuous and $g(y) \to \infty$ as $|y| \to \infty$, there exists a minimizing point of $g$. Let $S$ be such a point. Also let $s$ be the smallest scalar $z$ for which $z \leqslant S$ and $g(S) + K = g(z)$. For all $y$ with $y < s$ we have from the definition of $K$-convexity

$$K + g(S) \geqslant g(s) + \frac{S - s}{s - y}[g(s) - g(y)].$$

Since $K + g(S) - g(s) = 0$ we obtain $g(s) - g(y) \leqslant 0$. Since $y < s$ and $s$ is the smallest scalar for which $g(S) + K = g(s)$ we must have $g(s) < g(y)$ and (ii) is proved. Now for $y_1 < y_2 < s$ we have

$$K + g(S) \geqslant g(y_2) + \frac{S - y_2}{y_2 - y_1}[g(y_2) - g(y_1)].$$

Also from (ii),

$$g(y_2) > g(S) + K = g(s),$$

and by adding these two inequalities we obtain

$$0 > \frac{S - y_2}{y_2 - y_1} [g(y_2) - g(y_1)],$$

from which $g(y_1) > g(y_2)$ thus proving (iii). To prove (iv) we note that it holds for $y = z$ as well as for either $y = S$ or $y = s$. There remain two other possibilities, $S < y < z$ and $s < y < S$. If $S < y < z$, then by $K$-convexity

$$K + g(z) \geqslant g(y) + \frac{z - y}{y - S} [g(y) - g(S)] \geqslant g(y),$$

and (iv) is proved. If $s < y < S$, then by $K$-convexity

$$g(s) = K + g(S) \geqslant g(y) + \frac{S - y}{y - s} [g(y) - g(s)],$$

from which

$$\left[1 + \frac{S - y}{y - s}\right] g(s) \geqslant \left[1 + \frac{S - y}{y - s}\right] g(y),$$

and $g(s) \geqslant g(y)$. Noting that

$$g(z) + K \geqslant g(S) + K = g(s),$$

it follows that $g(z) + K \geqslant g(y)$. Thus (iv) is proved for this case as well. Q.E.D.

Consider now the function $G_{N-1}$ of (38):

$$G_{N-1}(y) = cy + L(y).$$

Clearly $G_{N-1}$ is a convex function and hence by part (a) of the previous lemma it is also $K$-convex. We have, from the analysis of the case where $K = 0$,

$$J_{N-1}(x) = \begin{cases} K + G_{N-1}(S_{N-1}) - cx & \text{for} \quad x < s_{N-1}, \\ G_{N-1}(x) - cx & \text{for} \quad x \geqslant s_{N-1}, \end{cases} \tag{41}$$

where $S_{N-1}$ minimizes $G_{N-1}(y)$ and $s_{N-1}$ is the smallest value of $y$ for which $G_{N-1}(y) = K + G_{N-1}(S_{N-1})$. Notice that since $K > 0$ we have $s_{N-1} \neq S_{N-1}$

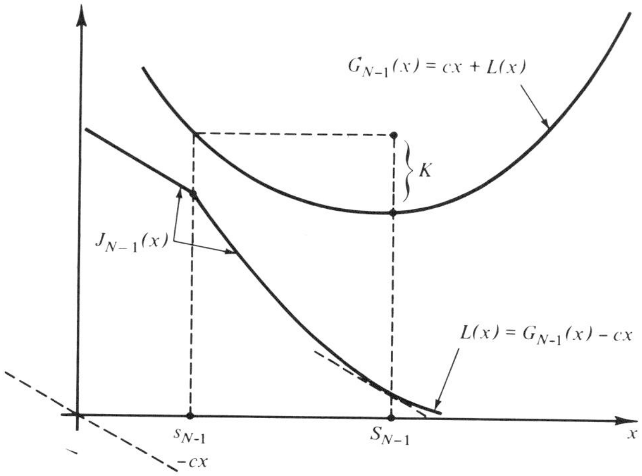


FIGURE 3.4

and furthermore the slope of $G_{N-1}$ at $s_{N-1}$ is negative. As a result the left slope of $J_{N-1}$ at $s_{N-1}$ is greater than the right slope as shown in Fig. 3.4 and $J_{N-1}$ is not convex. However, we will show that $J_{N-1}$ is $K$-convex based on the fact that $G_{N-1}$ is $K$-convex. To this end we must verify that

$$K + J_{N-1}(y+z) \geqslant J_{N-1}(y) + z\left[\frac{J_{N-1}(y) - J_{N-1}(y-b)}{b}\right] \qquad \forall z \geqslant 0, \quad b > 0, \quad y. \tag{42}$$

We distinguish three cases:

*Case* 1 $\quad y \geqslant s_{N-1}$ If $y - b \geqslant s_{N-1}$, then in this region of values of $z, b, y$ the function $J_{N-1}$, by (41), is the sum of a $K$-convex function and a linear function. Hence by part (b) of the lemma it is $K$-convex and (42) holds. If $y - b < s_{N-1}$, then in view of (41) we can write (42) as

$$K + G_{N-1}(y+z) - c(y+z) \geqslant G_{N-1}(y) - cy + z\left[\frac{G_{N-1}(y) - cy - G_{N-1}(s_{N-1}) + c(y-b)}{b}\right],$$

or equivalently

$$K + G_{N-1}(y+z) \geqslant G_{N-1}(y) + z\left[\frac{G_{N-1}(y) - G_{N-1}(s_{N-1})}{b}\right]. \tag{43}$$

Now if $y$ is such that $G_{N-1}(y) \geqslant G_{N-1}(s_{N-1})$, then by $K$-convexity of $G_{N-1}$ we have

$$K + G_{N-1}(y + z) \geqslant G_{N-1}(y) + z\left[\frac{G_{N-1}(y) - G_{N-1}(s_{N-1})}{y - s_{N-1}}\right]$$
$$\geqslant G_{N-1}(y) + z\left[\frac{G_{N-1}(y) - G_{N-1}(s_{N-1})}{b}\right].$$

Thus (43) and hence also (42) holds. If $y$ is such that $G_{N-1}(y) < G_{N-1}(s_{N-1})$, then we have

$$K + G_{N-1}(y + z) \geqslant K + G_{N-1}(S_{N-1}) = G_{N-1}(s_{N-1}) > G_{N-1}(y)$$
$$\geqslant G_{N-1}(y) + z\left[\frac{G_{N-1}(y) - G_{N-1}(s_{N-1})}{b}\right].$$

So for this case (43), and hence also (42), holds.

*Case* 2 $y \leqslant y + z \leqslant s_{N-1}$ In this region, by (41), the function $J_{N-1}$ is linear and hence (42) holds.

*Case* 3 $y < s_{N-1} < y + z$ For this case in view of (41) we can write (42) as

$$K + G_{N-1}(y + z) - c(y + z) \geqslant G_{N-1}(s_{N-1}) - cy$$
$$+ z\left[\frac{G_{N-1}(s_{N-1}) - cy - G_{N-1}(s_{N-1}) + c(y - b)}{b}\right]$$

or equivalently

$$K + G_{N-1}(y + z) \geqslant G_{N-1}(s_{N-1}),$$

which holds true by the definition of $s_{N-1}$.

We have thus proved that $K$-convexity and continuity of $G_{N-1}$ together with the fact that $G_{N-1}(y) \to \infty$ as $|y| \to \infty$ imply $K$-convexity of $J_{N-1}$. In addition, $J_{N-1}$ can be easily seen to be continuous. Now using the lemma it follows from (38) that $G_{N-2}$ is a $K$-convex function. Furthermore, by using the boundedness of $w_{N-2}$, it follows that $G_{N-2}$ is continuous and, in addition $G_{N-2}(y) \to \infty$ as $|y| \to \infty$. Repeating the argument above we obtain the fact that $J_{N-2}$ is $K$-convex and proceeding similarly we prove $K$-convexity and continuity of the functions $G_k$ for all $k$, as well as that $G_k(y) \to \infty$ as $|y| \to \infty$. At the same time [by using part (d) of the lemma] we prove optimality of the multiperiod $(s, S)$ policy of (39).

Optimality of policies of the $(s, S)$ type can be proved for a number of problems that constitute variations and generalizations of the problem considered in this section. Problems 4–6 treat the cases where forecasts on the uncertain demand become available during the control process, the case where there is a one-period time lag in delivery of inventory, and the case where unfilled demand is lost rather than backlogged (i.e., the system equation is $x_{k+1} = \max[0, x_k + u_k - w_k]$).

## 3.3 Dynamic Portfolio Analysis

Portfolio theory deals with the question of how to invest a certain amount of wealth among a collection of risky or riskless assets. The traditional and widely used approach to this problem [M4, S9], has been the so-called mean-variance approach examined in Section 1.3, whereby the investor is assumed to be maximizing the expected value of a utility function that depends on the mean and the variance of the rate of return of the investment. Since this approach cannot be readily generalized to the case where investment takes place over several periods of time and in addition is based on assumptions that may not be satisfied in a given practical situation, there has been considerable effort toward the development of alternative approaches for portfolio selection. In one such approach, the investor is assumed to be maximizing the expected utility of his final wealth. We shall discuss in this section some results related to this viewpoint. We will start with an analysis of a single-period model and then extend the results to the multiperiod case.

Let $x_0$ denote the initial wealth (measured in monetary units) of the investor and assume that there are $n$ risky assets, with corresponding random rates of return $e_1, e_2, \ldots, e_n$ among which the investor can allocate his wealth. The investor can also invest in a riskless asset offering a sure rate of return $s$. If we denote by $u_1, \ldots, u_n$ the corresponding amounts invested in the $n$ risky assets and by $(x_0 - u_1 - \cdots - u_n)$ the amount invested in the riskless asset, the final wealth of the decision maker is given by

$$x_1 = s(x_0 - u_1 - \cdots - u_n) + \sum_{i=1}^{n} e_i u_i,$$

or equivalently

$$x_1 = s x_0 + \sum_{i=1}^{n} (e_i - s) u_i. \tag{44}$$

The objective is to maximize over $u_1, \ldots, u_n$,

$$\underset{\substack{e_i \\ i=1,\ldots,n}}{E} \{U(x_1)\},$$

where $U$ is a utility function for the investor. We assume that the expected value above is well defined and finite for all $x_0$, $u_i$. We shall also assume that the domain of $U$ is an open set $D$ within which $U$ is concave and twice continuously differentiable. We shall not impose constraints on $u_1, \ldots, u_n$. This is necessary in order to obtain the results in convenient form. However, we shall assume that the probability distribution of $e_i$ and the minimizing values of $u_i$ are such that the resulting values of $x_1$ as given by (44) belong to the domain of $U$. A few additional assumptions will be made during the exposition of the results.

Let us consider the above problem for every value of initial wealth and denote by $u_i^* = \mu^{i*}(x_0)$, $i = 1, \ldots, n$, the optimal amounts to be invested in the $n$ risky assets, when the initial wealth is $x_0$.

We say that the portfolio $\{\mu^{1*}(x_0), \ldots, \mu^{n*}(x_0)\}$ is *partially separated* if

$$\mu^{i*}(x_0) = \alpha^i h(x_0), \qquad i = 1, \ldots, n, \tag{45}$$

where $\alpha^i$, $i = 1, \ldots, n$, are fixed constants and $h(x_0)$ is a function of $x_0$ (which is the same for all $i$).

When partial separation holds, the ratios of amounts invested in the risky assets are fixed and independent of the initial wealth, i.e.,

$$\mu^{i*}(x_0)/\mu^{j*}(x_0) = \alpha^i/\alpha^j \qquad \text{for} \quad 1 \leqslant i, j \leqslant n, \quad \alpha^j \neq 0.$$

Actually in the cases we shall examine when partial separation holds, the portfolio $\{\mu^{1*}(x_0), \ldots, \mu^{n*}(x_0)\}$ will be shown to consist of affine (linear plus constant) functions of $x_0$ that have the form

$$\mu^{i*}(x_0) = \alpha^i[a + bsx_0], \qquad i = 1, \ldots, n, \tag{46}$$

where $a$ and $b$ are constants characterizing the utility function $U$.

In the special case where $a = 0$ in (46) we say that the optimal portfolio is *completely separated* in the sense that the ratios of the amounts invested in both the risky asset *and* the riskless asset are fixed and independent of initial wealth. We now show that when the utility function satisfies

$$-U'(x_1)/U''(x_1) = a + bx_1 \qquad \forall x_1, \tag{47}$$

where $U'$ and $U''$ denote the first and second derivatives of $U$, respectively, and $a$ and $b$ are some scalars, then the optimal portfolio is given by

$$\mu^{i*}(x_0) = \alpha^i[a + bsx_0], \qquad i = 1, \ldots, n. \tag{48}$$

Furthermore, if $J(x_0)$ is the optimal value of the problem

$$J(x_0) = \min_{u_i} E\{U(x_1)\},$$

then we have

$$-J'(x_0)/J''(x_0) = (a/s) + bx_0 \qquad \forall x_0. \tag{49}$$

Let us assume that an optimal portfolio exists and is of the form

$$\mu^{i*}(x_0) = \alpha^i(x_0)[a + bsx_0],$$

where $\alpha^i(x_0)$, $i = 1, \ldots, n$, are some differentiable functions. We will prove that $d\alpha^i(x_0)/dx_0 = 0$ for all $x_0$ and hence the functions $\alpha^i$ must be constant.

We have for every $x_0$, by the optimality of $\mu^{i*}(x_0)$, for $i = 1, \ldots, n$,

$$dE\{U(x_1)\}/du_i = E\left\{U'\left[sx_0 + \sum_{j=1}^{n}(e_j - s)\alpha^j(x_0)(a + bsx_0)\right](e_i - s)\right\} = 0, \tag{50}$$

Differentiating the $n$ equations in (50) with respect to $x_0$ yields

$$E\left\{\begin{bmatrix}(e_1 - s)^2 \cdots (e_1 - s)(e_n - s)\\ \vdots \\ (e_n - s)(e_1 - s) \cdots (e_n - s)^2\end{bmatrix} U''(x_1)(a + bsx_0)\right\}\begin{bmatrix}\dfrac{d\alpha^1(x_0)}{dx_0}\\ \vdots \\ \dfrac{d\alpha^n(x_0)}{dx_0}\end{bmatrix}$$

$$= -\begin{bmatrix} E\left\{U''(x_1)(e_1 - s)s\left[1 + \sum_{i=1}^{n}(e_i - s)\alpha^i(x_0)b\right]\right\}\\ \vdots \\ E\left\{U''(x_1)(e_n - s)s\left[1 + \sum_{i=1}^{n}(e_i - s)\alpha^i(x_0)b\right]\right\}\end{bmatrix}. \tag{51}$$

Using relation (47) we have

$$U''(x_1) = -\frac{U'(x_1)}{a + b[sx_0 + \sum_{i=1}^{n}(e_i - s)\alpha^i(x_0)(a + bsx_0)]}$$

$$= -\frac{U'(x_1)}{(a + bsx_0)[1 + \sum_{i=1}^{n}(e_i - s)\alpha^i(x_0)b]}. \tag{52}$$

Substituting in (51) and using (50) we have that the right-hand side of (51) is the zero vector. The matrix on the left in (51), except for degenerate cases, can be shown to be nonsingular. Assuming that it is indeed nonsingular we obtain

$$d\alpha^i(x_0)/dx_0 = 0, \qquad i = 1, \ldots, n,$$

and $\alpha^i(x_0) = \alpha^i$, where $\alpha^i$ are some constants, thus proving (48).

We now turn our attention to proving relation (49). We have

$$J(x_0) = E\{U(x_1)\} = E\left\{U\left[s\left[1 + \sum_{i=1}^{n}(e_i - s)\alpha^i b\right]x_0 + \sum_{i=1}^{n}(e_i - s)\alpha^i a\right]\right\}$$

and hence

$$J'(x_0) = E\left\{U'(x_1)s\left[1 + \sum_{i=1}^{n}(e_i - s)\alpha^i b\right]\right\}, \tag{53}$$

$$J''(x_0) = E\left\{U''(x_1)s^2\left[1 + \sum_{i=1}^{n}(e_i - s)\alpha^i b\right]^2\right\}.$$

The last relation after some calculation and using (52) yields

$$J''(x_0) = -E\left\{U'(x_1)s\left[1 + \sum_{i=1}^{n}(e_i - s)\alpha^i b\right]\right\}s\Big/(a + bsx_0). \tag{54}$$

By combining (53) and (54) we obtain the desired result:

$$-J'(x_0)/J''(x_0) = (a/s) + bx_0.$$

The class of utility functions satisfying condition (47) is characterized by the fact that the inverse of the index of absolute risk aversion (sometimes referred to as the "risk-tolerance function") is affine in wealth. It can be easily shown that the following forms of utility functions satisfy this condition:

$$\begin{array}{rll} \text{exponential:} & -e^{-x/a} & \text{for} \quad b = 0, \\ \text{logarithmic:} & \ln(x + a) & \text{for} \quad b = 1, \\ \text{power:} & [1/(b - 1)](a + bx)^{1-(1/b)} & \text{otherwise.} \end{array} \tag{55}$$

Naturally in our portfolio problem only concave utility functions from this class are admissible. Furthermore if a utility function that is not defined over the whole real line is used, the problem should be formulated in a way that ensures that all possible values of the resulting final wealth are within the domain of definition of the utility function.

It is now easy to extend the one-period result of the preceding analysis to the case of a multiperiod model. We shall assume that the current wealth can be reinvested at the beginning of each of $N$ consecutive time periods. We denote

- $x_k$ the wealth of the investor at the beginning of the $k$th period,
- $u_i^k$ the amount invested at the beginning of the $k$th period in the $i$th risky asset,
- $e_i^k$ the rate of return of the $i$th risky asset during the $k$th period,
- $s_k$ the rate of return of the riskless asset during the $k$th period.

We have (in accordance with the single-period model) the system equation

$$x_{k+1} = s_k x_k + \sum_{i=1}^{n}(e_i^k - s_k)u_i^k, \qquad k = 0, 1, \ldots, N - 1. \tag{56}$$

We assume that the vectors $e^k = (e_1^k, \ldots, e_n^k)$, $k = 0, \ldots, N-1$, are independent random vectors with given probability distributions that result in finite expected values throughout the following analysis.

The objective is to maximize $E\{U(x_N)\}$, the expected utility of the terminal wealth $x_N$, where we assume that $U$ satisfies for all $x$

$$-U'(x)/U''(x) = a + bx.$$

Applying the DP algorithm to this problem we have

$$J_N(x_N) = U(x_N), \tag{57}$$

$$J_k(x_k) = \max_{u_1^k, \ldots, u_n^k} E\left\{J_{k+1}\left[s_k x_k + \sum_{i=1}^{n} (e_i^k - s_k)u_i^k\right]\right\}. \tag{58}$$

From the solution of the one-period problem we have that the optimal policy at the beginning of period $N-1$ is of the form

$$\mu_{N-1}^*(x_{N-1}) = \alpha_{N-1}[a + bs_{N-1}x_{N-1}],$$

where $\alpha_{N-1}$ is an appropriate $n$-dimensional vector. Furthermore, we have

$$-J_{N-1}'(x)/J_{N-1}''(x) = (a/s_{N-1}) + bx. \tag{59}$$

Hence applying the result of this section in (58) for the next to the last period we obtain the optimal policy

$$\mu_{N-2}^*(x_{N-2}) = \alpha_{N-2}[(a/s_{N-1}) + bs_{N-2}x_{N-2}],$$

where $\alpha_{N-2}$ is again an appropriate $n$-dimensional vector.

Proceeding similarly we have for the $k$th period

$$\mu_k^*(x_k) = \alpha_k[(a/s_{N-1} \cdots s_{k+1}) + bs_k x_k] \tag{60}$$

where $\alpha_k$, $k = 0, 1, \ldots, N-1$, are $n$-dimensional vectors that depend on the probability distributions of the rates of return $e_i^k$ of the risky assets and are determined by optimization of the expected value of the optimal "cost-to-go" functions $J_k$. These functions satisfy

$$-J_k'(x)/J_k''(x) = (a/s_{N-1} \cdots s_k) + bx, \qquad k = 0, 1, \ldots, N-1. \tag{61}$$

Thus one can see that the investor, when faced with the opportunity to reinvest sequentially his wealth, uses a policy similar to that of the single-period case. Carrying the analysis one step further, one can see that if the utility function $U$ is such that $a = 0$, i.e., $U$ has one of the forms

$$\begin{aligned} &\ln x && \text{for} \quad b = 1, \\ &[1/(b-1)](bx)^{1-(1/b)} && \text{for} \quad b \neq 0, \quad b \neq 1, \end{aligned}$$

then it follows from (60) that the investor acts at each stage $k$ as if he were faced with a *single-period* investment characterized by the rates of return $s_k, e_i^k, i = 1, \ldots, n,$ and the objective function $E\{U(x_{k+1})\}$. This policy whereby the investor can ignore the fact that he will have the opportunity to reinvest his wealth is called a *myopic policy* [M8].

Note that a myopic policy is also optimal when $s_k = 1, k = 0, \ldots, N-1$, which is the case when wealth is discounted at the rate of return of the riskless asset (going rate of interest for money). Furthermore, it can be proved that when $a = 0$ a myopic policy is optimal even in the more general case where the rates of return $s_k, k = 0, 1, \ldots, N-1$, are independent random variables [M8], and for the case where forecasts on the probability distributions of the rates of return $e_i^k$ of the risky assets become available during the investment process (see Problem 7).

It turns out that even for the more general case where $a \neq 0$ only a small amount of foresight is required on the part of the decision maker. It can be easily seen [compare (58)–(61)] that the optimal policy (60) at period $k$ is the same as the one that would be used if the investor were faced with a single-period problem whereby he would maximize over $u_i^k, i = 1, \ldots, n$,

$$E\{U(s_{N-1} \cdots s_{k+1} x_{k+1})\}$$

subject to $x_{k+1} = s_k x_k + \sum_{i=1}^n (e_i^k - s_k) u_i^k$. In other words, the investor maximizes the expected utility of wealth that results if amounts $u_i^k$ are invested in the risky assets in period $k$ and the resulting wealth $x_{k+1}$ is subsequently invested *exclusively* in the riskless asset during the remaining periods $k+1, \ldots, N-1$. This type of policy has been called a *partially myopic policy* [M8]. A partially myopic policy can also be shown to be optimal when forecasts on the probability distributions of the rates of return of the risky assets become available during the investment process (see Problem 7).

Another interesting aspect of the case where $a \neq 0$ is that when $s_k > 1$ for all $k$, then as the horizon becomes longer and longer ($N \to \infty$) the policy in the initial stages approaches a myopic policy [compare (60) and (61)]. Thus we can conclude that for $s_k > 1$ a partially myopic policy is asymptotically myopic as the horizon tends to infinity. This fact holds for an even larger class of utility functions than the class we have considered, as shown by Leland [L3].

In conclusion we mention that while the model we have examined ignores certain important aspects of the corresponding practical situation (constraints on investments, transaction costs, time correlation of rates of return, the possibility of intermediate consumption of wealth, market imperfections, etc.), it admits a closed-form solution for a large class of utility functions together with economic interpretations related to myopic and partially myopic policies that are undoubtedly of considerable interest. Thus while neither the particular model that we have examined nor its various refinements have achieved the preeminence and popularity of the mean-variance model, the corresponding analysis and results provide unquestionably important insights into the process of portfolio selection.

## 3.4 Optimal Stopping Problems—Examples

Optimal stopping problems of the type we will consider in this and subsequent sections typically involve a control space that consists of a finite number of elements (actions), one of which induces a termination (stopping) of the evolution of the system. Thus at each stage the controller observes the current state of the system and decides whether to continue the process (perhaps at a certain cost) or stop the process and incur a certain loss.

*Asset Selling Problem*

As a first example, consider a person having an asset (say a piece of land) for which he is offered a nonnegative amount of money from period to period. Let us assume that these random offers $w_0, w_1, \ldots, w_{N-1}$ are independent, identically distributed, and take values within some bounded interval. We consider a horizon of $N$ stages and assume that if the person accepts the offer, he can invest the money at a fixed rate of interest $r > 0$, and if he rejects the offer, he waits until the next period to consider the next offer. Offers rejected are not renewed and we initially assume that the last offer $w_{N-1}$ must be accepted if every prior offer has been rejected. The objective is to find a policy for accepting and rejecting offers that maximizes the revenue of the person at the $N$th period.

Let us try to embed this problem in the framework of the basic problem by defining the state space, control space, disturbance space, system equation, and cost functional. We consider as disturbance at time $k$ the random offer $w_k$ and as corresponding disturbance space the real line. The control space consists of two elements $u^1$, $u^2$, which correspond to the decisions "sell" and "do not sell," respectively. Concerning the state space we define it to be the real line, augmented with an additional state (call it $T$), which is a "termination state." The system moves into the termination state as soon as the asset is sold. By writing that the system is at a state $x_k \neq T$ at time $k$ we mean that the asset has not been sold as yet and the current offer under consideration is equal to $x_k$. By writing that the system is at state $x_k = T$ at time $k$ we mean that the asset has already been sold. With these conventions we may write a system equation of the form

$$x_{k+1} = f_k(x_k, u_k, w_k), \qquad k = 0, \ldots, N-1,$$
$$x_0 = 0,$$

where $x_k \in R \cup \{T\}$ and the function $f_k$ is defined via the relations

$$x_{k+1} = \begin{cases} T & \text{if } \quad u_k = u^1(\text{sell}) \quad \text{or} \quad x_k = T, \\ w_k & \text{otherwise.} \end{cases}$$

The corresponding reward function may be written

$$\underset{k=0,\ldots,N-1}{\underset{w_k}{E}} \left\{ g_N(x_N) + \sum_{k=0}^{N-1} g_k(x_k, u_k, w_k) \right\},$$

where

$$g_N(x_N) = \begin{cases} x_N & \text{if } \quad x_N \neq T, \\ 0 & \text{otherwise,} \end{cases}$$

$$g_k(x_k, u_k, w_k) = \begin{cases} (1 + r)^{N-k} x_k & \text{if } \quad x_k \neq T \quad \text{and} \quad u_k = u^1, \\ 0 & \text{otherwise.} \end{cases}$$

Based on this formulation we can write the corresponding DP algorithm over the states $x_k$:

$$J_N(x_N) = \begin{cases} x_N & \text{if } \quad x_N \neq \mathrm{T}, \\ 0 & \text{if } \quad x_N = T, \end{cases} \tag{62}$$

$$J_k(x_k) = \begin{cases} \max[(1 + r)^{N-k} x_k,\ E\{J_{k+1}(w_k)\}] & \text{if } \quad x_k \neq T, \\ 0 & \text{if } \quad x_k = T. \end{cases} \tag{63}$$

In Eq. (63) $(1 + r)^{N-k} x_k$ (where $x_k \neq T$) is the revenue resulting from decision $u^1$ (sell) when the offer under consideration is $x_k$, and $E\{J_{k+1}(w_k)\}$ represents the expected revenue corresponding to the decision $u^2$ (do not sell).

Actually the DP algorithm above could be derived by elementary reasoning without resorting to the elaborate formulation given earlier—something that we shall often do in the future. Our purpose for providing this formulation was simply to demonstrate to the reader the type of structure that one must adopt in order to embed a stopping problem into the framework of the basic problem.

Now from the DP algorithm (62) and (63) we obtain the following optimal policy for the case where $x_k \neq T$:

$$\begin{aligned} &\text{accept the offer } w_{k-1} = x_k \quad && \text{if} \quad (1 + r)^{N-k} x_k > E\{J_{k+1}(w_k)\}, \\ &\text{reject the offer } w_{k-1} = x_k \quad && \text{if} \quad (1 + r)^{N-k} x_k < E\{J_{k+1}(w_k)\}. \end{aligned}$$

When $(1 + r)^{N-k} x_k = E\{J_{k+1}(w_k)\}$ both acceptance and rejection are optimal.

The result can be put into a more convenient form by some further analysis. Let us introduce the functions

$$V_k(x_k) = \frac{1}{(1+r)^{N-k}} J_k(x_k) \qquad \forall x_k \neq T,$$

which represent discounted cost-to-go for the last $N - k$ stages. It can be easily seen that

$$V_N(x_N) = x_N, \tag{64}$$

$$V_k(x_k) = \max[x_k, (1+r)^{-1} E\{V_{k+1}(w_k)\}], \qquad k = 0, 1, \ldots, N-1. \tag{65}$$

By using the notation

$$\alpha_k = \frac{1}{1+r} E\{V_{k+1}(w_k)\},$$

the optimal policy is given by

$$\text{accept the offer } w_{k-1} = x_k \qquad \text{if} \quad x_k > \alpha_k,$$
$$\text{reject the offer } w_{k-1} = x_k \qquad \text{if} \quad x_k < \alpha_k,$$

while both acceptance and rejection are optimal for $x_k = \alpha_k$ (Fig. 3.5). Thus the optimal policy is completely determined by the sequence $\alpha_1, \ldots, \alpha_{N-1}$.

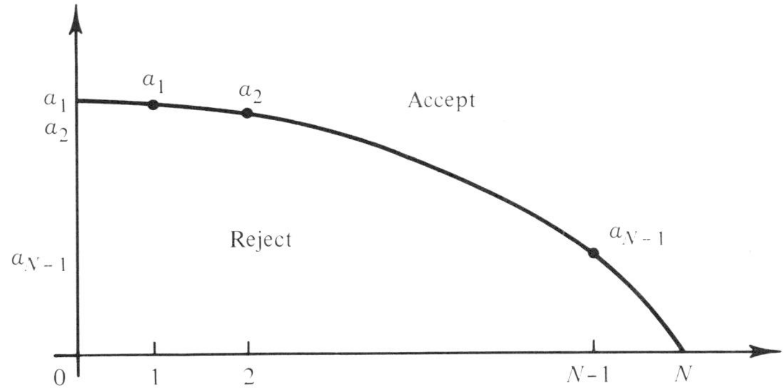


FIGURE 3.5

Now from the algorithm (64) and (65) we have

$$V_k(x_k) = \begin{cases} x_k & \text{if} \quad x_k > \alpha_k, \\ \alpha_k & \text{if} \quad x_k < \alpha_k. \end{cases}$$

Hence we obtain

$$\alpha_k = \frac{1}{1+r} E\{V_{k+1}(w_k)\} = \frac{1}{1+r} \int_0^{\alpha_{k+1}} \alpha_{k+1}\, dP(w_k) + \frac{1}{1+r} \int_{\alpha_{k+1}}^{\infty} w_k\, dP(w_k),$$

where the function $P$ is defined for all scalars $\lambda$ by

$$P(\lambda) = \text{Prob}\{w < \lambda\}.$$

Notice that the function $P$ is nondecreasing and continuous from the left for all $\lambda$. The difference equation for $\alpha_k$ above may also be written

$$\alpha_k = \frac{P(\alpha_{k+1})}{1+r}\alpha_{k+1} + \frac{1}{1+r}\int_{\alpha_{k+1}}^{\infty} w_k \, dP(w_k), \qquad k = 1, \ldots, N-1, \tag{66}$$

with $\alpha_N = 0$. Let us first show that the solution of the above difference equation is monotonically nonincreasing (as one would expect), i.e.,

$$0 \leqslant \alpha_{k+1} \leqslant \alpha_k, \qquad k = N-2, N-3, \ldots, \tag{67}$$

Indeed, we have

$$\alpha_{N-1} = \frac{1}{1+r}\int_0^{\infty} w_{N-1}\, dP(w_{N-1}) = \frac{E\{w_{N-1}\}}{1+r}$$

$$\begin{aligned}
\alpha_{N-2} &= \frac{\alpha_{N-1}}{1+r} P(\alpha_{N-1}) + \frac{1}{1+r}\int_{\alpha_{N-1}}^{\infty} w\, dP(w) \\
&= \frac{1}{1+r}\left(\int_0^{\alpha_{N-1}} \alpha_{N-1}\, dP(w) + \int_{\alpha_{N-1}}^{\infty} w\, dP(w)\right) \\
&\geqslant \frac{1}{1+r}\int_0^{\infty} w\, dP(w) = \alpha_{N-1}.
\end{aligned}$$

Assuming that $\alpha_{k+1} \geqslant \alpha_{k+2}$, we will show that $\alpha_k \geqslant \alpha_{k+1}$ and (67) will follow by induction. We have

$$\begin{aligned}
\alpha_k &= \frac{\alpha_{k+1}}{1+r} P(\alpha_{k+1}) + \frac{1}{1+r}\int_{\alpha_{k+1}}^{\infty} w\, dP(w) \\
&= \frac{\alpha_{k+1}}{1+r} P(\alpha_{k+1}) + \frac{1}{1+r}\int_{\alpha_{k+2}}^{\infty} w\, dP(w) - \frac{1}{1+r}\int_{\alpha_{k+2}}^{\alpha_{k+1}} w\, dP(w) \\
&\geqslant \frac{\alpha_{k+1}}{1+r} P(\alpha_{k+1}) + \frac{1}{1+r}\int_{\alpha_{k+2}}^{\infty} w\, dP(w) - \frac{\alpha_{k+1}}{1+r}[P(\alpha_{k+1}) - P(\alpha_{k+2})] \\
&= \frac{\alpha_{k+1}}{1+r} P(\alpha_{k+2}) + \frac{1}{1+r}\int_{\alpha_{k+2}}^{\infty} w\, dP(w) \\
&\geqslant \frac{\alpha_{k+2}}{1+r} P(\alpha_{k+2}) + \frac{1}{1+r}\int_{\alpha_{k+2}}^{\infty} w\, dP(w) = \alpha_{k+1}
\end{aligned}$$

and $\alpha_k \geqslant \alpha_{k+1}$ proving (67).

Now since we have

$$0 \leqslant \frac{P(\alpha)}{1+r} \leqslant \frac{1}{1+r} < 1 \qquad \forall \alpha \geqslant 0,$$

$$0 \leqslant \frac{1}{1+r} \int_{\alpha_{k+1}}^{\infty} w \, dP(w_k) \leqslant \frac{E\{w_k\}}{1+r} \qquad \forall k,$$

it can be easily seen, using the monotonicity property (67), that the sequence $\{\alpha_k\}$ generated (backwards) by the difference equation (66) converges (as $k \to -\infty$) to a constant $\bar{\alpha}$ satisfying

$$\bar{\alpha}(1+r) = P(\bar{\alpha})\bar{\alpha} + \int_{\bar{\alpha}}^{\infty} w \, dP(w).$$

This equation is obtained from (66) by taking limits as $k \to -\infty$ and by using the continuity from the left of the function $P$.

Thus when the horizon tends to become longer and longer (i.e., $N \to \infty$) the optimal policy for every fixed $k \geqslant 1$ approximates the stationary policy:

$$\text{accept the offer } w_{k-1} = x_k \quad \text{if} \quad x_k > \bar{\alpha},$$
$$\text{reject the offer } w_{k-1} = x_k \quad \text{if} \quad x_k < \bar{\alpha}.$$

The optimality of such a policy for the corresponding infinite horizon problem will be shown in Chapter 6.

*Purchasing with a Deadline*

Let us consider another problem of similar nature. Assume that a certain quantity of raw material is required by a certain time. If the price of this material fluctuates, then there arises the problem of deciding, given the price at any time, whether to purchase at that price or wait a further period, during which the price may go up or down.

Let us assume that successive prices are independent (the case of correlated prices will be examined later) and that the cumulative distribution $P(w)$ of the purchase cost $w$ of the raw material is the same for each time period. The problem is to decide, given the current purchase price, whether to purchase or not to purchase the amount of raw material needed. The purchase must be made within $N$ time periods.

This problem has obvious similarities with the previous problem. Let us denote by

$$x_{k+1} = w_k$$

the price prevailing in the beginning of period $k+1$. We have similarly as before the DP algorithm

$$J_N(x_N) = x_N,$$
$$J_k(x_k) = \min[x_k, E\{J_{k+1}(w_k)\}],$$

and the optimal policy is given by

$$\begin{aligned} &\text{purchase} && \text{if} \quad x_k < E\{J_{k+1}(w_k)\} = \alpha_k, \\ &\text{do not purchase} && \text{if} \quad x_k > E\{J_{k+1}(w_k)\} = \alpha_k. \end{aligned}$$

We have similarly that the critical numbers $\alpha_1, \alpha_2, \ldots, \alpha_{N-1}$ can be obtained from the discrete-time equation

$$\alpha_k = \alpha_{k+1}[1 - P(\alpha_{k+1})] + \int_0^{\alpha_{k+1}} w \, dP(w),$$

$$\alpha_{N-1} = \int_0^{\infty} wP(w) = E\{w\}.$$

Consider now a variation of this problem whereby we do not assume that the successive prices $w_0, \ldots, w_{N-1}$ are independent but rather that they are correlated and can be represented as

$$w_k = y_{k+1} \qquad k = 0, 1, \ldots, N-1,$$

with

$$y_{k+1} = \lambda y_k + \xi_k, \qquad y_0 = 0,$$

where $\lambda$ is a scalar with $0 \leqslant \lambda < 1$ and $\xi_0, \xi_1, \ldots, \xi_{N-1}$ are independent identically distributed random variables taking positive values with given probability distribution. As discussed in Section 2.3 the DP algorithm under these circumstances takes the form

$$J_N(x_N) = x_N,$$
$$J_k(x_k) = \min[x_k, E\{J_{k+1}(\lambda x_k + \xi_k)\}],$$

where the cost associated with the purchasing decision is $x_k$ and the cost associated with the waiting decision is $E\{J_{k+1}(\lambda x_k + \xi_k)\}$.

We shall show that in this case the optimal policy is also of the same type as the one for independent prices. Indeed, we have

$$J_{N-1}(x_{N-1}) = \min[x_{N-1}, \lambda x_{N-1} + \bar{\xi}],$$

where $\bar{\xi} = E\{\xi_{N-1}\}$. As shown in Fig. 3.6 an optimal policy at time $N-1$ is given by

$$\begin{aligned} &\text{purchase} && \text{if} \quad x_{N-1} < \alpha_{N-1}, \\ &\text{do not purchase} && \text{if} \quad x_{N-1} > \alpha_{N-1}, \end{aligned}$$

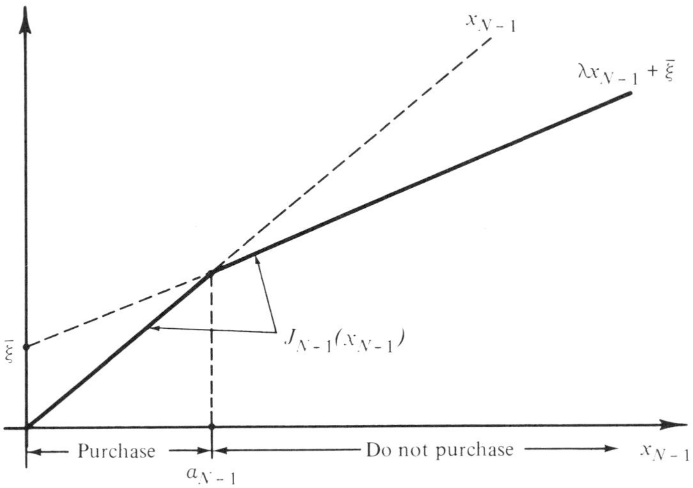


FIGURE 3.6

where $\alpha_{N-1}$ is defined from the equation $\alpha_{N-1} = \lambda\alpha_{N-1} + \bar{\xi}$, i.e.,

$$\alpha_{N-1} = \frac{1}{1-\lambda}\bar{\xi}.$$

Note that

$$J_{N-1}(x) \leqslant J_N(x) \qquad \forall x,$$

and that $J_{N-1}$ is concave and increasing in $x$. Using this fact in the DP algorithm one may easily show that

$$J_k(x) \leqslant J_{k+1}(x) \qquad \forall x, \quad k = 1, \ldots, N-1,$$

and that $J_k$ is concave and increasing in $x$ for all $k$. Furthermore, in view of the fact that $\bar{\xi} = E\{\xi_k\} > 0$ for all $k$, one can show that

$$E\{J_{k+1}(\xi_k)\} > 0, \qquad k = 0, 1, \ldots, N-2.$$

These facts imply (as shown in Fig. 3.7) that the optimal policy for every period $k$ is of the form

$$\begin{array}{rl} \text{purchase} & \text{if} \quad x_k < \alpha_k, \\ \text{do not purchase} & \text{if} \quad x_k > \alpha_k, \end{array}$$

where the scalar $\alpha_k$ is obtained as the unique positive solution of the equation

$$x = E\{J_{k+1}(\lambda x + \xi_k)\}.$$

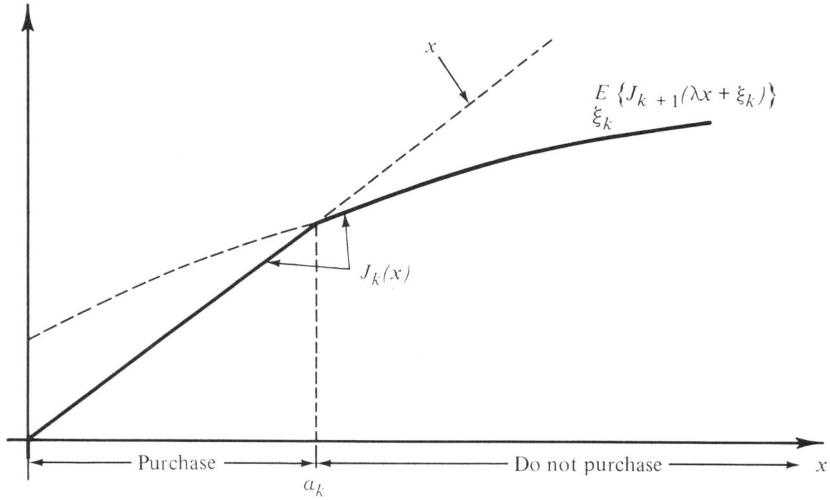


FIGURE 3.7

Note that $J_k(x) \leqslant J_{k+1}(x)$ implies

$$\alpha_{k-1} \leqslant \alpha_k \leqslant \alpha_{k+1} \qquad \forall k,$$

and hence (as one would expect) the threshold price below which one should purchase is lower in the early stages of the process and increases as the deadline comes nearer.

A number of variations on the theme of this section are of interest (see Problems). The main point to be recalled is that when only a finite number of actions $(u^1, \ldots, u^n)$ are possible at each stage, the optimal control law $\mu_k^*(x_k)$ is determined by a partition of the underlying state space $S_k$ (minus the termination state) into $n$ subsets $S_k^1, \ldots, S_k^n$, with $\bigcup_{i=1}^n S_k^i = S_k$, each subset $S_k^i$ being associated with the corresponding action $u^i$. In other words, the control law for all $k$ has the form

$$\mu_k^*(x_k) = u^i \qquad \text{if} \quad x_k \in S_k^i, \quad i = 1, \ldots, n.$$

In the examples we considered the state space was the positive half-line, and there were two actions available (accept–reject, or purchase–do not purchase). The partition of the state space $S_k$ was determined in each case by the single number $\alpha_k$.

## 3.5 Notes

The certainty equivalence principle for dynamic linear–quadratic problems was first discussed by Simon [S10]. His work was preceded by that of Theil [T2], who considered a single-period case, and Holt *et al.* [H11], who considered a deterministic case. Similar problems were considered somewhat later (and apparently independently) by Kalman and Koepcke [K2], Gunckel and Franklin [G2], and Joseph and Tou [J5]. Since their work, the literature on linear–quadratic problems has grown tremendously. The special issue on the linear–quadratic problem of the *IEEE Transactions on Automatic Control* [I1] contains most of the pertinent theory and variations thereof, together with hundreds of references. For a multidimensional version of Problem 3 see the paper by Jacobson [J1]. The result of Problem 14 is also due to Jacobson [J2].

The literature on inventory control stimulated by the pioneering paper of Arrow *et al.* [A6] is also voluminous. The 1966 survey paper by Veinott [V3] contains 118 references. An important work summarizing most of the research up to 1958 is the book by Arrow *et al*, [A7]. The ingenious line of argument for proving the optimality of $(s, S)$ policies in the case of nonzero fixed costs is due to Scarf [S6]. However, his original proof contains some minor flaws.

Most of the material in Section 3.3 is taken from the paper by Mossin [M8]. Some other interesting papers in the same area are by Hakansson [H1–H4], Kamin [K4], Levhari and Srinivasan [L4], and Phelps [P5].

## Problems

**1.** Show the optimality of the control law given by (30)–(33) for the linear–quadratic problem where the matrices $A_k$, $B_k$ are random.

**2.** *Linear–Quadratic Problems with Forecasts* Consider the linear–quadratic problem first examined in Section 3.1 ($A_k$, $B_k$: known) for the case where at the beginning of period $k$ there is available a forecast $y_k \in \{1, 2, \ldots, n\}$ consisting of an accurate prediction that $w_k$ will be selected in accordance with a particular probability distribution $P_{k|y_k}$ (cf. Section 2.3). The vectors $w_k$ need not have zero mean under the distributions $P_{k|y_k}$. Show that the optimal control law is of the form

$$\mu_k(x_k, y_k) = -(B_k' K_{k+1} B_k + R_k)^{-1} B_k' K_{k+1}[A_k x_k + E\{w_k | y_k\}] + \alpha_k,$$

where the matrices $K_i$, $i = 1, \ldots, N$, are given by the Riccati equation (5) and (6) and $\alpha_k$ are appropriate vectors in $R^n$.

**3.** Consider (for simplicity) a scalar linear system

$$x_{k+1} = a_k x_k + b_k u_k + w_k, \qquad k = 0, 1, \ldots, N-1,$$

where $a_k, b_k \in R$ are given and $w_k$ is for each $k$ a Gaussian disturbance with zero mean and variance $\sigma^2$. Show that the control law $\{\mu_0^*, \mu_1^*, \ldots, \mu_{N-1}^*\}$

that minimizes the risk-sensitive cost functional

$$E\left\{\exp\left[x_N^2 + \sum_{k=0}^{N-1}(x_k^2 + ru_k^2)\right]\right\}, \qquad r > 0,$$

is linear in the state variable. We assume no control constraints and independent disturbances. We also assume that the optimal value is finite for every $x_0$. Show by example that the Gaussian assumption is essential for the result to hold.

**4.** Consider an inventory control problem similar to the multistage inventory problem of Section 3.2. The only difference is that at the beginning of each period $k$ the decision maker, in addition to knowing the current inventory level $x_k$, receives an accurate forecast that the demand $w_k$ will be selected in accordance with one out of two possible probability distributions $P_l$, $P_s$ (large demand, small demand). The a priori probability of a large demand forecast is known (cf. Section 2.3).

(a) Obtain the optimal inventory ordering policy for the case of a single-period problem.

(b) Extend the result to the $N$-period case.

(c) Extend the result to the case of any finite number of possible distributions.

Consider also the inventory control problem where the purchase costs $c_k$, $k = 0, 1, \ldots, N-1$, are not known at the beginning of the process but instead they are independent random variables with a priori known probability distributions. The exact value of the cost $c_k$, however, becomes known to the decision maker at the beginning of the $k$th period, so that the inventory purchasing decision at time $k$ is made with exact knowledge of the cost $c_k$. Characterize the optimal ordering policy for this case.

**5.** Consider the multiperiod inventory model of Section 3.2 for the case where there is a one-period time lag between order and delivery of inventory, i.e., the system equation is of the form

$$x_{k+1} = x_k + u_{k-1} - w_k, \qquad k = 1, \ldots, N-1,$$
$$x_1 = x_0 - w_0.$$

Show that the optimal policy for this problem is an $(s, S)$ policy.

**6.** Consider the inventory problem under the assumption that unfilled demand at each stage is not backlogged but rather is lost, i.e., the system equation is $x_{k+1} = \max[0, x_k + u_k - w_k]$ instead of $x_{k+1} = x_k + u_k - w_k$. Show that a multiperiod $(s, S)$ policy is optimal.

*Abbreviated Proof* (S. Shreve) Let $J_N(x) = 0$ and for all $k$

$$G_k(y) = cy + E\{h \max[0, y - w_k] + p \max[0, w_k - y] + J_{k+1}(\max[0, y - w_k])\},$$

$$J_k(x) = -cx + \min_{u \geqslant 0}[K\delta(u) + G_k(x + u)],$$

where $\delta(0) = 0$, $\delta(u) = 1$ for $u > 0$. The result will follow if we can show that $G_k$ is $K$-convex, continuous, and $G_k(y) \to \infty$ as $|y| \to \infty$. The difficult part is to show $K$-convexity since $K$-convexity of $G_{k+1}$ does not imply $K$-convexity of $E\{J_{k+1}(\max[0, y - w])\}$. It will be sufficient to show that $K$-convexity of $G_{k+1}$ implies $K$-convexity of

$$H(y) = p \max[0, -y] + J_{k+1}(\max[0, y]), \tag{68}$$

or equivalently that

$$K + H(y + z) \geqslant H(y) + z\,\frac{H(y) - H(y - b)}{b} \qquad \forall z \geqslant 0, \quad b > 0, \quad y. \tag{69}$$

By $K$-convexity of $G_{k+1}$ we have for appropriate scalars $s_{k+1}$ and $S_{k+1}$ such that $G_{k+1}(S_{k+1}) = \min_y G_{k+1}(y)$ and $K + G_{k+1}(S_{k+1}) = G_{k+1}(s_{k+1})$:

$$J_{k+1}(x) = -cx + \begin{cases} G_{k+1}(x) & \text{if} \quad s_{k+1} \leqslant x, \\ K + G_{k+1}(S_{k+1}) & \text{if} \quad x < s_{k+1}, \end{cases} \tag{70}$$

and $J_{k+1}$ is $K$-convex by the theory of Section 3.2.

*Case* 1 $0 \leqslant y - b < y \leqslant y + z$ For this region (69) follows from $K$-convexity of $J_{k+1}$.

*Case* 2 $y - b < y \leqslant y + z \leqslant 0$ In this region $H$ is linear and hence $K$-convex.

*Case* 3 $y - b < y \leqslant 0 \leqslant y + z$ In this region (69) may be written [in view of (68)] as $K + J_{k+1}(y + z) \geqslant J_{k+1}(0) - p(y + z)$. We will show that

$$K + J_{k+1}(z) \geqslant J_{k+1}(0) - pz \qquad \forall z \geqslant 0. \tag{71}$$

If $0 \leqslant s_{k+1} \leqslant z$, then using (70) and the fact $p > c$,

$$K + J_{k+1}(z) = K - cz + G_{k+1}(z) \geqslant K - pz + G_{k+1}(S_{k+1}) = J_{k+1}(0) - pz.$$

If $0 \leqslant z \leqslant s_{k+1}$, then using (70) and the fact $p > c$,

$$K + J_{k+1}(z) = 2K - cz + G_{k+1}(S_{k+1}) \geqslant K - pz + G_{k+1}(S_{k+1}) = J_{k+1}(0) - pz.$$

If $s_{k+1} \leqslant 0 \leqslant z$, then using (70), the fact $p > c$, and part (iv) of the lemma in Section 3.2,

$$K + J_{k+1}(z) = K - cz + G_{k+1}(z) \geqslant G_{k+1}(0) - pz = J_{k+1}(0) - pz.$$

Thus (71) is proved and (69) follows for the case under consideration.

*Case* 4 $\quad y - b < 0 < y \leqslant y + z \quad$ Then $0 < y < b$. If

$$[H(y) - H(0)]/y \geqslant [H(y) - H(y - b)]/b, \tag{72}$$

then since $H$ agrees with $J_{k+1}$ on $[0, \infty)$ and $J_{k+1}$ is $K$-convex,

$$K + H(y + z) \geqslant H(y) + z\frac{H(y) - H(0)}{y} \geqslant H(y) + z\frac{H(y) - H(y - b)}{b},$$

where the last step follows from (72). If

$$[H(y) - H(0)]/y < [H(y) - H(y - b)]/b,$$

then we have

$$H(y) - H(0) < (y/b)[H(y) - H(y - b)] = (y/b)[H(y) - H(0) + p(y - b)].$$

It follows that

$$(1 - (y/b))[H(y) - H(0)] < (y/b)p(y - b) = -py(1 - (y/b)),$$

and since $b > y$,

$$H(y) - H(0) < -py. \tag{73}$$

Now we have, using the definition of $H$, (71), and (73),

$$\begin{aligned} H(y) + z\frac{H(y) - H(y - b)}{b} &= H(y) + z\frac{H(0) - py - H(0) + p(y - b)}{b} \\ &= H(y) - pz < H(0) - p(y + z) \\ &\leqslant K + H(y + z). \end{aligned}$$

Hence (69) is proved for this case as well. Q.E.D.

**7.** Consider the dynamic portfolio problem of Section 3.3 for the case where at each period $k$ there is a forecast that the rates of return of the risky assets for that period will be selected in accordance with a particular probability distribution as in Section 2.3. Show that a partially myopic policy is optimal.

**8.** Consider a problem involving the linear system

$$x_{k+1} = A_k x_k + B_k u_k, \qquad k = 0, 1, \ldots, N - 1,$$

where the $n \times n$ matrices $A_k$ are given and the $n \times m$ matrices $B_k$ are independent random matrices and have given probability distributions that do not depend on $x_k, u_k$. The problem is to find the optimal control law $\{\mu_0^*(x_0), \ldots, \mu_{N-1}^*(x_{N-1})\}$ that maximizes the cost functional $E\{U(c'x_N)\}$, where $c$ is a given $n$-dimensional vector. We assume that $U: R \to R$ is a concave utility function satisfying for all $y$

$$-U'(y)/U''(y) = a + by,$$

and that the control is unconstrained. Show that the control law consists of affine functions of the current state.

*Hint* Reduce the problem to a one-dimensional problem, and use the results of Section 3.3.

**9.** Consider the asset-selling problem of Section 3.4 for the case where successive offers are not lost but can be accepted at any subsequent time period. Determine the general form of the optimal policy.

**10.** Suppose that an individual wants to sell his house and an offer comes in at the beginning of each day. We assume that successive offers are independent and an offer is $x_j$ with probability $p_j$, $j = 1, \ldots, n$, where $x_j$ are given non-negative scalars. Any offer not immediately accepted is not lost but may be accepted at any later date. Also, a maintenance cost $c$ is incurred for each day that the house remains unsold. The objective is to maximize the price at which the house is sold minus the maintenance costs. Consider the problem when there is a deadline to sell the house within $N$ days and characterize the optimal policy.

**11.** *Capacity Expansion Problem* Consider a problem of expanding the capacity of a facility for production of a single type of nonstorable good or service over $N$ time periods. Let us denote by $x_k$ the production capacity at the beginning of the $k$th period and by $u_k \geq 0$ the addition to capacity during the $k$th period. Thus capacity evolves according to

$$x_{k+1} = x_k + u_k, \qquad k = 0, 1, \ldots, N-1.$$

The demand at the $k$th period is denoted $w_k$ and has a known probability distribution that does not depend on either $x_k$ or $u_k$. Also successive demands are assumed to be independent and bounded. We denote:

| | |
|---|---|
| $C_k(u_k)$ | expansion cost associated with adding capacity $u_k$ |
| $P_k(x_k + u_k - w_k)$ | penalty cost associated with capacity $x_k + u_k$ and demand $w_k$, |
| $S(x_N)$ | salvage value of final capacity $x_N$. |

Thus the cost functional takes the form

$$\underset{\substack{w_k \\ k=0,1,\ldots,N-1}}{E} \left\{ -S(x_N) + \sum_{k=0}^{N-1} [C_k(u_k) + P_k(x_k + u_k - w_k)] \right\}.$$

(a) Derive the DP algorithm for solving this problem.

(b) Assume that $S$ is a concave function with $\lim_{x\to\infty} dS(x)/dx = 0$, $P_k$ are convex functions, and the expansion cost $C_k$ is of the form

$$C_k(u) = \begin{cases} K + c_k u & \text{if} \quad u > 0, \\ 0 & \text{if} \quad u = 0, \end{cases}$$

where $K \geqslant 0$, $c_k > 0$ for all $k$. Show that the optimal policy is of the $(s, S)$ type assuming $c_k y + E\{P_k(y - w_k)\} \to \infty$ as $|y| \to \infty$.

(c) In (b) assume that forecasts on $w_k$ as in Problem 5 are available. Derive the optimal policy for this case.

**12.** *A Gambling Problem* A gambler enters a game whereby he may at any time $k$ stake any amount $u_k \geqslant 0$ that does not exceed his current fortune $x_k$ (defined to be his initial capital plus his gain or minus his loss thus far). He wins his stake back and as much more with probability $p$, where $\frac{1}{2} < p < 1$, and he loses his stake with probability $(1 - p)$. Show that the gambling strategy that maximizes $E\{\ln x_N\}$, where $x_N$ denotes his fortune after $N$ plays, is to stake at each time $k$ an amount $u_k = (2p - 1)x_k$.

(*Note* The problem is related to the portfolio problem of Section 3.3.)

**13.** *Optimal Termination of Sampling* A collection of $N \geqslant 2$ objects is observed randomly and sequentially one at a time. The observer may either select the current object observed, in which case the selection process is terminated, or reject the object and proceed to observe the next. The observer can rank each object relative to those he has already observed and the objective is to maximize the probability of selecting the "best" object according to some criterion. It is assumed that no two objects can be judged to be equal. Let $r^*$ be the smallest positive integer $r$ such that

$$\frac{1}{N-1} + \frac{1}{N-2} + \cdots + \frac{1}{r} \leqslant 1.$$

Show that an optimal policy requires that the first $r^*$ objects be observed. If the $r^*$th object has rank 1 relative to the others already observed, it should be selected, otherwise the observation process should be continued until an object of rank 1 relative to those already observed is found.

*Hint* We assume that if the $r$th object has rank 1 relative to the previous $(r - 1)$ objects, then the probability that it is best is $r/N$. For $k \geqslant r^*$ let $J_k(0)$ be the maximal probability of finding the best object assuming $k$ objects have been selected and the $k$th object is not best relative to the previous $(k - 1)$ objects. Show that

$$J_k(0) = \frac{k}{N}\left(\frac{1}{N-1} + \cdots + \frac{1}{k}\right).$$

**14.** *A Class of Nonlinear Quadratic Problems* Within the framework of the basic problem consider the case of a quadratic cost functional of the form

$$\frac{1}{2} \mathop{E}_{w_k} \left\{ x_N' Q_N x_N + \sum_{k=0}^{N-1} (x_k' Q_k x_k + u_k' R_k u_k) \right\}$$

and an $n$-dimensional nonlinear system of the form

$$x_{k+1} = A_k x_k + B_k u_k + m_k + f_k(x_k, u_k, w_k), \qquad k = 0, 1, \ldots, N-1,$$

where $A_k$, $B_k$, $m_k$, and $f_k$ are given. We assume for all $k$, $x_k$, $u_k$,

$$\mathop{E}_{w_k} \{ f_k(x_k, u_k, w_k) | x_k, u_k \} = 0,$$

and that the covariance matrix

$$F(x_k, u_k) = \mathop{E}_{w_k} \{ f_k(x_k, u_k, w_k) f_k(x_k, u_k, w_k)' | x_k, u_k \}$$

is a general quadratic function of $x_k$, $u_k$ for all $k$, i.e., $F_k$ has a representation of the form

$$F_k(x_k, u_k) = P_k^0 + \sum_{i=1}^{n'} P_k^i (\tfrac{1}{2} x_k' W_k^i x_k + u_k' N_k^i x_k + \tfrac{1}{2} u_k' M_k^i u_k + x_k' g_k^i + u_k' h_k^i),$$

where $n' = [n(n+1)/2]$, $P_k^i$, $W_k^i$, $N_k^i$, $M_k^i$ are given matrices of appropriate dimensions, $g_k^i$, $h_k^i$ are given vectors, and $P_k^i$, $W_k^i$, $M_k^i$ are symmetric. For any square matrix $L$ denote by $\operatorname{tr}(L)$ the trace of $L$, i.e., the sum of the diagonal elements of $L$. Define for all $k$

$$W_k = \frac{1}{2} \sum_{i=1}^{n'} \operatorname{tr}(S_{k+1} P_k^i) W_k^i, \qquad N_k = \frac{1}{2} \sum_{i=1}^{n'} \operatorname{tr}(S_{k+1} P_k^i) N_k^i,$$

$$M_k = \frac{1}{2} \sum_{i=1}^{n'} \operatorname{tr}(S_{k+1} P_k^i) M_k^i,$$

$$g_k = \frac{1}{2} \sum_{i=1}^{n'} \operatorname{tr}(S_{k+1} P_k^i) g_k^i, \qquad h_k = \frac{1}{2} \sum_{i=1}^{n'} \operatorname{tr}(S_{k+1} P_k^i) h_k^i,$$

$$\tilde{R}_k = R_k + B_k' S_{k+1} B_k + M_k, \qquad \tilde{A}_k = B_k' S_{k+1} A_k + N_k,$$

$$\tilde{m}_k = B_k' S_{k+1} m_k + B_k' d_{k+1} + h_k,$$

where

$$S_k = Q_k + A_k' S_{k+1} A_k + W_k - \tilde{A}_k' \tilde{R}_k^{-1} \tilde{A}_k, \qquad S_N = Q_N,$$

$$d_k = A_k' S_{k+1} m_k + A_k' d_{k+1} + g_k - \tilde{A}_k \tilde{R}_k^{-1} m_k, \qquad d_N = 0,$$

$$e_k = \tfrac{1}{2} \operatorname{tr}(S_{k+1} P_k^0) + \tfrac{1}{2} m_k' S_{k+1} m_k + d_{k+1}' m_k + e_{k+1} - \tfrac{1}{2} \tilde{m}_k' \tilde{R}_k^{-1} \tilde{m}_k, \qquad e_N = 0.$$

Show that if the matrices $\tilde{R}_k$ defined above are positive definite for all $k$, then the control law $\{\mu_0^*, \mu_1^*, \ldots, \mu_{N-1}^*\}$ defined by

$$\mu_k^*(x_k) = -\tilde{R}_k^{-1}(\tilde{A}_k x_k + \tilde{m}_k), \qquad k = 0, 1, \ldots, N-1,$$

is optimal.

*Hint* Show by induction that the cost-to-go functions obtained from the DP algorithm are of the form

$$J_k(x_k) = \tfrac{1}{2}x_k' S_k x_k + d_k' x_k + e_k, \qquad k = 0, \ldots, N-1.$$

*Chapter 4*

# Problems with Imperfect State Information

## 4.1 Reduction to the Perfect State Information Case

We consider now an important class of sequential optimization problems that is characterized by a situation often appearing in practice. Again we have the discrete-time dynamic system considered in the basic problem of Chapter 2 that we wish to control. However, we assume that the state of the system is no longer known at each stage to the controller (imperfect state information). Instead the controller receives some information at each stage about the value of the current state. Borrowing from engineering terminology, we shall loosely describe this information as noisy measurements of a function of the system state. The inability of the controller to observe the exact state could be due to physical inaccessibility of some of the state variables, or to inaccuracies of the sensors or procedures used for measurement. For example, in problems of control of chemical processes it may be physically impossible to measure exactly some of the state variables. In other cases it may be very costly to obtain the exact value of the state even though it may be physically possible to do so. In such problems it may be more efficient to base decisions on inaccurate information concerning the system state, which may be obtained at substantially less cost. In other situations, such as hypothesis testing problems, often of interest in statistics, the exact value of the state (true hypothesis) may be found only in an asymptotic sense after an infinite number of measurements (samples) has been obtained. Given that measurements are costly, a problem often posed is to determine when to terminate sampling so as to minimize observation costs plus the cost of error in the estimation of the system state. Thus in such situations not only is it impossible to determine the exact state of the system but the measurement process is the focal point of the decision problem itself.

Problems with imperfect state information are, generally speaking, considerably more complex than problems with perfect state information both from the analytical and computational point of view. Conceptually, however, they are no different from problems with perfect state information, and in fact, as we will demonstrate shortly, they can be reduced to problems of perfect state information by means of a simple reformulation.

First let us state the basic problem with imperfect state information with which we will be concerned.

BASIC PROBLEM WITH IMPERFECT STATE INFORMATION Consider the basic problem of Section 2.1, where the controller instead of having perfect knowledge of the state has access to observations $z_k$ of the form

$$z_0 = h_0(x_0, v_0), \qquad z_k = h_k(x_k, u_{k-1}, v_k), \qquad k = 1, 2, \ldots, N-1. \tag{1}$$

The observation $z_k$ belongs to a given observation space $Z_k$; $v_k$ is a random observation disturbance belonging to a given space $V_k$, and is characterized by given probability measures $P_{v_0}(\cdot \mid x_0)$, $P_{v_k}(\cdot \mid x_k, u_{k-1})$, $k = 1, \ldots, N-1$, which may depend explicitly on the state $x_k$ and the control $u_{k-1}$ but not on prior observation disturbances $v_0, \ldots, v_{k-1}$ or any of the disturbances $w_0, \ldots, w_k$. The initial state $x_0$ is also random and characterized by a given probability measure $P_{x_0}$. The probability measure $P_{w_k}(\cdot \mid x_k, u_k)$ of $w_k$ is given and may depend explicitly on $x_k$ and $u_k$ but not on prior disturbances $w_0, \ldots, w_{k-1}$. The control $u_k$ is constrained to take values from a given nonempty subset $U_k$ of the control space $C_k$. It is assumed that this subset does not depend on $x_k$.

Let us denote by $I_k$ the information available to the controller at time $k$ and call it the *information vector*. We have

$$\begin{aligned} I_k &= (z_0, z_1, \ldots, z_k, u_0, u_1, \ldots, u_{k-1}), \qquad k = 1, 2, \ldots, N-1, \\ I_0 &= z_0. \end{aligned} \tag{2}$$

We consider the class of control laws (or policies), which consist of a finite sequence of functions $\pi = \{\mu_0, \mu_1, \ldots, \mu_{N-1}\}$ where each function $\mu_k$ maps the information vector $I_k$ into the control space $C_k$ and

$$\mu_k(I_k) \in U_k \qquad \forall I_k, \quad k = 0, \ldots, N-1.$$

Such control laws are termed *admissible*. The problem is to find an admissible control law $\pi = \{\mu_0, \mu_1, \ldots, \mu_{N-1}\}$ that minimizes the cost functional

$$J_\pi = \underset{\substack{x_0, w_k, v_k \\ k=0,\ldots,N-1}}{E} \left\{ g_N(x_N) + \sum_{k=0}^{N-1} g_k[x_k, \mu_k(I_k), w_k] \right\} \tag{3}$$

subject to the system equation

$$x_{k+1} = f_k[x_k, \mu_k(I_k), w_k], \qquad k = 0, 1, \ldots, N-1,$$

and the measurement equation

$$z_0 = h_0(x_0, v_0),$$
$$z_k = h_k[x_k, \mu_{k-1}(I_{k-1}), v_k], \qquad k = 1, 2, \ldots, N-1.$$

The real-valued functions

$$g_N: S_N \to R, \qquad g_k: S_k \times C_k \times D_k \to R, \qquad k = 0, 1, \ldots, N-1,$$

are given.

Notice the difference from the case of perfect state information. Whereas before we were seeking a rule that would specify the control $u_k$ to be applied for each state $x_k$ and time $k$, now we are looking for a rule that tells us the control to be applied for every possible information vector $I_k$ (or state of information), i.e., for every sequence of measurements received and controls employed up to time $k$. This difference, at least at the conceptual level, is actually more apparent than real as will be seen in what follows.

Similarly, as for the basic problem of Section 2.1, the following sequence of events is envisioned in the problem above once an admissible policy $\pi = \{\mu_0, \mu_1, \ldots, \mu_{N-1}\}$ is adopted:

*Stage* 0 (1) The initial state $x_0$ is generated according to the given probability measure $P_{x_0}$.

(2) The observation disturbance $v_0$ is generated according to the probability measure $P_{v_0}(\cdot \mid x_0)$.

(3) The controller observes $z_0 = h_0(x_0, v_0)$ and applies $u_0 = \mu_0(I_0)$, where $I_0 = z_0$.

(4) The input disturbance $w_0$ is generated according to the probability measure $P_{w_0}(\cdot \mid x_0, \mu_0(I_0))$.

(5) The cost $g_0[x_0, \mu_0(I_0), w_0]$ is incurred.

(6) The next state $x_1$ is generated according to the system equation $x_1 = f_0[x_0, \mu_0(I_0), w_0]$.

*Stage* $k$ (1) The observation disturbance $v_k$ is generated according to the probability measure $P_{v_k}(\cdot \mid x_k, u_{k-1})$.

(2) The controller observes $z_k = h_k[x_k, \mu_{k-1}(I_{k-1}), v_k]$ and applies $u_k = \mu_k(I_k)$, where $I_k = (z_0, \ldots, z_k, u_0, \ldots, u_{k-1})$.

(3) The input disturbance $w_k$ is generated according to the probability measure $P_{w_k}(\cdot \mid x_k, \mu_k(I_k))$.

(4) The cost $g_k[x_k, \mu_k(I_k), w_k]$ is incurred and added to previous costs.

(5) The next state $x_{k+1}$ is generated according to the system equation $x_{k+1} = f_k[x_k, \mu_k(I_k), w_k]$.

*Last Stage* $(N - 1)$ (1) The observation disturbance $v_{N-1}$ is generated according to $P_{v_{N-1}}(\cdot \mid x_{N-1}, u_{N-2})$.

(2) The controller observes $z_{N-1} = h_{N-1}[x_{N-1}, \mu_{N-2}(I_{N-2}), v_{N-1}]$ and applies $u_{N-1} = \mu_{N-1}(I_{N-1})$, where $I_{N-1} = (z_0, \ldots, z_{N-1}, u_0, \ldots, u_{N-2})$.

(3) The input disturbance $w_{N-1}$ is generated according to the probability measure $P_{w_{N-1}}(\cdot \mid x_{N-1}, \mu_{N-1}(I_{N-1}))$.

(4) The cost $g_{N-1}[x_{N-1}, \mu_{N-1}(I_{N-1}), w_{N-1}]$ is incurred and added to previous costs.

(5) The final state $x_N$ is generated according to

$$x_N = f_{N-1}[x_{N-1}, \mu_{N-1}(I_{N-1}), w_{N-1}].$$

(6) The terminal cost $g_N(x_N)$ is incurred and added to previous costs.

Again the above process is well defined and the stochastic variables are generated by means of a precisely formulated probabilistic mechanism. Similarly, however, as for the perfect information case the cost functional

$$g_N(x_N) + \sum_{k=0}^{N-1} g_k[x_k, \mu_k(I_k), w_k]$$

is not in general a well-defined random variable in the absence of additional assumptions and structure. Again we shall bypass a rigorous formulation of the problem in view of the introductory nature of the text.† However, we mention that the cost functional can be viewed as a well-defined random variable if the space of the initial state $S_0$ and the disturbance spaces $D_k$ and $V_k$, $k = 0, 1, \ldots, N - 1$, are finite or countable sets. As we shall shortly demonstrate, the imperfect state information problem defined above may be converted by reformulation into a perfect state information problem. Once this reformulation is considered the problem may be rigorously posed in the manner described in Section 2.1.

We now provide an example of a problem that fits the general framework introduced in this section:

† For a recent treatment of the delicate mathematical questions involved in a rigorous and general analysis of the problem of this chapter we refer the reader to the monograph by Striebel [S18a].

EXAMPLE 1 A machine can be in one of two states denoted 0 and 1. State 0 corresponds to a machine in proper condition (good state) and state 1 to a machine in improper condition (bad state). If the machine is operated for one unit time, it stays in state 0 with probability $\frac{2}{3}$ provided it started at 0, and it stays in state 1 with probability 1 if it started at 1, as Fig. 4.1 shows. The

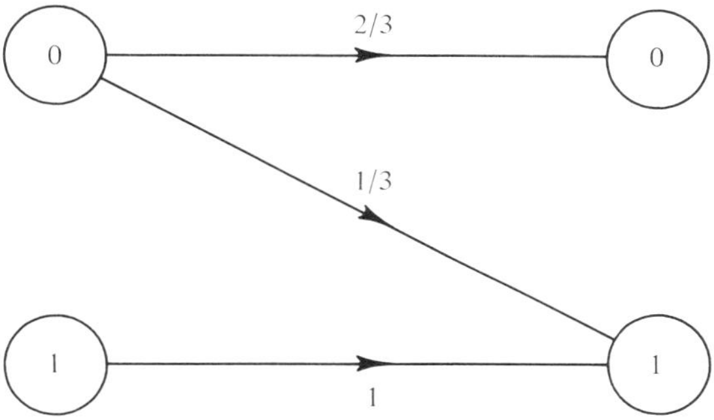


FIGURE 4.1

machine is operated for a total of three units of time and starts at state 0. At the end of the first and second unit of time the machine is inspected and there are two possible inspection outcomes denoted $G$ (probably good state) and $B$ (probably bad state). If the machine is in state $x = 0$, the inspection outcome is $G$ with probability $\frac{3}{4}$; if the machine is in $x = 1$, the inspection outcome is $B$ with probability $\frac{3}{4}$:

$$P(G|x = 0) = \tfrac{3}{4}, \quad P(B|x = 0) = \tfrac{1}{4}, \qquad P(G|x = 1) = \tfrac{1}{4}, \quad P(B|x = 1) = \tfrac{3}{4}.$$

After each inspection one of two possible actions can be taken:

*Action C* continue operation of the machine.

*Action S* stop the machine, do a complete and accurate inspection, and if the machine is in state 1 bring it back to the proper state 0.

There is a cost of 2 units for using a machine in state 1 for one time unit and zero cost for using a machine in state 0 for one time unit. There is also a cost of 1 unit for taking action $S$.

The problem is to determine the policy that minimizes the expected costs over the three time periods. In other words, we want to find the optimal course of action after the result of the first inspection is known, and after the results of the first and second inspection (and, of course, the action taken after the first inspection) are known.

It is not difficult to see that this example falls within the general framework of the problem of this section. The state space consists of the two states 0 and 1,

$$\text{state space} = \{0, 1\},$$

and the control space consists of the two actions

$$\text{control space} = \{C, S\}.$$

The system evolution may be described by introducing a system equation

$$x_{k+1} = w_k, \qquad k = 0, 1,$$

where for $k = 0, 1$ the probability distribution of $w_k$ is given by

$$P(w_k = 0 \mid x_k = 0, u_k = C) = \tfrac{2}{3}, \qquad P(w_k = 1 \mid x_k = 0, u_k = C) = \tfrac{1}{3},$$

$$P(w_k = 0 \mid x_k = 1, u_k = C) = 0, \qquad P(w_k = 1 \mid x_k = 1, u_k = C) = 1,$$

$$P(w_k = 0 \mid x_k = 0, u_k = S) = \tfrac{2}{3}, \qquad P(w_k = 1 \mid x_k = 0, u_k = S) = \tfrac{1}{3},$$

$$P(w_k = 0 \mid x_k = 1, u_k = S) = \tfrac{2}{3}, \qquad P(w_k = 1 \mid x_k = 1, u_k = S) = \tfrac{1}{3}.$$

We denote by $x_0, x_1, x_2$ the state of the machine at the end of the first, second, and third time unit, respectively. Also we denote by $u_0$ the action taken after the first inspection (end of first time unit) and by $u_1$ the action taken after the second inspection (end of second time unit). The probability distribution of $x_0$ is

$$P(x_0 = 0) = \tfrac{2}{3}, \qquad P(x_0 = 1) = \tfrac{1}{3}.$$

Concerning the observation process we do not have perfect state information since the inspections do not reveal the state of the machine with certainty. Rather the result of each inspection may be viewed as a measurement on the state of the system taking the form

$$z_k = v_k, \qquad k = 0, 1,$$

where for $k = 0, 1$ the probability distribution of $v_k$ is given by

$$P(v_k = G \mid x_k = 0) = \tfrac{3}{4}, \qquad P(v_k = B \mid x_k = 0) = \tfrac{1}{4},$$

$$P(v_k = G \mid x_k = 1) = \tfrac{1}{4}, \qquad P(v_k = B \mid x_k = 1) = \tfrac{3}{4}.$$

The cost resulting from a sequence of states $x_0, x_1$ and actions $u_0, u_1$ is

$$g(x_0, u_0) + g(x_1, u_1),$$

where

$$g(0, C) = 0, \quad g(0, S) = 1, \qquad g(1, C) = 2, \quad g(1, S) = 1.$$

The information vector at times 0 and 1 is

$$I_0 = z_0, \qquad I_1 = (z_0, z_1, u_0),$$

and we seek functions $\mu_0(I_0)$, $\mu_1(I_1)$ that minimize

$$\underset{\substack{x_0, w_0, w_1\\ v_0, v_1}}{E} \{g[x_0, \mu_0(I_0)] + g[x_1, \mu_1(I_1)]\}$$

$$= \underset{\substack{x_0, w_0, w_1\\ v_0, v_1}}{E} \{g[x_0, \mu_0(z_0)] + g[x_1, \mu_1(z_0, z_1, \mu_0(z_0))]\}. \tag{4}$$

We shall provide a complete solution of this problem once we introduce the related DP algorithm.

Let us now show how the general problem of this section can be reformulated into the framework of the basic problem with perfect state information. Similarly, as in the discussion of the state augmentation device of Section 2.3, it is intuitively clear that we should define a new system the state of which at time $k$ consists of all variables the knowledge of which can be of benefit to the controller when making the $k$th decision. Thus a first candidate as the state of the new system is the information vector $I_k$. Indeed we will show that this choice is appropriate.

We have by definition [cf. Eq. (2)] for every $k$,

$$I_{k+1} = (I_k, z_{k+1}, u_k), \quad k = 0, 1, \ldots, N-2, \qquad I_0 = z_0. \tag{5}$$

*These equations can be viewed as describing the evolution of a system of the same nature as the one considered in the basic problem of Section* 2.1. The state of the system is $I_k$, the control $u_k$, and $z_{k+1}$ can be viewed as a random disturbance. Furthermore, we have, simply by the definition of $I_k$,

$$P(z_{k+1} \in \bar{Z}_{k+1} | I_k, u_k) = P(z_{k+1} \in \bar{Z}_{k+1} | I_k, u_k, z_0, z_1, \ldots, z_k),$$

for any event $\bar{Z}_{k+1}$ (a subset of $Z_{k+1}$). Thus the probability measure of $z_{k+1}$ depends explicitly only on the state $I_k$ and control $u_k$ of the new system (5) and not on the prior "disturbances" $z_k, \ldots, z_1$. It should be noted that the probability measure of $z_{k+1}$ is completely determined from the statistics of the problem. We have for any event $\bar{Z}_{k+1}$

$$P(z_{k+1} \in \bar{Z}_{k+1} | I_k, u_k) = P[h_{k+1}(x_{k+1}, u_k, v_{k+1}) \in \bar{Z}_{k+1} | I_k, u_k],$$

and the probability on the right is completely determined from the statistics of the problem (i.e., the probability measures of $w_k, \ldots, w_0, v_{k+1}, \ldots, v_0, x_0$) and $I_k$ and $u_k$.

Now we turn to the reformulation of the cost functional in terms of the variables of the new system. We have for any admissible control law $\pi = \{\mu_0, \mu_1, \ldots, \mu_{N-1}\}$

$$
\begin{aligned}
J_\pi &= \underset{\substack{x_0, w_k, v_k \\ k=0,\ldots,N-1}}{E} \left\{ g_N(x_N) + \sum_{k=0}^{N-1} g_k[x_k, \mu_k(I_k), w_k] \right\} \\
&= \underset{\substack{z_k \\ k=0,\ldots,N-1}}{E} \left\{ \sum_{k=0}^{N-1} \tilde{g}_k[I_k, \mu_k(I_k)] \right\}, \qquad (6)
\end{aligned}
$$

where the functions $\tilde{g}_k$, $k = 0, \ldots, N-1$, are defined by

$$
\begin{aligned}
&\tilde{g}_{N-1}[I_{N-1}, \mu_{N-1}(I_{N-1})] \\
&\quad = \underset{x_{N-1}, w_{N-1}}{E} \{g_N[f_{N-1}[x_{N-1}, \mu_{N-1}(I_{N-1}), w_{N-1}]] \\
&\qquad + g_{N-1}[x_{N-1}, \mu_{N-1}(I_{N-1}), w_{N-1}] \,|\, I_{N-1}, \mu_{N-1}(I_{N-1})\} \qquad (7)
\end{aligned}
$$

$$
\tilde{g}_k[I_k, \mu_k(I_k)] = \underset{x_k, w_k}{E} \{g_k[x_k, \mu_k(I_k), w_k] \,|\, I_k, \mu_k(I_k)\}. \qquad (8)
$$

The conditional probabilities $P(w_k \,|\, I_k, \mu_k(I_k))$, $k = 0, 1, \ldots, N-1$, are determined from $P(w_k \,|\, x_k, \mu_k(I_k))$ and $P(x_k \,|\, I_k)$.

Thus the basic problem with imperfect state information has been reformulated to a problem with perfect state information that involves system (5) and cost functional (6). By writing the DP algorithm for this latter problem we have

$$
J_{N-1}(I_{N-1}) = \inf_{u_{N-1} \in U_{N-1}} \tilde{g}_{N-1}(I_{N-1}, u_{N-1}),
$$

$$
J_k(I_k) = \inf_{u_k \in U_k} \left[ \tilde{g}_k(I_k, u_k) + \underset{z_{k+1}}{E} \{J_{k+1}(I_k, z_{k+1}, u_k) \,|\, I_k, u_k\} \right].
$$

Using (7) and (8) the algorithm can be written

$$
\begin{aligned}
J_{N-1}(I_{N-1}) = \inf_{u_{N-1} \in U_{N-1}} \Big[ &\underset{x_{N-1}, w_{N-1}}{E} \{g_N[f_{N-1}(x_{N-1}, u_{N-1}, w_{N-1})] \\
&+ g_{N-1}(x_{N-1}, u_{N-1}, w_{N-1}) \,|\, I_{N-1}, u_{N-1}\} \Big], \qquad (9)
\end{aligned}
$$

$$
\begin{aligned}
J_k(I_k) = \inf_{u_k \in U_k} \Big[ &\underset{x_k, w_k, v_{k+1}}{E} \{g_k(x_k, u_k, w_k) \\
&+ J_{k+1}[I_k, h_{k+1}[f_k(x_k, u_k, w_k), u_k, v_{k+1}], u_k] \,|\, I_k, u_k\} \Big]. \qquad (10)
\end{aligned}
$$

Equations (9) and (10) constitute the basic DP algorithm for the problem of this section. An optimal control law $\{\mu_0^*, \mu_1^*, \ldots, \mu_{N-1}^*\}$ is obtained by first solving the minimization problem in (9) for every possible value of the information vector $I_{N-1}$ to obtain $\mu_{N-1}^*(I_{N-1})$. Simultaneously $J_{N-1}(I_{N-1})$ is computed and used in the computation of $J_{N-2}(I_{N-2})$ via the minimization in (10), which is carried out for every possible value of $I_{N-2}$. Proceeding similarly one obtains $J_{N-3}(I_{N-3})$ and $\mu_{N-3}^*$ and so on until $J_0(I_0) = J_0(z_0)$ is computed. The optimal cost $J^*$ is then obtained from

$$J^* = \underset{z_0}{E}\{J_0(z_0)\}.$$

It can also be seen that the DP algorithm (9) and (10) is valid even if the probability measures of $w_k, v_k$ depend on prior disturbances. We now continue our earlier example:

EXAMPLE 1 (continued) We first use (9) to compute $J_1(I_1)$ for each of the eight possible information vectors $I_1 = (z_0, z_1, u_0)$. These vectors are

$$I_1 = (G, G, C), (G, G, S), (B, G, C), (B, G, S), (G, B, C), (G, B, S), (B, B, C), (B, B, S).$$

We shall compute for each $I_1$ the expected cost associated with $u_1 = C$ and $u_1 = S$ and select as optimal the action with the smallest cost. We have

$$\text{cost of } C = 2 \times P(x_1 = 1 \mid I_1), \qquad \text{cost of } S = 1,$$

and therefore

$$J_1(I_1) = \min[2P(x_1 = 1 \mid I_1), 1].$$

The probabilities $P(x_1 = 1 \mid I_1)$ may be computed by using Bayes' rule and the data of the problem. Some of the details will be omitted. We have:

*For* $I_1 = (G, G, S)$

$$P(x_1 = 1 \mid G, G, S) = \frac{P(x_1 = 1, G, G, S)}{P(G, G, S)} = \frac{\frac{1}{3} \times \frac{1}{4}}{\frac{2}{3} \times \frac{3}{4} + \frac{1}{3} \times \frac{1}{4}} = \frac{1}{7}.$$

Hence

$$\triangleright \qquad J_1(G, G, S) = \tfrac{2}{7}, \qquad \mu_1^*(G, G, S) = C. \qquad \triangleleft$$

*For* $I_1 = (B, G, S)$

$$P(x_1 = 1 \mid B, G, S) = P(x_1 = 1 \mid G, G, S) = \tfrac{1}{7},$$

$$\triangleright \qquad J_1(B, G, S) = \tfrac{2}{7}, \qquad \mu_1^*(B, G, S) = C. \qquad \triangleleft$$

*For* $I_1 = (G, B, S)$

$$P(x_1 = 1 | G, B, S) = \frac{P(x_1 = 1, G, B, S)}{P(G, B, S)} = \frac{\frac{1}{3} \times \frac{3}{4}}{\frac{2}{3} \times \frac{1}{4} + \frac{1}{3} \times \frac{3}{4}} = \frac{3}{5},$$

▷ $$J_1(G, B, S) = 1, \qquad \mu_1^*(G, B, S) = S.$$ ◁

*For* $I_1 = (B, B, S)$

$$P(x_1 = 1 | B, B, S) = P(x_1 = 1 | G, B, S) = \tfrac{3}{5},$$

▷ $$J_1(B, B, S) = 1, \qquad \mu_1^*(B, B, S) = S.$$ ◁

*For* $I_1 = (G, G, C)$

$$P(x_1 = 1 | G, G, C) = \frac{P(x_1 = 1, G, G, C)}{P(G, G, C)} = \frac{1}{5},$$

▷ $$J_1(G, G, C) = \tfrac{2}{5}, \qquad \mu_1^*(G, G, C) = C.$$ ◁

*For* $I_1 = (B, G, C)$

$$P(x_1 = 1 | B, G, C) = \tfrac{11}{23},$$

▷ $$J_1(B, G, C) = \tfrac{22}{23}, \qquad \mu_1^*(B, G, C) = C.$$ ◁

*For* $I_1 = (G, B, C)$

$$P(x_1 = 1 | G, B, C) = \tfrac{9}{13},$$

▷ $$J_1(G, B, C) = 1, \qquad \mu_1^*(G, B, C) = S.$$ ◁

*For* $I_1 = (B, B, C)$

$$P(x_1 = 1 | B, B, C) = \tfrac{33}{37},$$

▷ $$J_1(B, B, C) = 1, \qquad \mu_1^*(B, B, C) = S.$$ ◁

Summarizing the results for the last stage the optimal policy is to continue ($u_1 = C$) if the result of the last inspection was $G$, and to stop ($u_1 = S$) if the result of the last inspection was $B$.

*First Stage* Here we use (10) to compute $J_0(I_0)$ for each of the two possible information vectors $I_0 = (G)$, $I_0 = (B)$. We have

$$\begin{aligned} \text{cost of } C &= 2P(x_0 = 1 | I_0, C) + \mathop{E}_{z_1} \{J_1(I_0, z_1, C) | I_0, C\} \\ &= 2P(x_0 = 1 | I_0, C) + P(z_1 = G | I_0, C) J_1(I_0, G, C) \\ &\quad + P(z_1 = B | I_0, C) J_1(I_0, B, C), \end{aligned}$$

$$\begin{aligned} \text{cost of } S &= 1 + \mathop{E}_{z_1} \{J_1(I_0, z_1, S) | I_0, S\} \\ &= 1 + P(z_1 = G | I_0, S) J_1(I_0, G, S) + P(z_1 = B | I_0, S) J_1(I_0, B, S), \end{aligned}$$

and

$$J_0(I_0) = \min[2P(x_0 = 1 | I_0, C) + \underset{z_1}{E}\{J_1(I_0, z_1, C) | I_0, C\}, 1 + \underset{z_1}{E}\{J_1(I_0, z_1, S) | I_0, S\}].$$

*For* $I_0 = (G)$ Direct calculation yields

$$P(z_1 = G | G, C) = \tfrac{15}{28}, \qquad P(z_1 = B | G, C) = \tfrac{13}{28},$$
$$P(z_1 = G | G, S) = \tfrac{7}{12}, \qquad P(z_1 = B | G, S) = \tfrac{5}{12},$$
$$P(x_0 = 1 | G, C) = \tfrac{1}{7},$$

and hence

$$J_0(G) = \min[2 \times \tfrac{1}{7} + \tfrac{15}{28}J_1(G, G, C) + \tfrac{13}{28}J_1(G, B, C), 1 + \tfrac{7}{12}J_1(G, G, S) + \tfrac{5}{12}J_1(G, B, S)].$$

Using the values of $J_1$ obtained in the previous stage

$$J_0(G) = \min[2 \times \tfrac{1}{7} + \tfrac{15}{28} \times \tfrac{2}{5} + \tfrac{13}{28} \times 1, 1 + \tfrac{7}{12} \times \tfrac{2}{7} + \tfrac{5}{12} \times 1] = \min[\tfrac{27}{28}, \tfrac{19}{12}] = \tfrac{27}{28},$$

▷ $$J_0(G) = \tfrac{27}{28}, \qquad \mu_0^*(G) = C.$$ ◁

*For* $I_0 = (B)$ Direct calculation yields

$$P(z_1 = G | B, C) = \tfrac{23}{60}, \qquad P(z_1 = B | B, C) = \tfrac{37}{60},$$
$$P(z_1 = G | B, S) = \tfrac{7}{12}, \qquad P(z_1 = B | B, S) = \tfrac{5}{12},$$
$$P(x_0 = 1 | B, C) = \tfrac{3}{5},$$

and

$$J_0(B) = \min[2 \times \tfrac{3}{5} + \tfrac{23}{60}J_1(B, G, C) + \tfrac{37}{60}J_1(B, B, C), 1 + \tfrac{7}{12}J_1(B, G, S) + \tfrac{5}{12}J_1(B, B, S)].$$

Using the values of $J_1$ obtained in the previous stage

$$J_0(B) = \min[\tfrac{131}{60}, \tfrac{19}{12}] = \tfrac{19}{12},$$

▷ $$J_0(B) = \tfrac{19}{12}, \qquad \mu_0^*(B) = S.$$ ◁

Summarizing, the optimal policy for both stages is to continue if the result of the latest inspection is $G$ and stop otherwise.

The optimal cost is

$$J^* = P(G)J_0(G) + P(B)J_0(B).$$

Since $P(G) = \frac{7}{12}$, $P(B) = \frac{5}{12}$,

$$J^* = \frac{7}{12} \times \frac{27}{28} + \frac{5}{12} \times \frac{19}{12} = \frac{176}{144}.$$

In the above example the computation of the optimal policy and the optimal cost by means of the DP algorithm (9) and (10) was made possible by the great simplicity of the problem. It is easy to see that for a more complex problem and in particular one where the number of possible information vectors $I_k$ is large (or infinite) and the number of stages $N$ is also large, the computational requirements of the DP algorithm can be truly prohibitive. This is due to the fact that it is necessary to apply the algorithm over the space of the information vector $I_k$, and, even if the control and observation spaces are simple (one dimensional or finite), the space of the information vector $I_k$ may be very complex and may have large dimension particularly for large values of $k$. This fact makes the application of the algorithm very difficult or computationally impossible in many cases. Not only is it necessary to carry out the DP computation over spaces of large dimension but also the storage of the functions that constitute the optimal controller presents a serious problem since values of control input must be stored for each possible value of information vector $I_k$ and for every $k$. Furthermore, the measurements $z_k$ obtained during the control process must be continuously stored by the controller. This rather disappointing situation motivates efforts aimed at the reduction of the data, which is truly necessary for control purposes. In other words, it is of interest to look for quantities that ideally would be of smaller dimension (can be characterized by a smaller set of numbers) than the information vector $I_k$ and nonetheless contain all the information in $I_k$ that is necessary for control purposes. Such quantities are called *sufficient statistics* and are the subject of Section 4.2.

## 4.2 Sufficient Statistics

Referring to the DP algorithm of Eqs. (9) and (10) let us consider the following definition:

**Definition** Let $S_k(\cdot)$, $k = 0, 1, \ldots, N-1$, be functions mapping the information vector $I_k$ into some space $M_k$, $k = 0, 1, \ldots, N-1$. We shall say that the functions $S_0(\cdot), \ldots, S_{N-1}(\cdot)$ constitute a *sufficient statistic* with respect to the problem of the previous section if there exist functions $H_0, \ldots, H_{N-1}$ such that

$$H_k[S_k(I_k), u_k] = \underset{x_k, w_k, v_{k+1}}{E} \{g_k(x_k, u_k, w_k) + J_{k+1}[I_k, h_{k+1}[f_k(x_k, u_k, w_k), u_k, v_{k+1}], u_k] | I_k, u_k\}$$
$$\forall I_k, u_k \in U_k, \quad k = 0, 1, \ldots, N-2 \tag{11}$$

$$H_{N-1}[S_{N-1}(I_{N-1}), u_{N-1}] = \underset{x_{N-1}, w_{N-1}}{E} \{g_N[f_{N-1}(x_{N-1}, u_{N-1}, w_{N-1})] + g_{N-1}(x_{N-1}, u_{N-1}, w_{N-1}) | I_{N-1}, u_{N-1}\}$$
$$\forall I_{N-1}, u_{N-1} \in U_{N-1}, \tag{12}$$

where $J_k$ are the cost-to-go functions of the DP algorithm (9) and (10).

Since the minimization step of the DP algorithm (9) and (10) can be written in terms of the functions $H_k$ as

$$\inf_{u_k \in U_k} H_k[S_k(I_k), u_k], \tag{13}$$

it follows immediately from the above definition that *an optimal control law need only depend on the information vector $I_k$ via the sufficient statistic $S_k(I_k)$.* In other words if the minimization problem in (13) has a solution for every $I_k$ and $k$, there exists an optimal control law that can be written

$$\mu_k^*(I_k) = \bar{\mu}_k^*[S_k(I_k)], \qquad k = 0, 1, \ldots, N-1, \tag{14}$$

where $\bar{\mu}_k^*$ is an appropriate function determined from the minimization (13). Thus if the sufficient statistic is characterized by a set of fewer numbers than the information vector $I_k$, it may be easier to implement the control law in the form $u_k = \bar{\mu}_k^*[S_k(I_k)]$ and take advantage of the resulting data reduction.

While it is possible to show that many different functions constitute a sufficient statistic for the problem that we are considering [the identity function $S_k(I_k) = I_k$ is certainly one of them], we will focus attention on a particular one that is useful both from the analytical and the conceptual point of view in many cases. This sufficient statistic is the conditional probability measure of the state $x_k$, given the information vector $I_k$,

$$S_k(I_k) = P_{x_k|I_k}, \qquad k = 0, 1, \ldots, N-1.$$

This function $S_k$ maps the space of information vectors into the space of all probability measures on the state space. It is easy to see that the conditional probability measures $P_{x_k|I_k}$ indeed constitute a sufficient statistic.† This is

† Abusing mathematical language we make no distinction between the function $S_k$ and its value $S_k(I_k)$, calling them both a sufficient statistic.

evident from definition (11) and (12) and the fact that the probability measures of $w_k$ and $v_{k+1}$ depend explicitly only on $x_k$, $u_k$ and $x_{k+1}$, $u_k$, respectively, and not on prior disturbances.

Now the sufficient statistic $P_{x_k|I_k}$ is generated recursively in time and can be viewed as the state of a controlled discrete-time dynamic system. By using Bayes' rule we can write

$$P_{x_{k+1}|I_{k+1}} = \Phi_k(P_{x_k|I_k}, u_k, z_{k+1}), \qquad k = 0, \ldots, N-2, \tag{15}$$

where $\Phi_k$ is some function that can be determined from the data of the problem, $u_k$ is the control of the system, and $z_{k+1}$ plays the role of a random disturbance the statistics of which are known and depend explicitly on $P_{x_k|I_k}$ and $u_k$ only and not on $z_k, \ldots, z_0$. For the case where the state, control, observation, and disturbance spaces are finite sets, see Problem 7(a). In the case where these spaces are the real line and all random variables involved possess probability density functions, the conditional density $p(\cdot|I_{k+1})$ is generated from $p(\cdot|I_k)$, $u_k$, and $z_{k+1}$ by means of the equation

$$p(x_{k+1}|I_{k+1}) = p(x_{k+1}|I_k, u_k, z_{k+1}) = \frac{p(x_{k+1}, z_{k+1}|I_k, u_k)}{p(z_{k+1}|I_k, u_k)}$$

$$= \frac{p(x_{k+1}|I_k, u_k)p(z_{k+1}|u_k, x_{k+1})}{\int_{-\infty}^{\infty} p(x_{k+1}|I_k, u_k)p(z_{k+1}|u_k, x_{k+1})\, dx_{k+1}}.$$

In this equation all the probability densities appearing in the right-hand side may be expressed in terms of $p(\cdot|I_k)$, $u_k$, and $z_{k-1}$ alone. In particular, the density $p(\cdot|I_k, u_k)$ may be expressed through $p(\cdot|I_k)$, $u_k$, and the system equation $x_{k+1} = f_k(x_k, u_k, w_k)$ using the given density $p(\cdot|x_k, u_k)$ and the relation

$$p(w_k|I_k, u_k) = \int_{-\infty}^{\infty} p(x_k|I_k)p(w_k|x_k, u_k)\, dx_k.$$

The density $p(\cdot|u_k, x_{k+1})$ is expressed through the measurement equation $z_{k+1} = h_{k+1}(x_{k+1}, u_k, v_{k+1})$ using the given probability density

$$p(\cdot|u_k, u_{k+1})$$

By substituting these expressions in the equation for $p(x_{k+1}|I_{k+1})$ we obtain an equation of the form of (15). Other explicit examples of equations of the form of (15) will be given in subsequent sections. In any case one can see that the *system described by* (15) *is one that fits the framework of the basic problem.* Furthermore, the controller can calculate (at least in principle) at time $k$ the conditional probability measure $P_{x_k|I_k}$. Therefore, the controlled system (15) is one for which perfect information prevails for the controller.

Now the DP algorithm (9) and (10) can be written in terms of the sufficient statistic $P_{x_k|I_k}$ by making use of the new system equation (15) as follows:

$$\bar{J}_{N-1}(P_{x_{N-1}|I_{N-1}}) = \inf_{u_{N-1}\in U_{N-1}} \Bigg[ \mathop{E}_{x_{N-1},\, w_{N-1}} \{g_N[f_{N-1}(x_{N-1}, u_{N-1}, w_{N-1})] + g_{N-1}(x_{N-1}, u_{N-1}, w_{N-1}) \,|\, I_{N-1}, u_{N-1}\} \Bigg], \tag{16}$$

$$\bar{J}_k(P_{x_k|I_k}) = \inf_{u_k\in U_k} \Bigg[ \mathop{E}_{x_k,\, w_k,\, z_{k+1}} \{g_k(x_k, u_k, w_k) + \bar{J}_{k+1}[\Phi_k(P_{x_k|I_k}, u_k, z_{k+1})] \,|\, I_k, u_k\} \Bigg]. \tag{17}$$

Furthermore, this DP algorithm yields a control law of the form

$$u_k^* = \bar{\mu}_k^*(P_{x_k|I_k}), \qquad k = 0, 1, \ldots, N-1,$$

which results from the minimization of the right-hand side of (16) and (17). This control law yields in turn an optimal control law $\{\mu_0^*, \ldots, \mu_{N-1}^*\}$ for the basic problem with imperfect state information by writing

$$\mu_k^*(I_k) = \bar{\mu}_k^*(P_{x_k|I_k}), \qquad k = 0, 1, \ldots, N-1.$$

In addition, the optimal value of the problem is given by

$$J^* = \mathop{E}_{z_0} \{\bar{J}_0(P_{x_0|z_0})\},$$

where $\bar{J}_0$ is obtained by the last step of the algorithm (16) and (17) and the probability measure of $z_0$ is obtained from the statistics of $x_0$ and $v_0$ and the measurement equation $z_0 = h_0(x_0, v_0)$.

It should be evident to the perceptive reader that the development of the DP algorithm (16) and (17) was based on nothing more than a reduction of the basic problem with imperfect state information to a problem with perfect state information that involves system (15), the state of which is $P_{x_k|I_k}$, and an appropriately reformulated cost functional. Thus the analysis of this section is in effect an alternate reduction of the problem of Section 4.1 to the basic problem with perfect state information. A conclusion that can be drawn from the analysis is that *the conditional probability $P_{x_k|I_k}$ summarizes all the information that is necessary for control purposes at period k*. In the absence of perfect knowledge of the state *the controller can be viewed as controlling the* *"probabilistic state"* $P_{x_k|I_k}$ *so as to minimize the expected value of the cost-to-go conditioned on the information* $I_k$ *available.*

The two reductions into the basic problem with perfect state information that we gave in this and the previous section are by far the most useful such reductions. Naturally there arises the question as to which of the two DP algorithms (9) and (10) or (16) and (17) is the most efficient. This depends on the nature of the problem at hand and, in particular, on the dimensionality of the probability distribution $P_{x_k|I_k}$. If $P_{x_k|I_k}$ is characterized by a finite set of numbers and may be computed with relative ease, then it may be profitable to carry out the DP algorithm over the space of the sufficient statistic $P_{x_k|I_k}$. Such, for example, is the case where the state space for each time $k$ is a finite set $\{x^1, \ldots, x^n\}$ (i.e., the system is a controlled finite state Markov chain). In this case $P_{x_k|I_k}$ consists of the $n$ probabilities $P(x_k = x^i | I_k)$, $i = 1, \ldots, N$. Another case is when the probability distribution $P_{x_k|I_k}$ is Gaussian and is therefore completely characterized by its mean and covariance matrix. Examples of these cases will be given in subsequent sections. If $P_{x_k|I_k}$ cannot be characterized by a finite set of numbers, then the DP algorithm (9) and (10), which is carried over the space of the information vector $I_k$, may be preferable, although actual calculation of an optimal control law is possible only in very simple cases due to the dimensionality problem.

Let us now demonstrate algorithm (16) and (17) by means of an example.

EXAMPLE 1 (continued) In the two-state example of the previous section let us denote

$$p_1 = P(x_1 = 1 | I_1), \qquad p_0 = P(x_0 = 1 | I_0).$$

The equation relating $p_1, p_0, u_0, z_1$ [cf. Eq. (15)] is written

$$p_1 = \Phi_0(p_0, u_0, z_1).$$

One may verify by straightforward calculation that $\Phi_0$ is given by

$$p_1 = \Phi_0(p_0, u_0, z_1) = \begin{cases} \dfrac{1}{7} & \text{if} \quad u_0 = S, \quad z_1 = G, \\[2ex] \dfrac{3}{5} & \text{if} \quad u_0 = S, \quad z_1 = B, \\[2ex] \dfrac{1 + 2p_0}{7 - 4p_0} & \text{if} \quad u_0 = C, \quad z_1 = G, \\[2ex] \dfrac{3 + 6p_0}{5 + 4p_0} & \text{if} \quad u_0 = C, \quad z_1 = B. \end{cases}$$

Algorithm (16) and (17) may be written in terms of $p_0$, $p_1$, and $\Phi_0$ above as

$$\bar{J}_1(p_1) = \min[2p_1, 1],$$

$$\begin{aligned}\bar{J}_0(p_0) = \min[&2p_0 + P(z_1 = G \mid I_0, C)\bar{J}_1[\Phi_0(p_0, C, G)] \\ &+ P(z_1 = B \mid I_0, C)\bar{J}_1[\Phi_0(p_0, C, B)], \\ &1 + P(z_1 = G \mid I_0, S)\bar{J}_1[\Phi_0(p_0, S, G)] \\ &+ p(z_1 = B \mid I_0, S)\bar{J}_1[\Phi_0(p_0, S, B)]].\end{aligned}$$

The probabilities entering in the second equation may be expressed in terms of $p_0$ by straightforward calculation as

$$P(z_1 = G \mid I_0, C) = (7 - 4p_0)/12, \qquad P(z_1 = B \mid I_0, C) = (5 + 4p_0)/12,$$

$$P(z_1 = G \mid I_0, S) = 7/12, \qquad P(z_1 = B \mid I_0, S) = 5/12.$$

Using these values we have

$$\begin{aligned}\bar{J}_0(p_0) = \min\Bigg[&2p_0 + \frac{7 - 4p_0}{12}\bar{J}_1\left(\frac{1 + 2p_0}{7 - 4p_0}\right) \\ &+ \frac{5 + 4p_0}{12}\bar{J}_1\left(\frac{3 + 6p_0}{5 + 4p_0}\right), 1 + \frac{7}{12}\bar{J}_1\left(\frac{1}{7}\right) + \frac{5}{12}\bar{J}_1\left(\frac{3}{5}\right)\Bigg].\end{aligned}$$

Now by minimization in the equation defining $\bar{J}_1(p_1)$ we obtain an optimal control law for the last stage

$$\bar{\mu}_1^*(p_1) = \begin{cases} C & \text{if} \quad p_1 \leqslant \frac{1}{2}, \\ S & \text{if} \quad p_1 > \frac{1}{2}. \end{cases}$$

Also by substitution of $\bar{J}_1(p_1)$ and by carrying out the straightforward calculation we obtain

$$\bar{J}_0(p_0) = \begin{cases} 19/12 & \text{if} \quad \frac{3}{8} \leqslant p_0 \leqslant 1, \\ (7 + 32p_0)/12 & \text{if} \quad 0 \leqslant p_0 \leqslant \frac{3}{8}, \end{cases}$$

and an optimal control law for the first stage

$$\bar{\mu}_0^*(p_0) = \begin{cases} C & \text{if} \quad p_0 \leqslant \frac{3}{8}, \\ S & \text{if} \quad p_0 > \frac{3}{8}. \end{cases}$$

Note that

$$P(x_0 = 1 \mid z_0 = G) = \tfrac{1}{7}, \qquad P(x_0 = 1 \mid z_0 = B) = \tfrac{3}{5},$$

$$P(z_0 = G) = \tfrac{7}{12}, \qquad P(z_0 = B) = \tfrac{5}{12},$$

so that the formula for the optimal value

$$J^* = \underset{z_0}{E}\{\bar{J}_0(P_{x_0|z_0})\} = \frac{7}{12}\bar{J}_0\left(\frac{1}{7}\right) + \frac{5}{12}\bar{J}_0\left(\frac{3}{5}\right) = \frac{176}{144}$$

yields the same optimal value as the one obtained in the previous section by means of the DP algorithm (9) and (10).

We finally note that, regardless of its computational value, the representation of the optimal control law as a sequence of functions of the conditional probability $P_{x_k|I_k}$ of the form

$$\mu_k^*(I_k) = \bar{\mu}_k^*(P_{x_k|I_k}), \qquad k = 0, 1, \ldots, N-1,$$

is a conceptually useful device. An interesting interpretation of this equation is that the optimal controller is composed of two cascaded parts: an *estimator*, which uses at time $k$ the measurement $z_k$ and the control $u_{k-1}$ to generate the probability distribution $P_{x_k|I_k}$, and an *actuator*, which generates a control input to the system as a function of the probability distribution $P_{x_k|I_k}$ (Fig. 4.2).

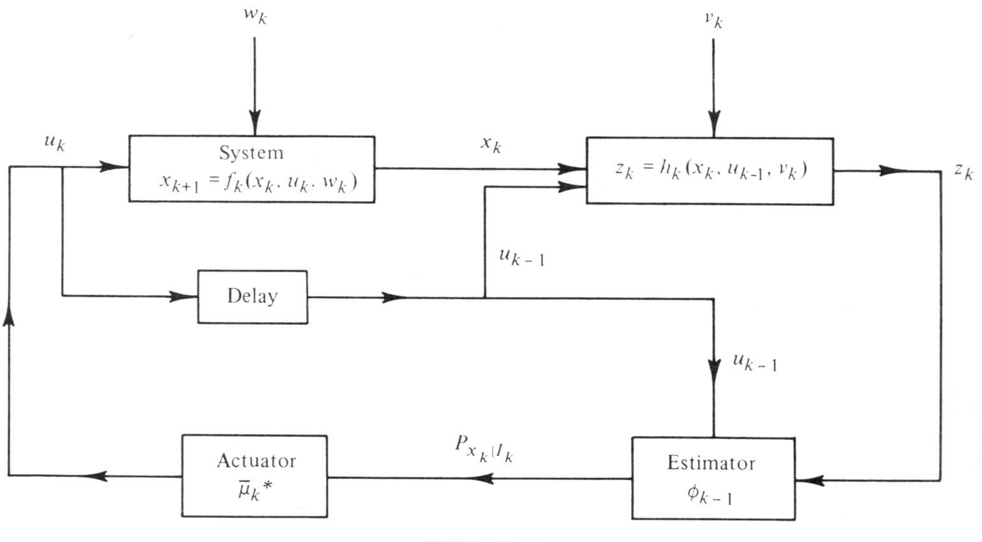


FIGURE 4.2

Aside from its conceptual and analytical importance this interpretation has formed the basis for various suboptimal control schemes that separate a priori the controller into an estimator and an actuator and attempt to design each part in a manner that seems "reasonable." Schemes of this type will be presented in the next chapter.

## 4.3 Linear Systems with Quadratic Cost Functionals —Separation of Estimation and Control

In this section we consider the imperfect state information version of the problem examined in Section 3.1, which involved a linear system and a quadratic cost functional. We have the same discrete-time linear system

$$x_{k+1} = A_k x_k + B_k u_k + w_k, \qquad k = 0, 1, \ldots, N-1, \tag{18}$$

and quadratic cost functional

$$E\left\{x_N' Q_N x_N + \sum_{k=0}^{N-1} (x_k' Q_k x_k + u_k' R_k u_k)\right\}, \tag{19}$$

but now the controller does not have access to the current system state. Instead it receives at the beginning of each period $k$ an observation of the form

$$z_k = C_k x_k + v_k, \qquad k = 0, 1, \ldots, N-1, \tag{20}$$

where $z_k \in R^s$, $C_k$ is a given $s \times m$ matrix, $k = 0, 1, \ldots, N-1$, and $v_k \in R^s$ is an observation noise vector with given probability distribution. Furthermore, the vectors $v_k$ are independent random vectors and are assumed to be independent from $w_k$ and $x_0$ as well. We make the same assumptions as in Section 3.1 concerning the input disturbances $w_k$, and we assume that the system matrices $A_k$, $B_k$ are known.

It is clear that this problem falls within the framework of the basic problem with imperfect state information of Section 4.1. We will analyze the problem by using the reduction and the DP algorithm of Section 4.1.

From Eqs. (9) and (10) we have

$$\begin{aligned} J_{N-1}(I_{N-1}) = \min_{u_{N-1}} \Big[ \underset{x_{N-1}, w_{N-1}}{E} &\{(A_{N-1}x_{N-1} + B_{N-1}u_{N-1} + w_{N-1})' Q_N \\ &\times (A_{N-1}x_{N-1} + B_{N-1}u_{N-1} + w_{N-1}) + x_{N-1}' Q_{N-1} x_{N-1} \\ &+ u_{N-1}' R_{N-1} u_{N-1} \,|\, I_{N-1}\} \Big]. \end{aligned}$$

Using the fact that $E\{w_{N-1} | I_{N-1}\} = E\{w_{N-1}\} = 0$, this expression can be written

$$\begin{aligned} J_{N-1}(I_{N-1}) = \underset{x_{N-1}}{E} &\{x_{N-1}'(A_{N-1}' Q_N A_{N-1} + Q_{N-1}) x_{N-1} \,|\, I_{N-1}\} \\ &+ \underset{w_{N-1}}{E} \{w_{N-1}' Q_{N-1} w_{N-1}\} \\ &+ \min_{u_{N-1}} [u_{N-1}'(B_{N-1}' Q_N B_{N-1} + R_{N-1}) u_{N-1} \\ &+ 2E\{x_{N-1} | I_{N-1}\}' A_{N-1}' Q_N B_{N-1} u_{N-1}]. \end{aligned} \tag{21}$$

The minimization yields the optimal control law for the last stage,

$$\begin{aligned} u_{N-1}^* &= \mu_{N-1}^*(I_{N-1}) \\ &= -(B_{N-1}'Q_N B_{N-1} + R_{N-1})^{-1} B_{N-1}'Q_N A_{N-1}\, E\{x_{N-1} \mid I_{N-1}\}, \end{aligned} \tag{22}$$

and upon substitution in (21) we obtain

$$\begin{aligned} J_{N-1}(I_{N-1}) = &\mathop{E}_{x_{N-1}} \{x_{N-1}' K_{N-1} x_{N-1} \mid I_{N-1}\} \\ &+ \mathop{E}_{x_{N-1}} \{[x_{N-1} - E\{x_{N-1} \mid I_{N-1}\}]' P_{N-1} [x_{N-1} \\ &- E\{x_{N-1} \mid I_{N-1}\}] \mid I_{N-1}\} + \mathop{E}_{w_{N-1}} \{w_{N-1}' Q_N w_{N-1}\}, \end{aligned} \tag{23}$$

where the matrices $K_{N-1}$ and $P_{N-1}$ are given by

$$\begin{aligned} P_{N-1} &= A_{N-1}' Q_N B_{N-1} (R_{N-1} + B_{N-1}' Q_N B_{N-1})^{-1} B_{N-1}' Q_N A_{N-1}, \\ K_{N-1} &= A_{N-1}' Q_N A_{N-1} - P_{N-1} + Q_{N-1}. \end{aligned}$$

Note that the control law (22) for the last stage is identical to the corresponding optimal control law for the problem of Section 3.1 except for the fact that $x_{N-1}$ is replaced by its conditional expectation $E\{x_{N-1} \mid I_{N-1}\}$. Notice also that expression (23) for the cost-to-go $J_{N-1}(I_{N-1})$ exhibits a corresponding similarity to the cost-to-go for the perfect information problem except for the fact that $J_{N-1}(I_{N-1})$ contains an additional middle term, which expresses a penalty due to estimation error.

Now the DP equation for period $N - 2$ is

$$\begin{aligned} J_{N-2}(I_{N-2}) &= \min_{u_{N-2}} \Big[ \mathop{E}_{\substack{x_{N-2} w_{N-2} \\ v_{N-1}}} \{x_{N-2}' Q_{N-2} x_{N-2} + u_{N-2}' R_{N-2} u_{N-2} \\ &\qquad + J_{N-1}(I_{N-1}) \mid I_{N-2}, u_{N-2}\} \Big] \\ &= \min_{u_{N-2}} \Big[ \mathop{E}_{x_{N-2} w_{N-2}} \{x_{N-2}' Q_{N-2} x_{N-2} + u_{N-2}' R_{N-2} u_{N-2} \\ &\qquad + (A_{N-2} x_{N-2} + B_{N-2} u_{N-2} + w_{N-2})' K_{N-1} (A_{N-2} x_{N-2} \\ &\qquad + B_{N-2} u_{N-2} + w_{N-2}) \mid I_{N-2}\} \Big] \\ &\qquad + \mathop{E}_{z_{N-1}} \Big\{ \mathop{E}_{x_{N-1}} \{[x_{N-1} - E\{x_{N-1} \mid I_{N-1}\}]' P_{N-1} [x_{N-1} \\ &\qquad - E\{x_{N-1} \mid I_{N-1}\}] \mid I_{N-1}\} \mid I_{N-2}, u_{N-2} \Big\} \\ &\qquad + \mathop{E}_{w_{N-1}} \{w_{N-1}' Q_N w_{N-1}\}. \end{aligned} \tag{24}$$

Note that we have excluded the next to last term from the minimization with respect to $u_{N-2}$. We have done so since this term will be shown to be independent of $u_{N-2}$—a fact due to the linearity of both the system *and* the measurement equation. This property follows as a special case of the following lemma.

**Lemma** Let $g: R^n \to R$ be any function such that the expected values appearing below are well defined and finite and consider the function

$$f_k(I_{k-1}, z_k, u_{k-1}) = \underset{x_k}{E}\{g[x_k - E\{x_k | I_{k-1}, z_k, u_{k-1}\}] | I_{k-1}, z_k, u_{k-1}\}.$$

Then for every $k = 1, 2, \ldots, N-1$, the function

$$\underset{z_k}{E}\{f_k(I_{k-1}, z_k, u_{k-1}) | I_{k-1}, u_{k-1}\}$$

does not depend on $u_{k-1}$, i.e.,

$$\underset{z_k}{E}\{f_k(I_{k-1}, z_k, u_{k-1}) | I_{k-1}, u_{k-1}\} = \underset{z_k}{E}\{f_k(I_{k-1}, z_k, \bar{u}_{k-1}) | I_{k-1}, \bar{u}_{k-1}\},$$

for all $I_{k-1}, u_{k-1}, \bar{u}_{k-1}$.

*Proof* Define for every $i = 0, 1, \ldots, N-1$, the vectors

$$\tilde{u}_i = \begin{bmatrix} u_0 \\ u_1 \\ \vdots \\ u_i \end{bmatrix}, \qquad \tilde{w}_i = \begin{bmatrix} w_0 \\ w_1 \\ \vdots \\ w_i \end{bmatrix}.$$

Then, in view of the linearity of the system equation, we have for all $i$,

$$x_{i+1} = \Gamma_i x_0 + \Delta_i \tilde{w}_i + E_i \tilde{u}_i$$

where $\Gamma_i, \Delta_i, E_i$ are known matrices of appropriate dimension. These matrices are obtained from the system matrices $A_i, B_i$ by straightforward calculation. Hence we have

$$f_k(I_{k-1}, z_k, u_{k-1}) = E\{g[\Gamma_{k-1}(x_0 - E\{x_0 | I_{k-1}, z_k, u_{k-1}\}) + \Delta_{k-1} \times (\tilde{w}_{k-1} - E\{\tilde{w}_{k-1} | I_{k-1}, z_k, u_{k-1}\})] | I_{k-1}, z_k, u_{k-1}\}.$$

The measurement equations may also be written

$$z_i = C_i \Gamma_{i-1} x_0 + C_i \Delta_{i-1} \tilde{w}_{i-1} + C_i E_{i-1} \tilde{u}_{i-1} + v_i, \qquad i = 1, 2, \ldots, N-1,$$

$$z_0 = C_0 x_0 + v_0.$$

Now in view of these expressions the conditional probability distribution of $(x_0, \tilde{w}_{k-1})$ given $(I_{k-1}, z_k, u_{k-1})$ depends on $I_{k-1}, z_k, u_{k-1}$ through the vectors $z_0, z_1 - C_1 E_0 \tilde{u}_0, \ldots, z_k - C_k E_{k-1} \tilde{u}_{k-1}$, and hence we have

$$f_k(I_{k-1}, z_k, u_{k-1}) = h_k(z_0, z_1 - C_1 E_0 \tilde{u}_0, \ldots, z_k - C_k E_{k-1} \tilde{u}_{k-1})$$

for some function $h_k$. Thus for any $I_{k-1}, u_{k-1}$ we have

$$\underset{z_k}{E}\{f_k(I_{k-1}, z_k, u_{k-1}) | I_{k-1}, u_{k-1}\}$$
$$= \underset{z_k}{E}\{h_k(z_0, z_1 - C_1 E_0 \tilde{u}_0, \ldots, z_k - C_k E_{k-1} \tilde{u}_{k-1}) | I_{k-1}, u_{k-1}\}.$$

Equivalently by using the equation

$$z_k - C_k E_{k-1} \tilde{u}_{k-1} = C_k \Gamma_{k-1} x_0 + C_k \Delta_{k-1} \tilde{w}_{k-1} + v_k$$

we obtain

$$\underset{z_k}{E}\{f_k(I_{k-1}, z_k, u_{k-1}) | I_{k-1}, u_{k-1}\}$$
$$= \underset{x_0, \tilde{w}_{k-1}, v_k}{E}\{h_k(z_0, z_1 - C_1 E_0 \tilde{u}_0, \ldots, z_{k-1}$$
$$- C_{k-1} E_{k-2} \tilde{u}_{k-2}, C_k \Gamma_{k-1} x_0 + C_k \Delta_{k-1} \tilde{w}_{k-1} + v_k) | I_{k-1}, u_{k-1}\}.$$

Since the last expression above does not depend on $u_{k-1}$ the result follows. Q.E.D.

Returning now to our problem, the minimization in Eq. (24) yields using a similar type of argument as for the last stage

$$u_{N-2}^* = \mu_{N-2}^*(I_{N-2}) = -(R_{N-2} + B_{N-2}' K_{N-1} B_{N-2})^{-1}$$
$$\times B_{N-2}' K_{N-1} A_{N-2} \, E\{x_{N-2} | I_{N-2}\},$$

and proceeding similarly we obtain the optimal control law for every stage

▷ $$\mu_k^*(I_k) = L_k \, E\{x_k | I_k\} \qquad (25)$$ ◁

where

▷ $$L_k = -(R_k + B_k' K_{k+1} B_k)^{-1} B_k' K_{k+1} A_k,$$ ◁

and where the matrices $K_k$ are given recursively by the matrix Riccati equation

$$K_N = Q_N$$

▷ $$K_k = A_k'[K_{k+1} - K_{k+1} B_k (R_k + B_k' K_{k+1} B_k)^{-1} B_k' K_{k+1}] A_k + Q_k.$$ ◁

Generalizing an observation made earlier, we note that the optimal control law (25) is identical to the optimal control law for the corresponding perfect state information problem of Section 3.1 except for the fact that the state $x_k$ is now replaced by its conditional expectation $E\{x_k | I_k\}$.

It is interesting to note that the optimal controller can be decomposed into the two parts shown in Fig. 4.3, an estimator, which uses the data to generate

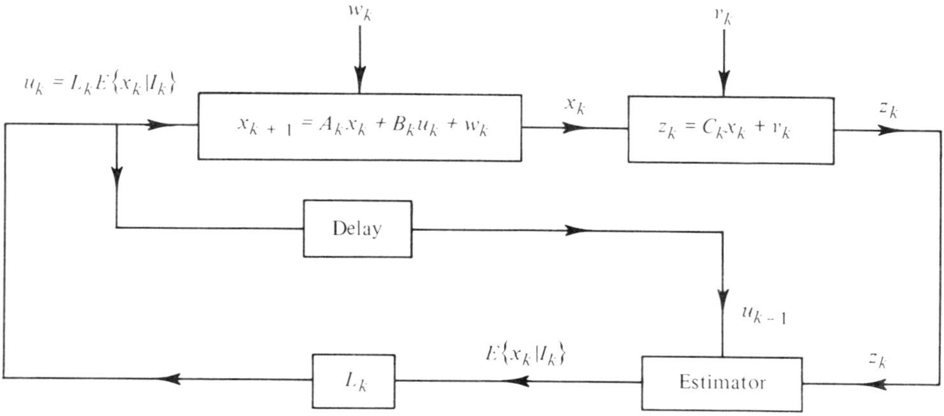


FIGURE 4.3

the conditional expectation $E\{x_k | I_k\}$, and an actuator, which multiplies $E\{x_k | I_k\}$ by the gain matrix $L_k$ and applies the control input $u_k = L_k E\{x_k | I_k\}$ to the system. Furthermore, the gain matrix $L_k$ is independent of the statistics of the problem and in particular is the same as the one that would be used if we were faced with the deterministic problem where $w_k$ and $x_0$ would be fixed and equal to their expected values. On the other hand it is known (see Section 1.3 and the Appendix to this chapter) that the estimate $\hat{x}(I)$ of a random $n$-vector $x$ given some information (random vector) $I$, which minimizes a positive definite quadratic form of the expected value of the estimation error

$$\underset{x, I}{E}\{[x - \hat{x}(I)]' M [x - \hat{x}(I)]\}, \qquad M = \text{positive definite symmetric},$$

is precisely the conditional expectation $E_x\{x | I\}$. Thus the *estimator portion of the optimal controller is an optimal solution of the problem of estimating the state $x_k$ assuming no control takes place, while the actuator portion is an optimal solution of the control problem assuming perfect state information prevails.* This interesting property, which shows that the two portions of the optimal controller can be designed independently as optimal solutions of an estimation and a control problem, has been called the *separation theorem for linear systems and quadratic criteria* and occupies a central position in modern automatic control theory.

Another interesting observation is that the optimal controller applies at each time $k$ the control that would be applied if we were faced with the deterministic problem of minimizing the cost-to-go

$$x_N' Q_N x_N + \sum_{i=k}^{N-1} (x_i' Q_i x_i + u_i' R_i u_i)$$

and the input disturbances $w_k, w_{k+1}, \ldots, w_{N-1}$ and current state $x_k$ were known and fixed at their conditional expected values, which are zero and $E\{x_k \mid I_k\}$, respectively. This is another form of the *certainty equivalence principle*, which was referred to in Section 3.1. For a generalization of this fact to the case of correlated disturbances see Problem 1, the hint to which provides the clue for a short proof of the separation theorem.

*Implementation Aspects—Steady-State Controller—Stability*

As explained in the perfect information case, the linear form of the actuator portion of the optimal control law is particularly attractive for implementation. In the imperfect information case, however, we are faced with the additional problem of constructing an estimator that produces the conditional expectation $E\{x_k \mid I_k\}$. The implementation of such an estimator is by no means easy in general. However, in one important special case, *the case where the disturbances $w_k$, $v_k$, and the initial state $x_0$ are Gaussian random vectors*, a convenient implementation of the estimator is possible by means of the celebrated Kalman filtering algorithm [K1]. This algorithm provides the conditional expectation $E\{x_k \mid I_k\}$, which due to the Gaussian nature of the uncertainties turns out to be a linear function of the information vector $I_k$,† i.e., the measurements $z_0, z_1, \ldots, z_k$ and the controls $u_0, u_1, \ldots, u_{k-1}$. The computations, however, are organized recursively so that only the most recent measurement $z_k$ and control $u_{k-1}$ are needed at time $k$, together with $E\{x_{k-1} \mid I_{k-1}\}$ in order to produce $E\{x_k \mid I_k\}$. The form of the algorithm is the following (see the Appendix at the end of this chapter for a derivation of the algorithm):

$$E\{x_{k+1} \mid I_{k+1}\} = A_k E\{x_k \mid I_k\} + B_k u_k + \Sigma_{k+1|k+1} C'_{k+1} N_{k+1}^{-1} [z_{k+1} - C_{k+1}(A_k E\{x_k \mid I_k\} + B_k u_k)] \qquad k = 0, 1, \ldots, N-1, \tag{26}$$

$$E\{x_0 \mid I_0\} = E\{x_0\} + \Sigma_{0|0} C'_0 N_0^{-1} [z_0 - C_0 E\{x_0\}], \tag{27}$$

where the matrices $\Sigma_{k|k}$ are precomputable and given recursively by

$$\Sigma_{k+1|k+1} = \Sigma_{k+1|k} - \Sigma_{k+1|k} C'_{k+1} (C_{k+1} \Sigma_{k+1|k} C'_{k+1} + N_{k+1})^{-1} C_{k+1} \Sigma_{k+1|k}, \tag{28}$$

$$\Sigma_{k+1|k} = A_k \Sigma_{k|k} A'_k + M_k, \qquad k = 0, 1, \ldots, N-1, \tag{29}$$

† Actually the conditional expectation $E\{x_k \mid I_k\}$ can be shown to be a linear function of $I_k$ for a more general class of probability distributions of $x_0$, $w_k$, $v_k$ that includes the Gaussian distribution as a special case, the class of so-called *spherically invariant distributions* [V6, B24].

with

$$\Sigma_{0|0} = S - SC_0'(C_0 SC_0' + N_0)^{-1}C_0 S.$$

In this equation $M_k$, $N_k$, and $S$ are the covariance matrices of $w_k$, $v_k$, and $x_0$, respectively, and we assume that $w_k$ and $v_k$ have zero mean, i.e.,

$$\begin{aligned} E\{w_k\} &= E\{v_k\} = 0, \\ M_k &= E\{w_k w_k'\}, \qquad N_k = E\{v_k v_k'\}, \qquad k = 0, 1, \ldots, N-1, \\ S &= E\{[x_0 - E\{x_0\}][x - E\{x_0\}]'\}. \end{aligned}$$

In addition the matrices $N_k$ are assumed to be positive definite.

Thus under the additional assumption of Gaussian uncertainties the implementation of the optimal controller is quite convenient. Furthermore, if the system and measurement equations and the disturbance statistics are stationary and the horizon tends to infinity, it can be shown under mild assumptions (see Section 3.1 and the Appendix to this chapter) that both the gain matrices $L_k$ and the matrices $\Sigma_{k|k}$ of the Kalman filtering algorithm (26)–(29) tend to steady-state matrices $L$ and $\bar{\Sigma}$, a fact that can be utilized to simplify greatly the implementation of the optimal controller by replacing the time-varying gain matrix $(\Sigma_{k+1|k+1}C_{k+1}'N_{k+1}^{-1})$ by a constant matrix.

Let us provide the equations for this steady-state implementation of the control law. Time indices will be dropped where they are superfluous due to stationarity. The control law takes the stationary form

$$\mu^*(I_k) = L\hat{x}_k,$$

where we use the notation

$$E\{x_k | I_k\} = \hat{x}_k, \qquad L = -(R + B'KB)^{-1}B'KA,$$

and $K$ is the unique positive semidefinite symmetric solution of the algebraic Riccati equation

$$K = A'[K - KB(R + B'KB)^{-1}B'K]A + Q.$$

By the theory of Section 3.1 this solution exists provided *the pair* $(A, B)$ *is controllable and the pair* $(A, F)$ *is observable where* $F$ *is a matrix such that* $Q = F'F$. The conditional expected value $\hat{x}_k$ is generated by the "steady-state" Kalman filtering equation (see the Appendix)

$$\hat{x}_{k+1} = (A + BL)\hat{x}_k + \bar{\Sigma}C'N^{-1}[z_{k+1} - C(A + BL)\hat{x}_k],$$

where $\bar{\Sigma}$ is given by

$$\bar{\Sigma} = \Sigma - \Sigma C'(C\Sigma C' + N)^{-1}C\Sigma,$$

and $\Sigma$ is the unique positive semidefinite symmetric solution of the algebraic Riccati equation

$$\Sigma = A[\Sigma - \Sigma C'(C\Sigma C' + N)^{-1}C\Sigma]A' + M.$$

As indicated in the Appendix, it follows from the theory of Section 3.1 that this solution exists provided *the pair* $(A, C)$ *is observable and the pair* $(A, D)$ *is controllable, where* $D$ *is a matrix such that* $M = DD'$.

Now by substituting $z_{k+1}$ with its value

$$z_{k+1} = Cx_{k+1} + v_{k+1} = CAx_k + CBL\hat{x}_k + Cw_k + v_{k+1},$$

the system equation and the Kalman filter equation can be written

$$\begin{aligned} x_{k+1} &= Ax_k + BL\hat{x}_k + w_k, \\ \hat{x}_{k+1} &= \bar{\Sigma}C'N^{-1}CAx_k + (A + BL - \bar{\Sigma}C'N^{-1}CA)\hat{x}_k \\ &\quad + \bar{\Sigma}C'N^{-1}(Cw_k + v_{k+1}). \end{aligned}$$

These equations may be viewed as representing a $2n$-dimensional dynamic system whose state is the vector $[x_k', \hat{x}_k']'$ and which is perturbed by vectors that depend on $w_k$ and $v_{k+1}$. From the practical point of view it is important that this system is stable, i.e., the $2n \times 2n$ matrix

$$\begin{bmatrix} A & BL \\ \bar{\Sigma}C'N^{-1}CA & A + BL - \bar{\Sigma}C'N^{-1}CA \end{bmatrix}$$

is a stable matrix. Indeed this can be shown under the preceding controllability and observability assumptions (see the Appendix in this chapter). Thus the combined estimation and control scheme based on approximation of the optimal controller by a stationary controller results in a stable system as well as an implementation that is attractive from the engineering point of view.

It is to be noted that in the case where the random vectors $w_k$, $v_k$, $x_0$ are not Gaussian the controller

$$\bar{\mu}^*(I_k) = L_k\hat{x}(I_k), \qquad k = 0, 1, \ldots, N-1, \tag{30}$$

is often used, where $\hat{x}(I_k)$ is the linear least-squares estimate of the state $x_k$ given the information vector $I_k$ (see the Appendix). It can be shown (see Problem 2) that the controller (30) is optimal in the restricted class of all controllers, which consists only of linear functions of the information vectors $I_k$. One can show (see the Appendix) that the estimate $\hat{x}_k(I_k)$ is again provided by the Kalman filtering algorithm (26)–(29) simply by replacing $E\{x_k | I_k\}$ by $\hat{x}_k(I_k)$, and thus the control law (30) can be conveniently implemented. Thus in the case of non-Gaussian uncertainties the control law (30) may represent an attractive suboptimal alternative to the optimal control law (25).

Nonetheless, using the control law (30) may result in a very substantial increase in the resulting value of the cost functional over the optimal value.

Finally a case that deserves special mention involves the linear system (18) and linear measurement equation (20) and a *nonquadratic* cost functional. Under the assumption that the random vectors are Gaussian and independent it can easily be shown by using the sufficient statistic algorithm of Section 4.2 that an optimal control law can be implemented in the form

$$\mu^*(I_k) = \bar{\mu}_k^*[E\{x_k | I_k\}].$$

In other words, the control input need only depend on the conditional expected value of the state. This result holds due to the fact that the conditional probability measure $P_{x_k|I_k}$ is Gaussian (due to linearity and Gaussian uncertainties) and furthermore the covariance matrix corresponding to $P_{x_k|I_k}$ is precomputable (via the Kalman filtering algorithm mentioned earlier) and cannot be influenced by the control input.

## 4.4 Finite State Markov Chains—A Problem of Instruction

Among problems with imperfect state information the class that involves a finite state system (finite state space) deserves particular attention. Not only is it a class of problems of general interest but also it admits a computationally tractable solution via the sufficient statistic algorithm of Section 4.2. As discussed in that section, the basic problem with imperfect state information can be reformulated into a problem with perfect state information which involves a system the state of which is the conditional probability measure $P_{x_k|I_k}$. When the state space contains only a finite number of states $x^1, x^2, \ldots, x^n$, the conditional probability measures $P_{x_k|I_k}$ are characterized by the $n$-tuple

$$P_k = \{p_k^1, \ldots, p_k^n\},$$

where

$$p_k^i = P(x_k = x^i | I_k), \qquad i = 1, 2, \ldots, n.$$

Thus for each period the state space for the reformulated problem is the simplex

$$\left\{(p^1, \ldots, p^n) | p^i \geq 0, \sum_{i=1}^{n} p^i = 1\right\},$$

and the control input need only be a function of the current conditional probabilities $p^i, i = 1, \ldots, n$. In this way substantial data reduction is achieved since all the necessary information provided by the measurements is summarized in these conditional probabilities. The probabilistic state $P_k$

can be computed at each time by the optimal controller on the basis of the previous state $P_{k-1}$, the control $u_{k-1}$, and the new measurement $z_k$. The DP algorithm (16) and (17) can be used for the calculation of the optimal control law.

Whenever the control and measurement spaces are also finite sets, it turns out that the cost-to-go functions $\bar{J}_k(P_k)$, $k = 0, 1, \ldots, N-1$, in algorithm (16) and (17) have a particularly simple form, a fact first shown by Smallwood and Sondik [S13]. One may show that

$$\bar{J}_k(P_k) = \min[P_k'\alpha_k^1, P_k'\alpha_k^2, \ldots, P_k'\alpha_k^{m_k}],$$

where $\alpha_k^1, \alpha_k^2, \ldots, \alpha_k^{m_k}$ are some $n$-dimensional vectors and $P_k'\alpha_k^j$, $j = 1, \ldots, m_k$, denotes the inner product of $P_k$ and $\alpha_k^j$. In other words, the functions $\bar{J}_k$ are "*piecewise linear*" *and concave* over $\{(p^1, \ldots, p^n) \mid p^i \geqslant 0, \sum_{i=1}^n p_i = 1\}$. The demonstration of this fact is straightforward but tedious and is outlined in the hint to Problem 7. The piecewise linearity of $\bar{J}_k$ is, however, an important property since $\bar{J}_k$ may be completely characterized by the vectors $\alpha_k^1, \ldots, \alpha_k^{m_k}$. These vectors may be computed through the DP algorithm by means of special procedures, which in addition yield an optimal policy. We will not describe these procedures here. They can be found in references [S13] and [S14]. Instead we shall demonstrate the DP algorithm by means of examples. The first of these examples, a problem of instruction, is considered in this section. The second, a hypothesis testing problem, is treated in the next section and is of importance in statistics.

### *A Problem of Instruction*

Consider a problem of instruction where the objective is to teach the student a certain simple item. The student may be at the beginning of each period in one of two possible states

$x^1$ item learned,
$x^2$ item not learned.

At the beginning of each period the instructor must make one of two decisions

$u^1$ terminate the instruction,
$u^2$ continue the instruction for one period and at the end of the period conduct a test the outcome of which gives an indication as to whether the student has learned the item.

The test has two possible outcomes

$z^1$ student gives a correct answer,
$z^2$ student gives an incorrect answer.

The transition probabilities from one state to the next if instruction takes place are given by

$$P(x_{k+1} = x^1 \mid x_k = x^1) = 1, \qquad P(x_{k+1} = x^2 \mid x_k = x^1) = 0,$$

$$P(x_{k+1} = x^1 \mid x_k = x^2) = t, \qquad P(x_{k+1} = x^2 \mid x_k = x^2) = 1 - t, \quad 0 < t < 1.$$

The outcome of the test depends probabilistically on the state of knowledge of the student as follows:

$$P(z_k = z^1 \mid x_k = x^1) = 1, \qquad P(z_k = z^2 \mid x_k = x^1) = 0,$$

$$P(z_k = z^1 \mid x_k = x^2) = r, \qquad P(z_k = z^2 \mid x_k = x^2) = 1 - r, \quad 0 < r < 1.$$

Concerning the cost structure we have that the cost of instruction and testing is $I$ per period and the cost of terminating the instruction is 0 and $C > 0$ if the student has learned or has not learned the item, respectively. The objective is to find the instruction–termination policy for each period $k$, as a function of the test information accrued up to that period, which minimizes the total expected cost, assuming that there is a maximum of $N$ periods of instruction.

It is easy to reformulate this problem into the framework of the basic problem with imperfect state information. We can define a corresponding system and measurement equation by introducing input and observation disturbances $w$ and $v$ with probability distributions expressing the probabilistic mechanism of the process similarly as was described in the example of Section 4.1. Subsequently we can use the sufficient statistic algorithm of Section 4.2 and conclude that the decision whether to terminate or continue instruction at period $k$ should depend on the conditional probability that the student has learned the item given the test results so far. This probability is denoted

$$p_k = P(x_k = x^1 \mid z_0, z_1, \ldots, z_k).$$

In addition we can use the DP algorithm (16) and (17) defined over the space of the sufficient statistic $p_k$ to obtain an optimal policy. However, rather than proceeding with this elaborate reformulation we prefer to argue and obtain directly this DP algorithm.

Concerning the evolution of the conditional probability $p_k$ (assuming instruction occurs) we have by Bayes' rule

$$\begin{aligned} p_{k+1} &= P(x_{k+1} = x^1 \mid z_0, \ldots, z_{k+1}) \\ &= \frac{P(x_{k+1} = x^1, z_{k+1} \mid z_0, \ldots, z_k)}{P(z_{k+1} \mid z_0, \ldots, z_k)} \\ &= \frac{P(z_{k+1} \mid z_0, \ldots, z_k, x_{k+1} = x^1) P(x_{k+1} = x^1 \mid z_0, \ldots, z_k)}{\sum_{i=1}^{2} P(x_{k+1} = x^i \mid z_0, \ldots, z_k) P(z_{k+1} \mid z_0, \ldots, z_k, x_{k+1} = x^i)}. \end{aligned} \tag{31}$$

From the probabilistic descriptions given we have

$$P(z_{k+1} | z_0, \ldots, z_k, x_{k+1} = x^1) = P(z_{k+1} | x_{k+1} = x^1) = \begin{cases} 1 & \text{if } z_{k+1} = z^1, \\ 0 & \text{if } z_{k+1} = z^2, \end{cases}$$

$$P(z_{k+1} | z_0, \ldots, z_k, x_{k+1} = x^2) = P(z_{k+1} | x_{k+1} = x^2) = \begin{cases} r & \text{if } z_{k+1} = z^1, \\ 1 - r & \text{if } z_{k+1} = z^2, \end{cases}$$

$$P(x_{k+1} = x^1 | z_0, \ldots, z_k) = p_k + (1 - p_k)t,$$

$$P(x_{k+1} = x^2 | z_0, \ldots, z_k) = (1 - p_k)(1 - t).$$

Substitution in (31) yields

$$p_{k+1} = \Phi(p_k, z_{k+1}), \tag{32}$$

where the function $\Phi$ is defined by

$$\Phi(p_k, z_{k+1}) = \begin{cases} \dfrac{p_k + (1 - p_k)t}{p_k + (1 - p_k)t + (1 - p_k)(1 - t)r} & \text{if } z_{k+1} = z^1, \\ 0 & \text{if } z_{k+1} = z^2, \end{cases}$$

or equivalently

$$\Phi(p_k, z_{k+1}) = \begin{cases} \dfrac{1 - (1 - t)(1 - p_k)}{1 - (1 - t)(1 - r)(1 - p_k)} & \text{if } z_{k+1} = z^1, \\ 0 & \text{if } z_{k+1} = z^2. \end{cases} \tag{33}$$

A cursory examination of this equation shows that, as expected, the conditional probability $p_{k+1}$ that the student has learned the item increases with every correct answer and drops to zero with every incorrect answer. We mention also that Eq. (32) is a special case of Eq. (15) of Section 4.2. The dependence of the function $\Phi$ on the control $u_k$ is not explicitly shown since there is only one possible action aside from termination.

We turn now to the development of the DP algorithm for the problem. At the end of the $N$th period, assuming instruction has continued to that period, the expected cost is

$$\bar{J}_N(p_N) = (1 - p_N)C. \tag{34}$$

At the end of period $N - 1$, the instructor has calculated the conditional probability $p_{N-1}$ that the student has learned the item and wishes to decide whether to terminate instruction and incur an expected cost $(1 - p_{N-1})C$ or continue the instruction and incur an expected cost $I + E_{z_N} \{\bar{J}_N(p_N)\}$. This leads to the following equation for the optimal expected cost-to-go:

$$\bar{J}_{N-1}(p_{N-1}) = \min\left[(1 - p_{N-1})C,\ I + \underset{z_N}{E}\ \{\bar{J}_N[\Phi(p_{N-1}, z_N)]\}\right].$$

Similarly the algorithm is written for every stage $k$ by replacing $N$ above by $k + 1$:

$$\bar{J}_k(p_k) = \min\left[(1 - p_k)C,\ I + \underset{z_{k+1}}{E}\ \{\bar{J}_{k+1}[\Phi(p_k, z_{k+1})]\}\right]. \tag{35}$$

Now using expression (33) for the function $\Phi$ and the probabilities

$$\begin{aligned} P(z_{k+1} = z^1 \mid p_k) &= p_k + (1 - p_k)[(1 - t)r + t] \\ &= 1 - (1 - t)(1 - r)(1 - p_k), \\ P(z_{k+1} = z^2 \mid p_k) &= 1 - p_k - (1 - p_k)[(1 - t)r + t] \\ &= (1 - t)(1 - r)(1 - p_k), \end{aligned}$$

Eq. (35) is written

$$\bar{J}_k(p_k) = \min[(1 - p_k)C,\ I + A_k(p_k)], \tag{36}$$

where

$$\begin{aligned} A_k(p_k) = {}& [1 - (1 - t)(1 - r)(1 - p_k)]\bar{J}_{k+1}\left[\frac{1 - (1 - t)(1 - p_k)}{1 - (1 - t)(1 - r)(1 - p_k)}\right] \\ & + (1 - t)(1 - r)(1 - p_k)\bar{J}_{k+1}(0). \end{aligned} \tag{37}$$

In particular, by using (34), (36), and (37) we have by straightforward calculation

$$\begin{aligned} \bar{J}_{N-1}(p_{N-1}) &= \min[(1 - p_{N-1})C,\ I + A_{N-1}(p_{N-1})] \\ &= \min[(1 - p_{N-1})C,\ I + (1 - t)(1 - p_{N-1})C]. \end{aligned}$$

Thus as shown in Fig. 4.4 if

$$I + (1 - t)C < C, \tag{38}$$

there exists a scalar $\alpha_{N-1}$ with $0 < \alpha_{N-1} < 1$ that determines an optimal policy for the last period:

| | |
|---|---|
| continue instruction | if $p_{N-1} \leqslant \alpha_{N-1}$, |
| terminate instruction | if $p_{N-1} > \alpha_{N-1}$. |

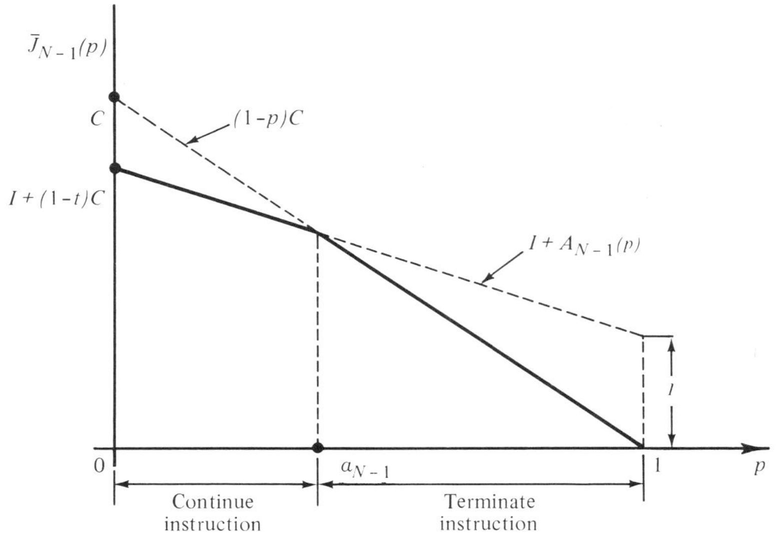


FIGURE 4.4

It is easy to show (see Problem 8) using (37) that under condition (38) the functions $A_k(p)$ are concave and "piecewise linear" (i.e., polyhedral) for each $k$ and satisfy for all $k$,

$$A_k(1) = 0. \tag{39}$$

Furthermore they satisfy for all $k$

$$A_k(p) \geqslant A_k(p') \qquad \text{if} \quad 0 \leqslant p \leqslant p' \leqslant 1, \tag{40}$$

$$A_{k-1}(p) \leqslant A_k(p) \leqslant A_{k+1}(p) \qquad \forall p \in [0, 1], \quad k = 1, 2, \ldots, N-2. \tag{41}$$

Thus from the DP algorithm (36) and Eqs. (39)–(41) we obtain that the optimal policy for each period is determined by the unique scalars $\alpha_k$, which are such that

$$(1 - \alpha_k)C = I + A_k(\alpha_k), \qquad k = 0, 1, \ldots, N-1.$$

An optimal policy for period $k$ is given by

$$\begin{aligned} &\text{continue instruction} && \text{if} \quad p_k \leqslant \alpha_k, \\ &\text{terminate instruction} && \text{if} \quad p_k > \alpha_k. \end{aligned}$$

Now since the functions $A_k(p)$ are monotonically nondecreasing with respect to $k$, it follows from Fig. 4.5 that

$$\alpha_{N-1} \leqslant \alpha_{N-2} \leqslant \cdots \leqslant \alpha_k \leqslant \alpha_{k-1} \leqslant \cdots \leqslant 1 - (I/C),$$

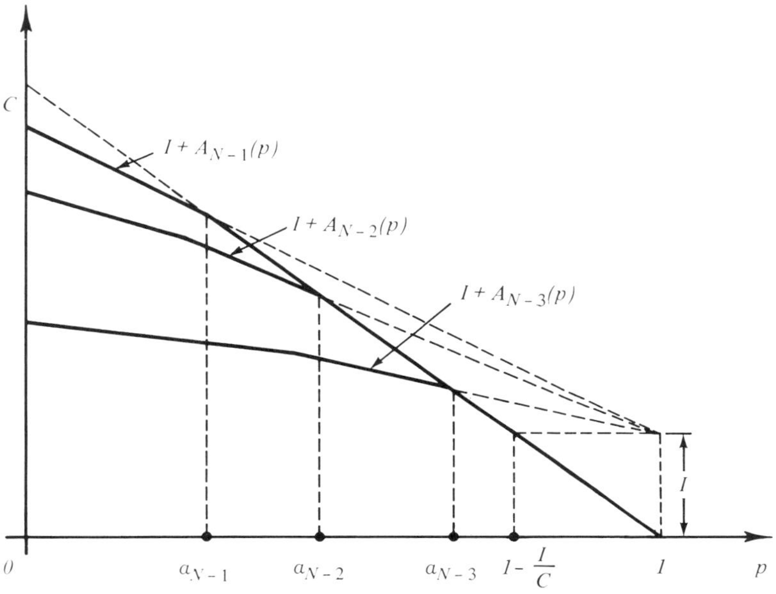


FIGURE 4.5

and therefore as $N \to \infty$ the corresponding sequence $\{\alpha_{N-i}\}$ converges to some scalar $\bar{\alpha}$ for every fixed $i$. Thus as the horizon gets longer, the optimal policy (at least for the initial stages) can be approximated by the stationary policy

$$\begin{aligned} &\text{continue instruction} \qquad &\text{if} \quad p_k \leqslant \bar{\alpha} \\ &\text{terminate instruction} \qquad &\text{if} \quad p_k > \bar{\alpha}. \end{aligned} \tag{42}$$

It turns out that this stationary policy has a very convenient implementation that does not require the calculation of the conditional probability at each stage. From Eq. (33) we have

$$p_{k+1} = \begin{cases} \dfrac{1-(1-t)(1-p_k)}{1-(1-t)(1-r)(1-p_k)} & \text{if} \quad z_{k+1} = z^1, \\ 0 & \text{if} \quad z_{k+1} = z^2. \end{cases}$$

Furthermore, $p_{k+1}$ increases over $p_k$ if a correct answer $z^1$ is given and drops to zero if an incorrect answer $z^2$ is given. Now define recursively the probabilities

$$\pi_1 = \Phi(0, z^1), \qquad \pi_2 = \Phi(\pi_1, z^1), \quad \ldots, \quad \pi_{k+1} = \Phi(\pi_k, z^1), \quad \ldots,$$

and let $n$ be the smallest integer for which $\pi_n > \bar{\alpha}$. It is clear that the stationary policy (42) can be implemented as follows:

| | |
|---|---|
| terminate instruction | if $n$ successive correct answers have been received, |
| continue instruction | otherwise. |

## 4.5 Hypothesis Testing—Sequential Probability Ratio Test

In this section we consider a hypothesis testing problem characteristic of a class of problems that are of central importance in statistical sequential analysis. The decision maker must sequentially and at each period either accept on the basis of past observations a certain hypothesis out of a given finite collection as being true and terminate experimentation, or he must delay his decision for at least one period and obtain, at a certain cost, an additional observation that provides information as to which hypothesis is the correct one. We will focus attention on the simplest and perhaps the most important case, where there are only two hypotheses. The approach can be easily generalized to the case of more hypotheses but the corresponding results are not as elegant.

Let $z_0, z_1, \ldots, z_{N-1}$ be a sequence of independent and identically distributed random variables taking values on a countable set $Z$. Suppose we know that the probability distribution of the $z_k$'s is either $f_0$ or $f_1$ and that we are trying to decide on one of these. Here for any element $z \in Z, f_0(z)$ $[f_1(z)]$ denotes the probability of $z$ occurring when $f_0$ $(f_1)$ is the true distribution. At time $k$ after observing $z_0, \ldots, z_k$ we may either stop observing and accept either $f_0$ or $f_1$, or we may take an additional observation at a cost $C > 0$. If we stop observing and make a choice, then we incur zero cost if our choice is correct, and costs $L_0, L_1$ if we choose incorrectly $f_0$ and $f_1$, respectively. We are given the a priori probability $p$ that the true distribution is $f_0$ and we assume that at most $N$ observations are possible.

It is easy to see that the problem described above can be formulated as a sequential optimization problem with imperfect state information involving a two-state Markov chain. The state space is $\{x^0, x^1\}$, where we use the notation

$x^0$ true density is $f_0$,
$x^1$ true density is $f_1$.

The system equation takes the simple form $x_{k+1} = x_k$ and we can write a measurement equation $z_k = v_k$, where $v_k$ is a random variable taking values in $Z$ with conditional probability distribution

$$P(v_k \mid x_k) = \begin{cases} f_0(v_k) & \text{if} \quad x_k = x^0, \\ f_1(v_k) & \text{if} \quad x_k = x^1. \end{cases}$$

Thus it is possible to reformulate the problem into the framework of the basic problem with imperfect state information and use the sufficient statistic DP algorithm of Section 4.2 for the analysis. This algorithm is defined over the interval [0, 1] of possible values of the conditional probability

$$p_k = P(x_k = x^0 \mid z_0, \ldots, z_k).$$

Similarly as in the previous section we shall obtain this algorithm directly.

The conditional probability $p_k$ is generated recursively according to the following equation (assuming $f_0(z) > 0, f_1(z) > 0$ for all $z \in Z$)

$$p_{k+1} = \frac{p_k f_0(z_{k+1})}{p_k f_0(z_{k+1}) + (1 - p_k) f_1(z_{k+1})}, \qquad k = 0, 1, \ldots, N-1, \tag{43}$$

$$p_0 = \frac{p f_0(z_0)}{p f_0(z_0) + (1 - p) f_1(z_0)}, \tag{44}$$

where $p$ is the a priori probability that the true distribution is $f_0$. The optimal expected cost for the last period is

$$\bar{J}_{N-1}(p_{N-1}) = \min[(1 - p_{N-1})L_0, p_{N-1}L_1], \tag{45}$$

where $(1 - p_{N-1})L_0$ is the expected cost for accepting $f_0$ and $p_{N-1}L_1$ is the expected cost for accepting $f_1$. Taking into account (43) and (44) we can obtain the optimal expected cost-to-go for the $k$th period from the equation

$$\bar{J}_k(p_k) = \min\Bigg[(1 - p_k)L_0, p_k L_1, C + \mathop{E}_{z_{k+1}} \left\{ \bar{J}_{k+1}\left[\frac{p_k f_0(z_{k+1})}{p_k f_0(z_{k+1}) + (1 - p_k) f_1(z_{k+1})}\right]\right\}\Bigg],$$

where the expectation over $z_{k+1}$ is taken with respect to the probability distribution

$$p(z_{k+1}) = p_k f_0(z_{k+1}) + (1 - p_k) f_1(z_{k+1}) \qquad \forall z_{k+1} \in Z.$$

Equivalently

$$\bar{J}_k(p_k) = \min[(1 - p_k)L_0, p_k L_1, C + A_k(p_k)], \qquad k = 0, 1, \ldots, N-2, \tag{46}$$

where

$$A_k(p_k) = \mathop{E}_{z_{k+1}} \left\{ \bar{J}_{k+1}\left[\frac{p_k f_0(z_{k+1})}{p_k f_0(z_{k+1}) + (1 - p_k) f_1(z_{k+1})}\right]\right\}. \tag{47}$$

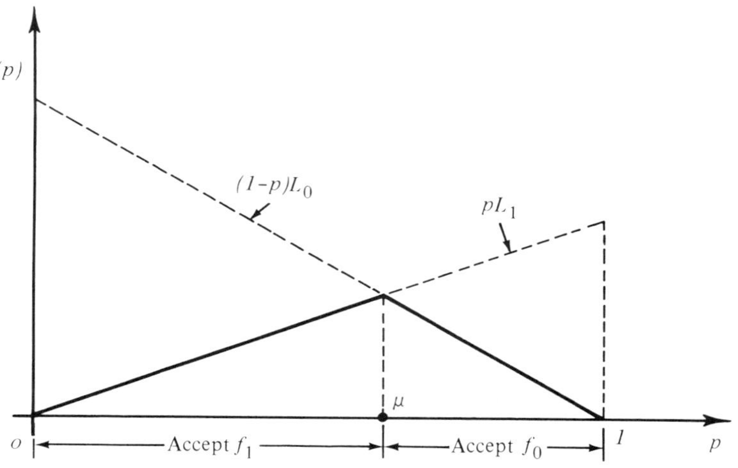


FIGURE 4.6

An optimal policy for the last period (see Fig. 4.6) is obtained from the minimization indicated in (45):

$$\begin{aligned} &\text{accept } f_0 \qquad \text{if} \quad p_{N-1} \geqslant \mu, \\ &\text{accept } f_1 \qquad \text{if} \quad p_{N-1} < \mu, \end{aligned}$$

where $\mu$ is determined from the relation $(1 - \mu)L_0 = \mu L_1$ or equivalently

$$\mu = L_0/(L_0 + L_1).$$

We now prove the following lemma.

**Lemma** The functions $A_k$: $[0, 1] \to R$ of (47) are concave and satisfy

$$A_k(0) = A_k(1) = 0 \qquad \forall k = 0, 1, \ldots, N-2,$$
$$A_{k-1}(p) \leqslant A_k(p) \qquad \forall p \in [0, 1], \quad k = 1, 2, \ldots, N-2.$$

*Proof* The last two relations are evident from (45)–(47). To prove concavity of $A_k$ in view of (45) and (46) it is sufficient to show that concavity of $\bar{J}_{k+1}$ implies concavity of $A_k$ through relation (47). Indeed assume that $\bar{J}_{k+1}$ is concave over [0, 1]. Let $z^1, z^2, z^3, \ldots$ denote the elements of the countable observation space $Z$. We have from (47) that

$$A_k(p) = \sum_{i=1}^{\infty} [pf_0(z^i) + (1-p)f_1(z^i)]\bar{J}_{k+1}\left[\frac{pf_0(z^i)}{pf_0(z^i) + (1-p)f_1(z^i)}\right].$$

Hence it is sufficient to show that concavity of $\bar{J}_{k+1}$ implies concavity of each of the functions

$$h_i(p) = [pf_0(z^i) + (1 - p)f_1(z^i)]\bar{J}_{k+1}\left[\frac{pf_0(z^i)}{pf_0(z^i) + (1 - p)f_1(z^i)}\right].$$

To show concavity of $h_i$ we must show that for every $\lambda \in [0, 1]$, $p_1, p_2 \in [0, 1]$ we have

$$\lambda h_i(p_1) + (1 - \lambda)h_i(p_2) \leqslant h_i[\lambda p_1 + (1 - \lambda)p_2].$$

Using the notation

$$\xi_1 = p_1 f_0(z^i) + (1 - p_1)f_1(z^i), \qquad \xi_2 = p_2 f_0(z^i) + (1 - p_2)f_1(z^i),$$

the inequality above is equivalent to

$$\frac{\lambda \xi_1}{\lambda\xi_1 + (1 - \lambda)\xi_2}\bar{J}_{k+1}\left[\frac{p_1 f_0(z^i)}{\xi_1}\right] + \frac{(1 - \lambda)\xi_2}{\lambda\xi_1 + (1 - \lambda)\xi_2}\bar{J}_{k+1}\left[\frac{p_2 f_0(z^i)}{\xi_2}\right]$$
$$\leqslant \bar{J}_{k+1}\left[\frac{(\lambda p_1 + (1 - \lambda)p_2)f_0(z^i)}{\lambda\xi_1 + (1 - \lambda)\xi_2}\right].$$

This relation, however, is implied by the concavity of $\bar{J}_{k+1}$. Q.E.D.

Using the lemma we obtain (see Fig. 4.7) that if

$$C + A_{N-2}[L_0/(L_0 + L_1)] < L_0 L_1/(L_0 + L_1),$$

then an optimal policy for each period $k$ is of the form

| | |
|---|---|
| accept $f_0$ | if $p_k \geqslant \alpha_k$, |
| accept $f_1$ | if $p_k \leqslant \beta_k$, |
| continue the observations | if $\beta_k < p_k < \alpha_k$, |

where the scalars $\alpha_k$, $\beta_k$ are determined from the relations

$$\beta_k L_1 = C + A_k(\beta_k) \qquad \forall k,$$
$$(1 - \alpha_k)L_0 = C + A_k(\alpha_k) \qquad \forall k.$$

Furthermore we have

$$\cdots \leqslant \alpha_{k+1} \leqslant \alpha_k \leqslant \alpha_{k-1} \leqslant \cdots \leqslant 1 - (C/L_0),$$
$$\cdots \geqslant \beta_{k+1} \geqslant \beta_k \geqslant \beta_{k-1} \geqslant \cdots \geqslant C/L_1.$$

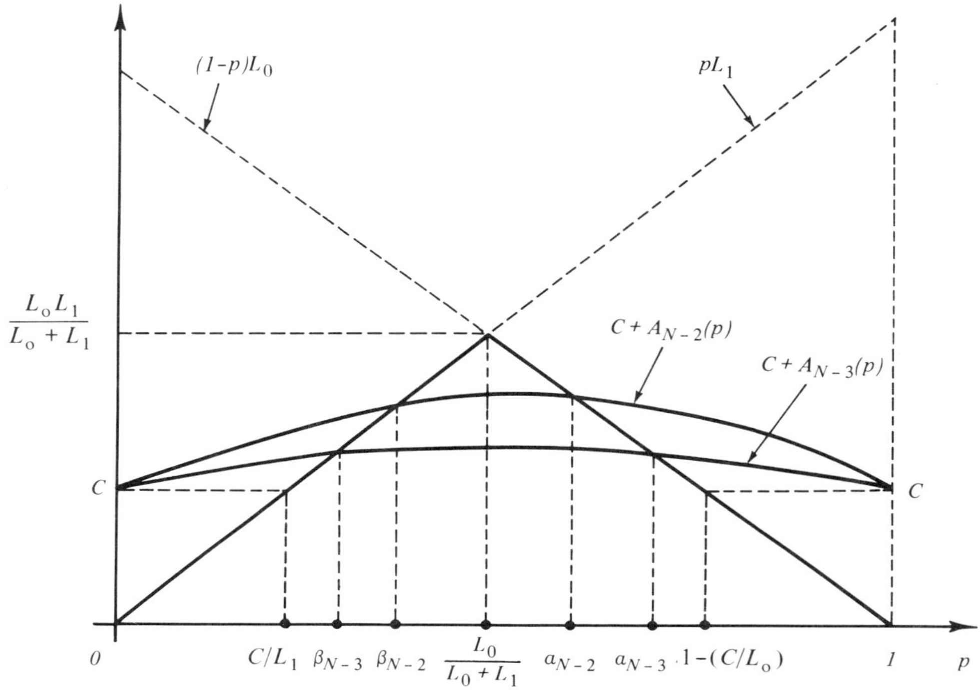


FIGURE 4.7

Hence as $N \to \infty$ the sequences $\{\alpha_{N-i}\}$, $\{\beta_{N-i}\}$ converge to scalars $\bar{\alpha}$, $\bar{\beta}$, respectively, for each fixed $i$, $i = 1, 2, \ldots$, and the optimal policy is approximated by the stationary policy

$$\begin{aligned} &\text{accept } f_0 && \text{if } \; p_k \geqslant \bar{\alpha}, \\ &\text{accept } f_1 && \text{if } \; p_k \leqslant \bar{\beta}, \\ &\text{continue the observations} && \text{if } \; \bar{\beta} < p_k < \bar{\alpha}. \end{aligned} \tag{48}$$

Now the conditional probability $p_k$ is given by

$$p_k = \frac{p f_0(z_0) f_0(z_1) \cdots f_0(z_k)}{p f_0(z_0) \cdots f_0(z_k) + (1 - p) f_1(z_0) \cdots f_1(z_k)}, \tag{49}$$

where $p$ is the a priori probability that $f_0$ is the true hypothesis. Using (49) the stationary policy (48) can be written in the form

$$\begin{aligned} &\text{accept } f_0 && \text{if } \; R_k \geqslant A = (1 - p)\bar{\alpha}/p(1 - \bar{\alpha}), \\ &\text{accept } f_1 && \text{if } \; R_k \leqslant B = (1 - p)\bar{\beta}/p(1 - \bar{\beta}), \\ &\text{continue the observations} && \text{if } \; B < R_k < A, \end{aligned} \tag{50}$$

where the *sequential probability ratio* $R_k$ is given by

$$R_k = f_0(z_0) \cdots f_0(z_k)/f_1(z_0) \cdots f_1(z_k).$$

Note that $R_k$ can be easily generated by means of the recursive equation

$$R_{k+1} = [f_0(z_{k+1})/f_1(z_{k+1})]R_k.$$

A sequential decision procedure of form (50) is called a "sequential probability ratio test." Procedures of this type were among the first formal methods of sequential analysis studied by Wald [W1] and have extensive applications in statistics. The optimality of policy (48) for a problem involving an unlimited number of observations will be shown in Section 7.2.

## 4.6 Sequential Sampling of a Large Batch

This section deals with a problem of inspection of a large batch of items. Samples from the batch are taken sequentially at a cost $C > 0$ per item and inspected for defects. On the basis of the number of defective and nondefective items inspected one must decide whether to classify the batch as defective or nondefective or continue sampling. Let $q$, with $0 \leqslant q \leqslant 1$, denote the quality of the batch, which represents the true proportion of defective items. We denote

- $a(q)$ cost of accepting a batch of quality $q$,
- $r(q)$ cost of rejecting a batch of quality $q$,
- $f(q)$ a priori probability density function of $q$.

We assume that $a(q)$ and $r(q)$ are nonnegative and bounded over $[0, 1]$. The problem is to find a sampling policy that minimizes the expected value of the sum of the sampling costs and the acceptance–rejection costs. We shall assume initially that the maximum number of samples that can be taken is $N$ and that sampling does not affect the composition of the batch. Subsequently we shall examine the limiting case where $N$ is very large and tends to infinity.

Clearly this problem is of a similar nature as the one of the previous section There are some important differences, however, which are perhaps worth going over. In both problems we have the same control space (accept, reject, or take another sample). The system and measurement equations in both problems can be taken to have the same form

$$x_{k+1} = x_k, \qquad z_k = v_k.$$

However, whereas in the problem of the previous section we had two hypotheses (corresponding to the two densities $f_0, f_1$) in the present problem essentially we have an infinity of hypotheses (one for each quality $q \in [0, 1]$).

Thus the state space is different in the two problems. The observation space is also different. In the previous section the observation space was countably infinite while in the present problem there are only two outcomes per sample (defective, nondefective), i.e., $v_k$ takes only the two values

$$\begin{array}{lll} \text{defective} & \text{with} & \text{probability } q, \\ \text{nondefective} & \text{with} & \text{probability } 1 - q. \end{array}$$

The differences described above induce an important change in the treatment of the present problem since it is not convenient to employ as a sufficient statistic the conditional probability density of $q$ given the measurements in view of infinite dimensionality. Fortunately enough there is another sufficient statistic, which turns out to be convenient in the present case thanks to the binary nature of the observation space. Assume that we are at time $k$ and consider the pair $(m, n)$, where $m + n = k$ and $m$ is the number of defective outcomes up to time $k$, and $n$ the number of nondefective outcomes up to time $k$. A little thought should convince the reader that all the information that the statistician needs to know at time $k$ for decision purposes is summarized in the pair $(m, n)$, and hence $(m, n)$ can be viewed as a sufficient statistic in accordance with the definition of Section 4.2. The verification of this fact is left to the reader. We proceed below to state and analyze the DP algorithm in terms of the sufficient statistic $(m, n)$. We obtain the algorithm directly rather than through a reformulation into the general problem of this chapter. Such a reformulation is of course possible.

Given $(m, n)$ the conditional probability density function of $q$ can be calculated to be

$$f(q \mid m, n) = \begin{cases} \dfrac{f(q)q^m(1 - q)^n}{\int_0^1 f(q)q^m(1 - q)^n \, dq} & \text{if } 0 \leqslant q \leqslant 1, \\ 0 & \text{otherwise.} \end{cases} \tag{51}$$

The conditional probability of obtaining a defective sample at time $k + 1$ given that $(m, n)$ with $m + n = k$ has occurred can be calculated to be

$$d(m, n) = \frac{\int_0^1 f(q)q^{m+1}(1 - q)^n \, dq}{\int_0^1 f(q)q^m(1 - q)^n \, dq} = \int_0^1 f(q \mid m, n)q \, dq, \tag{52}$$

and, of course, the conditional probability of a nondefective sample is $[1 - d(m, n)]$. Note that the scalars $f(q \mid m, n)$ and $d(m, n)$ may be computed a priori for each $(m, n)$.

Let us denote by $\bar{J}_k(m, n)$ the optimal cost-to-go of the process at time $k$ given that sampling has produced the pair $(m, n)$ with $m + n = k$. At this point the costs involved are

$$\text{accept} \qquad \int_0^1 a(q) f(q \mid m, n)\, dq,$$

$$\text{reject} \qquad \int_0^1 r(q) f(q \mid m, n)\, dq,$$

take another sample

$$C + d(m, n)\bar{J}_{k+1}(m+1, n) + [1 - d(m, n)]\bar{J}_{k+1}(m, n+1).$$

Let us denote by $g(m, n)$ the minimum of the acceptance and rejection costs given $(m, n)$:

$$g(m, n) = \min\left[\int_0^1 a(q) f(q \mid m, n)\, dq,\ \int_0^1 r(q) f(q \mid m, n)\, dq\right]. \tag{53}$$

The DP algorithm may now be written

$$\bar{J}_N(m, n) = g(m, n), \tag{54}$$

$$\begin{aligned}\bar{J}_k(m, n) = \min[g(m, n),\ & C + d(m, n)\bar{J}_{k+1}(m+1, n) \\ & + [1 - d(m, n)]\bar{J}_{k+1}(m, n+1)].\end{aligned} \tag{55}$$

Note that $\bar{J}_k$ is defined over the set $S_k$ of pairs of nonnegative integers $(m, n)$ with $m + n = k$. The sets $S_k$, which are, of course, finite, are shown in Fig. 4.8. Thus if the permissible number of samples $N$ is not very large, the sets $S_0, S_1, \ldots, S_N$ collectively contain relatively few points and the computation of the optimal policy via the DP algorithm (54) and (55) is relatively easy. On the other hand when $N$ is very large (and indeed there is a priori no reason why it should not be when the batch is large) computation (54) and (55) becomes inefficient. Nonetheless we will see that under an assumption often satisfied in practice one can limit the maximum number of samples to be taken without loss of optimality.

For any nonnegative integers $m, n, N$ with $m + n \leqslant N$ let us denote by $\bar{J}(m, n, N)$ the optimal "cost-to-go" $\bar{J}_k(m, n)$ $(k = m + n)$ obtained from algorithm (54) and (55) when the maximum sample size is $N$. In general, $\bar{J}(m, n, N)$ depends on $N$. If, however, the costs $a(q)$ and $r(q)$ and the probability density function $f(q)$ are such that for some integer $\bar{N}$

$$g(m, n) \leqslant C \qquad \text{for all } m, n \quad \text{with} \quad m + n \geqslant \bar{N}, \tag{56}$$

then from (55) we obtain

$$\bar{J}(m, n, N) = g(m, n) \qquad \text{for all } m, n \quad \text{with} \quad m + n \geqslant \bar{N}.$$

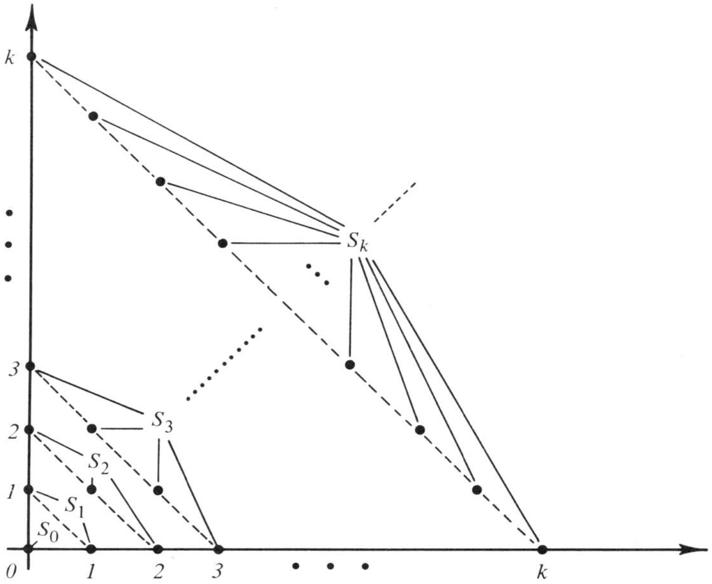


FIGURE 4.8

Under these circumstances acceptance or rejection takes place after *at most* $\bar{N}$ samples are taken, and the optimal sampling policy is the same for all $N \geqslant \bar{N}$. This policy can be obtained through the DP algorithm (54) and (55) provided the number of stages $N$ satisfies $N \geqslant \bar{N}$. As an example, let $a(q) = 0$ if $q < 0.75$; $a(q) = 1$ if $q \geqslant 0.75$; $r(q) = 0$ if $q > 0.25$; $r(q) = 1$ if $q \leqslant 0.25$; $f(q) = 1$, $C = 0.004$. Then one has

$$g(m, n) = \min\left[\frac{\int_{0.75}^{1} q^m(1-q)^n\,dq}{\int_0^1 q^m(1-q)^n\,dq}, \frac{\int_0^{0.25} q^m(1-q)^n\,dq}{\int_0^1 q^m(1-q)^n\,dq}\right]$$

and one can verify that $g(m, n) < 0.004 = C$ for $m + n \geqslant 10$. Thus it is sufficient to solve the problem for $N = 10$.

## 4.7 Notes

Sequential decision problems with imperfect state information and the idea of data reduction via a sufficient statistic have been considered very early particularly in the statistics literature (see, e.g., Blackwell and Girshick [B23] and the references quoted therein). In the area of stochastic control the sufficient statistic idea gained wide attention following the 1965 paper by Striebel [S18] (see also [S16, S19]) although the related facts were known to most researchers in the field by that time. For the analog of the sufficient statistic concept in sequential minimax problems see the work of Bertsekas and Rhodes [B18].

For literature on linear quadratic problems with imperfect state information see the references quoted for Section 3.1 and Witsenhausen's survey paper [W7]. The Kalman filtering algorithm [K1] is a well-known and widely used tool. Detailed discussions that supplement our exposition of the Appendix can be found in many textbooks [J3, L8, M6, M7, N1]. The result mentioned at the end of Section 4.3 was pointed out by Striebel [S18] (see also Bar-Shalom and Tse [B1]). The corresponding result for continuous-time systems has been shown under certain assumptions by Wonham [W11]. For linear quadratic problems with Gaussian uncertainties and observation cost in the spirit of Problem 6 see the works of Aoki and Li [A3] and Cooper and Nahi [C4]. Problems 1 and 2, which indicate the form of the certainty equivalence principle when the random disturbances are correlated, are based on an unpublished report by the author [B5].

The possibility of analysis of the problem of control of a finite state Markov chain with imperfect state information via the sufficient statistic algorithm of Section 4.2 was known for a long time. More recently it has been exploited by Eckles [E2], Smallwood and Sondik [S13], and Sondik [S14]. The proof of the "piecewise linearity" of the cost-to-go functions and an algorithm for their computation is given in references [S13] and [S14]. The instruction model described in Section 4.4 has been considered (with some variations) by a number of authors [A10, K5, G1, S12].

For a discussion of the sequential probability ratio test and related subjects see the book by DeGroot [D1], the lecture notes by Chernoff [C2], and the references quoted therein. The treatment provided here stems from Arrow *et al.* [A5]. The problem of Section 4.6 is treated by White [W3]. A similar class of problems that has received considerable attention is the class of the so-called two-armed bandit problems (see [D1]). A simple special case is treated in Problem 10.

## Problems

**1.** Consider the linear system (18) and measurement equation (20) of Section 4.3 and consider the problem of finding a control law $\{\mu_0^*(I_0), \ldots, \mu_{N-1}^*(I_{N-1})\}$ that minimizes the quadratic cost functional

$$E\left\{x_N' Q x_N + \sum_{k=0}^{N-1} u_k' R_k u_k\right\}.$$

Assume, however, that the random vectors $x_0, w_0, \ldots, w_{N-1}, v_0, \ldots, v_{N-1}$ are correlated and have given joint probability distribution and finite first and second moments. Show that the optimal control law is given by

$$\mu_k^*(I_k) = L_k E\{y_k | I_k\},$$

where the gain matrices $L_k$ are obtained from the recursive algorithm

$$\begin{aligned} L_k &= -(B_k' K_{k+1} B_k + R_k)^{-1} B_k' K_{k+1} A_k, \\ K_N &= Q, \\ K_k &= A_k'[K_{k+1} - K_{k+1} B_k (B_k' K_{k+1} B_k + R_k)^{-1} B_k' K_{k+1}] A_k, \\ &\qquad\qquad k = 0, 1, \ldots, N-1, \end{aligned}$$

and the vectors $y_k$ are given by

$$y_k = x_k + A_k^{-1} w_k + A_k^{-1} A_{k+1}^{-1} w_{k+1} + \cdots + A_k^{-1} \cdots A_{N-1}^{-1} w_{N-1}$$

(assuming the matrices $A_0, A_1, \ldots, A_{N-1}$ are invertible).

*Hint* Show that the cost functional can be written

$$E\left\{ y_0' K_0 y_0 + \sum_{k=0}^{N-1} (u_k - L_k y_k)' P_k (u_k - L_k y_k) \right\},$$

where

$$P_k = B_k' K_{k+1} B_k + R_k.$$

**2.** In Problem 1 show that the control law that minimizes the value of the cost functional among all control laws that consist of linear functions of the information vector $I_k$ is given by

$$\mu_k^*(I_k) = L_k \hat{y}_k(I_k),$$

where $\hat{y}_k(I_k)$ is the linear least squares estimate of $y_k$ given $I_k$, i.e., the linear function of $I_k$ that solves

$$\min_{y_k(\cdot),\, y_k(\cdot):\,\text{linear}} \; \underset{y_k, I_k}{E} \{[y_k - y_k(I_k)]'[y_k - y_k(I_k)]\}.$$

**3.** Consider Problems 1 and 2 with the difference that the measurements $z_k$ are received with a delay of $m \geqslant 1$ time periods, i.e., the information vector $I_k$ is given by

$$I_k = \begin{cases} (z_0, \ldots, z_{k-m}, u_0, \ldots, u_{k-1}) & \text{if } k \geqslant m, \\ (u_0, \ldots, u_{k-1}) & \text{if } k < m. \end{cases}$$

Show that the conclusion of both problems holds for this case as well.

**4.** Prove the result stated at the end of Section 4.3.

**5.** Consider a machine that can be in one of two states, good or bad. Suppose that the machine produces an item at the end of each period. The item produced is either good or bad depending on whether the machine is in a good or bad state, respectively. We suppose that once the machine is in a bad state, it remains in that state until it is replaced. If the machine is in a good state at the beginning of a certain period, then with probability $t$ it will be in the bad state at the end of the period. Once an item is produced we may inspect the item at a cost $I$, or not inspect. If an inspected item is found bad, the machine is replaced with a machine in good state at a cost $R$. The cost for producing a bad item is $C > 0$. Write a DP algorithm for obtaining an optimal inspection policy assuming an initial machine in good state, and a horizon of $N$ periods. Solve the problem for $t = 0.2$, $I = 1$, $R = 3$, $C = 2$, and $N = 8$. (The optimal policy is to inspect at the end of the third period and not inspect in any other period.) *Hint*: Search for a suitable sufficient statistic.

**6.** Consider the problem of estimating the common mean $x$ of a sequence of Gaussian random variables $z_1, z_2, \ldots, z_{N-1}$, where $z_k = x + v_k$ and $x, v_1, v_2, \ldots, v_{N-1}$ are independent Gaussian random variables with

$$E\{x\} = \mu, \qquad E\{(x - \mu)^2\} = \sigma_x^2, \qquad E\{v_k\} = 0, \qquad E\{v_k^2\} = \sigma^2.$$

At each period $k$ the sample $z_k$ may be observed at a cost $C > 0$ or else sampling is terminated and an estimate $\hat{x}(z_1, \ldots, z_{k-1})$ is selected as a function of samples $z_1, \ldots, z_{k-1}$ observed up to that period. The cost for termination after observing the $k$ first samples is

$$kC + \underset{x, z_1, \ldots, z_k}{E} \{[x - \hat{x}(z_1, \ldots, z_k)]^2\}.$$

The problem is to find the termination and estimation policy so as to minimize the cost. Show that the optimal termination time $\bar{k}$ is independent of the particular values $z_1, \ldots, z_k$ observed, and thus the problem may be solved by first determining $\bar{k}$ and then obtaining the optimal estimate $\hat{x}$ after taking $\bar{k}$ samples.

**7.** *Control of Finite-State Systems with Imperfect State Information* Consider a controlled dynamic system that at any time can be in any one of a finite number of states $x^1, x^2, \ldots, x^n$. When a control $u$ is applied, the transition probability of the system moving from state $x^i$ to state $x^j$ is denoted $p_{ij}(u)$. The control $u$ is chosen from a finite collection of controls $u^1, u^2, \ldots, u^m$. Following each state transition an observation is made by the controller. There is a finite number of possible observation outcomes denoted $z^1, z^2, \ldots, z^q$. The probability of the observation outcome $z^\theta$, $\theta = 1, \ldots, q$, occurring given that the current state is $j$ and the previous control was $u$ is denoted by either $r_j(u, \theta)$ or $r_j(u, z^\theta)$.

(a) Consider the process after $k + 2$ observations $z_0, z_1, \ldots, z_{k+1}$ have been made and controls $u_0, u_1, \ldots, u_k$ have been applied. Consider the conditional probability vectors

$$P_k = \{p_k^1, \ldots, p_k^n\}, \qquad P_{k+1} = \{p_{k+1}^1, \ldots, p_{k+1}^n\},$$

where

$$p_k^i = P(x_k = x^i \mid z_0, \ldots, z_k, u_0, \ldots, u_{k-1}), \qquad i = 1, \ldots, n,$$

$$p_{k+1}^i = P(x_{k+1} = x^i \mid z_0, \ldots, z_{k+1}, u_0, \ldots, u_k), \qquad i = 1, \ldots, n.$$

Show that

$$p_{k+1}^j = \frac{\sum_{i=1}^n p_k^i p_{ij}(u_k) r_j(u_k, z_{k+1})}{\sum_{j=1}^n \sum_{i=1}^n p_k^i p_{ij}(u_k) r_j(u_k, z_{k+1})}, \qquad j = 1, \ldots, n.$$

(b) Assume that there is a cost per stage denoted for the $k$th stage by $g_k(i, u, j)$, associated with the control $u$ and a transition from state $x^i$ to state $x^j$. Consider the problem of finding an optimal policy minimizing the sum of the costs per stage over $N$ periods. The terminal cost is denoted $g_N(x_N)$. Show that the corresponding DP algorithm is given by

$$\bar{J}_{N-1}(P_{N-1}) = \min_{u \in \{u^1, \ldots, u^m\}} \left[ \sum_{i=1}^n p_{N-1}^i \sum_{j=1}^n p_{ij}(u) [g_{N-1}(i, u, j) + g_N(x^j)] \right],$$

$$\bar{J}_k(P_k) = \min_{u \in \{u^1, \ldots, u^m\}} \Bigg[ \sum_{i=1}^n p_k^i \sum_{j=1}^n p_{ij}(u) [g_k(i, u, j)$$

$$+ \sum_{\theta=1}^q r_j(u, \theta) \bar{J}_{k+1} \left[ \frac{\sum_{i=1}^n p_k^i p_{i1}(u) r_1(u, \theta)}{\sum_{j=1}^n \sum_{i=1}^n p_k^i p_{ij}(u) r_j(u, \theta)}, \ldots, \frac{\sum_{i=1}^n p_k^i p_{in}(u) r_n(u, \theta)}{\sum_{j=1}^n \sum_{i=1}^n p_k^i p_{ij}(u) r_j(u, \theta)} \right] \Bigg] \Bigg]$$

(c) Show by induction that the functions $\bar{J}_{N-1}, \bar{J}_{N-2}, \ldots, \bar{J}_0$ are of the form

$$\bar{J}_k(P_k) = \min[P_k' \alpha_k^1, P_k' \alpha_k^2, \ldots, P_k' \alpha_k^{m_k}],$$

where $\alpha_k^1, \alpha_k^2, \ldots, \alpha_k^{m_k}$ are some vectors in $R^n$ and $P_k' \alpha_k^j$ denotes the inner product of $P_k$ and $\alpha_k^j$.

*Hint* If $\bar{J}_{k+1}$ is of the form above, show that

$$\bar{J}_{k+1} \left[ \frac{\sum_{i=1}^n p_k^i p_{i1}(u) r_1(u, \theta)}{\sum_{j=1}^n \sum_{i=1}^n p_k^i p_{ij}(u) r_j(u, \theta)}, \ldots, \frac{\sum_{i=1}^n p_k^i p_{in}(u) r_n(u, \theta)}{\sum_{j=1}^n \sum_{i=1}^n p_k^i p_{ij}(u) r_j(u, \theta)} \right]$$

$$= \frac{\min[P_k' \lambda_k^1(u, \theta), \ldots, P_k' \lambda_k^{m_{k+1}}(u, \theta)]}{\sum_{j=1}^n \sum_{i=1}^n p_k^i p_{ij}(u) r_j(u, \theta)},$$

where $\lambda_k^1(u, \theta), \ldots, \lambda_k^{m_{k+1}}(u, \theta)$ are some $n$-dimensional vectors depending on $u$ and $\theta$. Show next that

$$\bar{J}_k(P_k) = \min_{u \in \{u^1, \ldots, u^m\}} \left[ P_k' \bar{g}_k(u) + \sum_{\theta=1}^q \min[P_k' \lambda_k^i(u, \theta), \ldots, P_k' \lambda_k^{m_{k+1}}(u, \theta)] \right],$$

where $\bar{g}_k(u)$ is the $n$-vector with $i$th coordinate equal to $\sum_{j=1}^n p_{ij}(u) g_k(i, u, j)$.

Subsequently utilize the fact that any sum of the form

$$\min[P'_k\gamma_1, \ldots, P'_k\gamma_s] + \min[P'_k\tilde{\gamma}_1, \ldots, P'_k\tilde{\gamma}_{\tilde{s}}],$$

where $\gamma_1, \ldots, \gamma_s, \tilde{\gamma}_1, \ldots, \tilde{\gamma}_{\tilde{s}}$ are some $n$-dimensional vectors and $s$, $\tilde{s}$ are some positive integers is equal to

$$\min\{P'_k(\gamma_i + \tilde{\gamma}_j) \mid i = 1, \ldots, s, j = 1, \ldots, \tilde{s}\}.$$

**8.** Consider the functions $\bar{J}_k(p_k)$ of Eq. (35) in the instruction problem. Show inductively that each of these functions is piecewise linear and concave of the form

$$\bar{J}_k(p_k) = \min[\alpha_k^1 + \beta_k^1 p_k, \alpha_k^2 + \beta_k^2 p_k, \ldots, \alpha_k^{m_k} + \beta_k^{m_k} p_k],$$

where $\alpha_k^1, \ldots, \alpha_k^{m_k}, \beta_k^1, \ldots, \beta_k^{m_k}$ are suitable scalars.

**9.** Consider a hypothesis testing problem involving $n$ hypotheses. The a priori probability of hypothesis $i$, $i = 1, 2, \ldots, n$, being true is $p_i$. The cost of accepting erroneously hypothesis $i$ is $L_i$. At each sampling period an experiment is performed that has two possible outcomes denoted 0 and 1. The conditional probabilities of outcome 0 given that hypothesis $i$ holds true is known and denoted $\alpha_j$. We assume $\alpha_i \neq \alpha_j$ for $i \neq j$. At each period one may perform another experiment at the cost $C$ or terminate experimentation and accept one of the hypotheses. Provide a suitable DP algorithm for finding the optimal sampling plan when the maximum number of possible experiments is $N$.

**10.** *Two-Armed Bandit Problem* A person is offered $N$ free plays to be distributed as he pleases between two slot machines A and B. Machine A pays $\alpha$ dollars with known probability $s$ and nothing with probability $(1 - s)$. Machine B pays $\beta$ dollars with probability $p$ and nothing with probability $(1 - p)$. The person does not know $p$ but instead has an a priori cumulative probability distribution $F(p)$ of $p$. The problem is to find a playing policy that maximizes expected profit. Let $(m + n)$ denote the number of plays in $B$ after $k$ free plays $(m + n \leqslant k)$ and let $m$ denote the number of successes and $n$ the number of failures. Show that a DP algorithm that may be used to solve this problem is given by

$$\bar{J}_{N-1}(m, n) = \max\{s\alpha, p(m, n)\beta\}, \qquad m + n \leqslant N - 1,$$

$$\begin{aligned}\bar{J}_k(m, n) = \max\{&s[\alpha + \bar{J}_{k+1}(m, n)] + (1 - s)\bar{J}_{k+1}(m, n), \\ &p(m, n)[\beta + \bar{J}_{k+1}(m + 1, n)] + [1 - p(m, n)]\bar{J}_{k+1}(m, n + 1)\}, \\ &\qquad\qquad m + n \leqslant k,\end{aligned}$$

where

$$p(m, n) = \frac{\int_0^1 p^{m+1}(1 - p)^n \, dF(p)}{\int_0^1 p^m(1 - p)^n \, dF(p)}.$$

Solve the problem for $N = 6$, $\alpha = \beta = 1$, $s = 0.6$, $dF(p)/dp = 1$ for $0 \leqslant p \leqslant 1$. (The answer is to play machine B for the following pairs $(m, n)$: (0, 0), (1, 0), (2, 0), (3, 0), (4, 0), (5, 0), (2, 1), (3, 1), (4, 1). Otherwise machine A should be played.)

**11.** A person is offered 2 to 1 odds in a coin tossing game where he wins whenever a tail occurs. However, he suspects that the coin is biased and has an a priori cumulative probability distribution $F(p)$ for the probability $p$ that a head occurs at each toss. The problem is to find an optimal policy of deciding whether to continue or stop participating in the game given the outcomes of the game so far. A maximum of $N$ tossings is allowed. Indicate how such a policy can be found by means of DP.

## Appendix Least-Squares Estimation—The Kalman Filter

In this appendix we present the basic principles of least-squares estimation and their application in the problem of estimating the state of a linear discrete-time dynamic system using measurements that are linear in the state variable.

The basic problem is roughly the following. There are two random vectors $x$ and $y$ taking values in Euclidean spaces $R^n$ and $R^m$, respectively. The two vectors are related through their joint probability distribution so that the value of one of the two provides information about the value of the other. For example, $x$ may be the velocity of an automobile and $y$ the result of an inexact measurement of this velocity, which is equal to $x$ plus a random error that is, say, uniformly distributed over some interval. Now the situation that we examine is one where we get to know the value of $y$ and we would like to estimate the value of $x$ so that the estimation error is as small as possible in some sense. The criterion that we use is minimization of the expected value of the squared error between $x$ and its estimate, which explains the term least-squares estimation. This criterion is reasonable as well as convenient from the analytical point of view.

We begin with the problem of finding the least-squares estimate of a random vector $x$ given the value of the measured random vector $y$. Next we consider the problem of finding the least-squares estimate of the random vector $x$ within the class of all estimates that are *linear* in the measured vector $y$. Finally the results are applied to a special case where there is an underlying linear dynamic system the current state of which we would like to estimate using measurements that are obtained sequentially in time. Due to the special structure of this problem the computation of the state estimate can be organized conveniently in a recursive algorithm—the Kalman filter. Throughout the exposition we will not make a notational distinction between a random vector and particular sample values that it assumes. The precise meaning should be clear from the context.

## A.1 *Least-Squares Estimation*

Consider two jointly distributed random vectors $x$ and $y$ taking values in $R^n$ and $R^m$, respectively. In any particular occurrence of $x$ and $y$ we expect that the value of $y$ provides information that may be used to update our a priori estimate or guess of $x$. For example, while prior to knowing $y$ our estimate of $x$ may have been the expected value $E\{x\}$, once the value of $y$ is known we may wish to form an udpated estimate $x(y)$ of the value of $x$. This updated estimate depends, of course, on the value of $y$ and thus we are in effect interested in a rule that gives us the estimate for each possible value of $y$, i.e., we are interested in a function $x(\cdot)$, where $x(y)$ is the estimate of $x$ given $y$. Such a function $x(\cdot): R^m \to R^n$ we shall call an *estimator*. We are seeking an estimator that is optimal in some sense and the criterion we shall employ is based on minimization of

$$\underset{x,y}{E}\{\|x - x(y)\|^2\} = \underset{x,y}{E}\{[x - x(y)]'[x - x(y)]\}, \tag{57}$$

where $\|\cdot\|$ denotes the usual norm in $R^n(\|z\|^2 = z'z$ for $z \in R^n)$. An estimator that minimizes the measure of error above over all $x(\cdot): R^n \to R^m$ will be called a *least-squares estimator* and will be denoted $\hat{x}^*(\cdot)$. It is clear that $\hat{x}^*(\cdot)$ is a least-squares estimator if we have for every $y \in R^m$

$$\underset{x}{E}\{\|x - \hat{x}^*(y)\|^2 \,|\, y\} = \min_{z \in R^m} \underset{x}{E}\{\|x - z\|^2 \,|\, y\}, \tag{58}$$

where the expectations are taken with respect to the conditional distribution of $x$ given $y$ for fixed values of $y$.

The following proposition was proved in Section 1.3. We repeat the simple proof here. We shall assume in the proposition as well as throughout the Appendix that *all the expected values appearing are well defined and finite.*

**Proposition A.1** The least-squares estimator $\hat{x}^*(\cdot)$ is given by

$$\hat{x}^*(y) = \underset{x}{E}\{x \,|\, y\} \qquad \forall y \in R^m. \tag{59}$$

*Proof* We have for every fixed $z \in R^n$

$$\underset{x}{E}\{\|x - z\|^2 \,|\, y\} = E\{\|x\|^2 \,|\, y\} - 2z' \underset{x}{E}\{x \,|\, y\} + \|z\|^2.$$

This expression is minimized for $z = E_x\{x \,|\, y\}$ and using (58) the result follows. Q.E.D.

A.2 *Linear Least-Squares Estimation*

While the least-squares estimate is simply the conditional expectation $E_x\{x \mid y\}$, in general the function $E_x\{x \mid \cdot\}$ may be a complicated nonlinear function of $y$. As a result its practical computation and realization in a given application may be very cumbersome. This fact motivates us to consider estimators within a restricted class that may be realized with relative ease. An important such class is the class of *linear*† estimators, i.e., estimators of the form

$$x(y) = Ay + b, \tag{60}$$

where $A$ is an $n \times m$ matrix and $b$ is an $n$-dimensional vector. It is thus reasonable to consider the problem of finding a linear estimator of form (60) that minimizes the expected squared error (57). An estimator

$$\hat{x}(y) = \hat{A}y + \hat{b}$$

where $\hat{A}, \hat{b}$ minimize

$$\underset{x,\, y}{E}\{\|x - Ay - b\|^2\} = \underset{x,\, y}{E}\{(x - Ay - b)'(x - Ay - b)\}$$

over all $n \times m$ matrices $A$ and vectors $b \in R^n$ will be called a *linear least-squares estimator.*

Prior to proceeding with the derivation of the linear least-squares estimator we show that when $x$, $y$ are jointly Gaussian random vectors the conditional expectation $E_x\{x \mid y\}$ is a linear function of $y$ (plus a constant vector) and as a result for this case a linear least-squares estimator is also a least-squares estimator.

Consider the random (column) vector $z$

$$z = \begin{bmatrix} x \\ y \end{bmatrix}$$

taking values in $R^{n+m}$, and assume that $z$ is Gaussian with mean

$$\bar{z} = E\{z\} = \begin{bmatrix} E\{x\} \\ E\{y\} \end{bmatrix} = \begin{bmatrix} \bar{x} \\ \bar{y} \end{bmatrix} \tag{61}$$

and covariance matrix

$$\Sigma = E\{(z - \bar{z})(z - \bar{z})'\} = \begin{bmatrix} E\{(x - \bar{x})(x - \bar{x})'\} & E\{(x - \bar{x})(y - \bar{y})'\} \\ E\{(y - \bar{y})(x - \bar{x})'\} & E\{(y - \bar{y})(y - \bar{y})'\} \end{bmatrix}$$

$$= \begin{bmatrix} \Sigma_{xx} & \Sigma_{xy} \\ \Sigma_{yx} & \Sigma_{yy} \end{bmatrix}. \tag{62}$$

† A more precise term is "linear plus constant" or equivalently "affine" estimators.

We shall assume that $\Sigma$ is a positive definite symmetric $(n + m) \times (n + m)$ matrix so that it possesses an inverse. This assumption is made for convenience. The result to be proved holds, however, without this assumption. Since $z$ is Gaussian its probability density function is of the form (see, e.g., [M6], [P2])

$$p(z) = p(x, y) = c \exp[-\tfrac{1}{2}(z - \bar{z})'\Sigma^{-1}(z - \bar{z})],$$

where $c$ is given by

$$c = (2\pi)^{-(n+m)/2}(\det \Sigma)^{-1/2},$$

with $\det \Sigma$ denoting the determinant of $\Sigma$. Similarly the (marginal) probability density functions of $x$ and $y$ are of the form

$$p(x) = c_1 \exp[-\tfrac{1}{2}(x - \bar{x})'\Sigma_{xx}^{-1}(x - \bar{x})],$$
$$p(y) = c_2 \exp[-\tfrac{1}{2}(y - \bar{y})'\Sigma_{yy}^{-1}(y - \bar{y})],$$

where $c_1$ and $c_2$ are appropriate constants. By Bayes' rule the conditional probability density function of $x$ conditioned on $y$ is given by

$$\begin{aligned} p(x \mid y) &= p(x, y)/p(y) \\ &= (c/c_2) \exp\{-\tfrac{1}{2}[(z - \bar{z})'\Sigma^{-1}(z - \bar{z}) - (y - \bar{y})'\Sigma_{yy}^{-1}(y - \bar{y})]\}. \end{aligned} \tag{63}$$

It is now easy to see that there exist a positive definite $n \times n$ matrix $D$, an $n \times m$ matrix $A$, a vector $b \in R^n$, and a scalar $s$ such that

$$\begin{aligned} (z - \bar{z})'\Sigma^{-1}(z - \bar{z}) - (y - \bar{y})'\Sigma_{yy}^{-1}(y - \bar{y}) \\ = (x - Ay - b)'D^{-1}(x - Ay - b) + s. \end{aligned} \tag{64}$$

This is evident since by substitution of the expressions for $\bar{z}$ and $\Sigma$ of (61) and (62), the left part of (64) becomes a quadratic form in $x$ and $y$, which can be put in the form indicated in the right side of (64). In fact, by computing the inverse of $\Sigma$ using the partitioned matrix inversion formula (Appendix A) one may verify that $A$, $b$, $D$, and $s$ in (64) have the form

$$A = \Sigma_{xy}\Sigma_{yy}^{-1}, \qquad b = \bar{x} - \Sigma_{xy}\Sigma_{yy}^{-1}\bar{y}, \qquad D = \Sigma_{xx} - \Sigma_{xy}\Sigma_{yy}^{-1}\Sigma_{yx}, \qquad s = 0.$$

Now it follows from (64) and (63) that the conditional expectation $E_x\{x \mid y\}$ is of the form $Ay + b$, where $A$ is some $n \times m$ matrix and $b \in R^n$. Thus we have proved the following proposition.

**Proposition A.2** If $x$, $y$ are jointly Gaussian random vectors, then the least-squares estimate $E_x\{x \mid y\}$ of $x$ given $y$ is also a linear least-squares estimate of $x$ given $y$.

We note that this proposition holds not only for a Gaussian distribution of $(x, y)$ but also for a much wider class of distributions, which contains the Gaussian distribution as a special case (see [V6], [B24]).

We now turn to the characterization of the linear least-squares estimator.

**Proposition A.3** Let $x, y$ be random vectors taking values in $R^n$ and $R^m$, respectively, with given joint probability distribution. The expected values and covariance matrices of $x, y$ are assumed to exist and are denoted

$$E\{x\} = \bar{x} \qquad E\{y\} = \bar{y}, \tag{65}$$

$$E\{(x - \bar{x})(x - \bar{x})'\} = \Sigma_{xx}, \qquad E\{(y - \bar{y})(y - \bar{y})'\} = \Sigma_{yy}, \tag{66}$$

$$E\{(x - \bar{x})(y - \bar{y})'\} = \Sigma_{xy}, \qquad E\{(y - \bar{y})(x - \bar{x})'\} = \Sigma_{yx} = \Sigma'_{xy}, \tag{67}$$

We assume also that the inverse $\Sigma_{yy}^{-1}$ exists. Then the linear least-squares estimator of $x$ given $y$ has the form

$$\hat{x}(y) = \bar{x} + \Sigma_{xy}\Sigma_{yy}^{-1}(y - \bar{y}). \tag{68}$$

The corresponding error covariance matrix is given by

$$\underset{x,y}{E}\{[x - \hat{x}(y)][x - \hat{x}(y)]'\} = \Sigma_{xx} - \Sigma_{xy}\Sigma_{yy}^{-1}\Sigma_{yx}. \tag{69}$$

*Proof* The linear least-squares estimator is defined as

$$\hat{x}(y) = \hat{A}y + \hat{b},$$

where $\hat{A}$, $\hat{b}$ minimize the function $f(A, b) = E_{x,y}\{\|x - Ay - b\|^2\}$ over $A$ and $b$. Taking the derivatives of $f(A, b)$ with respect to $A$ and $b$ and setting them equal to zero we have

$$0 = \partial f/\partial A|_{\hat{A},\hat{b}} = 2\underset{x,y}{E}\{(\hat{b} + \hat{A}y - x)y'\}, \tag{70}$$

$$0 = \partial f/\partial b|_{\hat{A},\hat{b}} = 2\underset{x,y}{E}\{\hat{b} + \hat{A}y - x\}. \tag{71}$$

From (71) we have

$$\hat{b} = \bar{x} - \hat{A}\bar{y}, \tag{72}$$

and substitution in (70) yields

$$\underset{x,y}{E}\{y[\hat{A}(y - \bar{y}) - (x - \bar{x})]'\} = 0. \tag{73}$$

We now use the identity

$$\underset{x,y}{E}\{-\bar{y}[\hat{A}(y - \bar{y}) - (x - \bar{x})]'\} = -\bar{y}\underset{x,y}{E}\{\hat{A}(y - \bar{y}) - (x - \bar{x})\}' = 0. \tag{74}$$

Adding (74) and (73) yields

$$\underset{x,y}{E}\{(y - \bar{y})[\hat{A}(y - \bar{y}) - (x - \bar{x})]'\} = 0,$$

or equivalently

$$\Sigma_{yy}\hat{A}' - \Sigma_{yx} = 0,$$

from which

$$\hat{A} = \Sigma'_{yx}\Sigma_{yy}^{-1} = \Sigma_{xy}\Sigma_{yy}^{-1}. \tag{75}$$

From (72) and (75) we obtain

$$\hat{x}(y) = \hat{A}y + \hat{b} = \bar{x} + \Sigma_{xy}\Sigma_{yy}^{-1}(y - \bar{y}),$$

which was to be proved. Equation (69) follows immediately upon substitution of the expression for $\hat{x}(y)$ obtained above. Q.E.D.

We list below some of the properties of the least-squares estimator as corollaries.

**Corollary A.3.1** There holds

$$\bar{x} = \underset{x}{E}\{x\} = \underset{y}{E}\{\hat{x}(y)\}.$$

*Proof* Immediate from (68). Q.E.D.

**Corollary A.3.2** The estimation error $[x - \hat{x}(y)]$ is uncorrelated with both $y$ and $\hat{x}(y)$, i.e.,

$$\underset{x,y}{E}\{y[x - \hat{x}(y)]'\} = 0, \qquad \underset{x,y}{E}\{\hat{x}(y)[x - \hat{x}(y)]'\} = 0.$$

*Proof* The first equality is immediate from (70). The second equality is evident once we write $\hat{x}(y)$ in the form $\hat{A}y + \hat{b}$ and use the first equality and Corollary A.3.1. Q.E.D.

Corollary A.3.2 is sometimes called the *orthogonal projection principle*. It states a property that characterizes the linear least-squares estimate and forms the basis for alternative treatments of the least-squares estimation problem using the so-called projection theorem (see [L8], Chapter 4).

**Corollary A.3.3** Consider in addition to $x$ and $y$ the random vector $z$ defined by

$$z = Cx,$$

where $C$ is a $p \times m$ given matrix. Then the linear least-squares estimate $\hat{z}(y)$ of $z$ given $y$ has the form

$$\hat{z}(y) = C\hat{x}(y),$$

and the corresponding error covariance matrix is given by

$$\underset{z,y}{E}\{[z - \hat{z}(y)][z - \hat{z}(y)]'\} = C\underset{x,y}{E}\{[x - \hat{x}(y)][x - \hat{x}(y)]'\}C'.$$

*Proof* We have $E\{z\} = \bar{z} = C\bar{x}$ and

$$\Sigma_{zz} = \underset{z}{E}\{(z - \bar{z})(z - \bar{z})'\} = C\Sigma_{xx}C',$$

$$\Sigma_{zy} = \underset{z,\,y}{E}\{(z - \bar{z})(y - \bar{y})'\} = C\Sigma_{xy},$$

$$\Sigma_{yz} = \Sigma'_{zy} = \Sigma_{yx}C'.$$

By Proposition A.3 we have

$$\hat{z}(y) = \bar{z} + \Sigma_{zy}\Sigma_{yy}^{-1}(y - \bar{y}) = C\bar{x} + C\Sigma_{xy}\Sigma_{yy}^{-1}(y - \bar{y}) = C\hat{x}(y),$$

$$\underset{x,\,y}{E}\{[z - \hat{z}(y)][z - \hat{z}(y)]'\} = \Sigma_{zz} - \Sigma_{zy}\Sigma_{yy}^{-1}\Sigma_{yz} = C(\Sigma_{xx} - \Sigma_{xy}\Sigma_{yy}^{-1}\Sigma_{yx})C'$$

$$= C\underset{x,\,y}{E}\{[x - \hat{x}(y)][x - \hat{x}(y)]'\}C'. \quad \text{Q.E.D.}$$

**Corollary A.3.4** Consider in addition to $x$ and $y$ an additional random vector $z$ taking values in $R^p$ of the form

$$z = Cy + u, \tag{76}$$

where $C$ is a $p \times m$ with $p \leqslant m$ (nonrandom) given matrix with full rank and $u$ is a (nonrandom) given vector in $R^p$. Then the linear least-squares estimate $\hat{x}(z)$ of $x$ given $z$ has the form

$$\hat{x}(z) = \bar{x} + \Sigma_{xy}C'(C\Sigma_{yy}C')^{-1}(z - C\bar{y} - u), \tag{77}$$

and the corresponding error covariance matrix is given by

$$\underset{x,\,z}{E}\{[x - \hat{x}(z)][x - \hat{x}(z)]'\} = \Sigma_{xx} - \Sigma_{xy}C'(C\Sigma_{yy}C')^{-1}C\Sigma_{yx}. \tag{78}$$

*Proof* We have by direct calculation

$$E\{z\} = \bar{z} = C\bar{y} + u, \tag{79a}$$

$$E\{(z - \bar{z})(z - \bar{z})'\} = \Sigma_{zz} = C\Sigma_{yy}C', \tag{79b}$$

$$E\{(z - \bar{z})(x - \bar{x})'\} = \Sigma_{zx} = C\Sigma_{yx}, \tag{79c}$$

$$E\{(x - \bar{x})(z - \bar{z})'\} = \Sigma_{xz} = \Sigma_{xy}C'. \tag{79d}$$

From Proposition A.3 we have

$$\hat{x}(z) = \bar{x} + \Sigma_{xz}\Sigma_{zz}^{-1}(z - \bar{z}), \tag{80a}$$

$$\underset{x,\,z}{E}\{[x - \hat{x}(z)][x - \hat{x}(z)]'\} = \Sigma_{xx} - \Sigma_{xz}\Sigma_{zz}^{-1}\Sigma_{zx}, \tag{80b}$$

where $\Sigma_{zz} = C\Sigma_{yy}C'$ has an inverse since $\Sigma_{yy}$ is invertible and $C$ has full rank. By substituting relations (79) into (80) the result follows. Q.E.D.

Notice that the error covariance matrix $E_{x,z}\{[x - \hat{x}(z)][x - \hat{x}(z)]'\}$ does not depend on the vector $u$, i.e., the choice of $u$ cannot affect the quality of estimation.

**Corollary A.3.5** Consider in addition to $x$ and $y$ an additional random vector $z$ taking values in $R^p$, which is uncorrelated with $y$. Then the linear least-squares estimate $\hat{x}(y, z)$ of $x$ given $y$ *and* $z$ (i.e., given the composite vector $[y', z']'$) has the form

$$\hat{x}(y, z) = \hat{x}(y) + \hat{x}(z) - \bar{x}, \tag{81}$$

where $\hat{x}(y)$ and $\hat{x}(z)$ are the linear least-squares estimates of $x$ given $y$ and given $z$, respectively. Furthermore,

$$\mathop{E}_{x,y,z}\{[x - \hat{x}(y, z)][x - \hat{x}(y, z)]'\} = \Sigma_{xx} - \Sigma_{xy}\Sigma_{yy}^{-1}\Sigma_{yx} - \Sigma_{xz}\Sigma_{zz}^{-1}\Sigma_{zx}, \tag{82}$$

where

$$\Sigma_{xz} = \mathop{E}_{x,z}\{(x - \bar{x})(z - \bar{z})'\}, \qquad \Sigma_{zx} = \mathop{E}_{x,z}\{(z - \bar{z})(x - \bar{x})'\},$$

$$\Sigma_{zz} = \mathop{E}_{z}\{(z - \bar{z})(z - \bar{z})'\}, \qquad \bar{z} = \mathop{E}_{z}\{z\},$$

and it is assumed that $\Sigma_{zz}$ is invertible.

*Proof* Let

$$w = \begin{bmatrix} y \\ z \end{bmatrix}, \qquad \bar{w} = \begin{bmatrix} \bar{y} \\ \bar{z} \end{bmatrix}.$$

We have by (68) that

$$\hat{x}(w) = \bar{x} + \Sigma_{xw}\Sigma_{ww}^{-1}(w - \bar{w}). \tag{83}$$

Furthermore

$$\Sigma_{xw} = [\Sigma_{xy}, \Sigma_{xz}],$$

and since $y$ and $z$ are uncorrelated

$$\Sigma_{ww} = \begin{bmatrix} \Sigma_{yy} & 0 \\ 0 & \Sigma_{zz} \end{bmatrix}.$$

Substituting the above expressions in (83) we obtain

$$\hat{x}(w) = \bar{x} + \Sigma_{xy}\Sigma_{yy}^{-1}(y - \bar{y}) + \Sigma_{xz}\Sigma_{zz}^{-1}(z - \bar{z}) = \hat{x}(y) + \hat{x}(z) - \bar{x},$$

and (81) is proved. The proof of (82) is similar by using the relations above and (69). Q.E.D.

**Corollary A.3.6** Let $z$ be as in the previous corollary and assume that $y$ and $z$ are not necessary uncorrelated, i.e., we may have

$$\Sigma_{yz} = \Sigma'_{zy} = \underset{y,z}{E}\{(y - \bar{y})(z - \bar{z})'\} \neq 0.$$

Then

$$\hat{x}(y, z) = \hat{x}(y) + \hat{x}[z - \hat{z}(y)] - \bar{x}, \tag{84}$$

where $\hat{x}[z - \hat{z}(y)]$ denotes the linear least-squares estimate of $x$ given the random vector $[z - \hat{z}(y)]$ and $\hat{z}(y)$ is the linear least-squares estimate of $z$ given $y$. Furthermore,

$$\begin{aligned}
&\underset{x,y,z}{E}\{[x - \hat{x}(y, z)][x - \hat{x}(y, z)]'\} \\
&\quad = \underset{x,y}{E}\{[x - \hat{x}(y)][x - \hat{x}(y)]'\} - \underset{x,y,z}{E}\{(x - \bar{x})[z - \hat{z}(y)]'\} \\
&\qquad \times \left[\underset{y,z}{E}\{[z - \hat{z}(y)][z - \hat{z}(y)]'\}\right]^{-1} \underset{x,y,z}{E}\{[z - \hat{z}(y)](x - \bar{x})'\}. \qquad (85)
\end{aligned}$$

*Proof* By Corollary A.3.2 the random vectors $y$ and $[z - \hat{z}(y)]$ are uncorrelated. Given this observation the result follows by application of the previous corollary. Q.E.D.

Frequently one is faced with a situation whereby he wishes to estimate a vector of parameters $x \in R^n$ given a measurement vector $z \in R^m$ of the form

$$z = Cx + v,$$

where $C$ is a given $m \times n$ matrix and $v \in R^m$ is a random measurement error vector. The (a priori) probability distribution of $x$ and $v$ is given. The following corollary gives the linear least-squares estimate $\hat{x}(z)$ and its error covariance.

**Corollary A.3.7** Let $z$, $x$, $v$, $C$ be as above and assume that $x$ and $v$ are uncorrelated. Denote

$$E\{x\} = \bar{x}, \qquad E\{(x - \bar{x})(x - \bar{x})'\} = \Sigma_{xx},$$
$$E\{v\} = \bar{v}, \qquad E\{(v - \bar{v})(v - \bar{v})'\} = \Sigma_{vv},$$

and assume further that $\Sigma_{vv}$ is a positive definite matrix. Then

$$\hat{x}(z) = \bar{x} + \Sigma_{xx}C'(C\Sigma_{xx}C' + \Sigma_{vv})^{-1}(z - C\bar{x} - \bar{v}),$$

$$\underset{x,v}{E}\{[x - \hat{x}(z)][x - \hat{x}(z)]'\} = \Sigma_{xx} - \Sigma_{xx}C'(C\Sigma_{xx}C' + \Sigma_{vv})^{-1}C\Sigma_{xx}.$$

*Proof* Define $y = [x', v']' \in R^{n+m}$, $\bar{C} = [C, I]$, $\bar{y} = [\bar{x}', \bar{v}']'$. Then we have $z = \bar{C}y$ and by Corollary A.3.3

$$\hat{x}(z) = [I, 0]\hat{y}(z),$$

$$E\{[x - \hat{x}(z)][x - \hat{x}(z)]'\} = [I, 0]\,E\{[y - \hat{y}(z)][y - \hat{y}(z)]'\}\begin{bmatrix} I \\ 0 \end{bmatrix},$$

where $\hat{y}(z)$ is the linear least-squares estimate of $y$ given $z$. By applying Corollary A.3.4 with $u = 0$ and $x = y$ we have

$$\hat{y}(z) = \bar{y} + \Sigma_{yy}\bar{C}'(\bar{C}\Sigma_{yy}\bar{C}')^{-1}(z - \bar{C}\bar{y}),$$

$$E\{[y - \hat{y}(z)][y - \hat{y}(z)]'\} = \Sigma_{yy} - \Sigma_{yy}\bar{C}'(\bar{C}\Sigma_{yy}\bar{C}')^{-1}\bar{C}\Sigma_{yy}.$$

By using the equations

$$\Sigma_{yy} = \begin{bmatrix} \Sigma_{xx} & 0 \\ 0 & \Sigma_{vv} \end{bmatrix}$$

and $\bar{C} = [C, I]$ above and carrying out the straightforward calculation the result follows. Q.E.D.

## A.3 *State Estimation of Discrete-Time Dynamic Systems —The Kalman Filter*

Consider now a linear dynamic system of the type considered in Section 4.3 but without a control vector ($u_k = 0$)

$$x_{k+1} = A_k x_k + w_k, \qquad k = 0, 1, \ldots, N - 1, \tag{86}$$

where $x_k \in R^n$, $w_k \in R^n$ denote the state and random disturbance vectors, respectively, and the matrices $A_k$ are known (nonrandom). Consider also the measurement equation

$$z_k = C_k x_k + v_k, \qquad k = 0, 1, \ldots, N - 1, \tag{87}$$

where $z_k \in R^s$, $v_k \in R^s$ are the observation and observation noise vectors, respectively.

We assume that $x_0, w_0, w_1, \ldots, w_{N-1}, v_0, \ldots, v_{N-1}$ are mutually independent random vectors with given probability distributions. Furthermore, they have zero mean and finite second moments, i.e.,

$$E\{x_0\} = E\{w_k\} = E\{v_k\} = 0, \qquad k = 0, 1, \ldots, N - 1. \tag{88}$$

We use the notation

$$S = E\{x_0 x_0'\}, \qquad M_k = E\{w_k w_k'\}, \qquad N_k = E\{v_k v_k'\}, \tag{89}$$

and we shall assume that $N_k$ is a positive definite matrix for every $k$.

Consider now the problem of finding the linear least-squares estimate of the random vectors $x_{k+1}$ or $x_k$ given the values of $z_0, z_1, \ldots, z_k$, or equivalently given the random vector $Z_k = [z_0', \ldots, z_k']' \in R^{(k+1)s}$. Let us denote these estimates $\hat{x}_{k+1|k}$ and $\hat{x}_{k|k}$, respectively.

It is possible to provide with very little effort an equation for $\hat{x}_{k|k}$ by using the results already obtained. Indeed let us denote for each $k$

$$V_k = [v_0', v_1', \ldots, v_k']', \qquad r_{k-1} = [x_0', w_0', w_1', \ldots, w_{k-1}']'.$$

For each $i$ with $0 \leqslant i \leqslant k$ we have, by using the system equation,

$$x_{i+1} = L_i r_i,$$

where $L_i$ is the $n \times [n(i+1)]$ matrix

$$L_i = [A_i \cdots A_0, A_i \cdots A_1, \ldots, A_i, I].$$

As a result we may write

$$Z_k = \Phi_{k-1} r_{k-1} + V_k,$$

where $\Phi_{k-1}$ is an $[s(k+1)] \times (nk)$ matrix defined by

$$\Phi_{k-1} = \begin{bmatrix} C_0 & 0 \\ C_1 L_0 & 0 \\ \vdots & \vdots \\ C_{k-1} L_{k-2} & 0 \\ C_k L_{k-1} & \end{bmatrix},$$

where the zero matrices above have appropriate dimension. Thus the problem has been reformulated in such a way that we may use Corollary A.3.7, the equations above, and the data of the problem to compute

$$\hat{r}_{k-1}(Z_k) \qquad \text{and} \qquad E\{[r_{k-1} - \hat{r}_{k-1}(Z_k)][r_{k-1} - \hat{r}_{k-1}(Z_k)]'\}.$$

Subsequently we can obtain $\hat{x}_{k|k} = \hat{x}_k(Z_k)$ as well as the corresponding error covariance matrix by using Corollary A.3.3, i.e.,

$$\hat{x}_{k|k} = L_{k-1} \hat{r}_{k-1}(Z_k),$$

$$E\{[(x_k - \hat{x}_{k|k})(x_k - \hat{x}_{k|k})'\} \\ = L_{k-1} E\{[r_{k-1} - \hat{r}_{k-1}(Z_k)][r_{k-1} - \hat{r}_{k-1}(Z_k)]'\} L_{k-1}'.$$

These equations may in turn be used to yield $\hat{x}_{k+1|k}$ and an equation for the corresponding error covariance by again using Corollary A.3.3.

The conclusion from the above analysis is that one can provide in a straightforward manner equations for the least-squares estimate of the state $x_k$ and the corresponding error covariance by using the results already obtained earlier. However, these equations are very cumbersome to use when the number of measurements is large. Fortunately enough the sequential structure of the problem can be effectively exploited and the computations can be organized in a very convenient manner. The corresponding algorithm was originally proposed in the form given here by Kalman [K1] although related ideas were known much earlier. The main attractive feature of the Kalman filtering algorithm is that the estimate $\hat{x}_{k+1|k}$ can be obtained by means of a simple equation that involves the previous estimate $\hat{x}_{k|k-1}$ and the new measurement $z_k$ but *does not involve any of the past measurements* $z_0, z_1, \ldots, z_{k-1}$. In this way significant data reduction is achieved. We now proceed to derive the form of the algorithm.

Suppose that we have computed the estimate $\hat{x}_{k|k-1}$ together with the covariance matrix

$$\Sigma_{k|k-1} = E\{(x_k - \hat{x}_{k|k-1})(x_k - \hat{x}_{k|k-1})'\}. \tag{90}$$

At time $k$ we receive the additional measurement

$$z_k = C_k x_k + v_k.$$

We may use now Corollary A.3.6 to compute the linear least-squares estimate of $x_k$ given $Z_{k-1} = [z_0', z_1', \ldots, z_{k-1}']'$ *and* $z_k$. This estimate is denoted $\hat{x}_{k|k}$ and, by Corollary A.3.6, it is given by

$$\hat{x}_{k|k} = \hat{x}_{k|k-1} + \hat{x}_k[z_k - \hat{z}_k(Z_{k-1})] - E\{x_k\}, \tag{91}$$

where $\hat{z}_k(Z_{k-1})$ denotes the linear least-squares estimate of $z_k$ given $Z_{k-1}$ and $\hat{x}_k[z_k - \hat{z}_k(Z_{k-1})]$ denotes the linear least-squares estimate of $x_k$ given $[z_k - \hat{z}_k(Z_{k-1})]$. Now we have by (86)–(89) and Corollary A.3.3,

$$E\{x_k\} = 0, \qquad \hat{z}_k(Z_{k-1}) = C_k \hat{x}_{k|k-1}. \tag{92}$$

Also to calculate $\hat{x}_k[z_k - \hat{z}_k(Z_{k-1})]$ we use Corollary A.3.3 to obtain

$$E\{[z_k - \hat{z}_k(Z_{k-1})][z_k - \hat{z}_k(Z_{k-1})]'\} = C_k \Sigma_{k|k-1} C_k' + N_k, \tag{93}$$

$$\begin{aligned} E\{x_k[z_k - \hat{z}_k(Z_{k-1})]'\} &\\ = E\{x_k[C_k(x_k - \hat{x}_{k|k-1})]'\} &+ E\{x_k v_k'\} \\ = E\{(x_k - \hat{x}_{k|k-1})&(x_k - \hat{x}_{k|k-1})'\}C_k' \\ + E\{\hat{x}_{k|k-1}&(x_k - \hat{x}_{k|k-1})'\}C_k'. \end{aligned}$$

The last term on the right above is zero by Corollary A.3.2 so that using (90) we have

$$E\{x_k[z_k - \hat{z}_k(Z_{k-1})]'\} = \Sigma_{k|k-1} C_k'. \tag{94}$$

Using expressions (92)–(94) in Proposition A.3, we obtain

$$\hat{x}_k[z_k - \hat{z}_k(Z_{k-1})] = \Sigma_{k|k-1}C_k'(C_k\Sigma_{k|k-1}C_k' + N_k)^{-1}(z_k - C_k\hat{x}_{k|k-1}),$$

and (91) is written

$$\rhd \quad \hat{x}_{k|k} = \hat{x}_{k|k-1} + \Sigma_{k|k-1}C_k'(C_k\Sigma_{k|k-1}C_k' + N_k)^{-1}(z_k - C_k\hat{x}_{k|k-1}). \quad \lhd \qquad (95)$$

By using Corollary A.3.3 we also have

$$\rhd \qquad \hat{x}_{k+1|k} = A_k\hat{x}_{k|k}. \qquad \lhd \qquad (96)$$

Concerning the covariance matrix $\Sigma_{k+1|k}$ we have from the system equation (86) and (88), (89), and Corollary A.3.3:

$$\rhd \qquad \Sigma_{k+1|k} = A_k\Sigma_{k|k}A_k' + M_k, \qquad \lhd \qquad (97)$$

where

$$\Sigma_{k|k} = E\{(x_k - \hat{x}_{k|k})(x_k - \hat{x}_{k|k})'\}.$$

Now the error covariance matrix $\Sigma_{k|k}$ may be computed via Corollary A.3.6 similarly as $\hat{x}_{k|k}$ [cf. Eq. (91)]. We have from (85), (93), and (94) that

$$\rhd \quad \Sigma_{k|k} = \Sigma_{k|k-1} - \Sigma_{k|k-1}C_k'(C_k\Sigma_{k|k-1}C_k' + N_k)^{-1}C_k\Sigma_{k|k-1}. \qquad \lhd \qquad (98)$$

Equations (95)–(98) with the initial conditions [cf. Eqs. (88) and (89)]

$$\rhd \qquad \hat{x}_{0|-1} = 0, \qquad \Sigma_{0|-1} = S, \qquad \lhd \qquad (99)$$

constitute the *Kalman filtering algorithm*. This algorithm recursively generates the linear least-squares estimates $\hat{x}_{k+1|k}$ or $\hat{x}_{k|k}$ together with the associated error covariance matrices $\Sigma_{k+1|k}$ or $\Sigma_{k|k}$.

An alternative expression for Eq. (95) is

$$\rhd \quad \hat{x}_{k|k} = A_{k-1}\hat{x}_{k-1|k-1} + \Sigma_{k|k}C_k'N_k^{-1}(z_k - C_kA_{k-1}\hat{x}_{k-1|k-1}). \quad \lhd \qquad (100)$$

This expression is obtained from (95) and (96) by using the equality

$$\Sigma_{k|k}C_k'N_k^{-1} = \Sigma_{k|k-1}C_k'(C_k\Sigma_{k|k-1}C_k' + N_k)^{-1}.$$

This equality may be verified by using (98) to write

$$\begin{aligned}\Sigma_{k|k}C_k'N_k^{-1} &= [\Sigma_{k|k-1} - \Sigma_{k|k-1}C_k'(C_k\Sigma_{k|k-1}C_k' + N_k)^{-1}C_k\Sigma_{k|k-1}]C_k'N_k^{-1} \\ &= \Sigma_{k|k-1}C_k'[N_k^{-1} - (C_k\Sigma_{k|k-1}C_k' + N_k)^{-1}C_k\Sigma_{k|k-1}C_k'N_k^{-1}] \\ &= \Sigma_{k|k-1}C_k'(C_k\Sigma_{k|k-1}C_k' + N_k)^{-1},\end{aligned}$$

where the last step follows by writing

$$\begin{aligned}N_k^{-1} &= (C_k\Sigma_{k|k-1}C_k' + N_k)^{-1}(C_k\Sigma_{k|k-1}C_k' + N_k)N_k^{-1} \\ &= (C_k\Sigma_{k|k-1}C_k' + N_k)^{-1}(C_k\Sigma_{k|k-1}C_k'N_k^{-1} + I).\end{aligned}$$

When the system equation contains a control vector $u_k$ and has the form

$$x_{k+1} = A_k x_k + B_k u_k + w_k, \qquad k = 0, 1, \ldots, N-1,$$

and we seek the linear least-squares estimate $\hat{x}_{k|k}$ of $x_k$ given $z_0, z_1, \ldots, z_k$ and $u_0, u_1, \ldots, u_{k-1}$, it is easy to show that (100) takes the form

$$\begin{aligned}\hat{x}_{k|k} = {} & A_{k-1}\hat{x}_{k-1|k-1} + B_{k-1}u_{k-1} \\ & + \Sigma_{k|k} C_k' N_k^{-1}(z_k - C_k A_{k-1}\hat{x}_{k-1|k-1} - C_k B_{k-1} u_{k-1}).\end{aligned}$$

Equations (97)–(99) generating $\Sigma_{k|k}$ remain unchanged. Also if the mean of the initial state is nonzero, then the initial conditions (99) take the form

$$\hat{x}_{0|-1} = E\{x_0\}, \qquad \Sigma_{0|-1} = S.$$

Finally we note that (97) and (98) yield

$$\Sigma_{k+1|k} = A_k[\Sigma_{k|k-1} - \Sigma_{k|k-1}C_k'(C_k \Sigma_{k|k-1} C_k' + N_k)^{-1} C_k \Sigma_{k|k-1}]A_k' + M_k, \tag{101}$$

with the initial condition $\Sigma_{0|-1} = S$.

Equation (101) is a discrete-matrix Riccati equation of the type considered in Section 3.1. Thus when $A_k$, $C_k$, $N_k$, and $M_k$ are constant matrices

$$A_k = A, \qquad C_k = C, \qquad N_k = N, \qquad M_k = M, \qquad k = 0, 1, \ldots, N-1,$$

we have, by invoking the proposition proved there, that the solution of (101) tends to a positive definite matrix $\Sigma$ provided appropriate controllability and observability conditions hold. The conditions required are observability of the pair $(A, C)$ and controllability of the pair $(A, D)$, where $M = DD'$. When $k$ is large one may approximate the matrix $\Sigma_{k|k}$ in (110) by the constant matrix $\bar{\Sigma}$ to which $\Sigma_{k|k}$ converges as $k \to \infty$. We have from (98)

$$\bar{\Sigma} = \Sigma - \Sigma C'(C\Sigma C' + N)^{-1} C\Sigma,$$

and we may write (100) as

$$\hat{x}_{k|k} = A\hat{x}_{k-1|k-1} + \bar{\Sigma}C'N^{-1}(z_k - CA\hat{x}_{k-1|k-1}). \tag{102}$$

Since the "gain" matrix $(\bar{\Sigma}C'N^{-1})$ multiplying the "correction" term $(z_k - CA\hat{x}_{k-1|k-1})$ is independent of the index $k$, the implementation of the estimator (102) is considerably simplified.

## A.4 *Stability Aspects of the "Steady-State" Kalman Filtering Algorithm*

Let us consider now the stationary form of the Kalman filtering equations (95) and (96):

$$\hat{x}_{k+1|k} = A\hat{x}_{k|k-1} + A\Sigma C'(C\Sigma C' + N)^{-1}(z_k - C\hat{x}_{k|k-1}). \tag{103}$$

By using Eq. (103), the system equation

$$x_{k+1} = Ax_k + w_k,$$

and the measurement equation

$$z_k = Cx_k + v_k,$$

we obtain

$$e_{k+1} = [A - A\Sigma C'(C\Sigma C' + N)^{-1}C]e_k + w_k - A\Sigma C'(C\Sigma C' + N)^{-1}v_k, \tag{104}$$

where $e_k$ denotes for all $k$ the "one-step prediction" error

$$e_k = x_k - \hat{x}_{k|k-1}.$$

From the practical point of view it is important that the error equation (104) represents a stable system, i.e., the matrix

$$A - A\Sigma C'(C\Sigma C' + N)^{-1}C \tag{105}$$

is a stable matrix. This fact, however, is guaranteed under the observability and controllability assumptions given earlier since $\Sigma$ is the unique positive semidefinite symmetric solution of the algebraic Riccati equation

$$\Sigma = A[\Sigma - \Sigma C'(C\Sigma C' + N)^{-1}C\Sigma]A' + M$$

by the proposition proved in Section 3.1. Actually this proposition yields that the transpose of the matrix (105) is a stable matrix. This is, however, equivalent to the matrix (105) being a stable matrix, since for any matrix $D$ we have $D^k \to 0$ if and only if $D'^k \to 0$.

Having established the stability properties of the error equation (104) we now proceed to examine the stability properties of the equation governing the estimation error

$$\tilde{e}_k = x_k - \hat{x}_{k|k}.$$

We have by a straightforward calculation

$$\tilde{e}_k = [I - \Sigma C'(C\Sigma C' + N)^{-1}C]e_k - \Sigma C'(C\Sigma C' + N)^{-1}v_k. \tag{106}$$

By multiplying both sides of Eq. (104) by $[I - \Sigma C'(C\Sigma C' + N)^{-1}C]$ and using (106) we obtain

$$\begin{aligned}\tilde{e}_{k+1} &+ \Sigma C'(C\Sigma C' + N)^{-1}v_{k+1} \\ &= [A - \Sigma C'(C\Sigma C' + N)^{-1}CA][\tilde{e}_k + \Sigma C'(C\Sigma C' + N)^{-1}v_k] \\ &\quad + [I - \Sigma C'(C\Sigma C' + N)^{-1}C][w_k - A\Sigma C'(C\Sigma C' + N)^{-1}v_k],\end{aligned}$$

or equivalently

$$\begin{aligned}\tilde{e}_{k+1} &= [A - \Sigma C'(C\Sigma C' + N)^{-1}CA]\tilde{e}_k \\ &\quad + [I - \Sigma C'(C\Sigma C' + N)^{-1}C]w_k - \Sigma C'(C\Sigma C' + N)^{-1}v_{k+1}. \qquad (107)\end{aligned}$$

The stability of matrix (105) guarantees that the sequence $\{e_k\}$ generated by (104) tends to zero whenever the vectors $w_k$ and $v_k$ are identically zero for all $k$. Hence, by (106), the same is true for the sequence $\{\tilde{e}_k\}$. It follows from (107) that the matrix

$$A - \Sigma C'(C\Sigma C' + N)^{-1}CA \qquad (108)$$

is stable and hence the estimation error sequence $\{\tilde{e}_k\}$ is generated by a stable equation.

Let us consider now the stability properties of the $2n$-dimensional system of equations with state vector $[x_k', \hat{x}_k']'$:

$$x_{k+1} = Ax_k + BL\hat{x}_k, \qquad (109)$$

$$\hat{x}_{k+1} = \bar{\Sigma}C'N^{-1}CAx_k + (A + BL - \bar{\Sigma}C'N^{-1}CA)\hat{x}_k. \qquad (110)$$

This system was encountered at the end of Section 4.3.

We shall assume that the appropriate observability and controllability assumptions stated there are in effect. By using the equation

$$\bar{\Sigma}C'N^{-1} = \Sigma C'(C\Sigma C' + N)^{-1},$$

shown earlier, we obtain from (109) and (110) that

$$(x_{k+1} - \hat{x}_{k+1}) = [A - \Sigma C'(C\Sigma C' + N)^{-1}CA](x_k - \hat{x}_k).$$

In view of the stability of matrix (108) it follows that

$$\lim_{k\to\infty}(x_{k+1} - \hat{x}_{k+1}) = 0, \qquad (111)$$

for arbitrary initial states $x_0$ and $\hat{x}_0$. From (109) we obtain

$$x_{k+1} = (A + BL)x_k + BL(\hat{x}_k - x_k). \qquad (112)$$

Since in accordance with the theory of Sections 3.1 and 4.3 the matrix $(A + BL)$ is a stable matrix, it follows from (111) and (112) that we have

$$\lim_{k\to\infty} x_k = 0,$$

and hence by (111),

$$\lim_{k\to\infty} \hat{x}_k = 0.$$

Since the equations above hold for any initial states $x_0$ and $\hat{x}_0$ it follows that the system of equations (109) and (110) is a stable system, i.e., the matrix

$$\begin{bmatrix} A & BL \\ \bar{\Sigma}C'N^{-1}CA & A + BL - \bar{\Sigma}C'N^{-1}CA \end{bmatrix}$$

is a stable matrix.

### A.5 *Purely Linear Least-Squares Estimators*

Consider two jointly distributed random vectors $x$ and $y$ taking values in $R^n$ and $R^m$, respectively, as in Sections A.1 and A.2. Let us restrict attention to estimators of the form

$$x(y) = Ay,$$

which are purely linear (rather than linear plus constant). Similarly, as in Section A.2 we consider the problem of finding the optimal least-squares estimator within this restricted class, i.e., an estimator of the form

$$\hat{x}(y) = \hat{A}y$$

where $\hat{A}$ minimizes

$$f(A) = \underset{x,\,y}{E}\{\|x - Ay\|^2\}$$

over all $n \times m$ matrices $A$. We refer to such an estimator as a *purely linear least-squares estimator*. The derivation of the form of this estimator is very similar to the one of Section A.2.

By setting the derivative of $f$ with respect to $A$ equal to zero, we obtain

$$0 = \partial f/\partial A|_{\hat{A}} = 2\underset{x,\,y}{E}\{y(\hat{A}y - x)'\},$$

from which

$$\hat{A} = S_{xy}S_{yy}^{-1},$$

where

$$S_{xy} = E\{xy'\}, \qquad S_{yy} = E\{yy'\},$$

and we assume that $S_{yy}$ is invertible.

The purely linear least-squares estimator takes the form

$$\hat{x}(y) = S_{xy}S_{yy}^{-1}y. \tag{113}$$

The second moments of the corresponding error are given by

$$\underset{x,\,y}{E}\{[x - \hat{x}(y)][x - \hat{x}(y)]'\} = S_{xx} - S_{xy}S_{yy}^{-1}S_{yx}. \tag{114}$$

This equation is derived in an entirely similar manner as in Section A.2.

An advantage of the purely linear estimator is that it does not require knowledge of the means of $x$ and $y$. We only need to know the joint and individual second moments of $x$ and $y$. A price paid for this convenience, however, is that *the purely linear estimator yields in general biased estimates*, i.e., we may have

$$E\{\hat{x}(y)\} \neq E\{x\}.$$

By contrast, in the linear least-squares estimator examined earlier, we always have $E\{\hat{x}(y)\} = E\{x\}$ (cf. Corollary A.3.1).

As a final remark, we note that from Eqs. (68), (69) and (113), (114) it can be seen that the equations characterizing the purely linear estimator may be obtained from those of the linear estimator by setting $\bar{x} = 0$, $\bar{y} = 0$ and writing $S_{xx}$ and $S_{xy}$ in place of $\Sigma_{xx}$ and $\Sigma_{xy}$. As a result it is easy to see that there is an analog for the Kalman filtering algorithm corresponding to a purely linear least-squares estimator that is identical to the one described by Eqs. (95)–(99) and remains the same even if we do not assume that

$$E\{x_0\} = E\{w_k\} = E\{v_k\} = 0, \qquad k = 0, 1, \ldots,$$

provided that all given covariance matrices are replaced by the corresponding matrices of second moments.

## A.6 *Least-Squares Unbiased (Gauss–Markov) Estimators*

Let us assume that two random vectors $x \in R^n$ and $z \in R^m$ are related by means of the equation

$$z = Cx + v, \tag{115}$$

where $C$ is a given $m \times n$ matrix and $v$ a random measurement error vector uncorrelated with $x$, having known mean and covariance matrix

$$E\{v\} = \bar{v}, \qquad E\{(v - \bar{v})(v - \bar{v})'\} = \Sigma_{vv}. \tag{116}$$

The vector $z$ represents known measurements from which we wish to estimate the vector $x$. If the a priori probability distribution of $x$ is known then we may obtain a linear least-squares estimate of $x$ given $z$ by using the theory of Section A.2 (cf. Corollary A.3.7). In many cases, however, the probability distribution of $x$ is entirely unknown. In such cases it is possible to use the Gauss–Markov estimator, which we now describe.

Let us restrict attention to estimators of the form

$$x(z) = A(z - \bar{v}),$$

and seek an estimator of the form

$$\hat{x}(z) = \hat{A}(z - \bar{v}),$$

where $\hat{A}$ minimizes

$$f(A) = \underset{x,z}{E}\{\|x - A(z - \bar{v})\|^2\} \tag{117}$$

over all $n \times m$ matrices $A$. We have from (115)–(117) and the fact that $x$ and $v$ are uncorrelated,

$$\begin{aligned} f(A) &= \underset{x,v}{E}\{\|x - ACx - A(v - \bar{v})\|^2\} \\ &= \underset{x}{E}\{\|(I - AC)x\|^2\} + \underset{v}{E}\{(v - \bar{v})'A'A(v - \bar{v})\}, \end{aligned}$$

where $I$ is the $n \times n$ identity matrix. Since $f(A)$ depends on the unknown statistics of $x$, we see that the optimal matrix $\hat{A}$ will also depend on these statistics. We can circumvent this difficulty, however, by requiring that

$$AC = I.$$

Then our problem becomes

$$\begin{aligned} &\text{minimize} \quad E\{(v - \bar{v})'A'A(v - \bar{v})\} \\ &\text{subject to} \quad AC = I. \end{aligned} \tag{118}$$

Notice that the requirement $AC = I$ is equivalent to requiring that the estimator $x(z) = A(z - \bar{v})$ be *unbiased* in the sense that

$$E\{x(z)\} = E\{x\} = \bar{x} \qquad \forall \bar{x} \in R^n.$$

This can be seen by writing

$$E\{x(z)\} = E\{A(Cx + v - \bar{v})\} = AC\,E\{x\} = AC\bar{x} = \bar{x}.$$

There remains the task of solving problem (118). Let $a_i'$ denote the $i$th row of $A$. We have

$$\begin{aligned} (v - \bar{v})'A'A(v - \bar{v}) &= (v - \bar{v})'[a_1 \cdots a_n]\begin{bmatrix} a_1' \\ \vdots \\ a_n' \end{bmatrix}(v - \bar{v}) \\ &= \sum_{i=1}^{n}(v - \bar{v})'a_i a_i'(v - \bar{v}) = \sum_{i=1}^{n} a_i'(v - \bar{v})(v - \bar{v})'a_i. \end{aligned}$$

Hence, problem (118) can also be written

$$\begin{aligned} &\text{minimize} \quad \sum_{i=1}^{n} a_i'\Sigma_{vv}a_i \\ &\text{subject to} \quad C'a_i = e_i, \qquad i = 1, \ldots, n, \end{aligned}$$

where $e_i$ is the $i$th column of the identity matrix. The minimization can be carried out separately for each $i$, yielding

$$\hat{a}_i = \Sigma_{vv}^{-1} C (C' \Sigma_{vv}^{-1} C)^{-1} e_i, \qquad i = 1, \ldots, n,$$

and finally

$$\hat{A} = (C' \Sigma_{vv}^{-1} C)^{-1} C' \Sigma_{vv}^{-1},$$

where we assume that the inverses of $\Sigma_{vv}$ and $C' \Sigma_{vv}^{-1} C$ exist.

Thus, the Gauss–Markov estimator is given by

$$\hat{x}(z) = (C' \Sigma_{vv}^{-1} C)^{-1} C' \Sigma_{vv}^{-1} (z - \bar{v}). \tag{119}$$

Let us also calculate the corresponding error covariance matrix. We have

$$\begin{aligned} E\{[x - \hat{x}(z)][x - \hat{x}(z)]'\} &= E\{[x - \hat{A}(z - \bar{v})][x - \hat{A}(z - \bar{v})]'\} \\ &= E\{\hat{A}(v - \bar{v})(v - \bar{v})' \hat{A}'\} = \hat{A} \Sigma_{vv} \hat{A}' \\ &= (C' \Sigma_{vv}^{-1} C)^{-1} C' \Sigma_{vv}^{-1} \Sigma_{vv} \Sigma_{vv}^{-1} C (C' \Sigma_{vv}^{-1} C)^{-1}, \end{aligned}$$

and finally

$$E\{[x - \hat{x}(z)][x - \hat{x}(z)]'\} = (C' \Sigma_{vv}^{-1} C)^{-1}. \tag{120}$$

Finally, let us compare the Gauss–Markov estimator with the linear least-squares estimator of Corollary A.3.7. Whenever $\Sigma_{xx}$ is invertible the estimator of Corollary A.3.7 can also be written

$$\hat{x}(z) = \bar{x} + (\Sigma_{xx}^{-1} + C' \Sigma_{vv}^{-1} C)^{-1} C' \Sigma_{vv}^{-1} (z - C\bar{x} - \bar{v}). \tag{121}$$

This fact may be verified by straightforward calculation. By comparing Eqs. (119) and (121) we see that the Gauss–Markov estimator is obtained from the linear least-squares estimator by setting $\bar{x} = 0$ and $\Sigma_{xx}^{-1} = 0$, i.e., a zero mean and infinite covariance for the unknown random variable $x$. In this manner, the Gauss–Markov estimator may be viewed as a limiting form of the linear least-squares estimator. The error covariance matrix (120) of the Gauss–Markov estimator can also be obtained in the same manner from the error covariance matrix of the linear least-squares estimator.

## A.7 *Deterministic Least-Squares Estimation*

As in the case of Gauss–Markov estimation, let us consider vectors $x \in R^n$, $z \in R^m$, and $v \in R^m$ related by means of the equation

$$z = Cx + v,$$

where $C$ is a known $m \times n$ matrix. The vector $z$ represents known measurements from which we wish to estimate the vector $x$. However, we know nothing about the probability distribution of $x$ and $v$ and we are thus unable to utilize an estimator based on statistical considerations. Under these circumstances it is reasonable to select as our estimate the vector $\hat{x}$ that minimizes

$$f(x) = \|z - Cx\|^2,$$

i.e., the estimate that fits best the data in a least-squares sense. This estimate will, of course, depend on $z$ and will be denoted $\hat{x}(z)$.

By setting the gradient of $f$ at $\hat{x}(z)$ equal to zero, we obtain

$$\nabla f\,|_{\hat{x}(z)} = 2C'[C\hat{x}(z) - z] = 0,$$

from which

$$\hat{x}(z) = (C'C)^{-1}C'z, \tag{122}$$

provided $C'C$ is an invertible matrix.

An interesting observation is that the estimate (122) is the same as the Gauss–Markov estimate given by (119) provided the measurement error has zero mean and covariance matrix equal to the identity matrix, i.e., $\bar{v} = 0$, $\Sigma_{vv} = I$. In fact, if instead of $\|z - Cx\|^2$ we minimize

$$(z - \bar{v} - Cx)'\Sigma_{vv}^{-1}(z - \bar{v} - Cx),$$

then the deterministic least-squares estimate obtained would be identical to the Gauss–Markov estimate. If instead of $\|z - Cx\|^2$ we minimize

$$(x - \bar{x})'\Sigma_{xx}^{-1}(x - \bar{x}) + (z - \bar{v} - Cx)'\Sigma_{vv}^{-1}(z - \bar{v} - Cx),$$

then the estimate obtained would be identical to the linear least-squares estimate given by (121). Thus, we arrive at the interesting conclusion that the estimators obtained earlier on the basis of a stochastic optimization framework can also be obtained by minimization of a deterministic measure of fitness of estimated parameters to the data at hand.

*Chapter 5*

# Computational Aspects of Dynamic Programming—Suboptimal Control

## 5.1 The Curse of Dimensionality

Consider the DP algorithm for the basic problem

$$J_N(x_N) = g_N(x_N) \tag{1}$$

$$J_k(x_k) = \inf_{u_k \in U_k(x_k)} \underset{w_k}{E} \{g_k(x_k, u_k, w_k) + J_{k+1}[f_k(x_k, u_k, w_k)]\}. \tag{2}$$

As we have seen earlier it is possible in some cases of interest to obtain a closed-form solution to this algorithm or at least use the algorithm for the analysis of properties of the optimal policy. However, such cases tend to be the exception rather than the rule. In most cases it is necessary to solve numerically the DP equations in order to obtain an optimal policy. The computational requirements for doing so are often staggering to the point that for many problems a complete solution of the problem by DP is unthinkable at least with presently existing computers. The reason lies in what Bellman has called the "curse of dimensionality," which arises particularly when the state space is an infinite set. Consider, for example, a problem in which the state space is $R^n$ and the control space $R^m$. In order to obtain the function $J_{N-1}(x_{N-1})$ it is necessary first to discretize the state space. Taking, for example, 100 discretization points per axis results in a grid with $100^n$ points. For each of those points now, the minimization of the right-hand side of (2) must be carried out numerically. Each minimization is carried over $U_k(x_k)$ or over a grid of points covering $U_k(x_k)$—a nontrivial problem. Matters are further complicated by the requirement to carry out a numerical integration (the expectation operation) every time the function under minimization is evaluated, and by the possible presence of nondifferentiabilities introduced by discretization and interpolation. Computer storage also presents an acute problem. Thus for problems with Euclidean state and control spaces, DP can be applied only if the dimension of these spaces is very small. When the control space is one-dimensional, things are sometimes greatly simplified, since in this case effective one-dimensional minimization techniques such as the Fibonacci search [L10] may be used. Often the special structure of the problem can also be exploited to reduce the computational requirements. In other cases special computational algorithms can be used that take advantage of the particular features of the problem. This is particularly true of infinite-horizon problems.

Various devices have been suggested to help overcome the computational and storage problem, particularly for deterministic problems, such as the so-called coarse grid approach, the use of Lagrange multipliers, Legendre polynomial approximations, and specialized techniques [B4, K6, L2, L9, N2, W10]. These devices, though helpful for some problems, should be considered only as partial remedies, and in any case it is not our intention to discuss them at any length in this text. Instead we shall provide in the next section a simple discretization procedure for problems where an infinite state space is involved and we shall prove the validity of this procedure under certain reasonable assumptions. Subsequently we shall discuss various techniques for obtaining suboptimal control laws that are computationally efficient and hence of interest from a practical point of view.

## 5.2 Discretization Procedures and Their Convergence

In this section we consider a procedure for discretizing a DP algorithm defined over an infinite state space. The procedure is not necessarily the most efficient computationally. It is, however, used widely (perhaps with slight modifications) and is simple and straightfoward. The basic idea is to *approximate the "cost-to-go" functions of the DP algorithm as well as the corresponding policies by piecewise constant functions.* We consider problems where the state spaces are compact subsets of Euclidean spaces and we introduce certain continuity, compactness, and Lipschitz assumptions. Under these assumptions the discretization procedure is shown to be stable in the sense that it yields suboptimal policies whose performance approximates arbitrarily closely the optimal as the discretization grids become finer and finer. The analysis and proofs are straightforward but rather tedious and may be skipped by the less mathematically inclined or otherwise uninterested reader.

Consider the following DP algorithm:

$$J_N(x) = g_N(x), \qquad x \in S_N \subset R^{s_N} \tag{3}$$

$$J_k(x) = \sup_{u \in U_k(x)} \mathop{E}_{w} \{g_k(x, u, w) + J_{k+1}[f_k(x, u, w)]\},$$
$$x \in S_k \subset R^{s_k}, \quad k = 0, 1, \ldots, N-1. \tag{4}$$

This algorithm is associated with the basic problem with perfect state information involving the discrete-time dynamic system

$$x_{k+1} = f_k(x_k, u_k, w_k), \qquad k = 0, 1, \ldots, N-1, \tag{5}$$

with given initial state $x_0$ and the cost functional

$$E\left\{g_N(x_N) + \sum_{k=0}^{N-1} g_k(x_k, u_k, w_k)\right\}.$$

For the purpose of analytical convenience we are considering here maximization of the objective function rather than minimization. This, of course, does not affect the discretization procedures or the convergence results to be obtained. In the above equations the system state $x_k$ is an element of a Euclidean space $R^{s_k}$, $k = 0, 1, \ldots, N$. Algorithm (3) and (4) is defined over given *compact* subsets $S_k \subset R^{s_k}$, $k = 0, 1, \ldots, N-1$. The control input $u_k$ is an element of some space $C_k$, $k = 0, 1, \ldots, N-1$. In what follows we shall assume that $C_k$ is either a subset of a Euclidean space or a finite set.

The input disturbance $w_k$ is assumed to be an element of a set $W_k$, $k = 0, 1, \ldots, N-1$. We assume for simplicity that *each set $W_k$ has a finite number* (*say* $I_k$) *of elements*. This assumption is valid in many problems of interest, most notably in deterministic problems where the set $W_k$ consists of a single element. In problems where the sets $W_k$ are infinite, our assumption amounts to replacing the DP algorithm (3) and (4) by another algorithm whereby the expected value (integral) in Eq. (4) is approximated by a finite sum. For most problems of interest this finite sum approximation may be justified in the sense that the resulting error can be made arbitrarily small by taking a sufficiently large number of terms in the finite sum. The reader may also provide similar assumptions as the one made here under which the approximation is valid in the above sense and extend the result of this section to cases where $W_k$ are compact sets in Euclidean spaces. We shall denote the probabilities of the elements of $W_k$ by $p_k^i(x_k, u_k)$, $i = 1, \ldots, I_k$.

We shall consider two different sets of assumptions in addition to those already made. In the first set of assumptions the control space $C_k$ is assumed to be a finite set for each $k$. In the second set of assumptions the control space $C_k$ is assumed to be a Euclidean space in which case discretization of both the state space and the control space is required. The reader may easily extend our analysis and results to cases where the control space is the union or the Cartesian product of a finite set and a Euclidean space.

**Assumptions A**

**A.1** The control spaces $C_k$, $k = 0, 1, \ldots, N-1$, are finite sets and

$$U_k(x) = C_k \qquad \forall x \in S_k, \quad k = 0, 1, \ldots, N-1. \tag{6}$$

**A.2** The functions $f_k$ and $g_k$ satisfy the following conditions:

$$\|f_k(x, u, w) - f_k(x', u, w)\| \leqslant L_k\|x - x'\| \qquad \forall x, x' \in S_k, \quad u \in C_k, \quad w \in W_k, \quad k = 0, 1, \ldots, N-1, \tag{7}$$

$$|g_k(x, u, w) - g_k(x', u, w)| \leqslant M_k\|x - x'\| \qquad \forall x, x' \in S_k, \quad u \in C_k, \quad w \in W_k, \quad k = 0, 1, \ldots, N-1, \tag{8}$$

$$|g_N(x) - g_N(x')| \leqslant M_N\|x - x'\| \qquad \forall x, x' \in S_N, \tag{9}$$

where $M_N, M_k, L_k$, $k = 0, 1, \ldots, N-1$ are some positive constants and $\|\cdot\|$ denotes the usual Euclidean norm.

**A.3** The probabilities $p_k^i(x, u)$, $i = 1, 2, \ldots, I_k$, of the elements of the finite set $W_k = \{1, 2, \ldots, I_k\}$ satisfy the condition

$$|p_k^i(x, u) - p_k^i(x', u)| \leqslant N_k\|x - x'\| \qquad \forall x, x' \in S_k, \quad u \in C_k, \quad i \in W_k, \quad k = 0, 1, \ldots, N-1, \tag{10}$$

where $N_k$, $k = 0, \ldots, N-1$, are some positive constants. (This assumption is satisfied in particular if the probabilities $p_k^i$ do not depend on the state.)

**Assumptions B**

**B.1** The control space $C_k$, $k = 0, 1, \ldots, N-1$, is a compact subset of a Euclidean space. The sets $U_k(x)$ are compact for every $x \in S_k$, and in addition the set

$$U_k = \bigcup_{x \in S_k} U_k(x), \qquad k = 0, 1, \ldots, N-1, \tag{11}$$

is compact. Furthermore, the sets $U_k(x)$ satisfy

$$U_k(x) \subset U_k(x') + \{u \mid \|u\| \leqslant P_k\|x - x'\|\} \qquad \forall x, x' \in S_k, \quad k = 0, 1, \ldots, N-1, \tag{12}$$

where $P_k$ are some positive constants.†

† The advanced reader may verify that (12) is equivalent to assuming that the point-to-set map $x \to U_k(x)$ is Lipschitz continuous in the Hausdorff metric sense.

**B.2** The functions $f_k$, $g_k$ satisfy the following conditions:

$$\|f_k(x, u, w) - f_k(x', u', w)\| \leqslant \bar{L}_k(\|x - x'\| + \|u - u'\|) \qquad \forall x, x' \in S_k,$$
$$u, u' \in U_k, \quad w \in W_k, \quad k = 0, 1, \ldots, N-1, \tag{13}$$

$$|g_k(x, u, w) - g_k(x', u', w)| \leqslant \bar{M}_k(\|x - x'\| + \|u - u'\|) \qquad \forall x, x' \in S_k,$$
$$u, u' \in U_k, \quad w \in W_k, \quad k = 0, 1, \ldots, N-1, \tag{14}$$

$$|g_N(x) - g_N(x')| \leqslant \bar{M}_N\|x - x'\| \qquad \forall x, x' \in S_N, \tag{15}$$

where $\bar{M}_N$, $\bar{M}_k$, $\bar{L}_k$, $k = 0, 1, \ldots, N-1$, are some positive constants.

**B.3** The probabilities $p_k^i(x, u)$, $i = 1, \ldots, I_k$, of the elements of the finite set $W_k = \{1, 2, \ldots, I_k\}$ satisfy the condition

$$|p_k^i(x, u) - p_k^i(x', u')| \leqslant \bar{N}_k(\|x - x'\| + \|u - u'\|) \qquad \forall x, x' \in S_k,$$
$$u, u' \in U_k, \quad i \in W_k, \quad k = 0, 1, \ldots, N-1, \tag{16}$$

where $\bar{N}_k$, $k = 0, 1, \ldots, N-1$, are some positive constants.

Prior to considering discretization of the dynamic programming algorithm we establish the following property of the "cost-to-go" functions $J_k: S_k \to R$ of (3) and (4).

**Proposition 1** Under Assumptions A or B the functions $J_k: S_k \to R$, given by (3) and (4) satisfy

$$|J_k(x) - J_k(x')| \leqslant A_k\|x - x'\| \qquad \forall x, x' \in S_k, \quad k = 0, 1, \ldots, N, \tag{17}$$

where $A_k$, $k = 0, 1, \ldots, N$, are some positive constants.

*Proof* Under Assumptions A we have by (9) that (17) holds for $k = N$ with $A_N = M_N$. For $k = N-1$ we have that for each $x, x' \in S_{N-1}$,

$$\begin{aligned}
&|J_{N-1}(x) - J_{N-1}(x')| \\
&\quad = \Bigg| \max_{u \in C_{N-1}} \sum_{i=1}^{I_{N-1}} \{g_{N-1}(x, u, i) + J_N[f_{N-1}(x, u, i)]\} p_{N-1}^i(x, u) \\
&\qquad - \max_{u \in C_{N-1}} \sum_{i=1}^{I_{N-1}} \{g_{N-1}(x', u, i) + J_N[f_{N-1}(x', u, i)]\} p_{N-1}^i(x', u) \Bigg| \\
&\quad \leqslant \max_{u \in C_{N-1}} \left| \sum_{i=1}^{I_{N-1}} [g_{N-1}(x, u, i) p_{N-1}^i(x, u) - g_{N-1}(x', u, i) p_{N-1}^i(x', u)] \right| \\
&\qquad + \max_{u \in C_{N-1}} \Bigg| \sum_{i=1}^{I_{N-1}} [J_N[f_{N-1}(x, u, i)] p_{N-1}^i(x, u) \\
&\qquad - J_N[f_{N-1}(x', u, i)] p_{N-1}^i(x', u)] \Bigg| .
\end{aligned}$$

Now we use the fact that if $\alpha: S \to R$, $\beta: S \to R$ are real-valued functions over a compact subset $S$ of a Euclidean space satisfying for some constants $\mu_\alpha$, $\mu_\beta$, and all $t_1, t_2 \in S$,

$$|\alpha(t_1) - \alpha(t_2)| \leqslant \mu_\alpha \|t_1 - t_2\|, \qquad |\beta(t_1) - \beta(t_2)| \leqslant \mu_\beta \|t_1 - t_2\|,$$

then the product function $\alpha(\cdot)\beta(\cdot)$ satisfies for all $t_1, t_2 \in S$

$$|\alpha(t_1)\beta(t_1) - \alpha(t_2)\beta(t_2)| \leqslant \left[\mu_\alpha \max_{t\in S} |\beta(t)| + \mu_\beta \max_{t \in S} |\alpha(t)|\right] \|t_1 - t_2\|. \qquad (18)$$

Then the earlier estimate is strengthened to yield

$$|J_{N-1}(x) - J_{N-1}(x')| \leqslant A_{N-1}\|x - x'\| \qquad \forall x, x' \in S_{N-1},$$

where $A_{N-1}$ is given by

$$A_{N-1} = I_{N-1}(M_{N-1} + L_{N-1}A_N + B_{N-1}N_{N-1}),$$

$$B_{N-1} = \max\{|g_{N-1}(x, u, w)| \,|\, x \in S_{N-1}, u \in C_{N-1}, w \in W_{N-1}\}$$
$$+ \max\{|J_N[f_{N-1}(x, u, w)]| \,|\, x \in S_{N-1}, u \in C_{N-1}, w \in W_{N-1}\}.$$

Thus the result is proved for $k = N - 1$ and similarly it is proved for every $k$.

We turn now to proving the result under Assumptions B. Again the result holds for $k = N$ with $A_N = \overline{M}_N$. For $k = N - 1$ we have for each $x, x' \in S_{N-1}$,

$$|J_{N-1}(x) - J_{N-1}(x')|$$
$$= \left| \max_{u \in U_{N-1}(x)} \sum_{i=1}^{I_{N-1}} \{g_{N-1}(x, u, i) + J_N[f_{N-1}(x, u, i)]\} p^i_{N-1}(x, u) \right.$$
$$\left. - \max_{u \in U_{N-1}(x')} \sum_{i=1}^{I_{N-1}} \{g_{N-1}(x', u, i) + J_N[f_{N-1}(x', u, i)]\} p^i_{N-1}(x', u) \right|.$$

Now using (18), B.2, B.3, and the above equality it is straightforward to show that

$$|J_{N-1}(x) - J_{N-1}(x')|$$
$$\leqslant \max_{u \in U_{N-1}(x) \cup U_{N-1}(x')} \left| \sum_{i=1}^{I_{N-1}} [g_{N-1}(x, u, i) p^i_{N-1}(x, u) \right.$$
$$\left. - g_{N-1}(x', u, i) p^i_{N-1}(x', u)] \right|$$
$$+ \max_{u \in U_{N-1}(x) \cup U_{N-1}(x')} \left| \sum_{i=1}^{I_{N-1}} [J_{N-1}[f_{N-1}(x, u, i)] p^i_{N-1}(x, u) \right.$$
$$\left. - J_{N-1}[f_{N-1}(x', u, i)] p^i_{N-1}(x', u)] \right|$$
$$+ 2I_{N-1}(\overline{M}_{N-1} + \overline{L}_{N-1}A_N + \overline{B}_{N-1}\overline{N}_{N-1})P_{N-1}\|x - x'\|,$$

where

$$\bar{B}_{N-1} = \max\{|g_{N-1}(x, u, w)| \,|\, x \in S_{N-1}, u \in U_{N-1}, w \in W_{N-1}\}$$
$$+ \max\{|J_N[f_{N-1}(x, u, w)]| \,|\, x \in S_{N-1}, u \in U_{N-1}, w \in W_{N-1}\}. \quad (19)$$

Strengthening the above estimate and using (18) and our assumptions we obtain

$$|J_{N-1}(x) - J_{N-1}(x')| \leqslant A_{N-1}\|x - x'\|,$$

where

$$A_{N-1} = I_{N-1}(1 + 2P_{N-1})(\bar{M}_{N-1} + \bar{L}_{N-1}A_N + \bar{B}_{N-1}\bar{N}_{N-1}),$$

and the result is proved for $k = N - 1$. Similarly the result is proved under Assumptions B for all $k$. Q.E.D.

We now proceed to describe procedures for discretizing algorithm (3) and (4) under Assumptions A and B.

*Discretization Procedure under Assumptions A*

We partition each set $S_k$ into $n_k$ mutually disjoint sets $S_k^1, S_k^2, \ldots, S_k^{n_k}$ such that $S_k = \bigcup_{i=1}^{n_k} S_k^i$, and select arbitrary points $x_k^i \in S_k^i$, $i = 1, \ldots, n_k$. We approximate the DP algorithm (3) and (4) by the following algorithm, which is defined on the finite grids $G_k$, where

$$G_k = \{x_k^1, x_k^2, \ldots, x_k^{n_k}\}, \qquad k = 0, 1, \ldots, N. \quad (20)$$

We have

$$\hat{J}_N(x) = \begin{cases} g_N(x) & \text{if} \quad x \in G_N, & (21)\\ g_N(x_N^i) & \text{if} \quad x \in S_N^i, \quad i = 1, 2, \ldots, n_N, & (22)\end{cases}$$

$$\hat{J}_k(x) = \begin{cases} \max\limits_{u \in C_k} \mathop{E}\limits_{w} \{g_k(x, u, w) + \hat{J}_{k+1}[f_k(x, u, w)]\} \quad \text{if} \quad x \in G_k, & (23)\\ \hat{J}_k(x_k^i) \quad \text{if} \quad x \in S_k^i, \quad i = 1, 2, \ldots, n_k, \quad k = 0, 1, \ldots, N-1. & (24)\end{cases}$$

The algorithm above corresponds to computing the "cost-to-go" functions $\hat{J}_k$ on the finite grid by means of the DP algorithm (21) and (23), and extending their definition on the whole compact set $S_k$ by making them constant on each section $S_k^i$ of $S_k$ (see Fig. 5.1). Thus $\hat{J}_k$ may be viewed as a piecewise constant approximation of $J_k$. An alternative way of viewing the discretized algorithm (21) and (23) is to observe that it corresponds to a stochastic optimal control problem involving a certain finite state system (defined over the finite state spaces $G_0, \ldots, G_N$) and an appropriately reformulated cost functional.

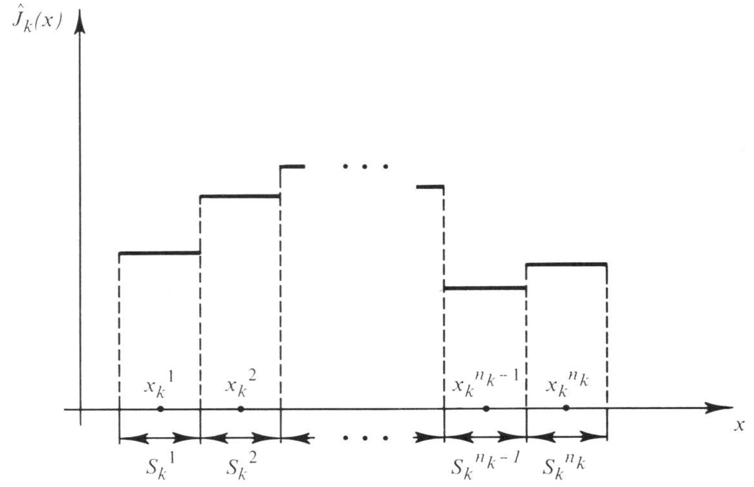


FIGURE 5.1

Carrying out the DP algorithm (21) and (23) involves a finite number of operations. Simultaneously we obtain an optimal control law as a sequence of functions $\hat{\mu}_k: G_k \to C_k$:

$$\hat{\mu}_0(x), \quad \hat{\mu}_1(x), \quad \ldots, \quad \hat{\mu}_{N-1}(x),$$

defined on the respective grids $G_k$, $k = 0, \ldots, N-1$, where $\hat{\mu}_k(x_k^i)$ maximizes the right-hand side of (23) when $x = x_k^i$, $i = 1, 2, \ldots, n_k$. We extend the definition of this control law over the whole state space by defining for every $x \in S_k$, $k = 0, 1, \ldots, N-1$,

$$\mu_k(x) = \hat{\mu}_k(x_k^i) \qquad \text{if} \quad x \in S_k^i, \quad i = 1, \ldots, n_k. \tag{25}$$

Thus we obtain a piecewise constant control law $\{\mu_0, \mu_1, \ldots, \mu_{N-1}\}$ defined over the whole space (see Fig. 5.2). The value of the cost functional corresponding to $\{\mu_0, \mu_1, \ldots, \mu_{N-1}\}$ is denoted $\tilde{J}_0(x_0)$ and is obtained by the last step of the algorithm

$$\tilde{J}_N(x) = g_N(x), \qquad x \in S_N \tag{26}$$

$$\tilde{J}_k(x) = \underset{w}{E}\{g[x, \mu_k(x), w] + \tilde{J}_{k+1}[f_k(x, \mu_k(x), w)]\}$$
$$x \in S_k, \quad k = 0, 1, \ldots, N-1. \tag{27}$$

Denote by $d_s$ the maximum "radius" of the sets $S_k^i$:

$$d_s = \max_{k=0,1,\ldots,N} \ \max_{i=1,\ldots,n_k} \ \sup_{x \in S_k^i} \|x - x_k^i\|. \tag{28}$$

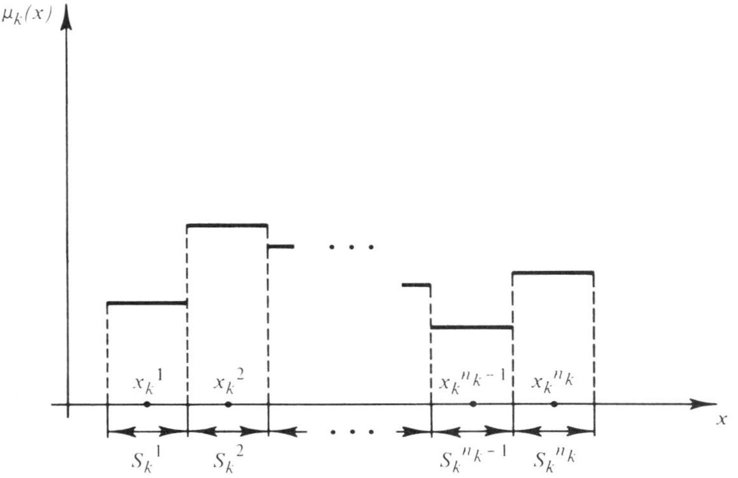


FIGURE 5.2

We shall be interested in whether $\hat{J}_k$ and $\tilde{J}_k$ converge in some sense to $J_k$ for each $k$ as $d_s$ tends to zero.

*Discretization Procedure under Assumptions B*

Here the state spaces $S_k$ are discretized in the same way as under Assumptions A. In addition, finite grids $H_k$ of points in $U_k$ are selected:

$$H_k = \{u_k^1, \ldots, u_k^{p_k}\} \subset U_k, \qquad k = 0, 1, \ldots, N-1. \tag{29}$$

We assume that

$$U_k(x_k^i) \cap H_k \neq \varnothing \qquad \forall i = 1, \ldots, n_k, \quad k = 0, 1, \ldots, N-1, \tag{30}$$

where $\varnothing$ denotes the empty set. We now approximate algorithm (3) and (4), by the following algorithm:

$$\hat{J}_N(x) = \begin{cases} g_N(x) & \text{if } x \in G_N, & (31) \\ g_N(x_N^i) & \text{if } x \in S_N^i, \quad i = 1, \ldots, n_N, & (32) \end{cases}$$

$$\hat{J}_k(x) = \begin{cases} \max\limits_{u \in U_k(x) \cap H_k} \underset{w}{E}\{g_k(x, u, w) + \hat{J}_{k+1}[f_k(x, u, w)]\} & \text{if } x \in G_k, & (33) \\ \hat{J}_k(x_k^i) \quad \text{if } x \in S_k^i, \quad i = 1, 2, \ldots, n_k, \quad k = 0, 1, \ldots, N-1. & & (34) \end{cases}$$

We obtain a control law $\{\hat{\mu}_0, \ldots, \hat{\mu}_{N-1}\}$ defined on the grids $G_k$ that can be extended over the whole space in a number of ways (see Problem 8). We consider the piecewise constant extension given by (25) which is admissible if $U_k(x)$ is constant, i.e., $U_k(x) \equiv U_k$ for some $U_k \subset C$. Even if it is not admissible, the extension (25) is often acceptable for practical purposes. The corresponding value $\tilde{J}_0(x_0)$ of the cost functional is given by (26) and (27).

Again we are interested in the question whether $\hat{J}_k$ and $\tilde{J}_k$ converge in some sense to $J_k$ for each $k$ as both $d_s$ and $d_c$ tend to zero, where

$$d_s = \max_{k=0,1,\ldots,N} \max_{i=1,\ldots,n_k} \sup_{x \in S_k^i} \|x - x_k^i\|, \tag{28'}$$

$$d_c = \max_{k=0,1,\ldots,N-1} \max_{i=1,\ldots,n_k} \max_{u \in U_k(x_k^i)} \min_{u' \in U_k(x_k^i) \cap H_k} \|u - u'\|. \tag{35}$$

This question is answered in the affirmative in the next proposition.

**Proposition 2** There exist positive constants $\alpha_k$, $\beta_k$, $k = 0, 1, \ldots, N$ (independent of the grids $G_0, \ldots, G_N, H_0, \ldots, H_{N-1}$ used in the discretization procedure) such that under Assumptions A

$$|J_k(x) - \hat{J}_k(x)| \leqslant \alpha_k d_s \qquad \forall x \in S_k, \quad k = 0, 1, \ldots, N, \tag{36}$$

$$|J_k(x) - \tilde{J}_k(x)| \leqslant \alpha_k d_s \qquad \forall x \in S_k, \quad k = 0, 1, \ldots, N, \tag{37}$$

and under Assumptions B

$$|J_k(x) - \hat{J}_k(x)| \leqslant \beta_k(d_s + d_c) \qquad \forall x \in S_k, \quad k = 0, 1, \ldots, N, \tag{38}$$

$$|J_k(x) - \tilde{J}_k(x)| \leqslant \beta_k(d_s + d_c) \qquad \forall x \in S_k, \quad k = 0, 1, \ldots, N, \tag{39}$$

where $J_k$, $\hat{J}_k$, $\tilde{J}_k$, $d_s$, $d_c$ are given by (3) and (4), (21)–(24) [or (31)–(34)], (26) and (27), (28) and (35), respectively.

*Proof* We first prove the proposition under Assumptions A. We have by (21) and (22) that $J_N(x) = \hat{J}_N(x)$ for all $x \in G_N$ while for any $x \in S_N^i$,

$$|J_N(x) - \hat{J}_N(x)| = |g_N(x) - g_N(x_N^i)| \leqslant M_N \|x - x_N^i\| \leqslant M_N d_s. \tag{40}$$

Hence (36) holds for $k = N$ with $\alpha_N = M_N$. Also $J_N(x) = \tilde{J}_N(x)\ \forall x \in S_N$, and hence (37) also holds for $k = N$. To prove (36) for $k = N - 1$ we have by (23) for any $i = 1, 2, \ldots, n_{N-1}$,

$$\begin{aligned}
&|J_{N-1}(x_{N-1}^i) - \hat{J}_{N-1}(x_{N-1}^i)| \\
&= \Big| \max_{u \in C_{N-1}} \underset{w}{E} \{g_{N-1}(x_{N-1}^i, u, w) + J_N[f_{N-1}(x_{N-1}^i, u, w)]\} \\
&\qquad - \max_{u \in C_{N-1}} \underset{w}{E} \{g_{N-1}(x_{N-1}^i, u, w) + \hat{J}_N[f_{N-1}(x_{N-1}^i, u, w)]\} \Big| \\
&\leqslant \max_{u \in C_{N-1}} \Big| \underset{w}{E} \{J_N[f_{N-1}(x_{N-1}^i, u, w)] - \hat{J}_N[f_{N-1}(x_{N-1}^i, u, w)]\} \Big| \leqslant \alpha_N d_s,
\end{aligned} \tag{41}$$

where the last step follows by (40). Also for any $x \in S^i_{N-1}$, $i = 1, \ldots, n_{N-1}$, we have using (41) and Proposition 1,

$$\begin{aligned}
&|J_{N-1}(x) - \hat{J}_{N-1}(x)| \\
&\quad = |J_{N-1}(x) - \hat{J}_{N-1}(x^i_{N-1})| \\
&\quad \leqslant |J_{N-1}(x) - J_{N-1}(x^i_{N-1})| + |J_{N-1}(x^i_{N-1}) - \hat{J}_{N-1}(x^i_{N-1})| \\
&\quad \leqslant A_{N-1}\|x - x^i_{N-1}\| + \alpha_N d_s \leqslant (A_{N-1} + \alpha_N) d_s.
\end{aligned}$$

Hence (36) holds for $k = N - 1$ with $\alpha_{N-1} = A_{N-1} + \alpha_N$, and similarly it is shown to hold for all $k$.

To prove (37) for $k = N - 1$ let $x \in S^i_{N-1}$. We have by (24) and the previous inequality,

$$\begin{aligned}
|J_{N-1}(x) - \tilde{J}_{N-1}(x)| &\leqslant |J_{N-1}(x) - \hat{J}_{N-1}(x)| + |\hat{J}_{N-1}(x) - \tilde{J}_{N-1}(x)| \\
&\leqslant (A_{N-1} + \alpha_N) d_s + |\hat{J}_{N-1}(x^i_{N-1}) - \tilde{J}_{N-1}(x)|. \quad (42)
\end{aligned}$$

For notational convenience we write $\mu_{N-1}(x) = \mu_{N-1}(x^i_{N-1}) = \mu^i_{N-1}$ [cf. (25)]. By using (23), (27), and (18), we have

$$\begin{aligned}
&|\hat{J}_{N-1}(x^i_{N-1}) - \tilde{J}_{N-1}(x)| \\
&\quad \leqslant \left| \sum_{j=1}^{I_{N-1}} g_{N-1}(x^i_{N-1}, \mu^i_{N-1}, j) p^j_{N-1}(x^i_{N-1}, \mu^i_{N-1}) \right. \\
&\qquad \left. - \sum_{j=1}^{I_{N-1}} g_{N-1}(x, \mu^i_{N-1}, j) p^j_{N-1}(x, \mu^i_{N-1}) \right| \\
&\qquad + \left| \sum_{j=1}^{I_{N-1}} \hat{J}_N[f_{N-1}(x^i_{N-1}, \mu^i_{N-1}, j)] p^j_{N-1}(x^i_{N-1}, \mu^i_{N-1}) \right. \\
&\qquad \left. - \sum_{j=1}^{I_{N-1}} \tilde{J}_N[f_{N-1}(x, \mu^i_{N-1}, j)] p^j_{N-1}(x, \mu^i_{N-1}) \right| \\
&\quad \leqslant I_{N-1}(M_{N-1} + N_{N-1} \max\{|g_{N-1}(x, u, w)| \mid x \in S_{N-1}, u \in C_{N-1}, \\
&\qquad w \in W_{N-1}\}) \|x - x^i_{N-1}\| + \left| \sum_{j=1}^{I_{N-1}} \{\hat{J}_N[f_{N-1}(x^i_{N-1}, \mu^i_{N-1}, j)] \right. \\
&\qquad \left. - J_N[f_{N-1}(x^i_{N-1}, \mu^i_{N-1}, j)]\} p^j_{N-1}(x^i_{N-1}, \mu^i_{N-1}) \right| \\
&\qquad + \left| \sum_{j=1}^{I_{N-1}} J_N[f_{N-1}(x^i_{N-1}, \mu^i_{N-1}, j)] p^j_{N-1}(x^i_{N-1}, \mu^i_{N-1}) \right. \\
&\qquad \left. - \sum_{i=1}^{I_{N-1}} J_N[f_{N-1}(x, \mu^i_{N-1}, j)] p^j_{N-1}(x, \mu^i_{N-1}) \right| \\
&\qquad + \left| \sum_{j=1}^{I_{N-1}} \{J_N[f_{N-1}(x, \mu^i_{N-1}, j)] - \tilde{J}_N[f_{N-1}(x, \mu^i_{N-1}, j)]\} p^j_{N-1}(x, \mu^i_{N-1}) \right|.
\end{aligned}$$

From the above inequality, Proposition 1, (36), (37) as proved for $k = N$, and conditions A.2 and A.3, we easily obtain

$$|\hat{J}_{N-1}(x_{N-1}^i) - \tilde{J}_{N-1}(x)| \leqslant \delta_N \|x - x_{N-1}^i\| \leqslant \delta_N d_s,$$

where $\delta_N$ is a positive scalar not depending on the grid $G_{N-1}$. Using the above inequality in (42),

$$|J_{N-1}(x) - \tilde{J}_{N-1}(x)| \leqslant (A_{N-1} + \alpha_N + \delta_N)d_s.$$

Thus (37) holds for $k = N - 1$ with $\alpha_{N-1} = A_{N-1} + \alpha_N + \delta_N$. Similarly (37) is shown to hold for all $k$.

We now turn to proving (38) and (39) under Assumptions B. Similarly, as in the case of Assumptions A, (38) and (39) hold for $k = N$ with $\beta_N = \overline{M}_N$. To prove (38) for $k = N - 1$ we have by (33), for any $i = 1, 2, \ldots, n_{N-1}$,

$$|J_{N-1}(x_{N-1}^i) - \hat{J}_{N-1}(x_{N-1}^i)|$$

$$= \Bigg| \max_{u \in U_{N-1}(x_{N-1}^i)} \mathop{E}_{w} \{g_{N-1}(x_{N-1}^i, u, w) + J_N[f_{N-1}(x_{N-1}^i, u, w)]\}$$

$$- \max_{u \in U_{N-1}(x_{N-1}^i) \cap H_{N-1}} \mathop{E}_{w} \{g_{N-1}(x_{N-1}^i, u, w) + \hat{J}_N[f_{N-1}(x_{N-1}^i, u, w)]\} \Bigg|.$$

We use the triangle inequality and strengthen it further as in (41), obtaining

$$|J_{N-1}(x_{N-1}^i) - \hat{J}_{N-1}(x_{N-1}^i)|$$

$$\leqslant \beta_N d_s + \Bigg| \max_{u \in U_{N-1}(x_{N-1}^i)} \mathop{E}_{w} \{g_{N-1}(x_{N-1}^i, u, w) + J_N[f_{N-1}(x_{N-1}^i, u, w)]\}$$

$$- \max_{u \in U_{N-1}(x_{N-1}^i) \cap H_{N-1}} \mathop{E}_{w} \{g_{N-1}(x_{N-1}^i, u, w) + J_N[f_{N-1}(x_{N-1}^i, u, w)]\} \Bigg|$$

$$\leqslant \beta_N d_s + I_{N-1}(\overline{M}_{N-1} + \overline{L}_{N-1} A_N + \overline{B}_{N-1} \overline{N}_{N-1}) d_c, \tag{43}$$

where $\overline{B}_{N-1}$ is given by (19). The last step in the above algebra is obtained in a straightforward manner by using Assumptions B, Proposition 1, and the definition of $d_c$. Also we have for any $x \in S_{N-1}^i$, $i = 1, \ldots, n_{N-1}$,

$$\begin{aligned}
|J_{N-1}(x) - \hat{J}_{N-1}(x)| &= |J_{N-1}(x) - \hat{J}_{N-1}(x_{N-1}^i)| \\
&\leqslant |J_{N-1}(x) - J_{N-1}(x_{N-1}^i)| + |J_{N-1}(x_{N-1}^i) - \hat{J}_{N-1}(x_{N-1}^i)| \\
&\leqslant A_{N-1}\|x - x_{N-1}^i\| + |J_{N-1}(x_{N-1}^i) - \hat{J}_{N-1}(x_{N-1}^i)| \\
&\leqslant A_{N-1} d_s + |J_{N-1}(x_{N-1}^i) - \hat{J}_{N-1}(x_{N-1}^i)|.
\end{aligned}$$

Using (43) we have that (38) holds for $k = N - 1$ with

$$\beta_{N-1} = \max\{\beta_N + A_{N-1}, I_{N-1}(\overline{M}_{N-1} + \overline{L}_{N-1} A_N + \overline{B}_{N-1} \overline{N}_{N-1})\}.$$

Similarly (38) is shown to hold for all $k$. Once (38) is proved, (39) follows in exactly the same manner as under Assumptions A. Q.E.D.

Proposition 2 demonstrates the validity (under mild continuity assumptions) of discretization procedures based on piecewise constant approximations of the functions obtained in the DP algorithm. The bound on the approximation error is proportional to the size of the discretization grid utilized and tends to zero as the grid becomes finer and finer.

## 5.3 Suboptimal Controllers and the Notion of Adaptivity

As discussed earlier, the numerical solution of many sequential optimization problems by DP is computationally impractical or infeasible due to the dimensionality problem. For this reason in practice one is often forced to use a suboptimal policy that can be more easily calculated and implemented than the optimal policy and hopefully does not result in a substantial increase of the value of the cost functional over the optimal value. There are a number of suboptimal approaches for sequential optimization and it is quite difficult to classify and analyze them in a unified manner. For example, one obvious approach is to simplify or modify the model so that the DP algorithm is either computationally feasible or possesses an analytical solution. Such simplifications include replacing nonlinear functions by linear and, perhaps, quadratic approximations, eliminating or aggregating certain variables in the model, neglecting small uncertainties and low-level correlations, etc. No general guidelines can be given for such an approach, and we will not be further concerned with it. On the other hand we shall discuss in the next two sections some approaches for suboptimal control that are of general applicability and are often considered in practice. Some other techniques are referred to in the last section. We first define the notion of *adaptivity*, which constitutes a desirable property for any suboptimal controller.

Let us take as our basic model the problem with imperfect state information of Section 4.1. This problem includes as a special case the problem with perfect state information of Section 2.1 except for the fact that the control constraint set $U_k$ is not state dependent (simply take the measurement equation to be of the form $z_k = x_k$ for all $k$). Now a good part of the difficulty in solving the problem by DP can be attributed to the fact that exact or inexact measurements of the state are taken during the process, and the control inputs depend on these measurements. If no measurements were taken, the problem would be reduced to finding a sequence $\{u_0^*, u_1^*, \ldots, u_{N-1}^*\}$, such that $u_k^* \in U_k$, $k = 0, \ldots, N-1$, which minimizes

$$\bar{J}(u_0, u_1, \ldots, u_{N-1}) = \underset{\substack{x_0, w_k \\ k=0,1,\ldots,N-1}}{E} \left\{ g_N(x_N) + \sum_{k=0}^{N-1} g_k(x_k, u_k, w_k) \right\} \tag{44}$$

subject to the constraints

$$x_{k+1} = f_k(x_k, u_k, w_k).$$

This is an essentially deterministic problem, which can be solved by deterministic optimization techniques. For example, if the state and control spaces are Euclidean, optimal control and mathematical programming methods can be efficiently used for solution of the problem. A control sequence $\{u_0^*, u_1^*, \ldots, u_{N-1}^*\}$ minimizing the cost functional (44) is called an *optimal open-loop control sequence*, and the optimal value $J_0^*$ of the cost (44) is called the *optimal open-loop value*:

$$J_0^* = \inf_{u_k \in U_k,\, k=0,\ldots,N-1} \bar{J}(u_0, u_1, \ldots, u_{N-1}).$$

Now if $J^*$ denotes the optimal value of the basic problem with imperfect state information of Section 4.1 we have

$$J^* \leqslant J_0^*.$$

This is true simply because the class of controllers that do not take into account the measurements (consist of functions that take a single constant value independent of the current information vector) is a strict subset of the class of admissible controllers for the problem of Section 4.1. The difference $(J_0^* - J^*)$ can be called the *value of information* supplied by the measurements. [When the cost functional includes directly costs for measurement, a more descriptive term for $(J_0^* - J^*)$ would be the *value of the option of taking measurements*.] The whole point for taking measurements (i.e., using feedback) is precisely to reduce the optimal cost from $J_0^*$ to $J^*$. Now it is evident that any suboptimal control law $\pi = \{\mu_0, \mu_1, \ldots, \mu_{N-1}\}$ that takes into account the measurements should be considered acceptable only if the corresponding value of the cost functional $J_\pi$ satisfies

$$J^* \leqslant J_\pi \leqslant J_0^*, \tag{45}$$

for otherwise the information supplied by the measurements is used with disadvantage rather than advantage.

**Definition** An admissible control law $\pi = \{\mu_0, \mu_1, \ldots, \mu_{N-1}\}$ that satisfies (45) will be called *quasi-adaptive*. If the right-hand side of (45) is satisfied with strict inequality, the control law $\pi$ will be called *adaptive*.

In other words an adaptive controller is one that uses the measurements with advantage. Of course, an optimal controller for the problem is quasi-adaptive but some of the suboptimal controllers commonly used in practice are not in general quasi-adaptive. One such example is the so-called naive feedback controller, which will be examined in the next section. In the same section we shall examine another suboptimal control scheme, the so-called open-loop feedback controller, which turns out to be always quasi-adaptive.

## 5.4 Naive Feedback and Open-Loop Feedback Controllers

The *naive feedback controller* (NFC) is a control scheme based on an idea that has been used with considerable success for many years. It is also called *certainty equivalent controller* and its conception and justification dates to the origins of feedback theory when feedback was employed as a device that compensated for uncertainties and noise in the system. Traditionally servomechanism engineers when faced with the design of a controller for an uncertain system, assumed away or neglected the uncertainty by fixing the uncertain quantitities at some typical values (for example, their expected values) and designed a feedback control scheme for the corresponding deterministic system on the basis of certain considerations (stability, optimality with respect to a criterion, etc.). They relied on the feedback mechanism to compensate for uncertainties and noise in the system. The NFC draws from that idea. It applies at each time the control input that would be optimal if all the uncertain quantities were fixed at their expected values, i.e., it acts as if a form of the certainty equivalence principle, discussed in Sections 1.3, 3.1, and 4.3, were holding.

We take as our model the basic problem with imperfect state information of Section 4.1 and we further assume that the probability measures of the input disturbances $w_k$ do not depend on $x_k$ and $u_k$. We assume that the state spaces and disturbance spaces are convex subsets of corresponding Euclidean spaces so that the expected values

$$\bar{x}_k = E\{x_k | I_k\}, \qquad \bar{w}_k = E\{w_k\},$$

belong to the corresponding state spaces and disturbance spaces.

The control input $\tilde{\mu}_k(I_k)$ applied by the NFC at each time $k$ is determined by the following rule:

(a) Given the information vector $I_k$, compute

$$\bar{x}_k = E\{x_k | I_k\}.$$

(b) Solve the deterministic problem of finding a control sequence $\{\tilde{u}_k, \tilde{u}_{k+1}, \ldots, \tilde{u}_{N-1}\}$ that minimizes

$$g_N(x_N) + \sum_{i=k}^{N-1} g_k(x_k, u_k, \bar{w}_k)$$

subject to the constraints†

$$u_i \in U_i, \qquad x_{i+1} = f_i(x_i, u_i, \bar{w}_i), \qquad i = k, k+1, \ldots, N-1, \qquad x_k = \bar{x}_k.$$

(c) Apply the control input

$$\tilde{\mu}_k(I_k) = \tilde{u}_k.$$

Note that if the current state $x_k$ is measured exactly (perfect state information), then step (a) is unnecessary. The deterministic optimization problem in step (b) must be solved at each time $k$ as soon as the initial state $\bar{x}_k = E\{x_k | I_k\}$ becomes known by means of an estimation (or perfect observation) procedure.‡ A total of $N$ such problems must be solved in any actual operation of the NFC. Each one of these problems, however, is a deterministic optimal control problem and is often of the type for which powerful deterministic optimization techniques such as steepest descent, conjugate gradient, Newton's method [L10] are applicable. Thus the NFC requires the solution of $N$ such problems in place of the solution of the DP algorithm required to obtain an optimal controller. Furthermore, the implementation of the NFC requires no storage of the type required for the optimal feedback controller—often a major advantage.

The implementation of the NFC given above requires the solution of $N$ optimal control problems in an "on-line" fashion, i.e., each of these problems is solved as soon as the necessary information $\bar{x}_k = E\{x_k | I_k\}$ becomes available during the actual control process. The alternative is to solve these problems a priori. This is accomplished by determining an optimal feedback controller for the deterministic optimal control problem obtained from the original problem by replacing all uncertain quantities by their expected values. It is easy to verify (based on the equivalence of open-loop and feedback implementation of optimal controllers for deterministic problems) that the implementation of the NFC given earlier is equivalent to the following implementation:

Let $\{\mu_0^d(x_0), \ldots, \mu_{N-1}^d(x_{N-1})\}$ be an optimal controller obtained from the DP algorithm for the deterministic problem

$$\text{minimize} \quad g_N(x_N) + \sum_{k=0}^{N-1} g_k[x_k, \mu_k(x_k), \bar{w}_k] \qquad \text{over all} \quad \{\mu_0, \ldots, \mu_{N-1}\}$$

$$\text{subject to} \quad \mu_k(x_k) \in U_k, \quad \forall x_k \in S_k, \qquad x_{k+1} = f_k[x_k, \mu_k(x_k), \bar{w}_k],$$
$$k = 0, 1, \ldots, N-1.$$

† We assume that there exists an optimal sequence $\{\tilde{u}_k, \tilde{u}_{k+1}, \ldots, \tilde{u}_{N-1}\}$. Furthermore, if multiple solutions exist for the minimization problem, some rule is used to select one of them in an unambiguous way.

‡ In practice, often one uses a suboptimal estimation scheme that generates an "approximate" value of $\bar{x} = E\{x_k | I_k\}$.

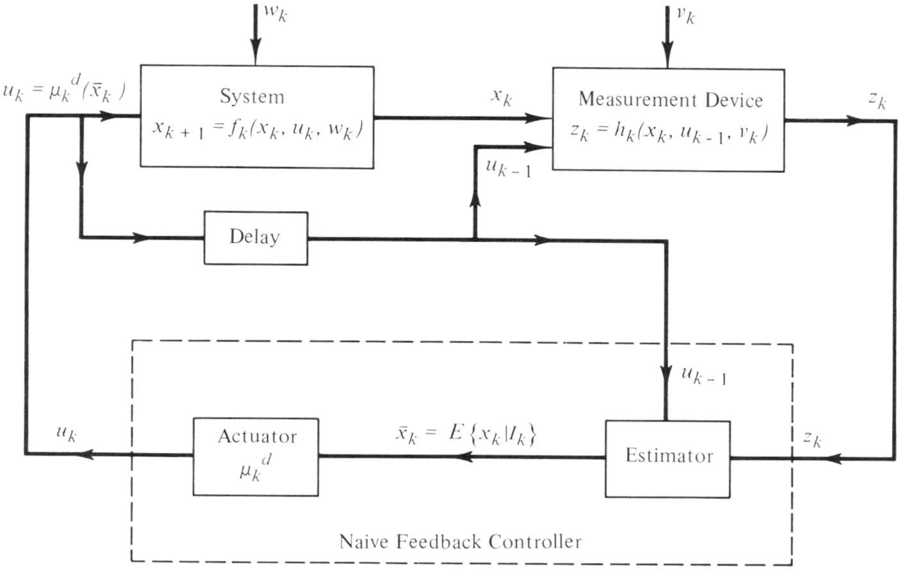


FIGURE 5.3

Then the control input $\tilde{\mu}_k(I_k)$ applied by the NFC at time $k$ is given by

$$\tilde{\mu}_k(I_k) = \mu_k^d[E\{x_k | I_k\}] = \mu_k^d(\bar{x}_k)$$

as shown in Fig. 5.3.

In other words *an alternative implementation of the NFC consists of finding a feedback controller* $\{\mu_0^d, \mu_1^d, \ldots, \mu_{N-1}^d\}$ *that is optimal for a corresponding deterministic problem, and subsequently using this controller for control of the uncertain system (modulo substitution of the state by its expected value) in the spirit of traditional servomechanism design.* Either of the definitions given for the NFC can serve as a basis for its implementation. Depending on the nature of the problem one method may be preferable to the other.

The NFC often performs well in practice and results in a value of the cost functional that is close to the optimal value. In fact for the linear-quadratic problems of Sections 3.1 and 4.3, it is identical to the optimal controller (certainty equivalence principle). However, it is possible that the NFC is not quasi-adaptive (performs strictly worse than the optimal open-loop controller), as the following intriguing example shows.

EXAMPLE Consider the following two-dimensional, two-stage linear system with scalar controls and disturbances:

$$x_{k+1} = \begin{bmatrix} x_{k+1}^1 \\ x_{k+1}^2 \end{bmatrix} = \begin{bmatrix} x_k^1 \\ x_k^2 \end{bmatrix} + \begin{bmatrix} 1 \\ 0 \end{bmatrix} u_k + \begin{bmatrix} \frac{1}{2} \\ \frac{1}{2}\sqrt{2} \end{bmatrix} w_k, \qquad k = 0, 1,$$

and with initial state $x_0^1 = x_0^2 = 0$. The controls $u_0, u_1$ are unconstrained and the disturbances $w_0, w_1$ are independent random variables with identical distribution. They take the values $+1$ and $-1$ each with probability $\frac{1}{2}$. Perfect state information prevails.

*Case 1* Consider the cost functional

$$\underset{w_0, w_1}{E} \{\|x_2\|^2\},$$

where $\|\cdot\|$ denotes the usual Euclidean norm. Since this is a quadratic cost functional and certainty equivalence holds, we expect that the NFC is an optimal controller for this problem. Indeed, let us first compute the optimal controller by the DP algorithm. We have

$$\begin{aligned} J_2(x_2) &= \|x_2\|^2, \\ J_1(x_1) &= \min_{u_1} \underset{w_1}{E} \{J_2[(x_1^1 + u_1 + \tfrac{1}{2}w_1, x_1^2 + \tfrac{1}{2}\sqrt{2}w_1)]\} \\ &= \min_{u_1} \underset{w_1}{E} \{(x_1^1 + u_1 + \tfrac{1}{2}w_1)^2 + (x_1^2 + \tfrac{1}{2}\sqrt{2}w_1)^2\} \\ &= \min_{u_1}[(x_1^1 + u_1)^2 + (x_1^2)^2 + \underset{w_1}{E} \{(\tfrac{1}{2}w_1)^2 + (\tfrac{1}{2}\sqrt{2}w_1)^2\}]. \end{aligned}$$

By minimization we obtain the optimal controller and "cost-to-go" for stage 1:

$$\mu_1^*(x_1) = -x_1^1,$$

and

$$J_1(x_1) = (x_1^2)^2 + \underset{w_1}{E} \{(\tfrac{1}{2}w_1)^2 + (\tfrac{1}{2}\sqrt{2}w_1)^2\} = (x_1^2)^2 + \tfrac{3}{4}.$$

Also

$$\begin{aligned} J_0(x_0) &= \min_{u_0} \underset{w_0}{E} \{J_1[(x_0^1 + u_0 + \tfrac{1}{2}w_0, x_0^2 + \tfrac{1}{2}\sqrt{2}w_0)]\} \\ &= \underset{w_0}{E} \{(x_0^2 + \tfrac{1}{2}\sqrt{2}w_0)^2 + \tfrac{3}{4}\} = (x_0^2)^2 + \tfrac{5}{4}. \end{aligned}$$

Thus we obtain the optimal cost corresponding to the initial state $x_0 = 0$:

$$J^* = J_0(0) = \tfrac{5}{4}.$$

Furthermore, applying any control $u_0$ at stage 0 is optimal.

The NFC may be obtained by means of the second implementation scheme described earlier whereby we consider the problem of obtaining an optimal feedback controller for the deterministic problem resulting when $w_0$ and $w_1$

are replaced by their expected values $\bar{w}_0 = \bar{w}_1 = 0$. The DP algorithm for this deterministic problem is given by

$$\tilde{J}_2(x_2) = \|x_2\|^2,$$
$$\tilde{J}_1(x_1) = \min_{u_1} \tilde{J}_2[(x_1^1 + u_1, x_1^2)] = (x_1^2)^2,$$
$$\tilde{J}_0(x_0) = \min_{u_1} \tilde{J}_1[(x_0^1 + u_0, x_0^2)] = (x_0^2)^2,$$

and one can easily see that the NFC is identical to the optimal controller. This is, of course, due to certainty equivalence. Thus if $\tilde{J}$ denotes the cost corresponding to the NFC, we have

$$J^* = \tilde{J} = \tfrac{5}{4}.$$

We also calculate the optimal open-loop value $J_0^*$, i.e., the optimal cost attained by an open-loop controller. Since (in view of $x_0 = 0$) the final state $x_2$ is given by

$$x_2 = \begin{bmatrix} u_0 + u_1 + \frac{1}{2}(w_0 + w_1) \\ \frac{1}{2}\sqrt{2}(w_0 + w_1) \end{bmatrix}$$

we have

$$J_0^* = \min_{u_0, u_1} \mathop{E}_{w_0, w_1} \{[u_0 + u_1 + \tfrac{1}{2}(w_0 + w_1)]^2 + \tfrac{1}{2}(w_0 + w_1)^2\}.$$

One may easily see that any pair $(u_0, u_1)$ with $u_0 + u_1 = 0$ attains the minimum above and computation of the expected value yields

$$J_0^* = \tfrac{3}{2}.$$

Thus for this problem we have

$$J^* = \tilde{J} < J_0^*,$$

and the NFC is adaptive.

*Case 2* Consider now the cost functional

$$\mathop{E}_{w_0, w_1} \{\|x_2\|\}.$$

The NFC is obtained from the DP algorithm for the corresponding deterministic problem. Straightforward calculation yields

$$\tilde{J}_2(x_2) = \|x_2\|,$$
$$\tilde{J}_1(x_1) = \min_{u_1} \tilde{J}_2[(x_1^1 + u_1, x_1^2)] = |x_1^2|,$$
$$\tilde{J}_0(x_0) = \min_{u_0} \tilde{J}_1[(x_0^1 + u_0, x_0^2)] = |x_0^2|.$$

The NFC is the same as for Case 1:

$$\tilde{\mu}_1(x_1) = -x_1^1, \qquad \tilde{\mu}_0(x_0) = \text{any scalar}.$$

This is due to the fact that minimizing $\|x_2\|$ is equivalent to minimizing $\|x_2\|^2$ in the absence of uncertainty. With this controller and given that $x_0 = 0$, the final state $x_2$ is equal to

$$x_2 = \begin{bmatrix} \frac{1}{2}w_1 \\ \frac{1}{2}\sqrt{2}(w_0 + w_1) \end{bmatrix}$$

and the corresponding value of the cost functional is

$$\tilde{J} = \mathop{E}_{w_0, w_1} \{[\tfrac{1}{4}w_1^2 + \tfrac{1}{2}(w_0 + w_1)^2]^{1/2}\} = 1.$$

The optimal open-loop value is obtained from

$$J_0^* = \min_{u_0, u_1} \mathop{E}_{w_0, w_1} \{([u_0 + u_1 + \tfrac{1}{2}(w_0 + w_1)]^2 + \tfrac{1}{2}(w_0 + w_1)^2)^{1/2}\}.$$

Any pair $(u_0, u_1)$ with $u_0 + u_1 = 0$ attains the minimum above and computation of the expected value yields

$$J_0^* = \tfrac{1}{2}\sqrt{3}.$$

Thus we have

$$J_0^* < \tilde{J}$$

and the NFC is not quasi-adaptive. Actually for this case one may verify that the optimal cost corresponding to an optimal feedback controller is

$$J^* = J_0^* = \tfrac{1}{2}\sqrt{3}$$

so that for this case the open-loop optimal value is equal to the optimal value.

This example is rather astonishing in view of the well-rooted conviction among engineers that feedback designs tend to compensate for uncertainties and noise in the system and hence are preferable to open-loop designs. The statement is true, of course, if one uses optimal feedback obtained through a problem formulation that explicitly takes into account the uncertainty. It is not necessarily true if one uses a feedback controller obtained from an optimization problem where uncertainties are not explicitly considered.

### *Open-Loop Feedback Controller*

The *open-loop feedback controller* (OLFC) is similar to the NFC except that it takes explicitly into account the uncertainty about $x_k, w_k, \ldots, w_{N-1}$ when calculating the control input to be applied at time $k$.

The OLFC applies at each time $k$ a control input $\bar{\mu}_k(I_k)$, which is determined by the following procedure:

(a) Given the information vector $I_k$ compute the conditional probability measure $P_{x_k|I_k}$.

(b) Find a control sequence $\{\bar{u}_k, \bar{u}_{k+1}, \ldots, \bar{u}_{N-1}\}$ that minimizes

$$\underset{x_k, w_k, \ldots, w_{N-1}}{E}\left\{g_N(x_N) + \sum_{i=k}^{N-1} g_i(x_i, u_i, w_i) \,|\, I_k\right\}$$

subject to the constraints

$$u_i \in U_i, \qquad x_{i+1} = f_i(x_i, u_i, w_i), \qquad i = k, \ldots, N-1.†$$

(c) Apply the control input

$$\bar{\mu}_k(I_k) = \bar{u}_k.$$

The operation of the OLFC can be interpreted as follows: At each time $k$ the controller uses the new measurement received to calculate the conditional probability $P_{x_k|I_k}$. However, it selects the control input as if no further measurements will be received in the future.

Similarly to the NFC, the OLFC requires the solution of $N$ optimal control problems in any actual operation of the system. Each of these problems may again be solved by deterministic optimal control or mathematical programming techniques. The computations are a little more complicated than those for the NFC since now the cost functional includes the expectation operation with respect to the uncertain quantities. The main difficulty in the implementation of the OLFC is the computation of $P_{x_k|I_k}$. In many cases one cannot compute $P_{x_k|I_k}$ exactly, in which case some "reasonable" approximation scheme must be used. Of course, if we have perfect state information, this potential problem does not arise.

We now prove that the OLFC is quasi-adaptive.

**Proposition** The value of the cost functional $J_{\bar{\pi}}$ corresponding to an OLFC $\bar{\pi} = \{\bar{\mu}_0, \bar{\mu}_1, \ldots, \bar{\mu}_{N-1}\}$ satisfies

$$J^* \leqslant J_{\bar{\pi}} \leqslant J_0^*. \tag{46}$$

*Proof*‡ We have

$$J_{\bar{\pi}} = \underset{z_0}{E}\{\bar{J}_0(I_0)\} = \underset{z_0}{E}\{\bar{J}_0(z_0)\}, \tag{47}$$

† Similarly as for the NFC, we assume that an optimal solution to this problem exists and ambiguities resulting from multiple solutions are resolved by some rule.

‡ We assume throughout the proof that all expected values appearing are well defined and finite and the minimum in (50) is attained for every $I_k$.

where the function $\bar{J}_0$ is obtained from the recursive algorithm

$$\bar{J}_{N-1}(I_{N-1}) = \underset{x_{N-1}, w_{N-1}}{E} \{g_N[f_{N-1}(x_{N-1}, \bar{\mu}_{N-1}(I_{N-1}), w_{N-1})] + g_{N-1}[x_{N-1}, \bar{\mu}_{N-1}(I_{N-1}), w_{N-1}] | I_{N-1}\}, \tag{48}$$

$$\bar{J}_k(I_k) = \underset{x_k, w_k, v_{k+1}}{E} \{g_k[x_k, \bar{\mu}_k(I_k), w_k] + \bar{J}_{k+1}[I_k, h_{k+1}[f_k(x_k, \bar{\mu}_k(I_k), w_k), \bar{\mu}_k(I_k), v_{k+1}], \bar{\mu}_k(I_k)] | I_k\}, \quad k = 0, 1, \ldots, N-1. \tag{49}$$

Consider also the functions $J_k^c(I_k)$, $k = 0, 1, \ldots, N-1$, defined by

$$J_k^c(I_k) = \min_{\substack{u_i \in U_i \\ i=k,\ldots,N-1}} \underset{\substack{x_k, w_i \\ x_{i+1} = f_i(x_i, u_i, w_i) \\ i=k,\ldots,N-1}}{E} \left\{ g_N(x_N) + \sum_{i=k}^{N-1} g_i(x_i, u_i, w_i) | I_k \right\}. \tag{50}$$

The minimization problem indicated in this equation is precisely the one that must be solved at time $k$ in order to calculate the control input $\bar{\mu}_k(I_k)$ of the OLFC. Clearly $J_k^c(I_k)$ can be interpreted as the calculated open-loop optimal cost from time $k$ to time $N$ when the current information vector is $I_k$. It is easy to see that we have

$$\underset{z_0}{E} \{J_0^c(z_0)\} \leqslant J_0^*. \tag{51}$$

We shall prove that

$$\bar{J}_k(I_k) \leqslant J_k^c(I_k) \qquad \text{for all} \quad I_k \quad \text{and} \quad k. \tag{52}$$

Then from (47), (51), and (52) it will follow that

$$J_{\bar{\pi}} \leqslant J_0^*,$$

which is the relation to be proved. We shall show (52) by induction.

By the definition of the OLFC and (50) we have

$$\bar{J}_{N-1}(I_{N-1}) = J_{N-1}^c(I_{N-1}) \qquad \forall I_{N-1},$$

and hence (52) holds for $k = N - 1$. Assume

$$\bar{J}_{k+1}(I_{k+1}) \leqslant J_{k+1}^c(I_{k+1}) \qquad \forall I_{k+1}. \tag{53}$$

Then from (49), (53), and (50) we have

$$
\begin{aligned}
\bar{J}_k(I_k) &= \underset{x_k, w_k, v_{k+1}}{E} \{g_k[x_k, \bar{\mu}_k(I_k), w_k] \\
&\quad + \bar{J}_{k+1}[I_k, h_{k+1}[f_k(x_k, \bar{\mu}_k(I_k), w_k), \bar{\mu}_k(I_k), v_{k+1}], \bar{\mu}_k(I_k)] \,|\, I_k\} \\
&\leqslant \underset{x_k, w_k, v_{k+1}}{E} \{g_k[x_k, \bar{\mu}_k(I_k), w_k] \\
&\quad + J^c_{k+1}[I_k, h_{k+1}[f_k(x_k, \bar{\mu}_k(I_k), w_k), \bar{\mu}_k(I_k), v_{k+1}], \bar{\mu}_k(I_k)] \,|\, I_k\} \\
&= \underset{x_k, w_k, v_{k+1}}{E} \Bigg\{ \min_{\substack{u_i \in U_i \\ i=k+1, \ldots, N-1}} \underset{\substack{x_{k+1}, w_i \\ x_{i+1} = f_i(x_i, u_i, w_i) \\ i=k+1, \ldots, N-1}}{E} \{g_k[x_k, \bar{\mu}_k(I_k), w_k] \\
&\quad + \sum_{i=k+1}^{N-1} g_i(x_i, u_i, w_i) + g_N(x_N) \,|\, I_{k+1}\} \,|\, I_k \Bigg\} \\
&\leqslant \min_{\substack{u_i \in U_i \\ i=k+1, \ldots, N-1}} \underset{\substack{x_k, w_k, w_i \\ x_{i+1} = f_i(x_i, u_i, w_i) \\ i=k+1, \ldots, N-1 \\ x_{k+1} = f_k[x_k, \bar{\mu}_k(I_k), w_k]}}{E} \Bigg\{ g_N(x_N) + g_k[x_k, \bar{\mu}_k(I_k), w_k] \\
&\quad + \sum_{i=k+1}^{N-1} g_i(x_i, u_i, w_i) \,|\, I_k \Bigg\} = J^c_k(I_k).
\end{aligned}
$$

The second inequality follows by interchanging expectation and minimization (notice that we have always $E\{\min[\cdot]\} \leqslant \min[E\{\cdot\}]$) and by "integrating out" $v_{k+1}$. The last equality follows from the definition of OLFC. Thus (52) is proved for all $k$ and the desired result is shown. Q.E.D.

It is worth noting that by (52) the calculated open-loop optimal cost from time $k$ to time $N$, $J^c_k(I_k)$, provides a readily obtainable performance bound for the OLFC.

The above proposition shows that the OLFC uses the measurements with advantage even though it selects at each period the present control input as if no further measurements will be taken in the future. Of course, this says nothing about how closely the resulting value of the cost functional approximates the optimal value. The general opinion, however, is that the OLFC is a fairly satisfactory mode of control for many problems, although future experience will undoubtedly shed more light on this question.

Concerning the comparison of the NFC and the OLFC one should note that even though the NFC may not be quasi-adaptive, it may be considerably easier to implement than the OLFC. On the other hand, the OLFC, in contrast to the NFC, can be used even for problems that are not defined over Euclidean spaces and is thus a more general form of suboptimal control. In general, the two controllers cannot be compared on the basis of their corresponding values of cost functional. The example and the proposition of this section show that the OLFC may perform better than the NFC. The example in Problem 4 shows that the opposite may also be true. Both controllers may exhibit some rather counterintuitive behavior due to their suboptimality. An example of such behavior is given in Problem 5.

## 5.5 Partial Open-Loop Feedback Controllers and the Efficient Utilization of Forecasts

As discussed in Section 5.4, the OLFC utilizes past measurements in the computation of the conditional probability measure $P_{x_k|I_k}$ but calculates the control input $\bar{\mu}_k(I_k)$ on the basis that no measurements will be taken in the future. A form of suboptimal control that is intermediate between the optimal feedback controller and the OLFC is provided by what we shall call the *partial open-loop feedback controller* (POLFC). This controller utilizes past measurements to compute $P_{x_k|I_k}$ but calculates the control input on the basis that *some* (but not all) of the measurements to be taken *will in fact be taken* in the future. This suboptimal controller, as we shall explain below, is characterized by better performance bounds than the OLFC and is particularly useful in problems where forecasts of future input disturbances become known during the control process.

Consider the basic problem with imperfect state information of Section 4.1. Let us assume that each measurement $z_k$ consists of two separate measurements $\bar{z}_k$, $\tilde{z}_k$, which are elements of appropriate spaces and are given by

$$z_k = \begin{bmatrix} \bar{z}_k \\ \tilde{z}_k \end{bmatrix} = \begin{bmatrix} \bar{h}_k(x_k, u_{k-1}, \bar{v}_k) \\ \tilde{h}_k(x_k, u_{k-1}, \tilde{v}_k) \end{bmatrix}, \qquad k = 0, 1, \ldots, N-1,$$

where $\bar{h}_k$, $\tilde{h}_k$ are given functions and $\bar{v}_k$, $\tilde{v}_k$ are random observation disturbances characterized by given probability measures that depend explicitly only on $x_k$ and $u_{k-1}$ and not on prior observation disturbances $\bar{v}_{k-1}, \tilde{v}_{k-1}, \ldots, \bar{v}_0, \tilde{v}_0$ or any of the input disturbances.

Roughly speaking, the POLFC computes the conditional probability measure $P_{x_k|I_k}$, where $I_k = (z_0, \ldots, z_k, u_0, \ldots, u_{k-1})$, and then calculates the optimal policy for the truncated optimization problem over the remaining periods $k, k+1, \ldots, N$ under the assumption that the measurements $\bar{z}_{k+1}, \ldots, \bar{z}_{N-1}$ *will* be taken, but the measurements $\tilde{z}_{k+1}, \ldots, \tilde{z}_{N-1}$ *will not.* The rationale for such a procedure is that in some cases it may be much easier to compute the optimal policy as a function of the measurements $\bar{z}_{k+1}, \ldots, \bar{z}_{N-1}$ only, rather than as a function of $\bar{z}_{k+1}, \ldots, \bar{z}_{N-1}$ *and* $\tilde{z}_{k+1}, \ldots, \tilde{z}_{N-1}$. Such situations occur, for example, when $\bar{z}_k$ represents an exact measurement of the current state and $\tilde{z}_k$ represents forecasts on future input disturbances, as will be explained later.

We define now precisely the POLFC, denoted $\{\bar{\mu}_0(I_0), \ldots, \bar{\mu}_{N-1}(I_{N-1})\}$, by specifying the rule by means of which the control input $\bar{\mu}_k(I_k)$ at time $k$ is calculated:

(a) Given the information $I_k$ compute the conditional probability measure $P_{x_k|I_k}$.

(b) Define the following *partial* information vectors:

$$
\begin{aligned}
I_{k+1}^k &= (\bar{z}_{k+1}, u_k), \\
I_{k+2}^k &= (\bar{z}_{k+1}, \bar{z}_{k+2}, u_k, u_{k+1}), \\
&\ \vdots \\
I_{N-1}^k &= (\bar{z}_{k+1}, \bar{z}_{k+2}, \ldots, \bar{z}_{N-1}, u_k, \ldots, u_{N-2}),
\end{aligned}
$$

and consider the problem of finding a control $u_k \in U_k$ and functions $\mu_i^k(I_i^k)$, such that $\mu_i^k(I_i^k) \in U_i$ for all $I_i^k$ and $i$, that minimize the cost functional

$$
\begin{aligned}
J^k(u_k, \mu_{k+1}^k, \ldots, \mu_{N-1}^k) = \underset{\substack{x_k, w_i \\ i=k,\ldots,N-1}}{E} \Bigg\{ & g_N(x_N) + \sum_{i=k+1}^{N-1} g_i[x_i, \mu_i^k(I_i^k), w_i] \\
& + g_k(x_k, u_k, w_k) \,|\, I_k \Bigg\}
\end{aligned}
$$

subject to the constraints

$$
\begin{aligned}
x_{k+1} &= f_k(x_k, u_k, w_k), \\
x_{i+1} &= f_i[x_i, \mu_i^k(I_i^k), w_i], \qquad i = k+1, \ldots, N-1.
\end{aligned}
$$

Let $\{\bar{u}_k, \bar{\mu}_{k+1}^k, \ldots, \bar{\mu}_{N-1}^k\}$ be an optimal policy for this problem.†

(c) Apply the control input

$$\bar{\mu}_k(I_k) = \bar{u}_k.$$

Note that if the measurements $\bar{z}_k$, $k = 0, \ldots, N-1$, are vacuous (provide no information about the current state as, for example, in the case where the values of the functions $\bar{h}_k$ do not depend on $x_k$), the POLFC is identical to the OLFC. If the measurements $\tilde{z}_k$ are vacuous, then the POLFC is identical with the optimal feedback controller.

Now let us consider the problem of minimizing the cost functional over all admissible controllers of the form $\{\mu_0(\bar{I}_0), \ldots, \mu_{N-1}(\bar{I}_{N-1})\}$, where the *partial* information vectors $\bar{I}_k$ are defined by

$$\bar{I}_0 = \{\bar{z}_0\}, \qquad \bar{I}_k = \{\bar{z}_0, \ldots, \bar{z}_k, u_0, \ldots, u_{k-1}\}.$$

† Similarly as for the NFC and OLFC we assume that such a policy exists and that a certain rule is used to resolve ambiguities resulting from multiple solutions.

This problem is obtained from the basic problem assuming that the measurements $\tilde{z}_k$ are not taken. If $\bar{J}^*$ is the corresponding optimal value, we clearly have

$$J^* \leqslant \bar{J}^* \leqslant J_0^*,$$

where $J^*$ and $J_0^*$ are the optimal value and optimal open-loop value of the basic problem.

The following proposition is analogous to the proposition of the previous section and admits an entirely similar proof.

**Proposition** The value of the cost functional $J_{\bar{\pi}}$ corresponding to a POLFC $\bar{\pi} = \{\bar{\mu}_0, \bar{\mu}_1, \ldots, \bar{\mu}_{N-1}\}$ satisfies

$$J^* \leqslant J_{\bar{\pi}} \leqslant \bar{J}^* \leqslant J_0^*.$$

It should be mentioned that while again it is not clear that the POLFC approximates closely in terms of performance the optimal controller, the upper bound $(\bar{J}^* - J^*)$ on the deviation of the value of the cost over the optimal is smaller than the corresponding bound $(J_0^* - J^*)$ for the OLFC. On the other hand, surprisingly, it is possible that the POLFC performs worse than the OLFC in a given situation. One example of such behavior has been constructed by E. Alleman—a student in one of the author's classes—but is too complicated to be presented here.

We now turn to the application of the POLFC idea to a specific class of problems involving forecasts on future input disturbances. Consider the basic problem with perfect state information where the probability measures of the input disturbances $w_k$ do not depend on $x_k$ or $u_k$, $k = 0, 1, \ldots, N-1$. Furthermore, assume that at each time the controller in addition to knowing the value of the current state $x_k$ has access to a forecast

$$\tilde{z}_k = (\tilde{z}_k^k, \tilde{z}_k^{k+1}, \ldots, \tilde{z}_k^{N-1}).$$

Each element $\tilde{z}_k^i$ of the forecast $\tilde{z}_k$ represents an appropriate noisy measurement of the future input disturbance $w_i$, $i = k, \ldots, N-1$. These forecasts can be utilized to update at each time the probability measures of future input disturbances.

It should be realized that the presence of the forecasts transforms the problem in effect into one of imperfect state information, which involves a system of the form

$$\tilde{x}_{k+1} = \tilde{f}_k(\tilde{x}_k, u_k), \qquad k = 0, 1, \ldots, N-1. \tag{54}$$

The state of this system is

$$\tilde{x}_k = (x_k, w_k, w_{k+1}, \ldots, w_{N-1}),$$

and the system function $\tilde{f}_k$ is defined in an obvious way from the system equation $x_{k+1} = f_k(x_k, u_k, w_k)$. The cost functional can also be easily reformulated in terms of the variables of system (54), and the measurement

$$z_k = \begin{bmatrix} x_k \\ \tilde{z}_k \end{bmatrix}$$

obtained by the controller at each time $k$ can be viewed as a measurement of state $\tilde{x}_k$ of system (54).

Now a sufficient statistic for the problem of imperfect state information outlined previously is provided by $(x_k, P_{w_k, \ldots, w_{N-1}|I_k})$, where $P_{w_k, \ldots, w_{N-1}|I_k}$ is the joint conditional probability measure of the future disturbances $w_k, \ldots, w_{N-1}$. In many cases $w_k, \ldots, w_{N-1}$ are independent under the conditional probability measure, in which case $P_{w_k, \ldots, w_{N-1}|I_k}$ is characterized by $P_{w_k|I_k}, \ldots, P_{w_{N-1}|I_{N-1}}$. We shall assume this independence in what follows. Thus the optimal feedback controller must be obtained from a DP algorithm that is carried over either the spaces of the information vectors $I_k$ or over the space of the sufficient statistic $(x_k, P_{w_k|I_k}, \ldots, P_{w_{N-1}|I_k})$. Clearly the computation of this optimal controller is extremely difficult except for special cases such as the one considered in Section 2.3.

When, however, the POLFC philosophy is adopted things may be considerably simplified. According to the scheme outlined earlier in this section, at each time $k$ the state $x_k$ and $P_{w_k|I_k}, \ldots, P_{w_{N-1}|I_{N-1}}$ are obtained, and the optimization problem of part (b) of the definition of the POLFC is solved to obtain the control input $\bar{\mu}_k(I_k)$. This problem consists of finding $\{\bar{\mu}_k(x_k), \bar{\mu}_{k+1}(x_{k+1}), \ldots, \bar{\mu}_{N-1}(x_{N-1})\}$, with $\bar{\mu}_i(x_i) \in U_i$, $i = k, \ldots, N-1$, that minimize

$$\underset{\substack{w_i \\ i=k,\ldots,N-1}}{E} \left\{ g_N(x_N) + \sum_{i=k}^{N-1} g_i[x_i, \bar{\mu}_i(x_i), w_i] \,|\, I_k \right\}$$

subject to the system equation constraints

$$x_{i+1} = f_i[x_i, \bar{\mu}_i(x_i), w_i], \qquad i = k, \ldots, N-1.$$

Now the above problem is one of perfect state information and may be much easier to solve than the original problem. For example, it may possess an analytical solution or be one dimensional in the state variable. In this way the implementation of the POLFC may be feasible while the implementation of the optimal controller may be impossible.

Summarizing the above discussion the POLFC for the perfect state information problem with forecasts operates as follows:

(a) An optimal feedback policy is computed at the beginning of the process on the basis of the a priori statistics of the input disturbances and assuming no forecasts will be available in the future.

(b) Each time a forecast becomes available the statistics of the future input disturbances are accordingly updated. The optimal feedback policy is revised to take into account the updated statistics but is computed on the basis that these statistics are final and will not be updated by means of future forecasts.

It is to be noted that this mode of operation does not require a detailed statistical modeling of the forecasting process. The only requirement is to be able to compute the updated probability measures $P_{w_k|I_k}, \ldots, P_{w_{N-1}|I_k}$ of future disturbances given the forecasts up to time $k$. This is particularly convenient in practice since forecasts may be of a complicated and unpredictable nature, such as marketing surveys, unforeseen "inside" information, unpredictable changes in the economic environment, and subjective feelings of the decision maker.

As an example, consider a dynamic inventory problem such as the one considered in Section 3.2 where forecasts on future demands become available during the process. The POLFC calculates an initial multiperiod $(s, S)$ policy based on the a priori probability distributions of future demands. As soon as new forecasts are obtained and these distributions are updated the current $(s, S)$ policy is abandoned in favor of a new one calculated on the basis of the updated demand distributions. The new $(s, S)$ policy is again updated as soon as new forecasts become available and so on. In this way a reasonable and implementable policy is adopted while the optimal policy would perhaps be impossible to calculate or implement.

As another example, consider an asset selling problem similar to the one of Section 3.4 where we assume that forecasts on future price offers become available during the sale process. The POLFC approach suggests the calculation and adoption of an initial policy, characterized by cutoff levels $\alpha_1, \ldots, \alpha_{N-1}$ for accepting and rejecting an offer, based on the initial probability distribution of offers. As soon as, say at time $k$, forecasts become available and the probability distribution of future offers is updated, new cutoff levels $\alpha_k, \ldots, \alpha_{N-1}$ are calculated, which specify a new policy to be followed until the next forecasts become available. The resulting expected revenue from using the POLFC is larger than that which would result if the decision maker were to stick to the initial policy and disregard the forecasts. It is safe to presume that many decision makers, guided by intuition rather than detailed planning, would in fact adopt a policy similar to the one suggested by the POLFC.

## 5.6 Control of Systems with Unknown Parameters —Self-Tuning Regulators

We have been dealing so far with systems having a known state equation. In practice, however, one is frequently faced with situations where the system equation contains parameters that are not known exactly. One possible approach, of course, is to conduct experiments and estimate the unknown parameters from input–output records of the system. This procedure, however, can be quite time consuming. Furthermore, it may be necessary to repeat the procedure if the parameters of the system change with time as is often the case in many industrial processes.

The alternative is to formulate the stochastic control problem in a way that unknown parameters are dealt with directly. It is very easy to show that problems involving unknown system parameters can be embedded within the framework of our basic problem with imperfect state information by using state augmentation. Indeed, let the system equation be of the form

$$x_{k+1} = f_k(x_k, \theta, u_k, w_k),$$

where $\theta$ is a vector of unknown parameters with a given a priori probability distribution. We introduce an additional state variable $y_k = \theta$ and obtain a system equation of the form

$$\begin{bmatrix} x_{k+1} \\ y_{k+1} \end{bmatrix} = \begin{bmatrix} f_k(x_k, y_k, u_k, w_k) \\ y_k \end{bmatrix}. \tag{55}$$

By defining $\tilde{x}_k = (x_k, y_k)$ as the new state, we obtain

$$\tilde{x}_{k+1} = \tilde{f}_k(\tilde{x}_k, u_k, w_k),$$

where $\tilde{f}_k$ is defined in an obvious manner from (55). The initial state is

$$\tilde{x}_0 = (x_0, \theta).$$

With a suitable reformulation of the cost functional, the resulting problem becomes one that fits our usual framework.

It is to be noted, however, that since $y_k = \theta$ is unobservable we are faced with a problem of imperfect state information even if the controller receives an exact measurement of the state $x_k$. Furthermore, the parameter vector $\theta$ usually enters the state equation in a manner that makes the augmented system (55) nonlinear. As a result, in the great majority of cases it is practically impossible to obtain an optimal controller by means of a DP algorithm. Suboptimal controllers are thus called for and in this section we discuss in some detail a form of the certainty equivalent controller that has been used with success in some problems involving linear systems with unknown parameters.

*Certainty Equivalent Control of Linear Systems with Unknown Parameters*

Consider a linear system of the form

$$x_{k+1} = A_k(\theta)x_k + B_k(\theta)u_k + w_k.$$

The matrices $A_k$ and $B_k$ depend on an unknown parameter vector $\theta$, which is an element of a Euclidean space. The functional form of $A_k(\cdot)$ and $B_k(\cdot)$ is known. The cost functional is quadratic of the form

$$E\left\{x_N' Q_N x_N + \sum_{k=0}^{N-1} (x_k' Q_k x_k + u_k' R_k u_k)\right\}.$$

The disturbances $w_k$ are independent and perfect state information prevails.

If the parameter vector $\theta$ were known, the corresponding optimal control law would be given by

$$\mu_k^*(x_k, \theta) = L_k(\theta)x_k,$$

where

$$L_k(\theta) = -[R_k + B_k(\theta)' K_{k+1}(\theta) B_k(\theta)]^{-1} B_k(\theta)' K_{k+1}(\theta) A_k(\theta), \qquad (56)$$

and $K_{k+1}(\theta)$ is given by the Riccati equation, which corresponds to $A_k(\theta)$, $B_k(\theta)$, $Q_k$, and $R_k$ (cf. Section 3.1).

When $\theta$ is not known we are faced with a problem of imperfect state information as explained earlier. The information vector is $I_k = \{x_0, \ldots, x_k, u_0, \ldots, u_{k-1}\}$ and the corresponding NFC (certainty equivalent controller) is given by

$$\tilde{\mu}_k(I_k) = L_k(\bar{\theta}_k)x_k,$$

where

$$\bar{\theta}_k = E\{\theta | I_k\}.$$

At each time $k$ the NFC computes the conditional expectation $\bar{\theta}_k$ and then calculates the gain $L_k(\bar{\theta}_k)$ from Eq. (56) once the matrix $K_{k+1}(\bar{\theta}_k)$ has been obtained from the corresponding Riccati equation. In practice the computation of $\bar{\theta}_k$ may present formidable difficulties and for this reason usually one computes an approximation of $\bar{\theta}_k$ by means of a conveniently implementable estimator. We now discuss in some detail a simple special case.

*Self-Tuning Regulators*

Consider a scalar controlled process described for $k = (m + n), (m + n + 1), \ldots,$ by the equation

$$y_{k+m+1} + a_1 y_{k+m} + \cdots + a_n y_{k+m+1-n} = b_1 u_k + \cdots + b_n u_{k-n+1} + e_k. \quad (57)$$

The scalars $y_k$, $u_k$ are considered to be the output and the input, respectively, of the process at time $k$. There is a delay of $m \geqslant 0$ time intervals between the time that the input is applied and the time at which it affects the process. The stochastic disturbances $e_k$ are independent and have zero mean and finite variance. The parameters $a_1, \ldots, a_n, b_1, \ldots, b_n$ are unknown and it is assumed that $b_1 \neq 0$. The algorithms to be presented can be easily modified if some of the parameters are unknown and the remaining are known.

The process described above when there is no control (i.e., $b_i = 0$) is known as an *autoregressive process of order n* and is a widely used model of linear processes. A more general model is obtained if we replace $e_k$ in Eq. (57) by $e_k + c_1 e_{k-1} + \cdots + c_n e_{k-n}$. Control algorithms similar to the ones that will be presented shortly can also be developed for such more general models. In order to keep the exposition simple, however, we shall restrict ourselves to the process (57).

Equation (57) may be written in a different form, which is more convenient for our purposes. We have from (57)

$$\begin{aligned} a_1(y_{k+m} + a_1 y_{k+m-1} + \cdots + a_n y_{k+m-n}) \\ = a_1(b_1 u_{k-1} + \cdots + b_n u_{k-n}) + a_1 e_{k-1}. \end{aligned} \tag{58}$$

We may eliminate $y_{k+m}$ by subtracting (58) from (57). We obtain

$$\begin{aligned} y_{k+m+1} &+ (a_2 - a_1^2) y_{k+m-1} + \cdots + (a_n - a_1 a_{n-1}) y_{k+m+1-n} - a_1 a_n y_{k+m-n} \\ &= b_1 u_k + (b_2 - a_1 b_1) u_{k-1} + \cdots + (b_n - a_1 b_{n-1}) u_{k-n+1} \\ &\quad - a_1 b_n u_{k-n} + e_k - a_1 e_{k-1}. \end{aligned}$$

Proceeding similarly we can obtain an equation of the form

$$y_{k+m+1} + \alpha_1 y_k + \cdots + \alpha_n y_{k-n+1} = \beta_0(u_k + \beta_1 u_{k-1} + \cdots + \beta_l u_{k-l}) + \tilde{e}_k \tag{59}$$

where $l = n + m - 1$, and $\tilde{e}_k$ is a linear combination of the disturbance vectors $e_i$. The coefficients $\alpha_1, \ldots, \alpha_n, \beta_0, \ldots, \beta_l$ in (59) are determined from the coefficients $a_1, \ldots, a_n, b_1, \ldots, b_n$ by means of the elimination process described above. Furthermore, it is easy to see that we have $\beta_0 = b_1 \neq 0$. Note that when there is no time delay (i.e., $m = 0$) Eqs. (57) and (59) coincide.

The problem that we are concerned with is to find a suitable control law

$$\mu_k(I_k) = \mu_k(y_0, \ldots, y_k, u_0, \ldots, u_{k-1})$$

depending on the past inputs and outputs, which are assumed to be observable without error. We utilize the quadratic cost functional

$$E\left\{\sum_{k=0}^{N} y_k^2\right\}. \tag{60}$$

Thus, the objective of control is to keep the output of the process near zero.

If the parameters $a_1, \ldots, a_n, b_1, \ldots, b_n$ in Eq (57) were known, the optimal control law could be obtained by reformulating our problem into the form of the basic problem with perfect state information via state augmentation, and by subsequently applying the DP algorithm. We shall bypass, however, this reformulation and obtain the optimal control law directly. Since the system is linear, the criterion is quadratic, and the disturbances are independent, certainty equivalence holds. The optimal control law is the same as the one that would be obtained if the disturbances $e_k$ are set equal to their expected value $\bar{e}_k = 0$, and the system equation is given by [cf. (57)]

$$y_{k+m+1} + \sum_{i=1}^{n} a_i y_{k+m+1-i} = \sum_{i=1}^{n} b_i u_{k+1-i},$$

or equivalently by [cf. (59)]

$$y_{k+m+1} + \alpha_1 y_k + \cdots + \alpha_n y_{k-n+1} = \beta_0(u_k + \beta_1 u_{k-1} + \cdots + \beta_l u_{k-l}).$$

By using the equation above in the cost functional (60) it follows immediately that the optimal control law for the corresponding deterministic problem is given by

$$\mu_k^{\mathrm{d}}(I_k) = -\sum_{i=1}^{l} \beta_i u_{k-i} + \frac{1}{\beta_0} \sum_{i=1}^{n} \alpha_i y_{k+1-i}.$$

By certainty equivalence the same control law is optimal for the stochastic problem assuming that the parameters $\alpha_1, \ldots, \alpha_n, \beta_0, \ldots, \beta_l$ are known. Note that this control law does not depend on the time index $k$ or the number of stages $N$.

When the parameters $\alpha_1, \ldots, \alpha_n, \beta_0, \ldots, \beta_l$ are unknown, then an approximate form of the NFC is given by

$$\tilde{\mu}_k(I_k) = -\sum_{i=1}^{l} \bar{\beta}_i^k u_{k-i} + \frac{1}{\bar{\beta}_0^k} \sum_{i=1}^{n} \bar{\alpha}_i^k y_{k+1-i}, \tag{61}$$

where $I_k = (y_0, \ldots, y_k, u_0, \ldots, u_{k-1})$ and $\bar{\alpha}_1^k, \ldots, \bar{\alpha}_n^k, \bar{\beta}_0^k, \ldots, \bar{\beta}_l^k$ are estimates of the parameters $\alpha_1, \ldots, \alpha_n, \beta_0, \ldots, \beta_l$ based on the inputs and outputs available up to time $k$.

If it were possible to construct an estimation scheme such that the estimates $\bar{\alpha}_i^k, \bar{\beta}_i^k$ converge to the true parameters $\alpha_i, \beta_i$, then the suboptimal controller defined by (61) would become asymptotically optimal. In fact, for this purpose it is sufficient that

$$\lim_{k\to\infty} \bar{\beta}_i^k = \beta_i, \qquad i = 1, \ldots, l, \tag{62}$$

$$\lim_{k\to\infty} \bar{\alpha}_i^k/\bar{\beta}_0^k = \alpha_i/\beta_0, \qquad i = 1, \ldots, n. \tag{63}$$

A suboptimal controller of the form (61) equipped with an estimation scheme for which (62) and (63) hold (with probability one) is referred to as a *self-tuning regulator*. Such regulators have many advantages. They are very simple to implement (provided the estimation scheme is simple), and they become optimal in the limit. Furthermore, not only they do not require a priori knowledge of the parameters of the process, but in addition they adjust appropriately when these parameters change—a frequent occurrence in many processes.

At the present time the analysis of the question as to when a regulator of the type described above has the self-tuning property is not complete. Future research will undoubtedly shed more light on this question. However, it appears that for many practical problems it is possible to construct simple self-tuning regulators.

Let us describe briefly a simple estimation scheme that can be used in conjunction with the control law (61). The scheme is based on the method of least squares and has been used with success in several applications. One may show that due to the employment of the feedback control law (61) it is possible that the least squares estimation scheme may encounter subtle difficulties if we try to estimate all the parameters $\alpha_1, \ldots, \alpha_n, \beta_0, \ldots, \beta_l$ (see Problem 7). On the other hand we note that for purposes of control we do not need to estimate both $\beta_0$ and $\alpha_i$, $i = 1, \ldots, n$. Rather, from (61) it follows that we need only estimate the ratios $\alpha_i/\beta_0$, $i = 1, \ldots, n$. This will be done by keeping $\beta_0$ fixed at some nonzero value $\bar{\beta}_0$, and by estimating $\alpha_i$, $i = 1, \ldots, n$, together with $\beta_i$, $i = 1, \ldots, l$. Let us define the $(n + l)$-dimensional column vectors

$$\theta = (\alpha_1, \ldots, \alpha_n, \beta_1, \ldots, \beta_l)',$$
$$z_k = (-y_k, \ldots, -y_{k-n+1}, \bar{\beta}_0 u_{k-1}, \ldots, \bar{\beta}_0 u_{k-l})',$$

where $\bar{\beta}_0$ has a fixed nonzero value assumed equal to $\beta_0$ for the purposes of estimation. Then Eq. (59) is written

$$y_k = \bar{\beta}_0 u_{k-m-1} + z'_{k-m-1}\theta + \tilde{e}_{k-m-1}.$$

The least squares estimate of $\theta$ at time $k$ is denoted by $\bar{\theta}_k$ and is the vector that minimizes

$$(\theta - \bar{\theta})'\bar{P}(\theta - \bar{\theta}) + \sum_{i=m+1+l}^{k} [y_i - \bar{\beta}_0 u_{i-m-1} - z'_{i-m-1}\theta]^2,$$

where $\bar{\theta}$ is an initial estimate of $\theta$ and $\bar{P}$ is a positive definite symmetric matrix. It is possible to show that the least squares estimate $\bar{\theta}_k$ can be generated recursively by the equations

$$\bar{\theta}_{k+1} = \bar{\theta}_k + g_k(y_k - \bar{\beta}_0 u_{k-m-1} - z'_{k-m-1}\bar{\theta}_k),$$

where the vectors $g_k \in R^{(n+1)}$ are given by

$$g_k = P_k z_{k-m-1}(1 + z'_{k-m-1} P_k z_{k-m-1})^{-1},$$

and the $(n + l) \times (n + l)$ matrices $P_k$ are given by

$$P_{k+1} = P_k - g_k g'_k (1 + z'_{k-m-1} P_k z_{k-m-1}).$$

The initial conditions are

$$\theta_{m+l} = \bar{\theta}, \qquad P_{m+l} = \bar{P}.$$

A proof of the validity of these equations may be found in most texts on parameter estimation theory (see, e.g., [M7]).

When estimation is carried out by the least squares method described above, the control law (61) takes the form

$$\tilde{\mu}_k(I_k) = (1/\bar{\beta}_0) z'_k \bar{\theta}_k.$$

This control law does not always have the self-tuning property

$$\lim_{k \to \infty} \frac{1}{\bar{\beta}_0} \bar{\theta}_k = \frac{1}{\beta_0} \theta.$$

However, both analysis and simulations indicate that if $\bar{\beta}_0$ is chosen judiciously and the order $n$ of the model and the size $m$ of the delay adequately represent those of the real process, then the self-tuning property can often be achieved. In practice one can settle on appropriate choices for $\bar{\beta}_0$, $n$, and $m$ by experimentation and simulation.

A detailed analysis of the regulator (61) equipped with the least squares estimation scheme described above would take us beyond the scope of the text and will not be undertaken. We mention that the main result, proved by Astrom and Wittenmark [A9] under a mild assumption on the process, is that *if the estimates $\bar{\theta}_k$ converge to some vector $\bar{\theta}$, then a self-tuning regulator is obtained, i.e.,*

$$\frac{1}{\beta_0} \theta = \frac{1}{\bar{\beta}_0} \bar{\theta}.$$

Since the estimates $\bar{\theta}_k$ do in fact converge in many cases, the result is quite reassuring. It is also a rather remarkable result in view of the fact that the fixed scalar $\bar{\beta}_0$ need not be equal to the true parameter $\beta_0$. Furthermore, even if $\bar{\beta}_0 = \beta_0$, it is by no means obvious that we should have $\bar{\theta} = \theta$ since the least squares method need not provide in the limit unbiased estimates of the true parameters. We shall demonstrate the result for a simple special case.

*First-Order Process*

Consider the process

$$y_{k+1} + \alpha y_k = \beta u_k + e_k. \tag{64}$$

Let $\bar{\beta}$ denote the fixed scalar used in the least squares estimation procedure (i.e., our earlier $\bar{\beta}_0$), and let $\bar{\alpha}_k$ denote the estimate of $\alpha$ at time $k$. Then $\bar{\alpha}_k$ minimizes over $\alpha$,

$$p_0(\alpha - \bar{\alpha}_0)^2 + \sum_{i=1}^{k} (y_i + \alpha y_{i-1} - \bar{\beta} u_{i-1})^2,$$

where $\bar{\alpha}_0$ is an initial estimate of $\alpha$ and $p_0$ some positive scalar. By setting the derivative at $\bar{\alpha}_k$ of the expression above equal to zero, we obtain

$$p_0(\bar{\alpha}_k - \bar{\alpha}_0) + \sum_{i=1}^{k} y_{i-1}(y_i + \bar{\alpha}_k y_{i-1} - \bar{\beta} u_{i-1}) = 0,$$

from which

$$\bar{\alpha}_k = \frac{p_0 \bar{\alpha}_0 - \sum_{i=1}^{k} y_{i-1}(y_i - \bar{\beta} u_{i-1})}{p_0 + \sum_{i=1}^{k} y_{i-1}^2}. \tag{65}$$

Now the controls $u_{i-1}$ are given by [cf. (61)]

$$u_{i-1} = (\bar{\alpha}_{i-1}/\bar{\beta}) y_{i-1}. \tag{66}$$

By substitution of (66) in (64), we obtain the form of the closed-loop system,

$$y_i = \left( \frac{\beta \bar{\alpha}_{i-1}}{\bar{\beta}} - \alpha \right) y_{i-1} + e_{i-1}. \tag{67}$$

By using Eqs. (66) and (67) in (65), we obtain

$$\bar{\alpha}_k = \frac{p_0 \bar{\alpha}_0 - \sum_{i=1}^{k} \{ y_{i-1}^2 [(\beta \bar{\alpha}_{i-1}/\bar{\beta}) - \alpha - \bar{\alpha}_{i-1}] + e_{i-1} y_{i-1} \}}{p_0 + \sum_{i=1}^{k} y_{i-1}^2}. \tag{68}$$

Let us assume that the estimates $\bar{\alpha}_k$ converge to some scalar $\bar{\alpha}$,

$$\lim_{k \to \infty} \bar{\alpha}_k = \bar{\alpha}. \tag{69}$$

Assume further that

$$\lim_{k \to \infty} \sum_{i=1}^{k} y_{i-1}^2 = \infty, \tag{70}$$

$$\lim_{k \to \infty} \frac{\sum_{i=1}^{k} e_{i-1} y_{i-1}}{p_0 + \sum_{i=1}^{k} y_{i-1}^2} = 0. \tag{71}$$

Relation (70) will hold with probability one if the variance of the disturbance $e_k$ is nonzero. Relation (71) can be expected to hold since the random variable $y_{i-1}$ depends only on the disturbances $e_{i-2}, e_{i-3}, \ldots,$ and is therefore independent of $e_{i-1}$. Since $e_{i-1}$ has zero mean it follows that when $y_{i-1}$ has finite first and second moments and we assume that expected values are equal to the limit of the corresponding average sums (an ergodic assumption), then (71) holds. Notice that relations (69) and (70) imply also

$$\lim_{k \to \infty} \frac{\sum_{i=1}^{k} y_{i-1}^2[(\beta/\bar{\beta}) - 1](\bar{\alpha}_{i-1} - \bar{\alpha})}{p_0 + \sum_{i=1}^{k} y_{i-1}^2} = 0. \tag{72}$$

Now Eq. (68) may be written

$$\bar{\alpha}_k = -\left(\frac{\beta\bar{\alpha}}{\bar{\beta}} - \alpha - \bar{\alpha}\right)\frac{\sum_{i=1}^{k} y_{i-1}^2}{p_0 + \sum_{i=1}^{k} y_{i-1}^2} - \frac{\sum_{i=1}^{k} y_{i-1}^2[(\beta/\bar{\beta}) - 1](\bar{\alpha}_{i-1} - \bar{\alpha})}{p_0 + \sum_{i=1}^{k} y_{i-1}^2}$$
$$+ \frac{p_0\bar{\alpha}_0 - \sum_{i=1}^{k} e_{i-1}y_{i-1}}{p_0 + \sum_{i=1}^{k} y_{i-1}^2}.$$

By taking the limit as $k \to \infty$ and using (69)–(72), we finally obtain $\beta\bar{\alpha}/\bar{\beta} - \alpha = 0$ or equivalently

$$\bar{\alpha}/\bar{\beta} = \alpha/\beta,$$

which is precisely the self-tuning property. Thus, we have proved that if the estimates $\bar{\alpha}_k$ converge [cf. (69)] a self-tuning regulator is obtained. The proof for more general processes follows a similar line of argument as the one given above.

## 5.7 Notes

The problems caused by large dimensionality have long been recognized as the principal computational drawback of DP. A great deal of effort by Bellman and his associates, and by others, has been directed toward finding effective techniques for alleviating these problems. A fairly extensive discussion of the computational aspects of DP can be found in the book by Nemhauser [N2]. Various techniques that are effective for solution of some deterministic dynamic optimization problems by DP can be found in references [K6], [L2], [L9], and [W10]. Finally we note that a class of two-stage stochastic optimization problems, called "stochastic programming problems," can be solved by using mathematical programming techniques (see Problems 1, 2, references [V1], [B7], [B10], and the references quoted therein). The discretization and convergence results of Section 5.2 are due to the author [B15].

The literature abounds with precise and imprecise definitions of adaptivity. The one adopted here is due to Witsenhausen [W4] and captures the idea that a controller should be called adaptive if it uses feedback with advantage.

The example of Section 5.3 showing that the naive feedback controller may perform worse than the optimal open-loop controller is due to Witsenhausen [Thau and Witsenhausen, T1]. For an interesting sensitivity property of the naive feedback controller see the paper by Malinvaud [M2]. The idea of open-loop feedback control is due to Dreyfus [D6], who demonstrated its superiority over open-loop control by means of some examples but did not give a general result. The result and its proof were given recently by the author [B8] in the context of minimax control. Suboptimal controllers other than the NFC and OLFC have been suggested by a number of authors [T3–T5, C1, D5, D7, C5, S1, S2, S15, S20, P6, A1, A9, B17, M9, W9]. The typical approach under imperfect state information is to separate the suboptimal controller into two parts—an estimator and an actuator—and use a "reasonable" design scheme for each part. The concept of the partial open-loop feedback controller is apparently new. Self-tuning regulators received wide attention following the paper by Aström and Wittenmark [A9]. For some recent analysis see Ljung and Wittenmark [L5].

Whenever a suboptimal controller is used in a practical situation it is desirable to know how close the resulting cost approximates the optimal. Tight bounds on the performance of the suboptimal controller are necessary but are usually quite hard to obtain. For some interesting results in this direction, see the papers by Witsenhausen [W5, W6].

As a final note we wish to emphasize that the problem of choice of a suitable suboptimal controller in a practical situation is by no means an easy one. It may be necessary to conduct extensive simulations and experiment with several schemes before settling on a reasonably practical and reliable mode of suboptimal control.

## Problems

**1.** The purpose of this problem is to show how certain stochastic control problems can be solved by (deterministic) mathematical programming techniques. Consider the basic problem of Chapter 2 for the case where there are only two stages ($N = 2$) and the disturbance set for the initial stage $D_0$ is a finite set $D_0 = \{w_0^1, \ldots, w_0^r\}$. The probability of $w_0^i$, $i = 1, \ldots, r$, is denoted $p_i$ and does not depend on $x_0$ or $u_0$. Verify that the optimal value function $J_0(x_0)$ given by

$$J_0(x_0) = \inf_{u_0 \in U_0(x_0)} \sum_{i=1}^{r} p_i[g_0(x_0, u_0, w_0^i) + \inf_{u_1 \in U_1[f_0(x_0, u_0, w_0^i)]} \mathop{E}_{w_1} \{g_1[f_0(x_0, u_0, w_0^i), u_1, w_1] + g_2[f_1[f_0(x_0, u_0, w_0^i), u_1, w_1]]\}]$$

is equal to the optimal value of the problem

$$\underset{\substack{u_0, z_i, u_1^i \\ i=1,\ldots,r}}{\text{minimize}} \quad \sum_{i=1}^{r} p_i[g_0(x_0, u_0, w_0^i) + z_i]$$

$$\text{subject to} \quad z_i \geqslant \mathop{E}_{w_1} \{g_1[f_0(x_0, u_0, w_0^i), u_1^i, w_1] + g_2[f_1[f_0(x_0, u_0, w_0^i), u_1^i, w_1]]\},$$

$$u_0 \in U_0(x_0), \qquad u_1^i \in U_1[f_0(x_0, u_0, w_0^i)].$$

Show also how a solution of the mathematical programming problem above may be used to yield an optimal control law.

**2.** Consider the problem of minimizing over $x$

$$g(x) + \mathop{E}_{r} \left\{ \min_{\substack{y \geqslant 0 \\ f(x) + Ay = r}} q'y \right\}$$

subject to $h_i(x) = 0$, $i = 1, \ldots, s$, $l_j(x) \leqslant 0$, $j = 1, \ldots, p$, where $x \in R^n$, $y \in R^m$, $q$ is a given vector in $R^m$, $r \in R^k$ is a random vector taking a finite number of values $r_1, \ldots, r_t$ with given probabilities $p_1, \ldots, p_t$, $g$, $h_i$, $l_j$ are given continuously differentiable real-valued functions, $f: R^n \to R^k$ is a continuously differentiable mapping, and $A$ is a given $k \times m$ matrix. Show that this problem may be viewed as a two-stage problem that fits the framework of the basic problem of Chapter 2. Show also how the problem can be converted to a deterministic problem that can be solved by standard mathematical programming techniques.

**3.** Consider a problem with perfect state information involving the $n$-dimensional linear system of Section 3.1:

$$x_{k+1} = A_k x_k + B_k u_k + w_k, \qquad k = 0, 1, \ldots, N-1,$$

and a cost functional of the form

$$\mathop{E}_{\substack{w_k \\ k=0,\ldots,N-1}} \{g_N(c'x_N)\},$$

where $c \in R^n$ is a given vector. Show that the DP algorithm for this problem can be carried over a one-dimensional state space.

**4.** Consider a two-stage problem with perfect state information involving the scalar system

$$x_0 = 1, \qquad x_1 = x_0 + u_0 + w_0, \qquad x_2 = f(x_1, u_1).$$

The control constraints are $u_0, u_1 \in \{0, -1\}$. The random variable $w_0$ takes the values $+1$ and $-1$ with equal probability $\frac{1}{2}$. The function $f$ is defined by

$$f(1, 0) = f(1, -1) = f(-1, 0) = f(-1, -1) = 0.5,$$
$$f(2, 0) = 0, \qquad f(2, -1) = 2, \qquad f(0, -1) = 0.6, \qquad f(0, 0) = 2.$$

The cost functional is

$$\underset{w_0}{E}\{x_2\}.$$

(a) Show that one possible OLFC for this problem is defined by

$$\bar{\mu}_0(x_0) = -1; \qquad \bar{\mu}_1(x_1) = \begin{cases} 0 & \text{if } x_1 = \pm 1, 2, \\ -1 & \text{if } x_1 = 0 \end{cases} \tag{73}$$

and the resulting value of the cost is 0.5.

(b) Show that one possible NFC for this problem is defined by

$$\tilde{\mu}_0(x_0) = 0; \qquad \tilde{\mu}_1(x_1) = \begin{cases} 0 & \text{if } x_1 = \pm 1, 2, \\ -1 & \text{if } x_1 = 0, \end{cases} \tag{74}$$

and the resulting value of the cost is 0.3. Show also that the NFC above is an optimal feedback controller.

**5.** Consider the system and cost functional of Problem 4 but with the difference that

$$f(0, -1) = 0.$$

(a) Show that the controller (73) of Problem 4 is both an OLFC and an NFC and that the corresponding value of the cost is 0.5.

(b) Assume that the control constraint set for the first stage is $\{0\}$ rather than $\{0, -1\}$. Show that the controller (74) of Problem 4 is both an OLFC and an NFC and that the corresponding value of the cost is 0.

**6.** Consider the linear quadratic problem of Section 4.3 where in addition to the linear measurements there are some nonlinear measurements received at each time $k$ of the form

$$\tilde{z}_k = h_k(x_k, \tilde{v}_k).$$

The random observation disturbances $\tilde{v}_k$ are independent random variables with given probability distributions. What is the POLFC that ignores the presence of $\tilde{z}_k$?

**7.** Consider the first-order process

$$y_{k+1} + \alpha y_k = \beta u_k + e_k$$

examined at the end of Section 5.6. Assume that a control law of the form $\tilde{\mu}_i(I_i) = (\bar{\alpha}/\bar{\beta})y_i$ is utilized and $\bar{\alpha}$, $\bar{\beta}$ minimize

$$\sum_{i=1}^{k} (y_i + \alpha y_{i-1} - \beta u_{i-1})^2.$$

Show that $[\bar{\alpha} + (\bar{\alpha}\gamma/\bar{\beta})]$, $(\bar{\beta} + \gamma)$ also minimize the expression above, where $\gamma$ is any scalar. (This problem indicates a possible instability in the least-squares estimation scheme if both $\alpha$ and $\beta$ are estimated.)

**8.** Consider the discretization procedure under Assumptions B of Section 5.2, and the control law $\{\hat{\mu}_0, \ldots, \hat{\mu}_{N-1}\}$ defined on the grids $G_k$ and obtained from (33). Extend this control law to the whole space by setting, for $x \in S_k^i$, $\mu_k(x)$ equal to a vector that minimizes $\|u - \hat{\mu}_k(x_k^i)\|$ over $u \in U_k(x)$. Show that Proposition 2 still holds.

*Part II*

# Control of Uncertain Systems over an Infinite Horizon

## General Remarks on Infinite Horizon Problems

The second part of this text is devoted to sequential optimization problems of the type considered in previous chapters but with two basic differences. First, the number of stages is assumed to be infinite, and second, the system is assumed to be stationary, i.e., the system equation and the random disturbance statistics do not change from one stage to the next. In addition, the cost per stage is assumed stationary (except perhaps for the presence of a discount factor).

The assumption of an infinite number of stages is, of course, a mathematical formalization since it is never satisfied in practice. Nonetheless, it constitutes a reasonable and analytically convenient approximation for problems involving a finite but very large number of stages. The assumption of stationarity is often satisfied in practice and in other cases it constitutes a reasonable approximation to a situation where the system parameters vary very slowly as time progresses. However, problems involving a nonstationary system or a nonstationary cost per stage can be reduced to the stationary case by means of a simple reformulation. This reformulation is discussed in Chapter 6 (Section 6.7).

Infinite horizon problems, as a general rule, require considerably more sophisticated mathematical analysis than their corresponding finite horizon counterparts. The analytical difficulties are of a twofold nature. First, the consideration of an infinite horizon introduces the need for analysis of limiting behavior, for example, the convergence of the DP algorithm and the corresponding optimal policies. This analysis is often nontrivial and at times reveals surprising possibilities. Second, a rigorous consideration of the probabilistic aspects of problems involving uncountable disturbance spaces requires the sophisticated machinery of measure-theoretic probability theory. The resulting analytical difficulties are considerably more severe than those of finite horizon problems and are far beyond the introductory scope of this text. For this reason and given that the need for precision is much greater in infinite horizon problems than in their finite horizon counterparts *we shall exclusively restrict ourselves to the case where the disturbance space is a countable set*. The advanced reader may consult the works of Blackwell [B20], Strauch [S17], and Hinderer [H9] for more general expositions.

It should be noted, however, that under the infinite horizon and stationarity assumptions one may obtain results of mathematical and conceptual elegance not to be found in finite horizon problems. Also the implementation of optimal policies for infinite horizon problems is often much simpler. For example, often the optimal policy can be selected to be stationary, i.e., the optimal rule for applying controls need not change from one time period to the next. In addition, in many cases of interest there are available powerful computational methods for calculation of such optimal policies.

Traditionally there have been three classes of infinite horizon problems of major interest:

(a) In the *discounted case* the cost functional takes the form

$$J_\pi(x_0) = \lim_{N\to\infty} \underset{\substack{w_k \\ k=0,1,\ldots}}{E} \left\{ \sum_{k=0}^{N-1} \alpha^k g[x_k, \mu_k(x_k), w_k] \right\},$$

where $J_\pi(x_0)$ denotes the cost associated with an initial state $x_0$ and a policy $\pi = \{\mu_0, \mu_1, \ldots\}$, and $\alpha$ is a scalar with $0 < \alpha < 1$, called the *discount factor*.

The discounted problem is by far the simplest infinite horizon problem particularly when the cost per stage $g(x, u, w)$ is bounded above and below. This case is examined in Sections 6.1–6.3. The case where $g$ may be unbounded either above or below is examined in Sections 6.4–6.6 and is very similar to the undiscounted case where $\alpha = 1$.

(b) In the *undiscounted case* the cost functional takes the form

$$J_\pi(x_0) = \lim_{N\to\infty} \underset{\substack{w_k \\ k=0,1,\ldots}}{E} \left\{ \sum_{k=0}^{N-1} g[x_k, \mu_k(x_k), w_k] \right\},$$

i.e., there is no discount factor ($\alpha = 1$). This case is the subject of Chapter 7.

(c) Minimization of $J_\pi(x_0)$ above makes sense, of course, if $J_\pi(x_0)$ is finite for at least some admissible policies $\pi$ and some initial states $x_0$. In many problems of interest it turns out that $J_\pi(x_0) = +\infty$, but the limit

$$\lim_{N\to\infty} (1/N) \underset{\substack{w_k \\ k=0,1,\ldots}}{E} \left\{ \sum_{k=0}^{N-1} g[x_k, \mu_k(x_k), w_k] \right\}$$

is finite for every policy $\pi = \{\mu_0, \mu_1, \ldots\}$ and initial state $x_0$. Under these circumstances it is reasonable to try to minimize the expression above, which may be viewed as an *average cost per stage* associated with policy $\pi$. Such problems are the subject of Chapter 8. However, for this case we shall restrict our attention mostly to problems involving a finite state and control space, i.e., the system considered is a controlled finite state Markov chain. Some results from the theory of Markov chains will be developed in Chapter 8 and Appendix D as necessary.

*Note*: A recent mathematically rigorous treatment of stochastic optimal control problems with uncountable disturbance spaces can be found in *Stochastic Optimal Control: The Discrete Time Case*, by D. P. Bertsekas and S. E. Shreve, Academic Press, New York, 1978.

*Chapter 6*

# Minimization of Total Expected Value-Discounted Cost

In this chapter we consider a class of infinite horizon problems that involves a discounted cost functional. The introduction of a discount factor is often justified, particularly when the cost per stage has a monetary interpretation. From the mathematical point of view the presence of the discount factor guarantees the finiteness of the cost functional provided costs per stage are uniformly bounded.

Let us define the problem we shall be considering in this chapter.

PROBLEM (D) Consider the stationary discrete-time dynamic system

$$x_{k+1} = f(x_k, u_k, w_k), \qquad k = 0, 1, 2, \ldots, \tag{1}$$

where the state $x_k$, $k = 0, 1, \ldots$, is an element of a space $S$, the control $u_k$, $k = 0, 1, \ldots$, an element of a space $C$, and the random disturbance $w_k$, $k = 0, 1, \ldots$, an element of a space $D$. It is assumed that $D$ is a countable set. The control $u_k$ is constrained to take values in a given nonempty subset $U(x_k)$ of $C$, which depends on the current state $x_k[u_k \in U(x_k)$, for all $x_k \in S$, $k = 0, 1, \ldots]$. The random disturbances $w_k$, $k = 0, 1, \ldots$, have identical statistics and are characterized by probabilities $P(\cdot \mid x_k, u_k)$ defined on $D$, where $P(w_k \mid x_k, u_k)$ is the probability of occurrence of $w_k$, when the current state and control are $x_k$ and $u_k$, respectively. The probability of $w_k$ may depend explicitly on $x_k$ and $u_k$ but not on values of prior disturbances $w_{k-1}, \ldots, w_0$.

Given an initial state $x_0$, the problem is to find a control law, or policy, $\pi = \{\mu_0, \mu_1, \ldots\}$ where $\mu_k: S \to C$, $\mu_k(x_k) \in U(x_k)$, for all $x_k \in S$, $k = 0, 1, \ldots$, that minimizes the cost functional†

$$J_\pi(x_0) = \lim_{N \to \infty} \underset{\substack{w_k \\ k=0,1,\ldots}}{E} \left\{ \sum_{k=0}^{N-1} \alpha^k g[x_k, \mu_k(x_k), w_k] \right\} \tag{2}$$

subject to the system equation constraint (1). The real-valued function $g: S \times C \times D \to R$ is given, and the discount factor $\alpha$ satisfies $0 < \alpha < 1$.

For any $x_0 \in S$ and policy $\pi$ the cost $J_\pi(x_0)$ given by (2) represents the limit of the (discounted) expected finite horizon costs and these costs are well defined as discussed in Section 2.1. Another possibility would be to minimize over $\pi$

$$\underset{\substack{w_k \\ k=0,1,\ldots}}{E} \left\{ \sum_{k=0}^{\infty} \alpha^k g[x_k, \mu_k(x_k), w_k] \right\}.$$

The rigorous introduction of such a cost functional would require the construction of a probability measure on a set of events in the space of all sequences $\{w_0, w_1, \ldots\}$ with elements in $D$ (see Kushner [K10]) and is well beyond the probabilistic framework of this text. However, we mention here that, under the assumptions we shall be using, the expression given above is equal to $J_\pi(x_0)$ as given by (2) for every $x_0 \in S$ and policy $\pi$. This may be proved by using the so-called monotone convergence theorem (see, e.g., Halmos [H5] and Royden [R5]), which allows the interchange of limit and expectation under assumptions that in our case are satisfied. It is also interesting to answer the question as to whether it is possible to reduce further the value of the cost functional by considering "history-remembering" policies of the form $\{\mu_0(I_0), \mu_1(I_1), \ldots\}$, where

$$I_k = (x_0, u_0, x_1, u_1, \ldots, u_{k-1}, x_k)$$

is the history of the system up to time $k$ (or information vector as in Section 4.1). The answer is negative within our framework (see Strauch [S17]), and as a result we shall not deal further with this possibility of enlarging the set of admissible policies.

It is to be noted that, while we allow an arbitrary state and control space, we have made a restrictive assumption in requiring that the disturbance space $D$ is a countable set. This assumption was mostly made in order to avoid

† In what follows we always assume that $g(x, u, w)$ is nonnegative or nonpositive for all $x, u, w$ and hence the limit is well defined as a real number or $\pm\infty$.

the extremely complicated mathematical difficulties associated with the need to restrict the class of admissible policies so that $\sum_{k=0}^{N-1} \alpha^k g[x_k, \mu_k(x_k), w_k]$ is guaranteed to be a well-defined random variable for each $N$. The countability assumption eliminates these difficulties as discussed in Section 2.1. Our assumption, however, is satisfied in many problems of interest, most notably for deterministic optimal control problems and problems of optimal control of a Markov chain with finite or countable number of states. Also for many problems of interest where our assumption is not satisfied, our main results still may be proved under other appropriate assumptions, usually by following the same line of argument as the one given here. For analysis related to problems involving uncountable disturbance sets the advanced reader may consult [B20], [B16], and [S17].

In the first three sections of this chapter we shall be operating under the following assumption:

**Assumption B** (Boundedness) The function $g$ in the cost functional (2) satisfies

$$0 \leqslant g(x, u, w) \leqslant M \qquad \forall (x, u, w) \in S \times C \times D, \tag{3}$$

where $M$ is some scalar.

Notice that (3) could be replaced by an inequality of the form

$$M_2 \leqslant g(x, u, w) \leqslant M_1,$$

where $M_1$, $M_2$ are arbitrary scalars, since addition of a constant $r$ to $g$ merely adds $(1 - \alpha)^{-1}r$ to the cost functional. The assumption above is not as restrictive as it may seem at first sight. First, it holds always for problems where the spaces $S$, $C$, and $D$ are finite sets. Second, the assumption of finiteness of $D$ is implicitly made each time a computational solution to the problem is sought. In addition, during the computations the effective state and control spaces will ordinarily be finite or bounded sets and relation (3) will usually hold. In other cases, it is often possible to reformulate the problem so that it is defined over bounded but arbitrarily large regions of the state space and the control space over which relation (3) holds. In any case the assumption will be somewhat relaxed in Section 6.4, where $g$ will be allowed to be unbounded either from below or from above.

Let us denote by $\Pi$ the set of all *admissible* policies $\pi$, i.e., the set of all sequences of functions $\pi = \{\mu_0, \mu_1, \ldots\}$ with $\mu_k : S \to C$, $\mu_k(x) \in U(x)$ for all $x \in S$, $k = 0, 1, \ldots$. Then the optimal value function $J^*$ given by

$$J^*(x) = \inf_{\pi \in \Pi} J_\pi(x) \qquad \forall x \in S$$

is well defined as a real-valued function under Assumption B. In fact, one may show that $J^*$ is uniformly bounded above and below.

A class of admissible policies of particular interest to us is the class of *stationary admissible policies* of the form $\pi = \{\mu, \mu, \ldots\}$, where $\mu: S \to C$, $\mu(x) \in U(x)$, $x \in S$. For such policies the rule for control selection is the same at every stage. The cost associated with an admissible stationary policy $\{\mu, \mu, \ldots\}$ and an initial state $x \in S$ will also be denoted by $J_\mu(x)$, i.e., for $\pi = \{\mu, \mu, \ldots\}$ we write

$$J_\mu(x_0) = J_\pi(x_0) = \lim_{N \to \infty} \underset{\substack{w_k \\ k=0,1,\ldots}}{E} \left\{ \sum_{k=0}^{N-1} \alpha^k g[x_k, \mu(x_k), w_k] \right\}.$$

Similarly as for $J^*$ we have that $J_\mu$ is well defined as a real-valued function under Assumption B. *A statement that the stationary policy $\{\mu^*, \mu^*, \ldots\}$ is optimal will mean throughout this chapter that $\{\mu^*, \mu^*, \ldots\}$ is admissible and we have $J^*(x) = J_{\mu^*}(x)$ for all $x \in S$.*

The next section gives a characterization of the optimal value function $J^*$ and provides some convergence and existence results. Section 6.2 describes computational methods for Problem (D), under the assumption that the state, control, and disturbance spaces are finite sets. The results obtained in Sections 6.1 and 6.2 are interpreted by means of the notion of a contraction mapping in Section 6.3. In Section 6.4 we relax Assumption B and consider costs per stage that are unbounded above or below. In Sections 6.5 and 6.6 we consider a problem involving a linear system and a discounted quadratic cost functional and an inventory control problem with discounted cost. Finally, in Section 6.7 we consider problems involving a nonstationary or periodic system and cost per stage. We show that such problems can be embedded within the framework of Problem (D), which involves a stationary system and cost per stage. Consequently we are able to obtain in a simple manner results for nonstationary problems that are analogous to those for the stationary case.

## 6.1 Convergence and Existence Results

Consider for every positive integer $N$ the following $N$-stage problem obtained from the infinite horizon problem defined earlier by means of truncation. This problem is to find a policy $\pi_N = \{\mu_0, \mu_1, \ldots, \mu_{N-1}\}$ with $\mu_k(x_k) \in U(x_k)$, $\forall x_k \in S$ that minimizes

$$J_{\pi_N}(x_0) = \underset{\substack{w_k \\ k=0,1,\ldots,N-1}}{E} \left\{ \sum_{k=0}^{N-1} \alpha^k g[x_k, \mu_k(x_k), w_k] \right\} \tag{4}$$

subject to the system equation constraints. The optimal value of this problem for each initial state $x_0$ is denoted $J_N^*(x_0)$ and is given by (cf. Chapter 2, Problem 6)

$$J_N^*(x_0) = J_N(x_0), \tag{5}$$

where for every $N$, $J_N(x_0)$ is given by the $N$th step of the algorithm

$$J_0(x) = 0 \qquad \forall x \in S \tag{6}$$

$$J_{k+1}(x) = \inf_{u \in U(x)} \mathop{E}_{w} \{g(x, u, w) + \alpha J_k[f(x, u, w)]\}$$

$$\forall x \in S, \quad k = 0, 1, \ldots, N-1. \tag{7}$$

The expectation above is, of course, taken with respect to the given distribution $P(\cdot \mid x, u)$, which depends on $x$, $u$.

Notice that in the DP algorithm (6) and (7) we have reversed the indexing of the optimal value functions so that now the algorithm proceeds from lower to higher values of indices $k$ in contrast with finite horizon algorithms. We have also dropped time indices where they are redundant due to stationarity. These notational conventions are convenient for infinite horizon problems and will be adopted throughout the remainder of the text.

We shall also denote for any functions $J: S \to R$, $\mu: S \to C$ with $\mu(x) \in U(x)$, $\forall x \in S$, for which the expected values below are well defined:

$$T(J)(x) = \inf_{u \in U(x)} \mathop{E}_{w} \{g(x, u, w) + \alpha J[f(x, u, w)]\}, \tag{8}$$

$$T_\mu(J)(x) = \mathop{E}_{w} \{g[x, \mu(x), w] + \alpha J[f(x, \mu(x), w)]\}. \tag{9}$$

In these relations $T(J)(\cdot)$ and $T_\mu(J)(\cdot)$ are functions defined on the state space $S$, and $T$, $T_\mu$ may be viewed as mappings that transform a function $J$ on $S$ into another function $[T(J)$ or $T_\mu(J)]$ on $S$. In terms of the notation above algorithm (6) and (7) can be written

$$J_0(x) = 0, \tag{10}$$

$$J_k(x) = T^k(J_0)(x), \qquad k = 0, 1, \ldots, \tag{11}$$

where $T^k$ is defined for all $k$ and $J: S \to R$ by

$$T^k(J)(x) = T[T^{k-1}(J)](x), \qquad T^0(J) = J,$$

i.e., $T^k$ is the composition of the mapping $T$ with itself $k$ times.

We have the following lemma.

**Lemma 1** For any bounded functions $J: S \to R$, $J': S \to R$, such that

$$J(x) \leqslant J'(x) \qquad \forall x \in S,$$

and for any function $\mu: S \to C$ with $\mu(x) \in U(x)$, $\forall x \in S$ we have

$$T^k(J)(x) \leqslant T^k(J')(x) \qquad \forall x \in S, \quad k = 0, 1, \ldots,$$
$$T_\mu^k(J)(x) \leqslant T_\mu^k(J')(x) \qquad \forall x \in S, \quad k = 0, 1, \ldots.$$

*Proof* For any $x$, $u$, $w$ we have $J[f(x, u, w)] \leqslant J'[f(x, u, w)]$, from which we obtain

$$\underset{w}{E}\{g(x, u, w) + \alpha J[f(x, u, w)]\} \leqslant \underset{w}{E}\{g(x, u, w) + \alpha J'[f(x, u, w)]\}.$$

From this relation we obtain $T(J)(x) \leqslant T(J')(x)$, $\forall x \in S$ by taking the infimum of both sides with respect to $u \in U(x)$, and the first inequality is proved for $k = 1$. Similarly, it is proved for all $k$. A similar argument proves also the second inequality. Q.E.D.

For any two functions $J: S \to R$, $J': S \to R$, we write

$$J \leqslant J' \quad \text{if} \quad J(x) \leqslant J'(x) \quad \forall x \in S.$$

With this notation Lemma 1 is stated as

$$J \leqslant J' \Rightarrow T^k(J) \leqslant T^k(J'), \qquad k = 1, 2, \ldots,$$
$$J \leqslant J' \Rightarrow T_\mu^k(J) \leqslant T_\mu^k(J'), \qquad k = 1, 2, \ldots.$$

Denote also by $e: S \to R$ the function taking the value 1 identically on $S$:

$$e(x) = 1 \qquad \forall x \in S. \tag{12}$$

We have from (8) and (9) for any function $J: S \to R$ and any scalar $r$,

$$T(J + re)(x) = T(J)(x) + \alpha r \qquad \forall x \in S, \tag{13}$$

$$T_\mu(J + re)(x) = T_\mu(J)(x) + \alpha r \qquad \forall x \in S. \tag{14}$$

The following proposition shows that the optimal value function $J_N(x_0)$ of the $N$-stage truncated problem converges to the optimal value of the infinite horizon problem. Not only that, but the proposition goes further and shows that the DP algorithm (6) and (7) and (10) and (11) converges to the optimal value function $J^*$ of the infinite horizon problem for an arbitrary bounded starting function $J_0$.

**Proposition 1** (Convergence of the DP Algorithm) The optimal value function $J^*$ of Problem (D) satisfies

$$0 \leqslant J^*(x) \leqslant M/(1 - \alpha) \qquad \forall x \in S, \tag{15}$$

where $M$ is the upper bound in (3). Furthermore, for any bounded function $J: S \to R$ there holds

$$J^*(x) = \lim_{k \to \infty} T^k(J)(x) \qquad \forall x \in S. \tag{16}$$

*Proof* From (3) we have for any initial state $x \in S$ and every policy $\{\mu_0, \mu_1, \ldots\}$,

$$\lim_{N \to \infty} E\left\{\sum_{k=0}^{N-1} \alpha^k g[x_k, \mu_k(x_k), w_k]\right\} \leqslant E\left\{\sum_{k=0}^{N-1} \alpha^k g[x_k, \mu_k(x_k), w_k]\right\} + M \sum_{k=N}^{\infty} \alpha^k.$$

By taking infima over $\{\mu_0, \mu_1, \ldots\}$ of both sides,

$$J^*(x) \leqslant J_N(x) + [\alpha^N/(1-\alpha)]M \qquad \forall x \in S, \quad N = 0, 1, \ldots,$$

where $J_N$ is defined for all $N$ by (6) and (7) [or (10) and (11)]. By combining this relation with (6) we obtain (15). Also in view of (3),

$$J_N(x) \leqslant J^*(x) \qquad \forall x \in S.$$

Combining the two inequalities above we obtain

$$J^*(x) = \lim_{N \to \infty} J_N(x) \qquad \forall x \in S. \tag{17}$$

Now for an arbitrary bounded function $J: S \to R$ let $r$ be a scalar such that

$$(J - re)(x) \leqslant 0, \qquad (J + re)(x) \geqslant 0, \qquad \forall x \in S,$$

where $e$ is the unit function defined by (12). We have by Lemma 1,

$$T^k(J - re)(x) \leqslant J_k(x) \leqslant T^k(J + re)(x) \qquad \forall x \in S, \tag{18}$$

and by (13)

$$T^k(J + re)(x) = T^k(J - re)(x) + 2\alpha^k r = T^k(J)(x) + \alpha^k r \qquad \forall x \in S. \tag{19}$$

Taking superior and inferior limits in (18) and using (17) we obtain

$$\begin{aligned} \limsup_{k \to \infty} T^k(J - re)(x) \leqslant J^*(x) \leqslant \limsup_{k \to \infty} T^k(J + re)(x), \\ \liminf_{k \to \infty} T^k(J - re)(x) \leqslant J^*(x) \leqslant \liminf_{k \to \infty} T^k(J + re)(x). \end{aligned} \tag{20}$$

On the other hand from (19) we have

$$\limsup_{k \to \infty} T^k(J - re)(x) = \limsup_{k \to \infty} T^k(J + re)(x) = \limsup_{k \to \infty} T^k(J)(x),$$

$$\liminf_{k \to \infty} T^k(J - re)(x) = \liminf_{k \to \infty} T^k(J + re)(x) = \liminf_{k \to \infty} T^k(J)(x).$$

Using the above relations in inequalities (20) we obtain

$$J^*(x) = \limsup_{k \to \infty} T^k(J)(x) = \liminf_{k \to \infty} T^k(J)(x)$$

and hence

$$J^*(x) = \lim_{k \to \infty} T^k(J)(x) \qquad \forall x \in S. \quad \text{Q.E.D.}$$

Now given any stationary policy $\pi = \{\mu, \mu, \ldots\}$ where $\mu: S \to C$ with $\mu(x) \in U(x)$, $\forall x \in S$, we can consider the problem that is the same as Problem (D) except for the fact that the control constraint set contains only one element for each state $x$, the control $\mu(x)$, i.e., a control constraint set of the form $\tilde{U}(x) = \{\mu(x)\}$, $\forall x \in S$. Clearly this problem falls within the framework of Problem (D) and since there is only one admissible control law (the policy $\{\mu, \mu, \ldots\}$) application of Proposition 1 yields the following corollary:

**Corollary 1.1** The value $J_\mu(x)$ of the cost functional (2) corresponding to an admissible stationary policy $\{\mu, \mu, \ldots\}$ when the initial state is $x$ satisfies

$$0 \leq J_\mu(x) \leq M/(1 - \alpha). \tag{21}$$

Furthermore, for any bounded function $J: S \to R$ there holds

$$J_\mu(x) = \lim_{k \to \infty} T_\mu^k(J)(x) \qquad \forall x \in S. \tag{22}$$

The next proposition shows that the function $J^*$ is the unique solution of a certain functional equation. This equation provides the means for obtaining a stationary optimal control law.

**Proposition 2** (Optimality Equation—Necessary and Sufficient Condition for Optimality) The optimal value function $J^*$ satisfies

$$J^*(x) = \inf_{u \in U(x)} \mathop{E}_{w} \{g(x, u, w) + \alpha J^*[f(x, u, w)]\} \qquad \forall x \in S \tag{23}$$

or equivalently

$$J^*(x) = T(J^*)(x) \qquad \forall x \in S.$$

Furthermore, $J^*$ is the unique bounded solution of the above functional equation. In addition, if $\mu^*: S \to C$ is a function such that $\mu^*(x) \in U(x)$, $\forall x \in S$, and $\mu^*(x)$ attains the infimum in the right-hand side of (23) for each $x \in S$, then the stationary policy $\{\mu^*, \mu^*, \ldots\}$ is optimal. Conversely if $\{\mu^*, \mu^*, \ldots\}$ is an optimal stationary policy, then $\mu^*(x)$ attains the infimum in the right-hand side of (23) for all $x \in S$.

*Proof* Let $J_0$ be the function that is identically zero on $S$ $[J_0(x) = 0, \forall x \in S]$. We have by (3) and (6), (7) [or (10) and (11)]

$$J_0(x) \leqslant T(J_0)(x) \leqslant \cdots \leqslant T^k(J_0)(x) \leqslant T^{k+1}(J_0)(x) \leqslant \cdots \leqslant J^*(x) \qquad \forall x \in S.$$

Hence for all $k$ and $x \in S$,

$$\begin{aligned} T^{k+1}(J_0)(x) &= \inf_{u \in U(x)} \mathop{E}_{w} \{g(x, u, w) + \alpha T^k(J_0)[f(x, u, w)]\} \\ &\leqslant \inf_{u \in U(x)} \mathop{E}_{w} \{g(x, u, w) + \alpha J^*[f(x, u, w)]\}. \end{aligned}$$

Taking the limit as $k \to \infty$ and using (16) we obtain

$$J^*(x) \leqslant \inf_{u \in U(x)} \mathop{E}_{w} \{g(x, u, w) + \alpha J^*[f(x, u, w)]\}, \tag{24}$$

or equivalently

$$J^*(x) \leqslant T(J^*)(x) \qquad \forall x \in S. \tag{25}$$

It follows from (25) that

$$J^*(x) \leqslant T(J^*)(x) \leqslant T^2(J^*)(x) \leqslant \cdots \leqslant T^k(J^*)(x) \leqslant T^{k+1}(J^*)(x) \leqslant \cdots.$$

By taking the limit as $k \to \infty$ and using $\lim_{k\to\infty} T^k(J^*)(x) = J^*(x)$ (Proposition 1), we obtain

$$J^*(x) \leqslant T(J^*)(x) \leqslant \cdots \leqslant T^k(J^*)(x) \leqslant \cdots \leqslant J^*(x).$$

Hence $J^*(x) = T(J^*)(x)$.

To show uniqueness simply observe that if $J_1^*, J_2^*$ were two bounded solutions of (23) we would have for all $k$

$$J_1^*(x) = T^k(J_1^*)(x), \qquad J_2^*(x) = T^k(J_2^*)(x) \qquad \forall x \in S.$$

By Proposition 1, however, we have

$$\lim_{k\to\infty} T^k(J_1^*)(x) = \lim_{k\to\infty} T^k(J_2^*)(x) = J^*(x) \qquad \forall x \in S.$$

Hence $J_1^* = J_2^* = J^*$. In order to prove the last part of the proposition let us state the following corollary, which follows from the part of Proposition 2 already proved by the same reasoning we used to obtain Corollary 1.1 from Proposition 1.

**Corollary 2.1** Let $\{\mu, \mu, \ldots\}$ be an admissible stationary policy. Then

$$J_\mu(x) = \mathop{E}_{w} \{g[x, \mu(x), w] + \alpha J_\mu[f(x, \mu(x), w)]\} \qquad \forall x \in S.$$

Furthermore, $J_\mu$ is the unique bounded solution of the above functional equation.

Now if $\mu^*(x)$ minimizes the right-hand side of (23) for each $x \in S$, then we have for all $x \in S$,

$$J^*(x) = \mathop{E}_{w} \{g[x, \mu^*(x), w] + \alpha J^*[f(x, \mu^*(x), w)]\}.$$

Hence by the uniqueness part of the corollary we must have $J^*(x) = J_{\mu^*}(x)$ for all $x \in S$, and it follows that $\{\mu^*, \mu^*, \ldots\}$ is optimal. Also if $\{\mu^*, \mu^*, \ldots\}$ is optimal, then we have $J^* = J_{\mu^*}$ and from the corollary $J_{\mu^*} = T_{\mu^*}(J_{\mu^*})$. Hence $J^* = T_{\mu^*}(J^*)$, which implies that $\mu^*(x)$ attains the infimum in (23) for all $x \in S$. Q.E.D.

Note that Proposition 2 implies the existence of an optimal stationary policy when the infimum in the right-hand side of (23) is attained for all $x \in S$. On the other hand, if the infimum is not attained, there arises the question whether one may approximate as closely as desired the optimal value $J^*(x)$ corresponding to an initial state $x$ by employing a stationary policy. The answer is affirmative (see Problem 12).

We finally show the following relation, which holds for any bounded function $J: S \to R$:

$$\sup_{x \in S} |T^k(J)(x) - J^*(x)| \leq \alpha^k \sup_{x \in S} |J(x) - J^*(x)|, \qquad k = 0, 1, \ldots.$$

This relation is a special case of the following result:

**Proposition 3** For any two bounded functions $J: S \to R$, $J': S \to R$, and for all $k = 0, 1, \ldots$ there holds

$$\sup_{x \in S} |T^k(J)(x) - T^k(J')(x)| \leq \alpha^k \sup_{x \in S} |J(x) - J'(x)|.$$

*Proof* It is sufficient to prove the result for $k = 1$ since repeated use of the inequality with $k = 1$ yields the desired result.

We have for any $x \in S$, $u \in U(x)$,

$$\begin{aligned} \underset{w}{E}\{g(x, u, w) + \alpha J[f(x, u, w)]\} &= \underset{w}{E}\{g(x, u, w) + \alpha J'[f(x, u, w)]\} \\ &\quad + \alpha \underset{w}{E}\{J[f(x, u, w)] - J'[f(x, u, w)]\} \\ &\leq \underset{w}{E}\{g(x, u, w) + \alpha J'[f(x, u, w)]\} \\ &\quad + \alpha \sup_{x \in S} |J(x) - J'(x)|. \end{aligned}$$

Taking the infimum of both sides over $u \in U(x)$, we obtain

$$T(J)(x) - T(J')(x) \leq \alpha \sup_{x \in S} |J(x) - J'(x)| \qquad \forall x \in S.$$

A similar argument shows that

$$T(J')(x) - T(J)(x) \leq \alpha \sup_{x \in S} |J(x) - J'(x)| \qquad \forall x \in S,$$

and hence

$$|T(J)(x) - T(J')(x)| \leq \alpha \sup_{x \in S} |J(x) - J'(x)| \qquad \forall x \in S.$$

By taking the supremum of the left side over $x \in S$ the result follows. Q.E.D.

As earlier, we have

**Corollary 3.1** For any two bounded functions $J: S \to R$, $J': S \to R$, and any admissible stationary policy $\{\mu, \mu, \ldots\}$, we have for all $k = 0, 1, \ldots,$

$$\sup_{x \in S} |T_\mu^k(J)(x) - T_\mu^k(J')(x)| \leqslant \alpha^k \sup_{x \in S} |J(x) - J'(x)|.$$

The main conclusion from the propositions established earlier is that the optimal value function $J^*$ is bounded and is the unique bounded solution of the functional equation (23). This equation yields an optimal stationary control law provided the infimum in its right-hand side is attained. Furthermore, the DP algorithm yields in the limit the function $J^*$ starting from an arbitrary bounded function $J$ and the rate of convergence is at least as fast as the rate of a convergent geometric progression (Proposition 3). Thus the DP algorithm may be used for actual computation of at least an approximation to $J^*$. This computational method together with some additional methods will be further examined in the next section. The remainder of this section is devoted to two examples, in which we do not state explicitly that the disturbance space is countable. The conclusions and results obtained are rigorous only for a countable disturbance space.

*Asset Selling Example*

Consider the asset selling problem of Section 3.4. When the problem is viewed over an infinite horizon it is essentially a discounted cost problem with discount factor $\alpha = 1/(1 + r)$ [cf. Eq. (3.65)]. If we assume that the offers $x$ are bounded, then the analysis of the present section is applicable and the optimal value function is the unique solution of the functional equation

$$J^*(x) = \max[x, (1 + r)^{-1} \underset{w}{E}\{J^*(w)\}].$$

The optimal policy is obtained from this equation and has the following form. If current offer $\geqslant (1 + r)^{-1} E_w \{J^*(w)\} = \bar{\alpha}$, sell and otherwise do not sell. The critical number $\bar{\alpha} = (1 + r)^{-1} E_w \{J^*(w)\}$ is obtained as in Section 3.4.

*Component Replacement Example*

A certain component of a machine tool can be in any one of a continuum of states, which we represent by the interval [0, 1]. At the beginning of each period the component is inspected, its current state $x \in [0, 1]$ determined, and a decision made whether or not to replace the component at a cost $R > 0$ by a new one at state $x = 0$. The expected cost of having the component at state $x$ for a single period is $C(x)$, where $C(\cdot)$ is a nonnegative bounded and increasing function of $x$ on [0, 1]. The conditional cumulative probability distribution $F(z|x)$ of the component being at a state less or equal to $z$ at the end of the period given that it was at state $x$ at the beginning of the period is known. Furthermore, for each $y \in [0, 1]$ we have

$$\int_y^1 dF(z|x_1) \leqslant \int_y^1 dF(z|x_2) \qquad \text{for} \quad 0 \leqslant x_1 \leqslant x_2 \leqslant 1.$$

This assumption implies that the component tends to turn worse gradually with usage, i.e., for each $y \in [0, 1]$ there is greater chance that the component will go to a final state in the interval $[y, 1]$ when at a worse initial state. Assuming a discount factor $\alpha \in (0, 1)$ and an infinite horizon, the problem is to determine the optimal replacement policy.

Except for the countability assumption, the problem clearly falls within the framework of this section and the optimal value function $J^*$ is the unique bounded solution of the functional equation

$$J^*(x) = \min\left[R + C(0) + \alpha \int_0^1 J^*(z)\, dF(z|0),\ C(x) + \alpha \int_0^1 J^*(z)\, dF(z|x)\right].$$

An optimal replacement policy is given by:

$$\text{Replace if} \quad R + C(0) + \alpha \int_0^1 J^*(z)\, dF(z|0) \leqslant C(x) + \alpha \int_0^1 J^*(z)\, dF(z|x).$$

Do not replace otherwise.

Now consider the DP algorithm

$$J_0(x) = 0,$$

$$T(J_0)(x) = \min[R + C(0), C(x)],$$

$$T^k(J_0)(x) = \min\left[R + C(0) + \alpha \int_0^1 T^{k-1}(J_0)(z)\, dF(z|0),\right.$$

$$\left. C(x) + \alpha \int_0^1 T^{k-1}(J_0)(z)\, dF(z|x)\right], \qquad k = 1, 2, \ldots.$$

Since $C(x)$ is increasing in $x$ we have that $T(J_0)(x)$ is nondecreasing in $x$ and in view of our assumption on the distributions $F(z|x)$ the same is true for $T^2(J_0)(x)$. Proceeding similarly it follows that $T^k(J_0)(x)$ is nondecreasing in $x$ and so is the limit

$$J^*(x) = \lim_{k\to\infty} T^k(J_0)(x).$$

It follows under our assumptions that the function $C(x) + \alpha \int_0^1 J^*(z)\, dF(z|x)$ is nondecreasing in $x$. This is simply a reflection of the intuitively clear fact

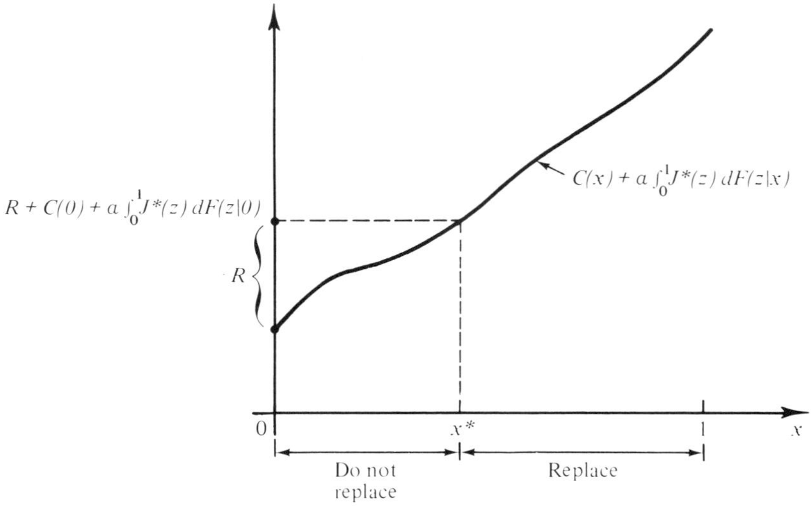


FIGURE 6.1

that the optimal cost cannot decrease as the initial state increases, i.e., we start at a worse initial state. Thus an optimal policy takes the form

$$\begin{array}{lll} \text{Replace} & \text{if} & x \geqslant x^* \\ \text{Do not replace} & \text{if} & x < x^*, \end{array}$$

where $x^*$ is the smallest scalar for which

$$R + C(0) + \alpha \int_0^1 J^*(z)\, dF(z\,|\,0) = C(x^*) + \alpha \int_0^1 J^*(z)\, dF(z\,|\,x^*),$$

as shown in Fig. 6.1.

## 6.2 Computational Methods—Successive Approximation, Policy Iteration, Linear Programming

This section presents three alternative approaches for solving the infinite horizon, discounted cost Problem (D). In order that these approaches be implementable in a computer it is necessary that the state space and control space be finite sets. When these spaces are infinite, it is necessary to replace them with finite sets by means of some discretization procedures. A discretization procedure analogous to the one given in Section 5.2 may be employed (see Bertsekas [B15] or Problem 2) and it possesses similar stability properties under analogous assumptions. The first approach, successive approximation, is essentially the DP algorithm and yields in the limit the optimal value function and an optimal policy as discussed in the previous section. Some variations aimed at accelerating the convergence are discussed in addition. The other two approaches, policy iteration and linear programming, terminate in a finite number of iterations (when the spaces involved are finite sets). However, they require solution of linear systems of equations or of a linear program of dimension as large as the number of points in the state space. When this dimension is very large their practicality is questionable.

Throughout this section we shall assume that *the spaces $S$, $C$, and $D$ in Problem (D) are finite sets*. Under these circumstances, Assumption B can be made to hold by adding a suitable constant to $g$ if necessary. When $S$ and $D$ are finite, the problem becomes one of control of a finite state Markov chain, and for this case one can represent the mappings $T$ and $T_\mu$ of (8) and (9) in a standard form, which is perhaps more convenient both from the point of view of conceptual understanding and from the point of view of computation.

Let $S$ consist of $n$ states denoted by $1, 2, \ldots, n$:

$$S = \{1, 2, \ldots, n\}.$$

Let us denote by $p_{ij}(u)$ *the transition probability*:

$$p_{ij}(u) = P(x_{k+1} = j \mid x_k = i, u_k = u) \qquad \forall i, j \in S, \quad u \in U(i).$$

Thus $p_{ij}(u)$ is the probability that the next state will be $j$ given that the current state is $i$ and control $u \in U(i)$ is applied. These transition probabilities may either be given a priori or they may be calculated from the system equation

$$x_{k+1} = f(x_k, u_k, w_k)$$

and the known probability distribution $P(\cdot \mid x, u)$ of the input disturbance $w_k$. Indeed, we have

$$p_{ij}(u) = P[W_{ij}(u) \mid i, u],$$

where $W_{ij}(u)$ is the (finite) set

$$W_{ij}(u) = \{w \in D \mid f(i, u, w) = j\}.$$

Now let us use the notation

$$\bar{g}(i, u) = \mathop{E}_{w}\{g(i, u, w)\} \qquad \forall i \in S, \quad u \in U(i),$$

where the expectation is taken with respect to the given probability distribution $P(\cdot \mid i, u)$. Then the basic expression

$$\mathop{E}_{w}\{g(i, u, w) + \alpha J[f(i, u, w)]\} \qquad \forall i \in S$$

may be written in terms of $p_{ij}(u)$ and $\bar{g}$ as

$$\bar{g}(i, u) + \alpha \sum_{j=1}^{n} p_{ij}(u)J(j) \qquad \forall i \in S.$$

As a result the mappings $T$ and $T_\mu$ of (8) and (9) can be written for any function $J: S \to R$ as

$$T(J)(i) = \inf_{u \in U(i)} \left[ \bar{g}(i, u) + \alpha \sum_{j=1}^{n} p_{ij}(u)J(j) \right], \qquad i = 1, 2, \ldots, n,$$

$$T_\mu(J)(i) = \bar{g}[i, \mu(i)] + \alpha \sum_{j=1}^{n} p_{ij}[\mu(i)]J(j), \qquad i = 1, 2, \ldots, n.$$

These expressions are often convenient to work with. Notice that the functions $J$, $T_\mu(J)$: $S \to R$, may be represented by the $n$-tuples (or $n$-dimensional vectors) of their values $J(1), \ldots, J(n)$, $T_\mu(J)(1), \ldots, T_\mu(J)(n)$

$$J = \begin{bmatrix} J(1) \\ \vdots \\ J(n) \end{bmatrix} \qquad T_\mu(J) = \begin{bmatrix} T_\mu(J)(1) \\ \vdots \\ T_\mu(J)(n) \end{bmatrix}$$

If for a $\mu: S \to C$ with $\mu(i) \in U(i)$, $i = 1, \ldots, n$, we form the transition probability matrix

$$P_\mu = \begin{bmatrix} p_{11}[\mu(1)] & \cdots & p_{1n}[\mu(1)] \\ \vdots & & \vdots \\ p_{n1}[\mu(n)] & \cdots & p_{nn}[\mu(n)] \end{bmatrix}$$

and consider the $n$-dimensional vector $\bar{g}_\mu$ defined by

$$\bar{g}_\mu = \begin{bmatrix} \bar{g}[1, \mu(1)] \\ \vdots \\ \bar{g}[n, \mu(n)] \end{bmatrix}$$

then we can write in vector notation

$$T_\mu(J) = \bar{g}_\mu + \alpha P_\mu J.$$

The value function $J_\mu$ corresponding to a stationary policy $\{\mu, \mu, \ldots\}$ is by Corollary 2.1 the unique solution of the equation

$$J_\mu = T_\mu(J_\mu) = \bar{g}_\mu + \alpha P_\mu J_\mu.$$

Hence for any admissible stationary policy $\{\mu, \mu, \ldots\}$ the corresponding function $J_\mu$ may be found as the unique $n$-dimensional vector that solves the system of $n$ linear equations

$$(I - \alpha P_\mu)J_\mu = \bar{g}_\mu,$$

or equivalently

$$J_\mu = (I - \alpha P_\mu)^{-1}\bar{g}_\mu,$$

where $I$ denotes the $n \times n$ identity matrix. The invertibility of the matrix $I - \alpha P_\mu$ is assured since we have proved that the system of equations representing the equation $J_\mu = T_\mu(J_\mu)$ has a unique solution for any possible value of the vector $\bar{g}_\mu$ (cf. Corollary 2.1).

*Successive Approximation*

Here we start with an arbitrary bounded function $J: S \to R$ and successively compute $T(J)$, $T^2(J)$, ..., where the mapping $T$ is defined by (8). By Proposition 1 we have

$$\lim_{k \to \infty} T^k(J)(x) = J^*(x) \qquad \forall x \in S.$$

Furthermore by Proposition 3, $|J^*(x) - T^k(J)(x)|$ is bounded by a multiple of a geometric progression for all $x \in S$. It is also of interest to note that the successive approximation method will yield an optimal policy after a finite number of iterations (see Problem 14). The successive approximation scheme can be considerably sharpened and improved by taking advantage of the special structure of the problem as we describe below.

Suppose that we have computed $J$, $T(J)$, $T^2(J)$, ..., $T^k(J)$. The following proposition provides upper and lower bounds for $J^*$, which are obtained from $J$, $T(J)$, ..., $T^k(J)$. These bounds converge monotonically to $J^*$.

**Proposition 4** Let $J: S \to R$. Then for all $x \in S$ and $k = 1, 2, \ldots,$

$$T^k(J)(x) + c_k \leqslant T^{k+1}(J)(x) + c_{k+1} \leqslant J^*(x) \leqslant T^{k+1}(J)(x) + \bar{c}_{k+1} \leqslant T^k(J)(x) + \bar{c}_k, \tag{26}$$

where for all $k = 1, 2, \ldots,$

$$c_k = \frac{\alpha}{1 - \alpha} \min_{x \in S}[T^k(J)(x) - T^{k-1}(J)(x)], \tag{27}$$

$$\bar{c}_k = \frac{\alpha}{1 - \alpha} \max_{x \in S}[T^k(J)(x) - T^{k-1}(J)(x)]. \tag{28}$$

Furthermore, the error bounds (26) are optimal in the following sense:

$$c_{k+1} = \max\{c \mid T^k(J) + (c/\alpha)e \leqslant T^{k+1}(J) + ce \leqslant J^*\}, \tag{29}$$

$$\bar{c}_{k+1} = \min\{\bar{c} \mid J^* \leqslant T^{k+1}(J) + \bar{c}e \leqslant T^k(J) + (\bar{c}/\alpha)e\}, \tag{30}$$

where $e$ is the unit function on $S$ [$e(x) = 1, \forall x \in S$].

*Proof* Denote

$$\gamma = \min_{x \in S}[T(J)(x) - J(x)].$$

We have

$$J + \gamma e \leqslant T(J). \tag{31}$$

Applying $T$ to both sides, using the monotonicity of $T$ and (13),

$$T(J) + \alpha\gamma e \leqslant T^2(J), \tag{32}$$

and because of (31),

$$J + (1 + \alpha)\gamma e \leqslant T(J) + \alpha\gamma e \leqslant T^2(J). \tag{33}$$

This process can be repeated, first applying $T$ to obtain

$$T(J) + (\alpha + \alpha^2)\gamma e \leqslant T^2(J) + \alpha^2\gamma e \leqslant T^3(J), \tag{34}$$

and then using (31) to write

$$J + (1 + \alpha + \alpha^2)\gamma e \leqslant T(J) + (\alpha + \alpha^2)\gamma e \leqslant T^2(J) + \alpha^2\gamma e \leqslant T^3(J). \tag{35}$$

After $k$ steps this results in the inequalities

$$\begin{aligned} J + \left(\sum_{i=0}^{k} \alpha^i\right)\gamma e &\leqslant T(J) + \left(\sum_{i=1}^{k} \alpha^i\right)\gamma e \\ &\leqslant T^2(J) + \left(\sum_{i=2}^{k} \alpha^i\right)\gamma e \leqslant \cdots \leqslant T^{k+1}(J). \end{aligned} \tag{36}$$

Taking the limit as $k \to \infty$ we obtain

$$J + (c_1/\alpha)e \leqslant T(J) + c_1 e \leqslant T^2(J) + \alpha c_1 e \leqslant J^*, \tag{37}$$

where $c_1$ is defined by (27). Replacing $J$ by $T^k(J)$ in this inequality, we have

$$T^{k+1}(J) + c_{k+1}e \leqslant J^*,$$

which is the second inequality in (26).

From (33) we have

$$\alpha\gamma \leqslant \min_{x \in S}[T^2(J)(x) - T(J)(x)],$$

and consequently

$$\alpha c_1 \leqslant c_2.$$

Using this in (37) yields

$$T(J) + c_1 e \leqslant T^2(J) + c_2 e,$$

and replacing $J$ by $T^{k-1}(J)$ we have the first inequality in (26).

To prove (29) we replace $J$ by $T^k(J)$ in (37) and obtain

$$T^k(J) + (c_{k+1}/\alpha)e \leqslant T^{k+1}(J) + c_{k+1}e \leqslant J^*,$$

which implies that $c_{k+1}$ is a member of the set on the right side of (29). If $c$ is any other member of this set, we have

$$c\left(\frac{1}{\alpha} - 1\right) \leqslant T^{k+1}(J)(x) - T^k(J)(x) \qquad \forall x \in S,$$

and so

$$\frac{1-\alpha}{\alpha} c \leqslant \min_{x \in S}[T^{k+1}(J)(x) - T^k(J)(x)] = \frac{1-\alpha}{\alpha} c_{k+1},$$

which shows $c \leqslant c_{k+1}$.

The last two inequalities in (26) and Eq. (30) follow by an analogous argument. Q.E.D.

Notice that the error bounds (26) may be easily computed as a by-product of the computations in the successive approximation method. In practice these bounds are extremely helpful and speed up the convergence considerably. Some properties of these bounds are given in Problem 7. Additional error bounds that are useful in successive approximation methods are given in Problems 3 and 4.

We now consider a computational example that illustrates the utility of the error bounds of Proposition 4.

EXAMPLE 1 Consider a problem where there are two states and two controls

$$S = \{1, 2\}, \qquad C = \{u^1, u^2\}.$$

FIGURE 6.2 (a) $u = u^1$. (b) $u = u^2$.

The transition probabilities corresponding to the controls $u^1$ and $u^2$ are as shown in Fig. 6.2, i.e., we have the transition probability matrices

$$P(u^1) = \begin{bmatrix} p_{11}(u^1) & p_{12}(u^1) \\ p_{21}(u^1) & p_{22}(u^1) \end{bmatrix} = \begin{bmatrix} \frac{3}{4} & \frac{1}{4} \\ \frac{3}{4} & \frac{1}{4} \end{bmatrix}$$

$$P(u^2) = \begin{bmatrix} p_{11}(u^2) & p_{12}(u^2) \\ p_{21}(u^2) & p_{22}(u^2) \end{bmatrix} = \begin{bmatrix} \frac{1}{4} & \frac{3}{4} \\ \frac{1}{4} & \frac{3}{4} \end{bmatrix}.$$

The transition costs are as follows:

$$g(1, u^1) = 2, \qquad g(1, u^2) = 0.5, \qquad g(2, u^1) = 1, \qquad g(2, u^2) = 3,$$

and the discount factor is $\alpha = 0.9$. The mapping $T$ is given by

$$T(J)(i) = \min\left\{ g(i, u^1) + \alpha \sum_{j=1}^{2} p_{ij}(u^1)J(j),\right.$$
$$\left. g(i, u^2) + \alpha \sum_{j=1}^{2} p_{ij}(u^2)J(j) \right\}, \qquad i = 1, 2.$$

The scalars $c_k$ and $\bar{c}_k$ of (27) and (28), respectively, are given by

$$c_k = \frac{\alpha}{1-\alpha} \min\{T^k(J)(1) - T^{k-1}(J)(1), T^k(J)(2) - T^{k-1}(J)(2)\},$$

$$\bar{c}_k = \frac{\alpha}{1-\alpha} \max\{T^k(J)(1) - T^{k-1}(J)(1), T^k(J)(2) - T^{k-1}(J)(2)\}.$$

The results of the successive approximation method starting with the zero function $J_0$ $[J_0(1) = J_0(2) = 0]$ are shown in Table 6.1 and illustrate the power and practicality of the error bounds.

TABLE 6.1

| $k$ | $T^k(J_0)(1)$ | $T^k(J_0)(2)$ | $T^k(J_0)(1) + c_k$ | $T^k(J_0)(1) + \bar{c}_k$ | $T^k(J_0)(2) + c_k$ | $T^k(J_0)(2) + \bar{c}_k$ |
|---|---|---|---|---|---|---|
| 0 | 0.00000 | 0.00000 | | | | |
| 1 | 0.50000 | 1.00000 | 5.00000 | 9.50000 | 5.50000 | 10.00000 |
| 2 | 1.28750 | 1.56250 | 6.35000 | 8.37500 | 6.62500 | 8.65000 |
| 3 | 1.84438 | 2.22063 | 6.85625 | 7.76750 | 7.23250 | 8.14375 |
| 4 | 2.41391 | 2.74459 | 7.12962 | 7.53969 | 7.46031 | 7.87038 |
| 5 | 2.89573 | 3.24692 | 7.23214 | 7.41667 | 7.58333 | 7.76786 |
| 6 | 3.34321 | 3.68517 | 7.28750 | 7.37054 | 7.62946 | 7.71250 |
| 7 | 3.73972 | 4.08583 | 7.30826 | 7.34563 | 7.65437 | 7.69174 |
| 8 | 4.09937 | 4.44362 | 7.31947 | 7.33628 | 7.66372 | 7.68053 |
| 9 | 4.42180 | 4.76689 | 7.32367 | 7.33124 | 7.66876 | 7.67633 |
| 10 | 4.71256 | 5.05727 | 7.32594 | 7.32935 | 7.67065 | 7.67406 |
| 11 | 4.97398 | 5.31886 | 7.32679 | 7.32833 | 7.67167 | 7.67321 |
| 12 | 5.20938 | 5.55418 | 7.32725 | 7.32794 | 7.67206 | 7.67275 |
| 13 | 5.42118 | 5.76602 | 7.32743 | 7.32774 | 7.67226 | 7.67257 |
| 14 | 5.61183 | 5.95665 | 7.32752 | 7.32766 | 7.67234 | 7.67248 |
| 15 | 5.78340 | 6.12823 | 7.32755 | 7.32762 | 7.67238 | 7.67245 |
| 16 | 5.93782 | 6.28265 | 7.32757 | 7.32760 | 7.67240 | 7.67243 |
| 17 | 6.07680 | 6.42163 | 7.32758 | 7.32759 | 7.67241 | 7.67242 |
| 18 | 6.20188 | 6.54670 | 7.32758 | 7.32759 | 7.67241 | 7.67242 |

Another possibility for accelerating the convergence of the successive approximation method is to modify the basic mapping $T$ in the following way.

Let us denote by $1, 2, \ldots, n$ the points of the state space $S$, i.e.,

$$S = \{1, 2, \ldots, n\}.$$

Given a function $J: S \to R$, define the mapping $F$ by

$$F(J)(x) = T_n(J)(x) \qquad \forall x \in S, \tag{38}$$

where the mapping $T_n$ is defined recursively by

$$T_0(J)(x) = J(x) \qquad \forall x \in S, \tag{39}$$

$$T_{i+1}(J)(x) = \begin{cases} \min\limits_{u \in U(x)} \mathop{E}\limits_{w} \{g(x, u, w) + \alpha T_i(J)[f(x, u, w)]\} \\ \qquad\qquad \text{if} \quad x = i + 1, \\ T_i(J)(x) \qquad \text{otherwise,} \end{cases} \tag{40}$$

or equivalently

$$T_{i+1}(J)(x) = \begin{cases} T[T_i(J)](x) & \text{if} \quad x = i + 1, \\ T_i(J)(x) & \text{otherwise.} \end{cases} \tag{41}$$

In words, for the state $x = i + 1$ we have that $T_{i+1}(J)(i + 1)$ is obtained from $T_i(J)$ by minimization over $u$ as in (40), but for all states $x \neq i + 1$, we have $T_{i+1}(J)(x) = T_i(J)(x)$. The function $F(J)$ is obtained from $J$ after minimization has been carried out for every state. The difference with the computation of $T(J)$ is that at each new minimization the "current" function $T_i(J)$ is used in (40) in place of $J$. A moment's reflection should convince the reader that the computation of $F(J)$ is as easy as the computation of $T(J)$. We may consider now the successive approximation method whereby we compute $J, F(J), F^2(J), \ldots$. The following propositions show that the method is valid in the sense that $F^k(J)(x) \to J^*(x)$ as $k \to \infty$ and in addition in many cases it is characterized by better convergence properties than the earlier method.

**Proposition 5** Let $J: S \to R$, $J': S \to R$, be two bounded functions. Then for any $k = 0, 1, \ldots,$

$$\max_{x \in S} |F^k(J)(x) - F^k(J')(x)| \leqslant \alpha^k \max_{x \in S} |J(x) - J'(x)|. \tag{42}$$

Furthermore we have

$$F(J^*)(x) = J^*(x) \qquad \forall x \in S, \tag{43}$$

$$\lim_{k \to \infty} F^k(J)(x) = J^*(x) \qquad \forall x \in S. \tag{44}$$

*Proof* It is sufficient to prove (42) for $k = 1$. We have by definition (38)–(41) of $F$ and Proposition 3 that

$$|T_1(J)(1) - T_1(J')(1)| \leqslant \alpha \max_{x \in S} |J(x) - J'(x)|.$$

Also using this inequality

$$|T_2(J)(1) - T_2(J')(1)| \leqslant \alpha \max_{x \in S} |J(x) - J'(x)|,$$

$$\begin{aligned} |T_2(J)(2) - T_2(J')(2)| &\leqslant \alpha \max\{|T_1(J)(1) - T_1(J')(1)|, \\ &\qquad |J(2) - J'(2)|, \ldots, |J(n) - J'(n)|\} \\ &\leqslant \alpha \max_{x \in S} |J(x) - J'(x)|. \end{aligned}$$

Proceeding similarly we have for every $i$ and $j \leqslant i$,

$$|T_i(J)(j) - T_i(J')(j)| \leqslant \alpha \max_{x \in S} |J(x) - J'(x)|,$$

and for $i = n$ the above relation is equivalent to (42) for $k = 1$. Relation (43) follows immediately from definition (38)–(41) and the fact that $J^* = T(J^*)$. Relation (44) follows immediately from (42) and (43). Q.E.D.

**Proposition 6** If $J: S \to R$ satisfies

$$J(x) \leqslant T(J)(x) \leqslant J^*(x) \qquad \forall x \in S,$$

then

$$T^k(J)(x) \leqslant F^k(J)(x) \leqslant J^*(x) \qquad \forall x \in S, \quad k = 1, 2, \ldots .$$

*Proof* The proof is immediate by using definition (38)–(41) and the monotonicity property of $T$. Q.E.D.

The preceding proposition provides the main motivation for employing the mapping $F$ in place of $T$ in the successive approximation method. A similar result may be proved for functions $J : S \to R$ satisfying $J^* \leqslant T(J) \leqslant J$ for all $x \in S$. Additional results applicable to the mapping $F$ and the related successive approximation method are given in Problems 3 and 4. We now provide a computational example.

EXAMPLE 1 (CONTINUED) Consider the example examined earlier in this section. The mapping $F$ is given by

$$F(J)(i) = T_2(J)(i), \qquad i = 1, 2,$$

where

$$T_1(J)(1) = \min\left\{g(1, u^1) + \alpha \sum_{j=1}^{2} p_{1j}(u^1)J(j),\ g(1, u^2) + \alpha \sum_{j=1}^{2} p_{1j}(u^2)J(j)\right\},$$

$$T_1(J)(2) = J(2),$$

$$T_2(J)(1) = T_1(J)(1),$$

$$T_2(J)(2) = \min\left\{g(2, u^1) + \alpha \sum_{j=1}^{2} p_{2j}(u^1)T_1(J)(j),\ g(2, u^2) + \alpha \sum_{j=1}^{2} p_{2j}(u^2)T_1(J)(j)\right\}.$$

One may show (Problem 3) that

$$\alpha^2 \leqslant [F(J + re)(i) - F(J)(i)]/r \leqslant \alpha, \qquad i = 1, 2, \quad r \neq 0,$$

and thus by the result of Problem 3 we have the error bounds

$$F^k(J)(i) + c_k \leqslant J^*(i) \leqslant F^k(J)(i) + \bar{c}_k, \qquad i = 1, 2,$$

where

$$c_k = \min\left\{\frac{\alpha^2}{1 - \alpha^2}\gamma_k, \frac{\alpha}{1 - \alpha}\gamma_k\right\}, \qquad \bar{c}_k = \max\left\{\frac{\alpha^2}{1 - \alpha^2}\bar{\gamma}_k, \frac{\alpha}{1 - \alpha}\bar{\gamma}_k\right\},$$

$$\gamma_k = \min[F^k(J)(1) - F^{k-1}(J)(1), F^k(J)(2) - F^{k-1}(J)(2)],$$

$$\bar{\gamma}_k = \max[F^k(J)(1) - F^{k-1}(J)(1), F^k(J)(2) - F^{k-1}(J)(2)].$$

TABLE 6.2

| $k$ | $F^k(J_0)(1)$ | $F^k(J_0)(2)$ | $F^k(J_0)(1) + c_k$ | $F^k(J_0)(1) + \bar{c}_k$ | $F^k(J_0)(2) + c_k$ | $F^k(J_0)(2) + \bar{c}_k$ |
|---|---|---|---|---|---|---|
| 0 | 0.00000 | 0.00000 | | | | |
| 2 | 1.51531 | 2.32373 | 5.71995 | 10.65312 | 6.52841 | 11.46159 |
| 4 | 3.16788 | 3.84690 | 6.14207 | 9.99341 | 6.82109 | 10.67243 |
| 6 | 4.35204 | 4.93594 | 6.47943 | 9.23554 | 7.06332 | 9.81943 |
| 8 | 5.19911 | 5.71495 | 6.72088 | 8.69239 | 7.23672 | 9.20823 |
| 10 | 5.80504 | 6.27219 | 6.89359 | 8.30387 | 7.36075 | 8.77102 |
| 12 | 6.23847 | 6.67080 | 7.01714 | 8.02594 | 7.44947 | 8.45827 |
| 14 | 6.54852 | 6.95594 | 7.10552 | 7.82714 | 7.51294 | 8.23456 |
| 16 | 6.77030 | 7.15990 | 7.16874 | 7.68493 | 7.55834 | 8.07453 |
| 18 | 6.92895 | 7.30580 | 7.21396 | 7.58320 | 7.59081 | 7.96006 |
| 20 | 7.04243 | 7.41017 | 7.24630 | 7.51043 | 7.61404 | 7.87817 |
| 22 | 7.12361 | 7.48482 | 7.26944 | 7.45838 | 7.63066 | 7.81960 |
| 24 | 7.18167 | 7.53823 | 7.28600 | 7.42115 | 7.64255 | 7.77370 |
| 26 | 7.22321 | 7.57643 | 7.29784 | 7.39451 | 7.65105 | 7.74773 |
| 28 | 7.25292 | 7.60375 | 7.30630 | 7.37546 | 7.65713 | 7.72629 |
| 30 | 7.27418 | 7.62330 | 7.31236 | 7.36183 | 7.66148 | 7.71095 |
| 32 | 7.28938 | 7.63728 | 7.31670 | 7.35208 | 7.66479 | 7.69998 |
| 34 | 7.30026 | 7.64728 | 7.31980 | 7.34511 | 7.66682 | 7.69213 |
| 36 | 7.30804 | 7.65444 | 7.32201 | 7.34012 | 7.66841 | 7.68652 |

The results of the successive approximation method starting with the zero function $J_0$ [$J_0(1) = J_0(2) = 0$] are shown in Table 6.2.

A comparison of Tables 6.1 and 6.2 reveals that the values $F^k(J_0)(1)$ and $F^k(J_0)(2)$ converge to $J^*(1)$ and $J^*(2)$ faster than $T^k(J_0)(1)$ and $T^k(J_0)(2)$ as predicted by Proposition 6. However, the error bounds $F^k(J_0)(i) + c_k$ and $F^k(J_0)(i) + \bar{c}_k$ converge much slower than the corresponding error bounds $T^k(J_0)(i) + c_k$ and $T^k(J_0)(i) + \bar{c}_k$. Thus the faster convergence property of the mapping $F$ is counterbalanced by the fact that the corresponding error bounds are not as tight as those associated with the mapping $T$. This disadvantage may be somewhat rectified by changing the successive approximation iteration so that instead of operating with $F$ on the current iterate function, say $J_k$, we operate on the average of the error bounding functions, i.e., by considering the iteration

$$J_k(i) = F(J_{k-1})(i) + \frac{c_k + \bar{c}_k}{2}, \qquad i = 1, 2,$$

where

$$c_k = \min\left\{\frac{\alpha^2}{1-\alpha^2}\gamma_k, \frac{\alpha}{1-\alpha}\gamma_k\right\}, \qquad \bar{c}_k = \max\left\{\frac{\alpha^2}{1-\alpha^2}\bar{\gamma}_k, \frac{\alpha}{1-\alpha}\bar{\gamma}_k\right\},$$

$$\gamma_k = \min[F(J_{k-1})(1) - J_{k-1}(1), F(J_{k-1})(2) - J_{k-1}(2)],$$

$$\bar{\gamma}_k = \max[F(J_{k-1})(1) - J_{k-1}(1), F(J_{k-1})(2) - J_{k-1}(2)],$$

TABLE 6.3

| $k$ | $J_k(1)$ | $J_k(2)$ | $F(J_{k-1})(1) + c_k$ | $F(J_{k-1})(1) + \bar{c}_k$ | $F(J_{k-1})(2) + c_k$ | $F(J_{k-1})(2) + \bar{c}_k$ |
|---|---|---|---|---|---|---|
| 0 | 0.00000 | 0.00000 | | | | |
| 1 | 7.58454 | 8.42204 | 2.63158 | 12.53750 | 3.46908 | 13.37500 |
| 2 | 8.37052 | 8.70077 | 6.08792 | 10.65312 | 6.41817 | 10.98338 |
| 3 | 7.24795 | 7.52229 | 6.72606 | 7.76983 | 7.00040 | 8.04418 |
| 4 | 7.19141 | 7.54122 | 6.85185 | 7.53098 | 7.20165 | 7.88079 |
| 5 | 7.34007 | 7.69411 | 7.28077 | 7.39936 | 7.63482 | 7.75341 |
| 6 | 7.34477 | 7.68881 | 7.29973 | 7.38982 | 7.64376 | 7.73386 |
| 7 | 7.32590 | 7.66956 | 7.31888 | 7.33292 | 7.66255 | 7.67658 |
| 8 | 7.32545 | 7.67038 | 7.31973 | 7.33117 | 7.66466 | 7.67611 |
| 9 | 7.32780 | 7.67277 | 7.32695 | 7.32865 | 7.67193 | 7.67362 |
| 10 | 7.32785 | 7.67366 | 7.32714 | 7.32856 | 7.67195 | 7.67338 |
| 11 | 7.32756 | 7.67237 | 7.32746 | 7.32766 | 7.67227 | 7.67247 |
| 12 | 7.32755 | 7.67238 | 7.32747 | 7.32764 | 7.67229 | 7.67247 |
| 13 | 7.32759 | 7.67242 | 7.32758 | 7.32760 | 7.67241 | 7.67243 |
| 14 | 7.32759 | 7.67242 | 7.32758 | 7.32760 | 7.67241 | 7.67243 |
| 15 | 7.32759 | 7.67241 | 7.32758 | 7.32759 | 7.67241 | 7.67241 |

and $J_0$ is arbitrary. The results of the computation for $J_0(1) = J_0(2) = 0$ are given in Table 6.3 and are considerably more favorable than those of Table 6.2. These computational results, however, are far too limited in scope to allow any general conclusions. Note that a similar modification of the successive approximation method based on the mapping $T$ does not lead to any improvement, as Problem 7 shows.

### *The Policy Iteration Algorithm*

The policy iteration algorithm (otherwise called policy improvement algorithm) operates as follows. An initial admissible stationary policy $\pi^0 = \{\mu^0, \mu^0, \ldots\}$ is adopted and the corresponding value function $J_{\mu^0} = J_{\pi^0}$ is calculated. Then an improved policy $\pi^1 = \{\mu^1, \mu^1, \ldots\}$ is computed, resulting in a decrease of the value of the cost, and the process is repeated.

The algorithm is based on Corollary 2.1 of Section 6.1 as well as the following proposition.

**Proposition 7** Let $\pi = \{\mu, \mu, \ldots\}$ be an admissible stationary policy and $J_\mu(x) = J_\pi(x)$ be the corresponding value of the cost functional (2) when the initial state is $x$. Let $\bar{\mu}: S \to C$, $\bar{\mu}(x) \in U(x)$, $\forall x \in S$, be a function satisfying

$$\underset{w}{E}\{g[x, \bar{\mu}(x), w] + \alpha J_\mu[f(x, \bar{\mu}(x), w)]\} = \min_{u \in U(x)} \underset{w}{E}\{g(x, u, w) + \alpha J_\mu[f(x, u, w)]\}. \tag{45}$$

Then if $J_{\bar{\mu}}(x)$ is the value of the cost corresponding to $\bar{\pi} = \{\bar{\mu}, \bar{\mu}, \ldots\}$ when the initial state is $x$, we have

$$J_{\bar{\mu}}(x) \leqslant J_{\mu}(x) \qquad \forall x \in S. \tag{46}$$

Furthermore, if the policy $\pi$ is not optimal, strict inequality holds in (46) for at least one state $x \in S$.

*Proof* From Corollary 2.1 and (45) we have for every $x \in S$,

$$\begin{aligned} J_{\mu}(x) &= \underset{w}{E}\{g[x, \mu(x), w] + \alpha J_{\mu}[f(x, \mu(x), w)]\} \\ &\geqslant \underset{w}{E}\{g[x, \bar{\mu}(x), w] + \alpha J_{\mu}[f(x, \bar{\mu}(x), w)]\} = T_{\bar{\mu}}(J_{\mu})(x). \end{aligned}$$

Applying repeatedly $T_{\bar{\mu}}$ on both sides of the above inequality and using the monotonicity of $T_{\bar{\mu}}$ and Corollary 1.1, we obtain

$$J_{\mu} \geqslant T_{\bar{\mu}}(J_{\mu}) \geqslant \cdots \geqslant T_{\bar{\mu}}^{k}(J_{\mu}) \geqslant \cdots \geqslant \lim_{k \to \infty} T_{\bar{\mu}}^{k}(J_{\mu}) = J_{\bar{\mu}},$$

proving (46). If $J_{\mu} = J_{\bar{\mu}}$, then from the inequality above we have $J_{\mu} = T_{\bar{\mu}}(J_{\mu})$ and from (45) we have $T_{\bar{\mu}}(J_{\mu}) = T(J_{\mu})$, so that $J_{\mu} = T(J_{\mu})$ and hence $J_{\mu} = J^*$ by Proposition 2. Hence $\pi = \{\mu, \mu, \ldots\}$ must be optimal. It follows that strict inequality holds in (46) for some $x \in S$ if $\pi$ is not optimal. Q.E.D.

**Policy Iteration Algorithm**

*Step 1* Guess an initial admissible stationary policy

$$\pi^0 = \{\mu^0, \mu^0, \ldots\}.$$

*Step 2* Given the admissible stationary policy

$$\pi^i = \{\mu^i, \mu^i, \ldots\},$$

compute the corresponding value function $J_{\mu^i}(x)$ using the successive approximation algorithm or the linear system of equations

$$(I - \alpha P_{\mu^i})J_{\mu^i} = \bar{g}_{\mu^i},$$

as described in the beginning of this section.

If $J_{\mu^i} = J_{\mu^{i-1}}$, stop—$\pi^i$ is optimal.

*Step 3* Obtain a new admissible policy $\pi^{i+1} = \{\mu^{i+1}, \mu^{i+1}, \ldots\}$ satisfying for all $x \in S$

$$\begin{aligned} &\underset{w}{E}\{g[x, \mu^{i+1}(x), w] + \alpha J_{\mu^i}[f(x, \mu^{i+1}(x), w)]\} \\ &\quad = \min_{u \in U(x)} \underset{w}{E}\{g(x, u, w) + \alpha J_{\mu^i}[f(x, u, w)]\} = T(J_{\mu^i})(x). \end{aligned}$$

Return to Step 2 and repeat the process.

Since the collection of all possible stationary policies is a finite collection (by the finiteness of $S$ and $C$) and an improved policy is generated at every iteration, it follows that the algorithm will find an optimal stationary policy in a finite number of iterations and thereby terminate. This property of the policy iteration algorithm is its main advantage over successive approximation, which in general converges in an infinite number of iterations. On the other hand, finding the exact value of $J_{\mu^i}$ in Step 2 of the algorithm requires solution of the system of linear equations representing $J_{\mu^i} = T_{\mu^i}(J_{\mu^i})$. The dimension of this system is equal to the number of points of the state space $S$ and thus when this number is very large the method is not very attractive.

We note that when the successive approximation method is used for carrying out Step 2 we may utilize error bounds similar to those described earlier in this section. We note also that one may construct a generalized policy iteration algorithm for the case where $S$, $C$, and $D$ are not necessarily finite sets (see Problem 13).

We now demonstrate the policy iteration algorithm by means of the example considered earlier in this section.

EXAMPLE 1 (CONTINUED)

*Step 1* Let us select an initial policy $\pi^0 = \{\mu^0, \mu^0, \ldots\}$, where

$$\mu^0(1) = u^1, \qquad \mu^0(2) = u^2.$$

*Step 2* We obtain $J_{\mu^0}$ through the equation $J_{\mu^0} = T_{\mu^0}(J_{\mu^0})$ or equivalently

$$J_{\mu^0}(1) = g(1, u^1) + \alpha p_{11}(u^1) J_{\mu^0}(1) + \alpha p_{12}(u^1) J_{\mu^0}(2),$$

$$J_{\mu^0}(2) = g(2, u^2) + \alpha p_{21}(u^2) J_{\mu^0}(1) + \alpha p_{22}(u^2) J_{\mu^0}(2).$$

Substituting the data of the problem,

$$J_{\mu^0}(1) = 2 + 0.9 \times \tfrac{3}{4} \times J_{\mu^0}(1) + 0.9 \times \tfrac{1}{4} \times J_{\mu^0}(2),$$

$$J_{\mu^0}(2) = 3 + 0.9 \times \tfrac{1}{4} \times J_{\mu^0}(1) + 0.9 \times \tfrac{3}{4} \times J_{\mu^0}(2).$$

Solving this system of linear equations for $J_{\mu^0}(1)$, $J_{\mu^0}(2)$ we obtain

$$J_{\mu^0}(1) \simeq 24.12, \qquad J_{\mu^0}(2) \simeq 25.96.$$

*Step 3* We now find $\mu^1(1)$, $\mu^1(2)$ satisfying $T_{\mu^1}(J_{\mu^0}) = T(J_{\mu^0})$. We have

$$\begin{aligned} T(J_{\mu^0})(1) &= \min\{2 + 0.9(\tfrac{3}{4} \times 24.12 + \tfrac{1}{4} \times 25.96), \\ &\qquad 0.5 + 0.9(\tfrac{1}{4} \times 24.12 + \tfrac{3}{4} \times 25.96)\} \\ &= \min\{24.12, 23.45\} = 23.45, \\ T(J_{\mu^0})(2) &= \min\{1 + 0.9(\tfrac{3}{4} \times 24.12 + \tfrac{1}{4} \times 25.96), \\ &\qquad 3 + 0.9(\tfrac{1}{4} \times 24.12 + \tfrac{3}{4} \times 25.96)\} \\ &= \min\{23.12, 25.95\} = 23.12. \end{aligned}$$

The minimizing controls are

$$\mu^1(1) = u^2, \qquad \mu^1(2) = u^1.$$

*Step 2* We obtain $J_{\mu^1}$ through the equation $J_{\mu^1} = T_{\mu^1}(J_{\mu^1})$:

$$J_{\mu^1}(1) = g(1, u^2) + \alpha p_{11}(u^2)J_{\mu^1}(1) + \alpha p_{12}(u^2)J_{\mu^1}(2),$$
$$J_{\mu^1}(2) = g(2, u^1) + \alpha p_{21}(u^1)J_{\mu^1}(1) + \alpha p_{22}(u^1)J_{\mu^1}(2).$$

Substitution of the data of the problem and solution of the system of equations yields

$$J_{\mu^1}(1) \simeq 7.33, \qquad J_{\mu^1}(2) \simeq 7.67.$$

*Step 3* We perform the minimization required to find $T(J_{\mu^1})$:

$$\begin{aligned} T(J_{\mu^1})(1) &= \min\{2 + 0.9(\tfrac{3}{4} \times 7.33 + \tfrac{1}{4} \times 7.67), \\ &\qquad 0.5 + 0.9(\tfrac{1}{4} \times 7.33 + \tfrac{3}{4} \times 7.67)\} \\ &= \min\{8.67, 7.33\} = 7.33, \\ T(J_{\mu^1})(2) &= \min\{1 + 0.9(\tfrac{3}{4} \times 7.33 + \tfrac{1}{4} \times 7.67), \\ &\qquad 3 + 0.9(\tfrac{1}{4} \times 7.33 + \tfrac{3}{4} \times 7.67)\} \\ &= \min\{7.67, 9.83\} = 7.67. \end{aligned}$$

Hence we have $J_{\mu^1} = T(J_{\mu^1})$, which implies that $\{\mu^1, \mu^1, \ldots\}$ is optimal and $J_{\mu^1} = J^*$:

$$\mu^*(1) = u^2, \quad \mu^*(2) = u^1, \qquad J^*(1) \simeq 7.33, \quad J^*(2) \simeq 7.67.$$

*Linear Programming*

As discussed earlier we have

$$J \leqslant T(J) \Rightarrow J \leqslant J^* = T(J^*).$$

Thus if we denote by $1, 2, \ldots, n$ the elements of the state space $S$, it is clear that $J^*(1), \ldots, J^*(n)$ solve the following maximization problem (in $\lambda_1, \ldots, \lambda_n$):

$$\max \sum_{i=1}^{n} \lambda_i$$

subject to

$$\lambda_i \leqslant T(J_\lambda)(i), \qquad i = 1, \ldots, n,$$

where the function $J_\lambda : S \to R$ is defined by

$$J_\lambda(i) = \lambda_i, \qquad i = 1, \ldots, n.$$

If we denote by $u^1, u^2, \ldots, u^m$ the elements of the control space, the problem above is written

$$\max \sum_{i=1}^{n} \lambda_i$$

subject to

$$\lambda_i \leqslant \underset{w}{E}\{g(i, u^k, w) + \alpha J_\lambda[f(i, u^k, w)] \mid i, u^k\}, \qquad i = 1, \ldots, n, \quad u^k \in U(i).$$

Using the notation given in the beginning of this section this linear program may be written in terms of transition probabilities $p_{ij}(u)$ as

$$\max \sum_{i=1}^{n} \lambda_i$$

subject to

$$\lambda_i \leqslant \bar{g}(i, u^k) + \alpha \sum_{j=1}^{n} p_{ij}(u^k)\lambda_j, \qquad i = 1, 2, \ldots, n, \quad u^k \in U(i)$$

This is a linear program with $n$ variables and as many as $n \times m$ constraints. As $n$ increases, its solution becomes more complex and for very large $n$ and $m$ (in the order of several hundreds) the linear programming approach becomes impractical.

For the example considered in this section the linear programming problem above takes the form

$$\text{maximize} \quad \lambda_1 + \lambda_2$$

subject to
$$\lambda_1 \leqslant 2 + 0.9(\tfrac{3}{4}\lambda_1 + \tfrac{1}{4}\lambda_2), \qquad \lambda_1 \leqslant 0.5 + 0.9(\tfrac{1}{4}\lambda_1 + \tfrac{3}{4}\lambda_2),$$
$$\lambda_2 \leqslant 1 + 0.9(\tfrac{3}{4}\lambda_1 + \tfrac{1}{4}\lambda_2), \qquad \lambda_2 \leqslant 3 + 0.9(\tfrac{1}{4}\lambda_1 + \tfrac{3}{4}\lambda_2).$$

## 6.3 Contraction Mappings

In this section we introduce the notion of a contraction mapping on the space of all bounded functions over the state space $S$. A basic fact about such mappings is that they possess a unique fixed point—a classical result of analysis. Furthermore the fixed point may be found in the limit by successive application of the mapping on any bounded function over $S$, i.e., by a method of successive approximation. It turns out that the mappings $T$, $T_\mu$, and $F$ considered in the past two sections [cf. (8), (9), (38)–(41)] are contraction mappings and that the corresponding results that we proved draw essentially their validity from the contraction property together with the monotonicity property of Lemma 1.

The connection with the theory of contraction mappings is very valuable from both the conceptual and the practical points of view. This is particularly so since it turns out that the contraction and monotonicity properties are present in several dynamic programming models other than Problem (D) and by themselves guarantee the validity of results and algorithms similar to those obtained in the past two sections. The corresponding theory is developed in some detail in Problem 4. Here we present only the notion of a contraction mapping and show that some of our earlier results concerning Problem 1 are special cases of a general result on fixed points.

Let $B(S)$ denote the set of all bounded real-valued functions on $S$. With every function $J\colon S \to R$ that belongs to $B(S)$ we associate the scalar

$$\|J\| = \sup_{x \in S} |J(x)|. \tag{47}$$

(For the benefit of the advanced reader we mention that the function $\|\cdot\|$ may be shown to be a norm on the linear space $B(S)$ and with this norm $B(S)$ becomes a complete normed linear space, i.e., a Banach space [R5]). The following definition and theorem are specializations to $B(S)$ of a more general notion and result (see, e.g., references [L5] and [L8]).

**Definition** A mapping $H\colon B(S) \to B(S)$ is said to be a *contraction mapping* if there exists a scalar $\rho < 1$ such that

$$\|H(J) - H(J')\| \leqslant \rho\|J - J'\|, \qquad \forall J, J' \in B(S),$$

where $\|\cdot\|$ is as in (47). It is said to be an *m-stage contraction mapping* if there exists a positive integer $m$ and some $\rho < 1$ such that

$$\|H^m(J) - H^m(J')\| \leqslant \rho\|J - J'\|, \qquad \forall J, J' \in B(S),$$

where $H^m$ denotes the composition $H \cdots H$ of $H$ with itself $m$ times.

The main result concerning contraction mappings is as follows.

**Contraction Mapping Fixed Point Theorem** If $H\colon B(S) \to B(S)$ is a contraction mapping or an $m$-stage contraction mapping, then there exists a unique fixed point of $H$, i.e., there exists a unique function $J^* \in B(S)$ such that

$$H(J^*) = J^*.$$

Furthermore, if $J$ is any function in $B(S)$ and $H^k$ is the composition $H \cdots H$ of $H$ with itself $k$ times, then

$$\lim_{k \to \infty} \|H^k(J) - J^*\| = 0,$$

i.e., the function $H^k(J)$ converges uniformly to the function $J^*$.

*Proof* See reference [L5] or [L8].

Now consider the mappings $T$ and $T_\mu$ defined by (8) and (9). Proposition 3 and Corollary 3.1 show that $T$ and $T_\mu$ are contraction mappings ($\rho = \alpha$). As a result, the fact that the successive approximation method converges to the unique fixed point of $T$ follows directly from the contraction mapping theorem. Notice also that by Proposition 5, the mapping $F$ defined by (38)–(41) is also a contraction mapping with $\rho = \alpha$ and the convergence result of Proposition 5 is again a special case of the fixed point theorem.

Several additional results on contraction mappings of the type considered in DP are pointed out in the problem section (Problem 4). In particular, the computational methods of the previous section are special cases of more general procedures that are applicable to such contraction mappings.

## 6.4 Unbounded Costs per Stage

In this section we consider Problem (D) but relax Assumption B by allowing costs per stage that are unbounded above or below. The complications resulting from relaxation of the boundedness assumption are substantial and the analysis required is considerably more sophisticated than the one under Assumption B. The main difficulty is that Proposition 1 and the results that depend on it need not be true anymore. We shall assume that one of the following two assumptions is in effect in place of Assumption B.

**Assumption P** (Positivity)† The function $g$ in the cost functional (2) satisfies

$$0 \leqslant g(x, u, w) \qquad \forall (x, u, w) \in S \times C \times D. \tag{48}$$

**Assumption N** (Negativity) The function $g$ in the cost functional (2) satisfies

$$g(x, u, w) \leqslant 0 \qquad \forall (x, u, w) \in S \times C \times D. \tag{49}$$

In problems where reward or utility per stage is nonnegative and total discounted expected reward is to be *maximized*, we may consider minimization of negative reward thus coming within the framework of Assumption N.

It is to be noted that (48) could be replaced by

$$M \leqslant g(x, u, w) \qquad \forall (x, u, w) \in S \times C \times D,$$

while (49) could be replaced by

$$g(x, u, w) \leqslant M \qquad \forall (x, u, w) \in S \times C \times D,$$

where $M$ is some scalar. When $g$ is either bounded above or below we may add a scalar to $g$ so that either (48) or (49) is satisfied. An optimal policy will not be affected by this change since, in view of the presence of the discount factor,

† Problems corresponding to Assumption P are sometimes referred to in the research literature as negative DP problems [S17]. In these problems the objective function is maximized and the reward per stage is negative. Similarly problems corresponding to Assumption N are sometimes referred to as positive DP problems [B21, S17].

the addition of a constant $r$ to $g$ merely adds $(1 - \alpha)^{-1}r$ to the cost associated with every policy.

One complication arising from allowing unbounded costs per stage is that the value $J_\pi(x_0)$ of the cost functional (2) for some initial states $x_0$ and some admissible policies $\pi = \{\mu_0, \mu_1, \ldots\}$ may be $+\infty$ (in the case of Assumption P) or $-\infty$ (in the case of Assumption N). Consider the following simple example.

EXAMPLE Let the system equation be

$$x_{k+1} = \beta x_k + u_k, \qquad k = 0, 1, 2, \ldots,$$

where $x_k, u_k \in R$, $k = 0, 1, \ldots$, and $\beta$ is a positive scalar. The control constraint is $|u_k| \leqslant 1$, i.e., $U(x) = \{u \mid |u| \leqslant 1\}$. Consider the cost functional

$$J_\pi(x_0) = \lim_{N\to\infty} \sum_{k=0}^{N-1} \alpha^k |x_k|,$$

where $\alpha < 1$ is the discount factor. Consider the policy $\tilde{\pi} = \{\tilde{\mu}, \tilde{\mu}, \ldots\}$, where $\tilde{\mu}(x) = 0$ for all $x \in R$. Then

$$J_{\tilde{\pi}}(x_0) = \lim_{N\to\infty} \sum_{k=0}^{N-1} \alpha^k \beta^k |x_0|,$$

and hence

$$J_{\tilde{\pi}}(x_0) = \infty \qquad \text{if} \quad x_0 \neq 0, \quad \alpha\beta \geqslant 1,$$

and $J_{\tilde{\pi}}(x_0)$ is finite otherwise. It is also possible to verify that when $\alpha\beta \geqslant 1$ the optimal value $J^*(x_0)$ is equal to $+\infty$ for $|x_0| > 1/(\beta - 1)$ and is finite for $|x_0| \leqslant 1/(\beta - 1)$.

We shall conduct our analysis with the notational understanding that the cost $J_\pi(x_0)$ corresponding to an initial state $x_0$ and a policy $\pi$ or the optimal cost $J^*(x_0)$ corresponding to an initial state $x_0$ may take the values $+\infty$ or $-\infty$ (depending on whether we operate under Assumption P or N). In other words, *we consider* $J_\pi(\cdot)$, $J^*(\cdot)$ *to be extended real-valued functions.*

The results to be presented provide characterizations of the optimal value function $J^*$ as well as optimal stationary policies. They also provide conditions under which the successive approximation method yields in the limit the optimal value function $J^*$. In the proofs we shall often need to interchange expectation and limit in various relations. This interchange is valid under the assumptions of the following theorem.

**Monotone Convergence Theorem** Let $P = (p_1, p_2, \ldots)$ be a probability distribution over a countable set $S$ denoted by $S = \{1, 2, \ldots\}$. Let $\{h_N\}$ be a sequence of extended real-valued functions on $S$ such that

$$0 \leqslant h_N(i) \leqslant h_{N+1}(i) \qquad \forall i, N = 1, 2, \ldots.$$

Let $h: S \to [0, +\infty]$ be the limit function

$$h(i) = \lim_{N\to\infty} h_N(i).$$

Then

$$\lim_{N\to\infty} \sum_{i=1}^{\infty} p_i h_N(i) = \sum_{i=1}^{\infty} p_i \lim_{N\to\infty} h_N(i) = \sum_{i=1}^{\infty} p_i h(i).$$

*Proof* We have

$$\sum_{i=1}^{\infty} p_i h_N(i) \leqslant \sum_{i=1}^{\infty} p_i h(i).$$

By taking the limit, we obtain

$$\lim_{N\to\infty} \sum_{i=1}^{\infty} p_i h_N(i) \leqslant \sum_{i=1}^{\infty} p_i h(i),$$

so that it remains to prove the reverse inequality. Now for every integer $M \geqslant 1$ we have

$$\lim_{N\to\infty} \sum_{i=1}^{\infty} p_i h_N(i) \geqslant \lim_{N\to\infty} \sum_{i=1}^{M} p_i h_N(i) = \sum_{i=1}^{M} p_i h(i),$$

and by taking the limit as $M \to \infty$ the reverse inequality follows. Q.E.D.

*Optimality Equation—Conditions for Optimality*

**Proposition 8** Under either Assumption P or N the optimal value function $J^*$ of Problem (D) satisfies

$$J^*(x) = \inf_{u\in U(x)} \mathop{E}_{w} \{g(x, u, w) + \alpha J^*[f(x, u, w)]\} \qquad \forall x \in S \tag{50a}$$

or in terms of the mapping $T$ of (8)

$$J^* = T(J^*). \tag{50b}$$

*Proof* Let $\pi = \{\mu_0, \mu_1, \ldots\}$ be an arbitrary admissible policy, and consider the value $J_\pi(x)$ of the cost functional corresponding to $\pi$ when the initial state is $x$. We have

$$J_\pi(x) = \mathop{E}_{w} \{g[x, \mu_0(x), w] + V_\pi[f(x, \mu_0(x), w)]\}, \tag{51}$$

where for all $z \in S$,

$$V_\pi(z) = \lim_{N\to\infty} \mathop{E}_{\substack{w_k \\ k=1,2,\ldots}} \left\{\sum_{k=1}^{N-1} \alpha^k g[x_k, \mu_k(x_k), w_k]\right\}.$$

In this equation we have $x_1 = z$, and $x_{k+1}$ is generated from $x_k$, $\mu_k(x_k)$, $w_k$ via the system equation (1). In other words, $V_\pi(z)$ is the cost from stage 1 to infinity using $\pi$ when the initial state $x_1$ is equal to $z$. We have clearly

$$V_\pi(z) \geqslant \alpha J^*(z) \qquad \forall z \in S.$$

Hence from (51)

$$\begin{aligned} J_\pi(x) &\geqslant \underset{w}{E}\{g[x, \mu_0(x), w] + \alpha J^*[f(x, \mu_0(x), w)]\} \\ &\geqslant \inf_{u \in U(x)} \underset{w}{E}\{g(x, u, w) + \alpha J^*[f(x, u, w)]\}. \end{aligned}$$

Taking the infimum over all admissible policies we have

$$\inf_\pi J_\pi(x) = J^*(x) \geqslant \inf_{u \in U(x)} \underset{w}{E}\{g(x, u, w) + \alpha J^*[f(x, u, w)]\} = T(J^*)(x).$$

Thus it remains to prove that the reverse inequality also holds.

Let $x_0 \in S$ be any initial state for which we have

$$\inf_{u \in U(x_0)} \underset{w}{E}\{g(x_0, u, w) + \alpha J^*[f(x_0, u, w)]\} > -\infty. \tag{52}$$

Notice that this inequality will always hold under P. Under (52), given any scalars $\varepsilon_1, \varepsilon_2 > 0$ let $\tilde{\pi} = \{\tilde{\mu}_0, \tilde{\mu}_1, \ldots\}$ be a policy such that

$$\begin{aligned} &\underset{w}{E}\{g[x_0, \tilde{\mu}_0(x), w] + \alpha J^*[f(x_0, \tilde{\mu}_0(x_0), w)]\} \\ &\qquad \leqslant \inf_{u \in U(x_0)} \underset{w}{E}\{g(x_0, u, w) + \alpha J^*[f(x_0, u, w)]\} + \varepsilon_1, \end{aligned} \tag{53}$$

and

$$\underset{w}{E}\{J_{\tilde{\pi}_1}[f(x_0, \tilde{\mu}_0(x_0), w)]\} \leqslant \underset{w}{E}\{J^*[f(x_0, \tilde{\mu}_0(x_0), w)]\} + \varepsilon_2, \tag{54}$$

where

$$\tilde{\pi}_1 = \{\tilde{\mu}_1, \tilde{\mu}_2, \ldots\}.$$

Such a policy clearly exists when the problem is deterministic, i.e., $D$ consists of a single element. It can also be shown to exist in the general case (see Problem 8). We have by (53) and (54)

$$\begin{aligned} J_{\tilde{\pi}}(x_0) &= \underset{w}{E}\{g[x_0, \tilde{\mu}_0(x_0), w] + \alpha J_{\tilde{\pi}_1}[f(x_0, \tilde{\mu}_0(x_0), w)]\} \\ &\leqslant \underset{w}{E}\{g[x_0, \tilde{\mu}_0(x_0), w] + \alpha J^*[f(x_0, \tilde{\mu}_0(x_0), w)]\} + \alpha\varepsilon_2 \\ &\leqslant \inf_{u \in U(x_0)} \underset{w}{E}\{g(x_0, u, w) + \alpha J^*[f(x_0, u, w)]\} + \varepsilon_1 + \alpha\varepsilon_2. \end{aligned}$$

Since

$$J^*(x_0) \leqslant J_{\tilde{\pi}}(x_0),$$

we obtain

$$J^*(x_0) \leqslant \inf_{u \in U(x_0)} \mathop{E}_{w} \{g(x_0, u, w) + \alpha J^*[f(x_0, u, w)]\} + \varepsilon_1 + \alpha\varepsilon_2,$$

Since $\varepsilon_1, \varepsilon_2 > 0$ are arbitrary, we obtain

$$J^*(x_0) \leqslant \inf_{u \in U(x_0)} \mathop{E}_{w} \{g(x_0, u, w) + \alpha J^*[f(x_0, u, w)]\} = T(J^*)(x_0), \quad (55)$$

for all initial states $x_0 \in S$ for which (52) holds.

If (under N) $x_0 \in S$ is such that

$$\inf_{u \in U(x_0)} \mathop{E}_{w} \{g(x_0, u, w) + \alpha J^*[f(x_0, u, w)]\} = -\infty,$$

then

$$\inf_{u \in U(x_0)} \mathop{E}_{w} \{g(x_0, u, w)\} = -\infty \qquad \text{or} \qquad \inf_{u \in U(x_0)} \mathop{E}_{w} \{J^*[f(x_0, u, w)]\} = -\infty.$$

In the first case we have clearly $J^*(x_0) = -\infty$ and (55) holds. In the second case given any $M > 0$ one may find a $\tilde{u} \in U(x_0)$ such that

$$\mathop{E}_{w} \{J^*[f(x_0, \tilde{u}, w)]\} < -M.$$

It follows, using the result of Problem 8, that we may also find a policy $\tilde{\pi}_1 = \{\tilde{\mu}_1, \tilde{\mu}_2, \ldots\}$ such that

$$\mathop{E}_{w} \{J_{\tilde{\pi}_1}[f(x_0, \tilde{u}, w)]\} < -M.$$

Consider a policy of the form $\tilde{\pi} = \{\tilde{\mu}_0, \tilde{\mu}_1, \tilde{\mu}_2, \ldots\}$, where $\tilde{\mu}_0(x_0) = \tilde{u}$. Then, since under N we have $E_w \{g[x_0, \tilde{\mu}_0(x_0), w]\} \leqslant 0$, it follows that

$$J^*(x_0) \leqslant J_{\tilde{\pi}}(x_0) = \mathop{E}_{w} \{g[x_0, \tilde{\mu}_0(x_0), w] + \alpha J_{\tilde{\pi}_1}[f(x_0, \tilde{\mu}_0(x_0), w)]\} < -\alpha M.$$

Since $M > 0$ is arbitrary, it follows that $J^*(x_0) = -\infty$ and thus inequality (55) is established for all $x_0 \in S$ for which (52) does not hold. Hence (55) holds for all $x_0 \in S$ and the proposition is proved. Q.E.D.

Similarly as in Corollaries 1.1, 2.1, and 3.1 we have:

**Corollary 8.1** Let $\pi = \{\mu, \mu, \ldots\}$ be an admissible stationary policy. Then under Assumption P or N we have

$$J_\mu(x) = \mathop{E}_{w} \{g[x, \mu(x), w] + \alpha J_\mu[f(x, \mu(x), w)]\}$$

or in terms of the mapping $T_\mu$ of (9)

$$J_\mu = T_\mu(J_\mu).$$

Contrary to the case of Assumption B the optimal value function $J^*$ under Assumption P or N need not be the unique solution of the functional equation

$$J(x) = T(J)(x) = \inf_{u \in U(x)} \underset{w}{E} \{g(x, u, w) + \alpha J[f(x, u, w)]\}. \tag{56}$$

Consider the following examples.

EXAMPLE 1 Let $S = [0, +\infty)$ (or $S = (-\infty, 0]$) and

$$g(x, u, w) = 0, \qquad f(x, u, w) = x/\alpha, \qquad \forall x, u, w.$$

Then for every $\beta$, the function $J$ given by

$$J(x) = \beta x \qquad \forall x \in S$$

is a solution of (56) and hence $T$ has an infinite number of fixed points in this case, although there is a unique fixed point within the class of bounded functions—the zero function $J_0(x) = 0$ $\forall x \in S$, which is the optimal value function for this problem. More generally it can be shown by using Proposition 9 that if there exists a bounded function that is a fixed point of $T$, then that function must be equal to the optimal value function $J^*$ (see Problem 15).

EXAMPLE 2 Let $S = [0, +\infty)$ (or $S = (-\infty, 0]$) and

$$g(x, u, w) = (1 - \sqrt{\alpha})x, \qquad f(x, u, w) = x/\sqrt{\alpha} \qquad \forall x, u, w.$$

Then the reader may verify that the functions $J(x) = x$, $J(x) = x + x^2$, and $J(x) = x - x^2$ are all solutions of (56).

It is to be noted also that when $\alpha = 1$ (a case to be examined in the next chapter), Eq. (56) may have an infinity of solutions even within the class of bounded functions. This is clear since if $\alpha = 1$ and $J(\cdot)$ is any solution of (56), then $J(\cdot) + r$, where $r$ is any scalar, is also a solution.

The optimal value function $J^*$, however, has the property that it is the smallest (under Assumption P) or largest (under Assumption N) fixed point of $T$ in the sense described in the following proposition.

**Proposition 9** (a) Under Assumption P if $\tilde{J}: S \to (-\infty, +\infty]$ is bounded below and satisfies $\tilde{J} = T(\tilde{J})$, then $J^* \leqslant \tilde{J}$.

(b) Under Assumption N if $\tilde{J}: S \to [-\infty, +\infty)$ is bounded above and satisfies $\tilde{J} = T(\tilde{J})$, then $\tilde{J} \leqslant J^*$.

*Proof* (a) Under Assumption P let $r$ be a scalar such that $\tilde{J}(x) + r \geqslant 0$ for all $x \in S$. For any sequence $\{\varepsilon_k\}$ with $\varepsilon_k > 0$, let $\tilde{\pi} = \{\tilde{\mu}_0, \tilde{\mu}_1, \ldots\}$ be an admissible policy for which we have for every $x \in S$ and $k$,

$$\underset{w}{E} \{g[x, \tilde{\mu}_k(x), w] + \alpha \tilde{J}[f(x, \tilde{\mu}_k(x), w)]\} \leqslant T(\tilde{J})(x) + \varepsilon_k. \tag{57}$$

Such a policy exists since $T(\tilde{J})(x) > -\infty$ for all $x \in S$. We have for any initial state $x_0 \in S$,

$$J^*(x_0) = \inf_{\pi} \lim_{N\to\infty} \underset{w_k}{E} \left\{ \sum_{k=0}^{N-1} \alpha^k g[x_k, \mu_k(x_k), w_k] \right\}$$

$$\leq \inf_{\pi} \liminf_{N\to\infty} \underset{w_k}{E} \left\{ \alpha^N [\tilde{J}(x_N) + r] + \sum_{k=0}^{N-1} \alpha^k g[x_k, \mu_k(x_k), w_k] \right\}$$

$$\leq \liminf_{N\to\infty} \underset{w_k}{E} \left\{ \alpha^N [\tilde{J}(x_N) + r] + \sum_{k=0}^{N-1} \alpha^k g[x_k, \tilde{\mu}_k(x_k), w_k] \right\}.$$

Now using (57) and the assumption $\tilde{J} = T(\tilde{J})$ we obtain

$$\underset{w_k}{E} \left\{ \alpha^N \tilde{J}(x_N) + \sum_{k=0}^{N-1} \alpha^k g[x_k, \tilde{\mu}_k(x_k), w_k] \right\}$$

$$= \underset{w_k}{E} \left\{ \alpha^N \tilde{J}[f(x_{N-1}, \tilde{\mu}_{N-1}(x_{N-1}), w_{N-1})] + \sum_{k=0}^{N-1} \alpha^k g[x_k, \tilde{\mu}_k(x_k), w_k] \right\}$$

$$\leq \underset{w_k}{E} \left\{ \alpha^{N-1} T(\tilde{J})(x_{N-1}) + \sum_{k=0}^{N-2} \alpha^k g[x_k, \tilde{\mu}_k(x_k), w_k] \right\} + \alpha^{N-1} \varepsilon_{N-1}$$

$$= \underset{w_k}{E} \left\{ \alpha^{N-1} \tilde{J}(x_{N-1}) + \sum_{k=0}^{N-2} \alpha^k g[x_k, \tilde{\mu}_k(x_k), w_k] \right\} + \alpha^{N-1} \varepsilon_{N-1}$$

$$\leq \underset{w_k}{E} \left\{ \alpha^{N-2} \tilde{J}(x_{N-2}) + \sum_{k=0}^{N-3} \alpha^k g[x_k, \tilde{\mu}_k(x_k), w_k] \right\} + \alpha^{N-2} \varepsilon_{N-2} + \alpha^{N-1} \varepsilon_{N-1}$$

$$\vdots$$

$$\leq \tilde{J}(x_0) + \sum_{k=0}^{N-1} \alpha^k \varepsilon_k.$$

Combining these inequalities we obtain

$$J^*(x_0) \leq \tilde{J}(x_0) + \lim_{N\to\infty} \left( \alpha^N r + \sum_{k=0}^{N-1} \alpha^k \varepsilon_k \right).$$

Since the sequence $\{\varepsilon_k\}$ is arbitrary (except for $\varepsilon_k > 0$) we may select $\{\varepsilon_k\}$ so that $\lim_{N\to\infty} \sum_{k=0}^{N-1} \alpha^k \varepsilon_k$ is arbitrarily close to zero, and the result follows.

(b) Under Assumption N, let $r$ be a scalar such that $\tilde{J}(x) + r \leqslant 0$ for all $x \in S$. We have for every initial state $x_0 \in S$,

$$J^*(x_0) = \inf_{\pi} \lim_{N\to\infty} \mathop{E}_{w_k} \left\{ \sum_{k=0}^{N-1} \alpha^k g[x_k, \mu_k(x_k), w_k] \right\}$$

$$\geqslant \inf_{\pi} \limsup_{N\to\infty} \mathop{E}_{w_k} \left\{ \alpha^N[\tilde{J}(x_N) + r] + \sum_{k=0}^{N-1} \alpha^k g[x_k, \mu_k(x_k), w_k] \right\}$$

$$\geqslant \limsup_{N\to\infty} \inf_{\pi} \mathop{E}_{w_k} \left\{ \alpha^N[\tilde{J}(x_N) + r] + \sum_{k=0}^{N-1} \alpha^k g[x_k, \mu_k(x_k), w_k] \right\}, \quad (58)$$

where the last inequality follows from the fact that for any sequence $\{h_N(\lambda)\}$ of functions of a parameter $\lambda$ we have

$$\inf_{\lambda} \limsup_{N\to\infty} h_N(\lambda) \geqslant \limsup_{N\to\infty} \inf_{\lambda} h_N(\lambda).$$

This inequality follows by writing

$$h_N(\lambda) \geqslant \inf_{\lambda} h_N(\lambda)$$

and by subsequently taking the limit superior of both sides and the infimum over $\lambda$ of the left-hand side.

Now we have by using the assumption $\tilde{J} = T(\tilde{J})$,

$$\inf_{\pi} \mathop{E}_{w_k} \left\{ \alpha^N \tilde{J}(x_N) + \sum_{k=0}^{N-1} \alpha^k g[x_k, \mu_k(x_k), w_k] \right\}$$

$$= \inf_{\pi} \mathop{E}_{w_k} \left\{ \sum_{k=0}^{N-2} \alpha^k g[x_k, \mu_k(x_k), w_k] \right.$$

$$+ \alpha^{N-1} \inf_{u_{N-1} \in U(x_{N-1})} \mathop{E}_{w_{N-1}} \{g(x_{N-1}, u_{N-1} w_{N-1})$$

$$\left. + \alpha \tilde{J}[f(x_{N-1}, u_{N-1}, w_{N-1})]\right\}$$

$$= \inf_{\pi} \mathop{E}_{w_k} \left\{ \alpha^{N-1} \tilde{J}(x_{N-1}) + \sum_{k=0}^{N-2} \alpha^k g[x_k, \mu_k(x_k), w_k] \right\}$$

$$\vdots$$

$$= \tilde{J}(x_0).$$

Using this equality in (58) we obtain

$$J^*(x_0) \geqslant \tilde{J}(x_0) + \lim_{N\to\infty} \alpha^N r = \tilde{J}(x_0). \quad \text{Q.E.D.}$$

Similarly as earlier we have the following corollary:

**Corollary 9.1** Let $\pi = \{\mu, \mu, \ldots\}$ be an admissible stationary policy.

(a) Under Assumption P if $J: S \to (-\infty, +\infty]$ is bounded below and satisfies $\tilde{J} = T_\mu(\tilde{J})$, then $J_\mu \leqslant \tilde{J}$.

(b) Under Assumption N if $\tilde{J}: S \to [-\infty, +\infty)$ is bounded above and satisfies $\tilde{J} = T_\mu(\tilde{J})$, then $\tilde{J} \leqslant J_\mu$.

Under Assumption N, Proposition 9 yields also the following corollary, which constitutes a necessary and sufficient condition for optimality of a stationary policy.

**Corollary 9.2** (Necessary and Sufficient Condition for Optimality under Assumption N) In order for an admissible stationary policy $\pi^* = \{\mu^*, \mu^*, \ldots\}$ to be optimal under Assumption N it is necessary and sufficient that

$$J_{\mu^*} = T_{\mu^*}(J_{\mu^*}) = T(J_{\mu^*}),$$

or equivalently

$$\begin{aligned} J_{\mu^*}(x) &= \underset{w}{E}\,\{g[x, \mu^*(x), w] + \alpha J_{\mu^*}[f(x, \mu^*(x), w)]\} \\ &\leqslant \underset{w}{E}\,\{g(x, u, w) + \alpha J_{\mu^*}[f(x, u, w)]\} \qquad \forall x \in S, \quad u \in U(x). \end{aligned}$$

*Proof* Assume that the above condition holds. Then since $J_{\mu^*}$ is a fixed point of $T$ we have by Proposition 9 that $J_{\mu^*} \leqslant J^*$, which implies that $\pi^*$ is optimal. Conversely if $\pi^*$ is optimal, we have $J^* = J_{\mu^*}$ and hence we obtain $T_{\mu^*}(J_{\mu^*}) = J_{\mu^*} = J^* = T(J^*) = T(J_{\mu^*})$, which proves the desired result. Q.E.D.

The sufficiency part of the above corollary need not be true under Assumption P, as the following example shows.

EXAMPLE Let $S = (-\infty, +\infty)$, $U(x) = (0, 1]$ for all $x \in S$,

$$g(x, u, w) = |x|, \qquad f(x, u, w) = \alpha^{-1}ux$$

for all $(x, u, w) \in S \times C \times D$. Let $\mu^*(x) = 1$ for all $x \in S$. Then $J_{\mu^*}(x) = +\infty$ if $x \neq 0$ and $J_{\mu^*}(0) = 0$. Furthermore we have $J_{\mu^*} = T_{\mu^*}(J_{\mu^*}) = T(J_{\mu^*})$ as the reader can easily verify. It is also easy to verify that $J^*(x) = |x|$ and hence the policy $\{\mu^*, \mu^*, \ldots\}$ is not optimal.

On the other hand under Assumption P we have a different optimality condition, which, in view of its importance, we state as a proposition.

**Proposition 10** (Necessary and Sufficient Condition for Optimality under Assumption P) In order for an admissible stationary policy $\pi^* = \{\mu^*, \mu^*, \ldots\}$ to be optimal under Assumption P it is necessary and sufficient that

$$J^* = T_{\mu^*}(J^*) = T(J^*),$$

or equivalently

$$J^*(x) = \underset{w}{E}\{g[x, \mu^*(x), w] + \alpha J^*[f(x, \mu^*(x), w)]\}$$
$$\leqslant \underset{w}{E}\{g(x, u, w) + \alpha J^*[f(x, u, w)]\} \qquad \forall x \in S, \quad u \in U(x).$$

*Proof* We have by Corollary 8.1 that $J_{\mu^*} = T_{\mu^*}(J_{\mu^*})$. If the above condition holds, i.e., $J^* = T_{\mu^*}(J^*)$, then we obtain from Corollary 9.1 that $J_{\mu^*} \leqslant J^*$, which implies optimality of $\pi^*$. Conversely if $\pi^*$ is optimal, we have $J^* = J_{\mu^*}$ and hence we obtain $T_{\mu^*}(J^*) = T_{\mu^*}(J_{\mu^*}) = J_{\mu^*} = J^* = T(J^*)$, which proves the desired result. Q.E.D.

Again the sufficiency part of the proposition need not be true under Assumption N as the following example shows.

EXAMPLE Let $S = C = (-\infty, 0]$, $U(x) = C$ for all $x \in S$,

$$g(x, u, w) = f(x, u, w) = u$$

for all $(x, u, w) \in S \times C \times D$. Then $J^*(x) = -\infty$ for all $x \in S$, and every admissible policy $\pi^* = \{\mu^*, \mu^*, \ldots\}$ satisfies the condition of the proposition above. On the other hand, for $\mu^*(x) = 0$ for all $x \in S$ we have $J_{\mu^*}(x) = 0$ for all $x \in S$ and hence $\{\mu^*, \mu^*, \ldots\}$ is not optimal.

It is worth noting that Proposition 10 implies the existence of an optimal stationary policy under Assumption P when $U(x)$ is a finite set for every $x \in S$. This result need not be true under Assumption N (an example for the related case where $\alpha = 1$ is given in Problem 10 of Chapter 7).

*The Successive Approximation Method*

We now turn to the question of whether it is possible to obtain the optimal value function $J^*$ (in the limit) by means of the DP algorithm. Let $J_0$ denote the zero function on $S$, i.e.,

$$J_0(x) = 0 \qquad \forall x \in S.$$

Then under Assumption P we have

$$J_0 \leqslant T(J_0) \leqslant T^2(J_0) \leqslant \cdots \leqslant T^k(J_0) \leqslant \cdots,$$

while under Assumption N we have

$$J_0 \geqslant T(J_0) \geqslant T^2(J_0) \geqslant \cdots \geqslant T^k(J_0) \geqslant \cdots.$$

In either case the limit function

$$J_\infty(x) = \lim_{k\to\infty} T^k(J_0)(x) \qquad \forall x \in S \tag{59}$$

is well defined provided we allow the possibility that $J_\infty$ can take the value $+\infty$ (under Assumption P) or $-\infty$ (under Assumption N). We shall be interested in the question whether

$$J_\infty = J^*. \tag{60}$$

This question is, of course, of computational interest but it is also of analytical interest since, if one knows that $J^* = \lim_{k\to\infty} T^k(J_0)$, then one can infer properties of the unknown function $J^*$ from properties of $T^k(J_0)$ that are functions defined in a concrete algorithmic manner.

When the costs per stage are bounded (i.e., under Assumption B) we proved that (60) always holds. It turns out that when costs per stage are unbounded it may happen that (60) fails to hold. While we shall prove that (60) holds under Assumption N, when Assumption P is in effect, the example of Problem 9 shows that it may happen that $J_\infty \neq J^*$. In what follows we shall provide additional assumptions that guarantee that (60) holds under Assumption P. We have the following proposition.

**Proposition 11** (a) Let Assumption P hold and assume that

$$J_\infty(x) = T(J_\infty)(x) \qquad \forall x \in S.$$

Then if $J : S \to R$ is any bounded function we have

$$\lim_{k\to\infty} T^k(J)(x) = J^*(x) \qquad \forall x \in S. \tag{61}$$

(b) Let Assumption N hold. Then if $J: S \to R$ is any bounded function, we have

$$\lim_{k\to\infty} T^k(J)(x) = J^*(x) \qquad \forall x \in S. \tag{62}$$

*Proof* (a) Since under Assumption P we have

$$J_0 \leqslant T(J_0) \leqslant \cdots \leqslant T^k(J_0) \leqslant \cdots \leqslant J^*,$$

it follows that $\lim_{k\to\infty} T^k(J_0) = J_\infty \leqslant J^*$. Since $J_\infty$ is also a fixed point of $T$ by assumption we obtain from Proposition 9 that $J^* \leqslant J_\infty$. It follows that

$$J_\infty = J^* \tag{63}$$

and hence (61) is proved for the case $J = J_0$. It remains to prove (61) for every bounded $J: S \to R$. The proof proceeds similarly as in the proof of Proposition 1. Let $r$ be a scalar such that

$$J - re \leqslant J_0, \qquad J_0 \leqslant J + re,$$

where $e$ is the unit function on $S$ $[e(x) = 1, \forall x \in S]$. We have for all $k$ and $x \in S$,

$$T^k(J - re)(x) \leqslant T^k(J_0)(x) \leqslant T^k(J + re)(x), \tag{64}$$

and

$$T^k(J + re)(x) = T^k(J - re)(x) + 2\alpha^k r = T^k(J)(x) + \alpha^k r. \tag{65}$$

It follows from these two relations that

$$\lim_{k\to\infty} T^k(J - re)(x) = \lim_{k\to\infty} T^k(J + re)(x) = \lim_{k\to\infty} T^k(J_0)(x) = J^*(x). \tag{66}$$

Since

$$\lim_{k\to\infty} T^k(J + re)(x) = \lim_{k\to\infty} T^k(J)(x) \qquad \forall x \in S, \tag{67}$$

we obtain (61).

(b) We shall prove that under Assumption N we have

$$J_\infty(x) = \lim_{k\to\infty} T^k(J_0)(x) = J^*(x).$$

Then (62) follows from (64)–(67). Under Assumption N we have

$$J_0 \geqslant T(J_0) \geqslant \cdots \geqslant T^k(J_0) \geqslant \cdots \geqslant J^*.$$

It follows that

$$J_\infty(x) = \lim_{k\to\infty} T^k(J_0)(x) \geqslant J^*(x) \qquad \forall x \in S. \tag{68}$$

Also from relation $T^k(J_0) \geqslant J_\infty$, by applying $T$ to both sides, we obtain $T^{k+1}(J_0) \geqslant T(J_\infty)$ and taking the limit as $k \to \infty$

$$J_\infty \geqslant T(J_\infty). \tag{69}$$

On the other hand, for any $x \in S$, $u \in U(x)$ we have

$$\underset{w}{E}\,\{g(x, u, w) + \alpha T^k(J_0)[f(x, u, w)]\} \geqslant J_\infty(x).$$

Taking the limit of the left side as $k \to \infty$ and using the fact that $T^k(J_0)$ converges monotonically to $J_\infty$ we have

$$\underset{w}{E}\,\{g(x, u, w) + \alpha J_\infty[f(x, u, w)]\} \geqslant J_\infty(x).$$

Taking the infimum over $u \in U(x)$ we obtain

$$T(J_\infty)(x) \geqslant J_\infty(x) \qquad \forall x \in S. \tag{70}$$

Combining (69) and (70),

$$J_\infty = T(J_\infty).$$

Hence by Proposition 9

$$J_\infty \leqslant J^*.$$

Combining this inequality with (68) we obtain $J_\infty = \lim_{k\to\infty} T^k(J_0) = J^*$ and the result follows. Q.E.D.

We now proceed to obtain conditions that guarantee that $J_\infty = T(J_\infty)$ under Assumption P. As part (a) of Proposition 11 shows, $J_\infty = T(J_\infty)$ is a sufficient condition for the equality $J_\infty = J^*$ to hold. It is also a necessary condition in view of relation

$$J_\infty \leqslant T(J_\infty) \leqslant J^*. \tag{71}$$

This last relation follows once we observe that

$$T^{k-1}(J_0) \leqslant T^k(J_0) \leqslant \cdots \leqslant J_\infty \leqslant J^*. \tag{72}$$

By applying $T$ throughout and using $J^* = T(J^*)$,

$$T^k(J_0) \leqslant T^{k+1}(J_0) \leqslant \cdots \leqslant T(J_\infty) \leqslant J^*. \tag{73}$$

By taking the limit as $k \to \infty$, (71) follows.

We prove two propositions providing conditions for $J_\infty = T(J_\infty)$. The first admits an easy proof but requires a restrictive assumption. The second is a little harder to prove but requires a much weaker assumption.

**Proposition 12** Let Assumption P hold and assume that the control constraint set $U(x)$ is a finite set for every $x \in S$. Then

$$J_\infty = T(J_\infty) = J^* = T(J^*).$$

*Proof* From (73) we have for all $x \in S$

$$T^{k+1}(J_0)(x) = \inf_{u \in U(x)} \mathop{E}_{w} \{g(x, u, w) + \alpha T^k(J_0)[f(x, u, w)]\} \leqslant T(J_\infty)(x), \tag{74}$$

and taking the limit in (74), $J_\infty \leqslant T(J_\infty)$. Suppose that there existed a state $\tilde{x} \in S$, such that

$$J_\infty(\tilde{x}) < T(J_\infty)(\tilde{x}). \tag{75}$$

Using the finiteness of $U(\tilde{x})$ let $u_k$ be the minimizing control in (74). Since $U(\tilde{x})$ is finite there must exist some $\tilde{u} \in U(\tilde{x})$ such that $u_k = \tilde{u}$ for all $k$ in some infinite subset $\mathscr{K}$ of the positive integers. By (74) we have for all $k \in \mathscr{K}$

$$T^{k+1}(J_0)(\tilde{x}) = \mathop{E}_{w} \{g(\tilde{x}, \tilde{u}, w) + \alpha T^k(J_0)[f(\tilde{x}, \tilde{u}, w)]\} \leqslant T(J_\infty)(\tilde{x}).$$

Taking the limit as $k \to \infty$, $k \in \mathscr{K}$, we obtain

$$\begin{aligned} J_\infty(\tilde{x}) &= \underset{w}{E}\{g(\tilde{x}, \tilde{u}, w) + \alpha J_\infty[f(\tilde{x}, \tilde{u}, w)]\} \\ &\leqslant T(J_\infty)(\tilde{x}) = \inf_{u \in U(\tilde{x})} \underset{w}{E}\{g(\tilde{x}, u, w) + \alpha J_\infty[f(\tilde{x}, u, w)]\}. \end{aligned}$$

It follows that $J_\infty(\tilde{x}) = T(J_\infty)(\tilde{x})$, contradicting (75). Q.E.D.

The following proposition strengthens Proposition 12 in that it requires a compactness rather than a finiteness assumption. We recall (see Appendix A) that a subset $X$ of an $n$-dimensional Euclidean space $R^n$ is said to be *compact* if every sequence $\{x_k\}$ with $x_k \in X$ contains a subsequence $\{x_k\}_{k \in \mathscr{K}}$ that converges to a point $x \in X$. Equivalently $X$ is compact if and only if it is closed and bounded. The empty set is (trivially) considered compact. Given any collection of compact sets, their intersection is a compact set (possibly empty). Given a sequence of nonempty compact sets $X_1, X_2, \ldots, X_k, \ldots$ such that

$$X_1 \supset X_2 \supset \cdots \supset X_k \supset X_{k+1} \supset \cdots,$$

their intersection $\bigcap_{k=1}^\infty X_k$ is both nonempty and compact. In view of this fact it follows that if $f: R^n \to [-\infty, +\infty]$ is a function such that the set

$$F_\lambda = \{x \in R^n \mid f(x) \leqslant \lambda\}$$

is compact for every $\lambda \in R$, then there exists a point $x^*$ minimizing $f$, i.e., there exists an $x^* \in R^n$ such that

$$f(x^*) = \inf_{x \in R^n} f(x).$$

To see this, take a sequence $\{\lambda_k\}$ such that $\lambda_k \to \inf_{x \in R^n} f(x)$ and $\lambda_k \geqslant \lambda_{k+1}$ for all $k$. If $\inf_{x \in R^n} f(x) < +\infty$ such a sequence exists and the sets

$$F_{\lambda_k} = \{x \in R^n \mid f(x) \leqslant \lambda_k\}$$

are nonempty and compact. Furthermore, $F_{\lambda_k} \supset F_{\lambda_{k+1}}$ for all $k$ and hence the intersection $\bigcap_{k=1}^\infty F_{\lambda_k}$ is also nonempty and compact. Let $x^*$ be any point in $\bigcap_{k=1}^\infty F_{\lambda_k}$. Then

$$f(x^*) \leqslant \lambda_k \qquad \forall k = 1, 2, \ldots,$$

and taking the limit as $k \to \infty$ we obtain $f(x^*) \leqslant \inf_{x \in R^n} f(x)$, proving that $x^*$ minimizes $f(x)$.

**Proposition 13** Let Assumption P hold and assume that the sets

$$U_k(x, \lambda) = \left\{ u \in U(x) \,\middle|\, \underset{w}{E}\{g(x, u, w) + \alpha T^k(J_0)[f(x, u, w)]\} \leqslant \lambda \right\} \tag{76}$$

are compact subsets of a Euclidean space for every $x \in S$, $\lambda \in R$, and for all $k$ greater than some integer $\bar{k}$. Then

$$J_\infty = T(J_\infty) = J^* = T(J^*). \tag{77}$$

Furthermore there exists a stationary optimal policy.

*Proof* Similarly as in Proposition 12 we have $J_\infty \leqslant T(J_\infty)$. Suppose that there existed a state $\tilde{x} \in S$ such that

$$J_\infty(\tilde{x}) < T(J_\infty)(\tilde{x}). \tag{78}$$

Clearly we must have $J_\infty(\tilde{x}) < +\infty$. For every $k \geqslant \bar{k}$ consider the sets

$$U_k[\tilde{x}, J_\infty(\tilde{x})] = \{u \in U(\tilde{x}) \mid \underset{w}{E}\{g(\tilde{x}, u, w) + \alpha T^k(J_0)[f(\tilde{x}, u, w)]\} \leqslant J_\infty(\tilde{x})\}.$$

Let also $u_k$ be a point attaining the infimum in

$$T^{k+1}(J_0)(\tilde{x}) = \inf_{u \in U(\tilde{x})} \underset{w}{E}\{g(\tilde{x}, u, w) + \alpha T^k(J_0)[f(\tilde{x}, u, w)]\} \leqslant J_\infty(\tilde{x}),$$

i.e., $u_k$ is such that

$$T^{k+1}(J_0)(\tilde{x}) = \underset{w}{E}\{g(\tilde{x}, u_k, w) + \alpha T^k(J_0)[f(\tilde{x}, u_k, w)]\} \leqslant J_\infty(\tilde{x}).$$

Such minimizing points $u_k$ exist by our compactness assumption. For every $k \geqslant \bar{k}$ consider the sequence $\{u_i\}_{i=k}^\infty$. Since $T^k(J_0) \leqslant T^{k+1}(J_0) \leqslant \cdots \leqslant J_\infty$ it follows that

$$\underset{w}{E}\{g(\tilde{x}, u_i, w) + \alpha T^k(J_0)[f(\tilde{x}, u_i, w)]\}$$
$$\leqslant \underset{w}{E}\{g(\tilde{x}, u_i, w) + \alpha T^i(J_0)[f(\tilde{x}, u_i, w)]\} \leqslant J_\infty(\tilde{x}) \qquad \forall i \geqslant k.$$

Hence $\{u_i\}_{i=k}^\infty \subset U_k[\tilde{x}, J_\infty(\tilde{x})]$, and since $U_k[\tilde{x}, J_\infty(\tilde{x})]$ is compact, all the limit points of $\{u_i\}_{i=k}^\infty$ belong to $U_k[\tilde{x}, J_\infty(\tilde{x})]$ and at least one such limit point exists. Hence the same is true of the limit points of the whole sequence $\{u_i\}_{i=\bar{k}}^\infty$. It follows that if $\tilde{u}$ is a limit point of $\{u_i\}_{i=\bar{k}}^\infty$, then

$$\tilde{u} \in \bigcap_{k=\bar{k}}^\infty U_k[\tilde{x}, J_\infty(\tilde{x})].$$

This implies by (76) that for all $k \geqslant \bar{k}$

$$J_\infty(\tilde{x}) \geqslant \underset{w}{E}\{g(\tilde{x}, \tilde{u}, w) + \alpha T^k(J_0)[f(\tilde{x}, \tilde{u}, w)]\} \geqslant T^{k+1}(J_0)(\tilde{x}).$$

Taking the limit as $k \to \infty$ we obtain

$$J_\infty(\tilde{x}) = \underset{w}{E}\{g(\tilde{x}, \tilde{u}, w) + \alpha J_\infty[f(\tilde{x}, \tilde{u}, w)]\}.$$

Since the right-hand side is greater or equal to $T(J_\infty)(\tilde{x})$, (78) is contradicted. Hence $J_\infty = T(J_\infty)$ and (77) is proved.

To show that there exists an optimal stationary policy observe that (77) and the last relation imply that $\tilde{u}$ attains the infimum in

$$J^*(\tilde{x}) = \inf_{u \in U(\tilde{x})} \mathop{E}_{w} \{g(\tilde{x}, u, w) + \alpha J^*[f(\tilde{x}, u, w)]\}$$

for a state $\tilde{x} \in S$ with $J^*(\tilde{x}) < +\infty$. For states $\tilde{x} \in S$ such that $J^*(\tilde{x}) = +\infty$ every $u \in U(\tilde{x})$ attains the infimum above. Hence by Proposition 10 an optimal stationary policy exists. Q.E.D.

The reader may verify by inspection of the proof that actually Proposition 13 may be proved under the weaker assumption that the sets $U_k[\tilde{x}, J_\infty(\tilde{x})]$ are compact for all $\tilde{x} \in S$ such that $J_\infty(\tilde{x}) < +\infty$, and all $k$ greater than some index $\bar{k}$.

Another fact that may be verified from the proof is that if $\mu_k(\tilde{x})$, $k = 0, 1, \ldots,$ attains the infimum in the relation

$$T^{k+1}(J_0)(\tilde{x}) = \inf_{u \in U(x)} \mathop{E}_{w} \{g(\tilde{x}, u, w) + \alpha T^k(J_0)[f(\tilde{x}, u, w)]\},$$

then if $\mu^*(\tilde{x})$ is a limit point of $\{\mu_k(\tilde{x})\}$, $\forall \tilde{x} \in S$, the policy $\pi^* = \{\mu^*, \mu^*, \ldots\}$ is optimal. Furthermore, $\{\mu_k(\tilde{x})\}$ has at least one limit point for every $\tilde{x} \in S$ for which $J^*(\tilde{x}) < +\infty$. Thus *the successive approximation method under the assumptions of Proposition* 12 *or* 13 *yields in the limit not only the optimal value function* $J^*$ *but also an optimal stationary policy.*

## 6.5 Linear Systems and Quadratic Cost Functionals

Consider the case where in Problem (D) the system is linear:

$$x_{k+1} = Ax_k + Bu_k + w_k, \qquad k = 0, 1, \ldots,$$

where $x_k \in R^n$, $u_k \in R^m$ for all $k$ and the matrices $A$, $B$ are known. As in Sections 3.1 and 4.3 we assume that the random disturbances $w_k$ are independent with zero mean and finite second moments. The cost functional is quadratic and has the form

$$J_\pi(x_0) = \lim_{N \to \infty} \mathop{E}_{\substack{w_k \\ k=0,\ldots,N-1}} \left\{ \sum_{k=0}^{N-1} \alpha^k [x_k' Q x_k + \mu_k(x_k)' R \mu_k(x_k)] \right\},$$

where $Q$ is a positive semidefinite symmetric $n \times n$ matrix and $R$ is a positive definite symmetric $m \times m$ matrix. The problem clearly falls under the framework of Assumption P.

Our approach will be to consider the ordinary DP algorithm, i.e., obtain the functions $T(J_0)$, $T^2(J_0), \ldots,$ as well as the pointwise limit function $J_\infty = \lim_{k \to \infty} T^k(J_0)$. Subsequently we show that $J_\infty$ satisfies $J_\infty = T(J_\infty)$

and hence, by Proposition 11, $J_\infty = J^*$. The optimal policy is then easily obtained from the optimal value function $J^*$ via the DP functional equation using Proposition 10.

As in Section 3.1 we have

$$J_0(x) = 0 \qquad \forall x \in R^n,$$

$$T(J_0)(x) = \inf_u (x'Qx + u'Ru) = x'Qx \qquad \forall x \in R^n,$$

$$T^2(J_0)(x) = \inf_u \underset{w}{E}\{x'Qx + u'Ru + \alpha(Ax + Bu + w)'Q(Ax + Bu + w)\}$$

$$= x'K_1x + \alpha \underset{w}{E}\{w'Qw\} \qquad \forall x \in R^n,$$

$$T^{k+1}(J_0)(x) = x'K_kx + \sum_{m=0}^{k-1} \alpha^{k-m} \underset{w}{E}\{w'K_m w\} \qquad \forall x \in R^n, \quad k = 1, 2, \ldots,$$

where the matrices $K_0, K_1, K_2, \ldots$ are given recursively by

$$K_0 = Q,$$

$$K_{k+1} = A'[\alpha K_k - \alpha^2 K_k B(\alpha B'K_k B + R)^{-1}B'K_k]A + Q, \qquad k = 0, 1, \ldots.$$

Now by writing $\tilde{R} = R/\alpha$ and $\tilde{A} = \sqrt{\alpha}A$ the preceding equation may be written

$$K_{k+1} = \tilde{A}'[K_k - K_k B(B'K_k B + \tilde{R})^{-1}B'K_k]\tilde{A} + Q,$$

and is of the form considered in Section 3.1. By making use of the result shown there we have

$$K_k \to K$$

provided the pairs $(\tilde{A}, B)$ and $(\tilde{A}, C)$, where $Q = C'C$, are controllable and observable, respectively. Since $\tilde{A} = \sqrt{\alpha}A$, controllability and observability of $(A, B)$ or $(A, C)$ are clearly equivalent to controllability and observability of $(\tilde{A}, B)$ or $(\tilde{A}, C)$. The matrix $K$ is positive definite and it is the unique solution of the equation

$$K = A'[\alpha K - \alpha^2 KB(\alpha B'KB + R)^{-1}B'K]A + Q \tag{79}$$

within the class of positive semidefinite symmetric matrices.

As a result of the preceding analysis we have that the pointwise limit of the functions $T^k(J_0)$ is given by

$$J_\infty(x) = \lim_{k\to\infty} T^k(J_0)(x) = x'Kx + c, \tag{80}$$

where

$$c = \lim_{k\to\infty} \sum_{m=0}^{k-1} \alpha^{k-m} \underset{w}{E}\{w'K_m w\}.$$

This limit is well defined since the scalars $c_k$, where

$$c_k = \sum_{m=0}^{k-1} \alpha^{k-m} \mathop{E}_{w} \{w'K_m w\},$$

satisfy the equation

$$c_{k+1} = \alpha c_k + \alpha \mathop{E}_{w} \{w'K_k w\}.$$

Since $\alpha < 1$ and $K_k \to K$ it follows that $c_k \to c$, where

$$c = [\alpha/(1 - \alpha)] \mathop{E}_{w} \{w'Kw\}. \tag{81}$$

Now using (79)–(81) one can easily verify that for all $x \in S$

$$J_\infty(x) = T(J_\infty)(x) = \min_u \left[ x'Qx + u'Ru + \alpha \mathop{E}_{w} \{J_\infty(Ax + Bu + w)\} \right] \tag{82}$$

and hence by Proposition 11, $J_\infty = J^*$. Another method for proving that $J_\infty = T(J_\infty)$ is to show that the assumption of Proposition 13 is satisfied, i.e., that the sets

$$U_k(x, \lambda) = \left\{ u \,\middle|\, \mathop{E}_{w} \{x'Qx + u'Ru + \alpha T^k(J_0)(Ax + Bu + w)\} \leqslant \lambda \right\}$$

are compact. This can be easily verified using the fact that $T^k(J_0)$ is a quadratic function and $R$ is positive definite. The optimal policy is obtained by minimization in (82) and has the form $\pi^* = \{\mu^*, \mu^*, \ldots\}$, where $\mu^*$ is given by

$$\mu^*(x) = -\alpha(\alpha B'KB + R)^{-1}B'KAx \qquad \forall x \in R^n.$$

The linearity and stationarity of this policy makes it very attractive for engineering applications. A number of generalized versions of the problem of this section, including the case of imperfect state information, are treated in the problem section.

## 6.6 Inventory Control

Let us consider an infinite horizon version of the inventory control problem of Section 3.2 where costs per stage are discounted. Inventory stock evolves according to the equation

$$x_{k+1} = x_k + u_k - w_k, \qquad k = 0, 1, \ldots. \tag{83}$$

Again we assume that the successive demands $w_k$ are independent and bounded and have identical probability distributions. We shall assume for simplicity that there is no fixed cost. A similar analysis may be carried out for the case of a nonzero fixed cost. The function to be minimized is given by

$$J_\pi(x_0) = \lim_{N\to\infty} \underset{\substack{w_k \\ k=0,1,\ldots}}{E} \left\{ \sum_{k=0}^{N-1} \alpha^k [c\mu_k(x_k) + p \max(0, w_k - x_k - \mu_k(x_k)) + h \max(0, x_k + \mu_k(x_k) - w_k)] \right\}. \tag{84}$$

The DP algorithm is given by

$$J_0(x) = 0,$$

$$T^{k+1}(J_0)(x) = \inf_{0 \leqslant u} \underset{w}{E} \{cu + p \max(0, w - x - u) + h \max(0, x + u - w) + \alpha T^k(J_0)(x + u - w)\}. \tag{85}$$

Let us first show that

$$J^*(x_0) = \inf_\pi J_\pi(x_0) < +\infty \qquad \forall x_0 \in S. \tag{86}$$

Indeed consider the policy $\tilde{\pi} = \{\tilde{\mu}, \tilde{\mu}, \ldots\}$, where $\tilde{\mu}$ is defined by

$$\tilde{\mu}(x) = \begin{cases} 0 & \text{if } x \geqslant 0, \\ -x & \text{if } x < 0. \end{cases}$$

Since $w_k$ is nonnegative and bounded it follows that the inventory stock $x_k$ when the policy $\tilde{\pi}$ is used satisfies

$$-w_{k-1} \leqslant x_k \leqslant \max[0, x_0], \qquad k = 1, 2, \ldots,$$

and is bounded. Hence $\tilde{\mu}(x_k)$ is also bounded. Hence the cost per stage incurred when $\tilde{\pi}$ is used is bounded, and in view of the presence of the discount factor we have

$$J_{\tilde{\pi}}(x_0) < +\infty \qquad \forall x_0 \in S.$$

Since $J^* \leqslant J_{\tilde{\pi}}$, (86) follows.

Next let us observe that under the assumption $c < p$, the functions $T^k(J_0)$ are real-valued convex functions. Indeed we have

$$J_0 \leqslant T(J_0) \leqslant \cdots \leqslant T^k(J_0) \leqslant \cdots \leqslant J^*, \tag{87}$$

which implies that $T^k(J_0)$ is real valued. Convexity follows easily by induction as shown in Section 3.2. Consider now the sets

$$U_k(x, \lambda) = \left\{ u \geqslant 0 \,\middle|\, \underset{w}{E} \{cu + p \max(0, w - x - u) + h \max(0, x + u - w) + \alpha T^k(J_0)(x + u - w)\} \leqslant \lambda \right\}. \tag{88}$$

These sets are bounded since the expected value above tends to $+\infty$ as $u \to +\infty$. Also the sets $U_k(x, \lambda)$ are closed since the expected value in (88) is a continuous function of $u$ [recall that $T^k(J_0)$ is a real-valued convex and hence continuous function]. Thus we may invoke Proposition 13 and assert that

$$J_\infty(x) = \lim_{k \to \infty} T^k(J_0)(x) = J^*(x) \qquad \forall x \in S.$$

It follows from the convexity of the functions $T^k(J_0)$ that the limit function $J^*$ *is a real-valued convex function.* Furthermore we have from Proposition 8 the optimality equation

$$J^*(x) = \inf_{u \geq 0} \mathop{E}_{w} \{cu + p \max(0, w - x - u) + h \max(0, x + u - w) + \alpha J^*(x + u - w)\}.$$

An optimal stationary policy $\pi^* = \{\mu^*, \mu^*, \ldots\}$ can be obtained from the above equation as in Sections 1.3 and 3.2. We have

$$\mu^*(x) = \begin{cases} S^* - x & \text{if} \quad x \leqslant S^*, \\ 0 & \text{otherwise}, \end{cases}$$

where $S^*$ is a minimizing point of

$$G^*(y) = cy + L(y) + \mathop{E}_{w} \{J^*(y - w)\},$$

with

$$L(y) = p \mathop{E}_{w} \{\max(0, w - y)\} + h \mathop{E}_{w} \{\max(0, y - w)\}.$$

It is easy to see that if $p > c$, we have $\lim_{|y| \to \infty} G^*(y) = +\infty$ so that such a minimizing point exists. Furthermore by utilizing the observation made at the end of Section 6.4 it follows that minimizing points $S^*$ of $G^*(y)$ may be obtained as limit points of sequences $\{S_k\}$, where for each $k$ the scalar $S_k$ minimizes

$$G_k(y) = cy + L(y) + \alpha \mathop{E}_{w} \{T^k(J_0)(y - w)\}$$

and is obtained by means of the successive approximation method.

In the case where there is a positive fixed cost ($K > 0$) the same line of argument may be used. Similarly we prove that $J^*$ is a real-valued $K$-convex function. A separate argument is necessary to prove that $J^*$ is also continuous (see references [B3] or [I2]). Once $K$-convexity and continuity of $J^*$ is established the optimality of a stationary $(s^*, S^*)$ policy follows from the equation

$$J^*(x) = \min_{u \geqslant 0} \mathop{E}_{w} \{C(u) + p \max(0, w - x - u) + h \max(0, x + u - w) + \alpha J^*(x + u - w)\},$$

where $C(u) = K + cu$ if $u > 0$ and $C(0) = 0$.

## 6.7 Nonstationary and Periodic Problems

The standing assumption so far in this chapter has been that the problem involves a stationary system and a stationary cost per stage (except for the presence of the discount factor). Problems where the system or the cost per stage are nonstationary arise occasionally in practice or in theoretical studies and are thus of some interest. It turns out that such problems can be embedded by means of a simple reformulation within the framework of Problem (D) for which stationarity prevails. Once this reformulation is considered, one easily obtains results analogous to those of Sections 6.1 and 6.4.

Consider a nonstationary system of the form

$$x_{k+1} = f_k(x_k, u_k, w_k), \qquad k = 0, 1, \ldots, \tag{89}$$

and a cost functional of the form

$$J_\pi(x_0) = \lim_{N \to \infty} \underset{\substack{w_k \\ k=0,1,\ldots,N-1}}{E} \left\{ \sum_{k=0}^{N-1} \alpha^k g_k[x_k, \mu_k(x_k), w_k] \right\}. \tag{90}$$

In the above equations, for each $k$, $x_k$ belongs to a space $S_k$, $u_k$ belongs to a space $C_k$ and satisfies $u_k \in U_k(x_k)$ for all $x_k \in S_k$, and $w_k$ belongs to a countable space $D_k$. The sets $S_k$, $C_k$, $U_k(x_k)$, $D_k$ may differ from one stage to the next. The random disturbances $w_k$ are characterized by probabilities $P_k(\cdot \mid x_k, u_k)$, which depend on $x_k$, $u_k$ as well as the time index $k$. The set of admissible policies $\Pi$ is the set of all sequences $\pi = \{\mu_0, \mu_1, \ldots\}$ with $\mu_k: S_k \to C_k$ with $\mu_k(x_k) \in U_k(x_k)$ for all $x_k \in S_k$ and $k = 0, 1, \ldots$. The functions $g_k : S_k \times C_k \times D_k \to R$ are given and are assumed to satisfy one of the following three assumptions, which are analogous to Assumptions B, P, and N considered earlier in this chapter:

**Assumption B′** The functions $g_k$ satisfy for all $k = 0, 1, \ldots,$

$$0 \leqslant g_k(x_k, u_k, w_k) \leqslant M \qquad \forall (x_k, u_k, w_k) \in S_k \times C_k \times D_k,$$

where $M$ is some scalar.

**Assumption P′** The functions $g_k$ satisfy for all $k = 0, 1, \ldots,$

$$0 \leqslant g_k(x_k, u_k, w_k) \qquad \forall (x_k, u_k, w_k) \in S_k \times C_k \times D_k.$$

**Assumption N′** The functions $g_k$ satisfy for all $k = 0, 1, \ldots,$

$$g_k(x_k, u_k, w_k) \leqslant 0 \qquad \forall (x_k, u_k, w_k) \in S_k \times C_k \times D_k.$$

We shall refer to the problem formulated above as the *nonstationary problem* (NSP).

Let us now convert the NSP to a stationary problem that fits within the framework of Problem (D). In order to simplify the notation *we shall assume that the state spaces* $S_i$, $i = 0, 1, \ldots,$ *the control spaces* $C_i$, $i = 0, 1, \ldots,$ *and the disturbance spaces* $D_i$, $i = 0, 1, \ldots,$ *are all mutually disjoint.* This assumption does not involve a loss of generality since, if necessary, we may relabel the elements of $S_i$, $C_i$, and $D_i$ without affecting the structure of the problem. Define now a new state space $S$, a new control space $C$, and a new (countable) disturbance space $D$ by

$$S = \bigcup_{i=0}^{\infty} S_i, \qquad C = \bigcup_{i=0}^{\infty} C_i, \qquad D = \bigcup_{i=0}^{\infty} D_i.$$

Introduce a new (stationary) system

$$\tilde{x}_{k+1} = f(\tilde{x}_k, \tilde{u}_k, \tilde{w}_k), \qquad k = 0, 1, \ldots, \tag{91}$$

where $\tilde{x}_k \in S$, $\tilde{u}_k \in C$, $\tilde{w}_k \in D$, and the system function $f: S \times C \times D \to S$ is defined by

$$f(\tilde{x}, \tilde{u}, \tilde{w}) = f_i(\tilde{x}, \tilde{u}, \tilde{w}) \qquad \text{if} \quad \tilde{x} \in S_i, \quad \tilde{u} \in C_i, \quad \tilde{w} \in D_i, \quad i = 0, 1, \ldots.$$

For triplets $(\tilde{x}, \tilde{u}, \tilde{w})$, where for some $i = 0, 1, \ldots$ we have $\tilde{x} \in S_i$ but $\tilde{u} \notin C_i$, or $\tilde{w} \notin D_i$, the definition of $f$ is immaterial—any definition is adequate for our purposes in view of the control constraints to be introduced. The control constraint is taken to be $\tilde{u} \in U(\tilde{x})$ for all $\tilde{x} \in S$, where $U(\cdot)$ is defined by

$$U(\tilde{x}) = U_i(\tilde{x}) \qquad \text{if} \quad \tilde{x} \in S_i, \quad i = 0, 1, \ldots.$$

The disturbance $\tilde{w}$ is characterized by probabilities $P(\tilde{w} \mid \tilde{x}, \tilde{u})$ such that

$$P(\tilde{w} \in D_i \mid \tilde{x} \in S_i, \tilde{u} \in C_i) = 1, \qquad i = 0, 1, \ldots,$$
$$P(\tilde{w} \notin D_i \mid \tilde{x} \in S_i, \tilde{u} \in C_i) = 0, \qquad i = 0, 1, \ldots.$$

Furthermore for any $w_i \in D_i$, $x_i \in S_i$, $u_i \in C_i$, $i = 0, 1, \ldots,$ we have

$$P(w_i \mid x_i, u_i) = P_i(w_i \mid x_i, u_i).$$

We also introduce a new cost functional

$$\tilde{J}_{\tilde{\pi}}(\tilde{x}_0) = \lim_{N \to \infty} \underset{\substack{\tilde{w}_k \\ k = 0, 1, \ldots, N-1}}{E} \left\{ \sum_{k=0}^{N-1} \alpha^k g[\tilde{x}_k, \tilde{\mu}_k(\tilde{x}_k), \tilde{w}_k] \right\}, \tag{92}$$

where the (stationary) cost per stage $g: S \times C \times D \to R$ is defined by

$$g(\tilde{x}, \tilde{u}, \tilde{w}) = g_i(\tilde{x}, \tilde{u}, \tilde{w}) \qquad \text{if} \quad \tilde{x} \in S_i, \quad \tilde{u} \in C_i, \quad \tilde{w} \in D_i, \quad i = 0, 1, \ldots.$$

For triplets $(\tilde{x}, \tilde{u}, \tilde{w})$, where for some $i = 0, 1, \ldots$ we have $\tilde{x} \in S_i$ but $\tilde{u} \notin C_i$ or $\tilde{w} \notin D_i$, any definition of $g$ is adequate provided $0 \leq g(\tilde{x}, \tilde{u}, \tilde{w}) \leq M$ for all $(\tilde{x}, \tilde{u}, \tilde{w})$ when Assumption B′ holds, $0 \leq g(\tilde{x}, \tilde{u}, \tilde{w})$ when P′ holds, and

$g(\tilde{x}, \tilde{u}, \tilde{w}) \leqslant 0$ when N′ holds. The set of admissible policies $\tilde{\Pi}$ for the new problem consists of all sequences $\tilde{\pi} = \{\tilde{\mu}_0, \tilde{\mu}_1, \ldots\}$ where $\tilde{\mu}_k: S \to C$ and $\tilde{\mu}_k(\tilde{x}) \in U(\tilde{x})$ for all $\tilde{x} \in S$ and $k = 0, 1, \ldots$.

The construction given above defines a problem that clearly fits the framework of Problem (D). We shall refer to this problem as the *stationary problem* (SP).

It is important to understand the nature of the intimate connection between the NSP and the SP formulated above. Let $\pi = \{\mu_0, \mu_1, \ldots\}$ be an admissible policy for the NSP. Also let $\tilde{\pi} = \{\tilde{\mu}_0, \tilde{\mu}_1, \ldots\}$ be an admissible policy for the SP such that

$$\tilde{\mu}_i(\tilde{x}) = \mu_i(\tilde{x}) \qquad \text{if} \quad \tilde{x} \in S_i, \quad i = 0, 1, \ldots. \tag{93}$$

Let $x_0 \in S_0$ be the initial state for the NSP and consider the same initial state for the SP, i.e., $\tilde{x}_0 = x_0 \in S_0$. Then the sequence of states $\{\tilde{x}_i\}$ generated in the SP will satisfy $\tilde{x}_i \in S_i$, $i = 0, 1, \ldots$, with probability one, i.e., the system will move from the set $S_0$ to the set $S_1$, then to $S_2$, etc., just as in the NSP. Furthermore, the probabilistic law of generation of states and costs is identical in the NSP and the SP. As a result it is easy to see that for any admissible policies $\pi$ and $\tilde{\pi}$ satisfying (93) and initial states $x_0$, $\tilde{x}_0$ satisfying $x_0 = \tilde{x}_0 \in S_0$, the sequence of generated states in the NSP and the SP is the same ($x_i = \tilde{x}_i$, $\forall i$) provided the generated disturbances $w_i$ and $\tilde{w}_i$ are also the same for all $i$ ($w_i = \tilde{w}_i$, $\forall i$). Furthermore, if $\pi$ and $\tilde{\pi}$ satisfy (93), we have $J_\pi(x_0) = \tilde{J}_{\tilde{\pi}}(\tilde{x}_0)$ if $x_0 = \tilde{x}_0 \in S_0$. Let us also consider the optimal value functions for the NSP and the SP

$$J^*(x_0) = \inf_{\pi \in \Pi} J_\pi(x_0), \qquad x_0 \in S_0,$$

$$\tilde{J}^*(\tilde{x}_0) = \inf_{\tilde{\pi} \in \tilde{\Pi}} \tilde{J}_{\tilde{\pi}}(\tilde{x}_0), \qquad \tilde{x}_0 \in S.$$

Then it follows from the construction of the SP that

$$\tilde{J}^*(\tilde{x}_0) = \tilde{J}^*(\tilde{x}_0, i) \qquad \text{if} \quad \tilde{x}_0 \in S_i, \quad i = 0, 1, \ldots, \tag{94}$$

where

$$\tilde{J}^*(\tilde{x}_0, i) = \inf_{\pi \in \Pi} \lim_{N \to \infty} \underset{k = i, \ldots, N-1}{\underset{w_k}{E}} \left\{ \sum_{k=i}^{N-1} \alpha^{k-i} g_k[x_k, \mu_k(x_k), w_k] \right\},$$

$$\text{if} \quad \tilde{x}_0 = x_i \in S_i, \quad i = 0, 1, \ldots. \tag{95}$$

Note that in this equation the right-hand side is defined in terms of the data of the NSP. As a special case of this equation we obtain

$$\tilde{J}^*(\tilde{x}_0) = \tilde{J}^*(\tilde{x}_0, 0) = J^*(x_0) \qquad \text{if} \quad \tilde{x}_0 = x_0 \in S_0. \tag{96}$$

Thus *the optimal value function* $J^*$ *of the NSP can be obtained from the optimal value function* $\tilde{J}^*$ *of the SP*. Furthermore, if $\tilde{\pi}^* = \{\tilde{\mu}_0^*, \tilde{\mu}_1^*, \ldots\}$ is an optimal policy for the SP, then the policy $\pi^* = \{\mu_0^*, \mu_1^*, \ldots\}$ defined by

$$\mu_i^*(x_i) = \tilde{\mu}_i^*(x_i) \qquad \forall x_i \in S_i, \quad i = 0, 1, \ldots, \tag{97}$$

is an optimal policy for the NSP. *Thus optimal policies for the SP yield optimal policies for the NSP via* (97). Another point to be noted is that *if Assumption B′ (P′, N′) is satisfied for the NSP, then Assumption B (P, N) introduced earlier in this chapter is satisfied for the SP.*

The observations above indicate clearly that one may analyze the NSP by means of the SP. In fact every result given in Sections 6.1 and 6.4 when applied to the SP yields a corresponding result for the NSP. We shall content ourselves with providing the form of the optimality equation for the NSP in the following proposition.

**Proposition 14** Under Assumption B′ (P′, N′) there holds

$$J^*(x_0) = \tilde{J}^*(x_0, 0) \qquad \forall x_0 \in S_0,$$

where for all $i = 0, 1, \ldots$ the functions $\tilde{J}^*(\cdot, i)$ map $S_i$ into $[0, \infty)$ ($[0, \infty]$, $[-\infty, 0]$), are given by (95), and satisfy

$$\tilde{J}^*(x_i, i) = \inf_{u_i \in U_i(x_i)} \mathop{E}_{w_i} \{g_i(x_i, u_i, w_i) + \alpha \tilde{J}^*[f_i(x_i, u_i, w_i), i + 1]\}$$

$$\forall x_i \in S_i, \quad i = 0, 1, \ldots. \tag{98}$$

Under Assumption B′ the functions $\tilde{J}^*(\cdot, i)$, $i = 0, 1, \ldots$, are the unique bounded solutions of the set of equations (98). Furthermore, under Assumption B′ or P′, if $\mu_i^*(x_i) \in U_i(x_i)$ attains the infimum in (98) for all $x_i \in S_i$, and $i$, then the policy $\pi^* = \{\mu_0^*, \mu_1^*, \ldots\}$ is optimal for the NSP.

*Proof* Apply Propositions 2, 8, and 10 to the SP. Then the result follows immediately by making use of definitions (94)–(96). Q.E.D.

Notice that an optimal policy for the NSP will normally be nonstationary even though the SP may possess an optimal stationary policy. Furthermore, such an optimal policy is in general impossible to obtain in practice even if the state spaces $S_i$, $i = 0, 1, \ldots$, are finite sets, in view of the fact that an infinite number of functions $\mu_i^*$ are involved. However, there are special cases where important simplifications occur. We proceed to examine two such cases.

*Eventually Stationary Problems*

Within the framework of the NSP consider the case where the spaces $S_i$, $C_i$, $D_i$, the sets $U_i(\cdot)$, the probability distributions $P_i(\cdot \mid x_i, u_i)$, and the functions $f_i$ and $g_i$ remain unchanged after some index $\bar{k}$, i.e.,

$$S_i = S_j = \bar{S}, \qquad C_i = C_j = \bar{C}, \qquad D_i = D_j = \bar{D} \qquad \forall i, j \geqslant \bar{k},$$

$$U_i(\cdot) = U_j(\cdot) = \bar{U}(\cdot), \qquad P_i(\cdot \,|\, x, u) = P_j(\cdot \,|\, x, u) \qquad \forall i, j \geqslant \bar{k},$$

$$f_i = f_j = \bar{f}, \qquad g_i = g_j = \bar{g}, \qquad \forall i, j \geqslant \bar{k}.$$

Notice that finite horizon problems may be embedded within the framework of the above problem by taking the function $\bar{g}$ identically equal to zero. We shall assume that the spaces $S_i, C_i, D_i, i = 0, 1, \ldots, \bar{k} - 1$, are mutually disjoint and disjoint from $\bar{S}, \bar{C}, \bar{D}$, respectively. Then we may define a new state space, control space, and disturbance space by

$$S = \bigcup_{i=0}^{\bar{k}-1} S_i \cup \bar{S}, \qquad C = \bigcup_{i=0}^{\bar{k}-1} C_i \cup \bar{C}, \qquad D = \bigcup_{i=0}^{\bar{k}-1} D_i \cup \bar{D},$$

and similarly as earlier we may obtain an equivalent stationary problem. The optimality equation for this problem reduces to the system of $(\bar{k} + 1)$ equations

$$\tilde{J}^*(x_i, i) = \inf_{u_i \in U_i(x_i)} \underset{w_i}{E} \{g_i(x_i, u_i, w_i) + \alpha \tilde{J}^*[f_i(x_i, u_i, w_i), i + 1]\}$$

$$\forall x_i \in S_i, \quad i = 0, 1, \ldots, \bar{k} - 1,$$

$$\tilde{J}^*(x) = \inf_{u \in \bar{U}(x)} \underset{w}{E} \{\bar{g}(x, u, w) + \alpha \tilde{J}^*[\bar{f}(x, u, w)]\} \qquad \forall x \in \bar{S},$$

where $\tilde{J}^*(x) = \tilde{J}^*(x, \bar{k})$ for all $x \in S_{\bar{k}} = \bar{S}$. If the infimum on the right-hand side of the above equations is attained for all $x_i, i = 0, 1, \ldots, \bar{k} - 1$, and $x$, then under Assumptions B′ and P′ there exist (eventually stationary) optimal policies of the form $\pi^* = \{\mu_0^*, \mu_1^*, \ldots, \mu_{\bar{k}}^*, \mu_{\bar{k}}^*, \ldots\}$.

*Periodic Problems*

Assume within the framework of the NSP that there exists an integer $p \geqslant 2$ (called the *period*) such that for all integers $i$ and $j$ with $|i - j| = \lambda p$, $\lambda = 1, 2, \ldots$, we have

$$S_i = S_j, \qquad C_i = C_j, \qquad D_i = D_j, \qquad U_i(\cdot) = U_j(\cdot),$$

$$f_i = f_j, \qquad g_i = g_j, \qquad P_i(\cdot \,|\, x, u) = P_j(\cdot \,|\, x, u) \qquad \forall (x, u) \in S_i \times C_i.$$

We assume that the spaces $S_i, C_i, D_i, i = 0, 1, \ldots, p - 1$, are mutually disjoint. Then we may define a new state space, control space, and disturbance space by

$$S = \bigcup_{i=0}^{p-1} S_i, \qquad C = \bigcup_{i=0}^{p-1} C_i, \qquad D = \bigcup_{i=0}^{p-1} D_i.$$

As done earlier, we may obtain an equivalent stationary problem. The optimality equation for this problem reduces to the system of $p$ equations

$$\tilde{J}^*(x_0, 0) = \inf_{u_0 \in U_0(x_0)} \underset{w_0}{E} \{g_0(x_0, u_0, w_0) + \alpha \tilde{J}^*[f_0(x_0, u_0, w_0), 1]\},$$

$$\tilde{J}^*(x_1, 1) = \inf_{u_1 \in U_1(x_1)} \underset{w_1}{E} \{g_1(x_1, u_1, w_1) + \alpha \tilde{J}^*[f_1(x_1, u_1, w_1), 2]\},$$

$$\vdots$$

$$\tilde{J}^*(x_{p-1}, p-1) = \inf_{u_{p-1} \in U_{p-1}(x_{p-1})} \underset{w_{p-1}}{E} \{g_{p-1}(x_{p-1}, u_{p-1}, w_{p-1}) + \alpha \tilde{J}^*[f_{p-1}(x_{p-1}, u_{p-1}, w_{p-1}), 0]\}. \quad (99)$$

These equations may be used to obtain (under Assumption B′ or P′) a periodic policy of the form $\{\mu_0^*, \ldots, \mu_{p-1}^*, \mu_0^*, \ldots, \mu_{p-1}^*, \ldots\}$ whenever the infimum of the right-hand side is attained for all $x_i$, $i = 0, 1, \ldots, p-1$.

Concerning the algorithmic solution of periodic problems, we mention that when all spaces involved are finite sets then an optimal policy may be found through a finite number of arithmetic operations by means of the policy iteration algorithm or linear programming. The form of these algorithms may be obtained by applying them to the corresponding SP, the state, control, and disturbance spaces of which are now the finite sets $S$, $C$, and $D$.

Finally, we provide the form of the successive approximation method with starting functions equal to zero:

$$\tilde{J}_0(x_i, i) = 0 \qquad \forall x_i \in S_i, \quad i = 0, 1, \ldots, p-1.$$

The $(k+1)$st iteration of the successive approximation method is given by

$$\tilde{J}_{k+1}(x_i, i) = \inf_{u_i \in U_i(x_i)} \underset{w_i}{E} \{g_i(x_i, u_i, w_i) + \alpha \tilde{J}_k[f_i(x_i, u_i, w_i), i+1]\}, \qquad i = 0, 1, \ldots, p-2,$$

$$\tilde{J}_{k+1}(x_{p-1}, p-1) = \inf_{u_{p-1} \in U_{p-1}(x_{p-1})} \underset{w_{p-1}}{E} \{g_{p-1}(x_{p-1}, u_{p-1}, w_{p-1}) + \alpha \tilde{J}_k[f_{p-1}(x_{p-1}, u_{p-1}, w_{p-1}), 0]\}. \quad (100)$$

Under Assumptions B′ and N′ we have (by applying Proposition 1 or 11 to the corresponding SP)

$$\lim_{k \to \infty} \tilde{J}_k(x_i, i) = \tilde{J}^*(x_i, i) \qquad \forall x_i \in S_i, \quad i = 0, \ldots, p-1,$$

while under Assumption P′ the same equations hold provided the sets

$$U_k(x_i, \lambda, i) = \left\{u_i \in U_i(x_i) \,\middle|\, \underset{w_i}{E}\{g_i(x_i, u_i, w_i) + \alpha \tilde{J}_k[f_i(x_i, u_i, w_i), i+1]\} \leqslant \lambda\right\}, \qquad i = 0, \ldots, p-2,$$

$$U_k(x_{p-1}, \lambda, p-1) = \left\{u_{p-1} \in U_{p-1}(x_{p-1}) \,\middle|\, \underset{w_{p-1}}{E}\{g_{p-1}(x_{p-1}, u_{p-1}, w_{p-1}) + \alpha \tilde{J}_k[f_{p-1}(x_{p-1}, u_{p-1}, w_{p-1}), 0]\} \leqslant \lambda\right\}, \tag{101}$$

are compact subsets of Euclidean spaces for all $x_i \in S_i$, $\lambda \in R$, and $k$ greater than some integer $\bar{k}$ (Proposition 13 applied to the SP). Under the same compactness condition an optimal periodic policy is guaranteed to exist.

## 6.8 Notes

The discounted problem with bounded cost per stage is by far the simplest and most well-behaved infinite horizon problem. This is due to the contraction property induced by the presence of the discount factor. Many authors have contributed to its analysis, most notably Bellman [B3], Howard [H15], and Blackwell [B20]. Contraction type properties were first exploited in a DP setting by Shapley [S8] in a paper on multistage games. The mapping $F$ of Section 6.2 and the corresponding algorithms are given by Kushner [K10], where the connection with Gauss–Seidel iterations is pointed out (see also Hastings [H7]). The linear programming approach of Section 6.2 was proposed by D'Epenoux [D3]. The error bounds given in Section 6.2 and Problem 3 are improvements on results of McQueen [M5] and Denardo [D2] (see [B14]). The convergence results and discretization procedures of Problem 2 are taken from Bertsekas [B15]. The essential structure of the discounted cost problem with bounded cost per stage was captured in the abstract framework introduced in an important paper by Denardo [D2] (see Problem 4). This framework contains many other interesting problems similar to Problem (D), such as the so-called Markov-renewal or semi-Markov decision problems [J4, H16, R4] or minimax discounted cost problems (see Problems 1 and 6). Denardo's framework relies strongly on contraction properties and is thus generally inapplicable to problems of the type examined in Section 6.4 and in Chapter 7. A related framework that does not employ contraction assumptions and is applicable to problems such as those of Section 6.4 and Chapter 7 was developed recently by the author (see Problem

9 in Chapter 7 and [B13] and [B16]). For analysis of discounted cost problems involving linear systems and convex cost functionals see references [B11] and [K8]. For analysis related to problems with imperfect state information see references [D9], [S5], and [S11]. The form of the generalized policy iteration algorithm of Problem 13 is apparently new.

Discounted cost problems with unbounded cost per stage are similar to undiscounted cost problems, which will be examined in the next section. Important works in this area are those of Dubins and Savage [D8], Blackwell [B21], and particularly Strauch [S17] (see Hinderer [H9] for an account). These authors considered explicitly the thorny measurability questions arising from uncountable disturbance spaces. The analysis of Section 6.4 is mostly a synthesis of results given in these references. Propositions 9a and 13 are new results [B13, B16]. The result of Proposition 13 can be generalized to the case where the sets $U_k(x, \lambda)$ of (76) are compact subsets of a Hausdorff topological space. For generalizations of the analysis of Section 6.4 see Problem 9 in Chapter 7 and [B13] and [B16].

It is to be noted that in our formulation of the problem of this chapter we have specified that the initial state $x_0$ is fixed and given. Thus whenever the optimal cost $J^*(x_0)$ corresponding to $x_0$ is finite, it can be attained within any $\varepsilon > 0$ by an admissible policy $\pi_\varepsilon(x_0)$, i.e., given any $\varepsilon > 0$ there exists an admissible $\pi_\varepsilon(x_0)$ such that

$$J_{\pi_\varepsilon(x_0)}(x_0) \leqslant J^*(x_0) + \varepsilon.$$

The policy $\pi_\varepsilon(x_0)$ will depend on $x_0$ and it does not necessarily follow that given any $\varepsilon > 0$ there exists an admissible policy $\pi_\varepsilon$ (independent of $x_0$) such that

$$J_{\pi_\varepsilon}(x_0) \leqslant J^*(x_0) + \varepsilon \qquad \forall x_0 \in S.$$

Neither does it follow that policy $\pi_\varepsilon$ can be taken to be stationary unless Assumption B is satisfied (Problem 12). A considerable amount of research has been directed toward clarifying these fine points and the advanced reader is referred to references [B16], [B21], [B22], [O2], and [S17] for related analysis and counterexamples (see also Problems 22–26).

The results on linear quadratic problems are well known (see, e.g., Kushner [K10]). The inventory control problem has been analyzed by Bellman [B3] (see also Iglehart [I2]). For some recent results see Kalymon [K3]. The line of argument adopted here is new.

The treatment of nonstationary and periodic problems by means of reduction to the stationary case is apparently new. Earlier works on the subject [F4, H9] do not take advantage of the possibility of this reduction. The results on periodic linear–quadratic problems and inventory control (Problems 18 and 20) seem to be new.

The formulation of Problem (D) excludes the possibility of constraints on the state $x_k$ of the form $x_k \in X \subset S$. Such constraints can be taken into account under Assumption P by adding to the cost per stage $g$ the indicator function $\delta(x \mid X)$ of the set $X$:

$$\delta(x \mid X) = \begin{cases} 0 & \text{if} \quad x \in X, \\ +\infty & \text{if} \quad x \in X. \end{cases}$$

This formulation, however, requires that $g$ can take the value $+\infty$. Nonetheless all the results of Section 6.4 shown under Assumption P may be proved for $g$ satisfying

$$0 \leqslant g(x, u, w) \leqslant +\infty \qquad \forall (x, u, w) \in S \times C \times D,$$

i.e., whenever $g$ is allowed to take the value $+\infty$ (Problem 10). For an analysis and treatment of state constraints see references [B9] and [B11] or Problem 13 in Chapter 7.

Finally, we note that even though the problem of this chapter excludes specifically the possibility of an uncountable disturbance space, it may still serve as the starting point of analysis of a problem with uncountable disturbance space. This can be done by reducing such a problem to a deterministic problem (i.e., one where the disturbance space consists of a single element) with state space a set of probability measures. The basic idea of this reduction is demonstrated in Problem 21. The advanced reader may consult [B16], [W8] and see how a related reduction can be effected for a very broad class of finite and infinite horizon problems.

## Problems

**1.** Provide analogs for the results and algorithms of Sections 6.1 and 6.2 for the problem of minimizing

$$J_\pi(x_0) = \lim_{N \to \infty} \sup_{\substack{w_k \in W[x_k, \mu_k(x_k)] \\ k = 0, 1, \ldots}} \sum_{k=0}^{N-1} \alpha^k g[x_k, \mu_k(x_k), w_k],$$

over all polices $\pi = \{\mu_0, \mu_1, \ldots\}$ with $\mu_k(x_k) \in U(x_k)\, \forall x_k \in S$, where $\alpha \in (0, 1)$, $g$ satisfies Assumption B, $x_k$ is generated by $x_{k+1} = f[x_k, \mu_k(x_k), w_k]$, and $W(x, u)$ is a given nonempty subset of $D$ for each $(x, u) \in S \times C$.

**2.** The purpose of this problem is to provide discretization procedures and related convergence results analogous to those of Section 5.2. Consider the functional equation for $J^*: S \to R$:

$$J^*(x) = \max_{u \in C} \underset{w}{E} \{g(x, u, w) + \alpha J^*[f(x, u, w)]\}$$

where $0 < \alpha < 1$ and $g, f, S, C$, and $w$ satisfy continuity, compactness, and finiteness assumptions analogous to Assumptions A of Section 5.2. Let $S^1, S^2, \ldots, S^n$ be mutually disjoint sets with $S = \bigcup_{i=1}^n S^i$, select arbitrary points $x^i \in S^i$, $i = 1, \ldots, n$, and consider the discretized functional equation

$$\hat{J}^*(x) = \begin{cases} \max_{u \in C} \underset{w}{E} \{g(x, u, w) + \alpha \hat{J}^*[f(x, u, w)]\} & \\ \qquad\qquad \text{if} \quad x = x^i, \quad i = 1, \ldots, n, \\ \hat{J}^*(x^i) \qquad \text{if} \quad x \in S^i, \quad i = 1, \ldots, n. \end{cases}$$

(a) Show that both equations have unique solutions $J^*$ and $\hat{J}^*$ within the class of all bounded functions $J: S \to R$ and furthermore

$$\lim_{d_s \to 0} \sup_{x \in S} |J^*(x) - \hat{J}^*(x)| = 0,$$

where

$$d_s = \max_{i=1,\ldots,n} \sup_{x \in S^i} \|x - x^i\|.$$

(b) Provide a discretization procedure and prove a similar result under assumptions analogous to Assumption B of Section 5.2.

*Hint*: Use the results already proved in Section 5.2.

**3.** Let $S$ be a set and $B(S)$ be the set of all bounded real-valued functions on $S$. Let $T: B(S) \to B(S)$ be a mapping with the following two properties:

(1) $T(J) \leqslant T(J')$ for all $J, J' \in B(S)$ with $J \leqslant J'$.
(2) For every scalar $r \neq 0$ and all $x \in S$

$$\alpha_1 \leqslant [T(J + re)(x) - T(J)(x)]/r \leqslant \alpha_2,$$

where $\alpha_1, \alpha_2$ are two scalars with $0 \leqslant \alpha_1 \leqslant \alpha_2 < 1$.

(a) Show that $T$ is a contraction mapping on $B(S)$ and hence for every $J \in B(S)$ we have

$$\lim_{k \to \infty} T^k(J)(x) = J^*(x) \qquad \forall x \in S,$$

where $J^*$ is the unique fixed point of $T$ in $B(S)$.

(b) Show that for all $J \in B(S)$, $x \in S$, and $k = 1, 2, \ldots,$

$$T^k(J)(x) + c_k \leqslant T^{k+1}(J)(x) + c_{k+1} \leqslant J^*(x)$$
$$\leqslant T^{k+1}(J)(x) + \bar{c}_{k+1} \leqslant T^k(J)(x) + \bar{c}_k,$$

where for all $k$

$$c_k = \min\left\{\frac{\alpha_1}{1-\alpha_1}\inf_{x\in S}[T^k(J)(x) - T^{k-1}(J)(x)],\right.$$
$$\left.\frac{\alpha_2}{1-\alpha_2}\inf_{x\in S}[T^k(J)(x) - T^{k-1}(J)(x)]\right\},$$
$$\bar{c}_k = \max\left\{\frac{\alpha_1}{1-\alpha_1}\sup_{x\in S}[T^k(J)(x) - T^{k-1}(J)(x)],\right.$$
$$\left.\frac{\alpha_2}{1-\alpha_2}\sup_{x\in S}[T^k(J)(x) - T^{k-1}(J)(x)]\right\}.$$

A geometric interpretation of these relations for the case where $S$ consists of a single element is provided in Fig. 6.3.

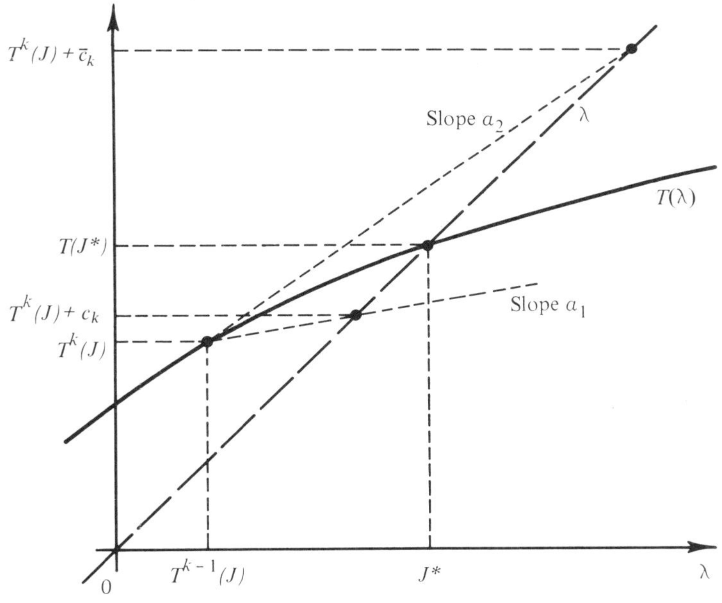


FIGURE 6.3

(c) Show that the mapping $F$ defined by (38)–(41) satisfies

$$\alpha^n \leqslant [F(J + re)(x) - F(J)(x)]/r \leqslant \alpha,$$

where $n$ is the number of elements in $S$.

*Hint*: Use a line of argument similar to the one of Section 6.2.

**4.** Let $S$ be a set, let $B(S)$ be the set of all bounded real-valued functions on $S$, let $C$ be another set, and for each $x \in S$ let $U(x)$ be a given nonempty subset of $C$. Assume that we are given a mapping $H: S \times C \times B(S) \to R$ having the monotonicity property

$$J \leqslant J' \Rightarrow H(x, u, J) \leqslant H(x, u, J') \qquad \forall J, J' \in B(S), \qquad (x, u) \in S \times C, \qquad \text{(M)}$$

and the contraction property

$$|H(x, u, J) - H(x, u, J')| \leqslant \alpha \|J - J'\| \qquad \forall J, J' \in B(S), \quad (x, u) \in S \times C, \quad \text{(C)}$$

where $\alpha$ is some scalar with $0 < \alpha < 1$ and

$$\|J - J'\| = \sup_{x \in S} |J(x) - J'(x)|.$$

Let $M$ denote the set of all functions $\mu: S \to C$ with $\mu(x) \in U(x)\ \forall x \in S$. For any $\mu \in M$ define the mapping $T_\mu$ by

$$T_\mu(J)(x) = H[x, \mu(x), J] \qquad \forall x \in S,$$

and the mapping $T$ by

$$T(J)(x) = \inf_{u \in U(x)} H(x, u, J) \qquad \forall x \in S.$$

We assume that $T(J)$, $T_\mu(J) \in B(S)$ for all $J \in B(S)$, $\mu \in M$. Show the following:

(a) There exists a unique function $J^* \in B(S)$, and for each $\mu \in M$ a unique function $J_\mu \in B(S)$ such that

$$J^* = T(J^*), \qquad J_\mu = T_\mu(J_\mu).$$

Furthermore,

$$\|J^* - T^k(J)\| \to 0, \qquad \|J_\mu - T_\mu^k(J)\| \to 0 \qquad \forall J \in B(S).$$

(b) There holds

$$J^*(x) = \inf_{\mu \in M} J_\mu(x) \qquad \forall x \in S,$$

and if for some $\mu \in M$ we have $T_\mu(J^*) = J^*$, then $J_\mu = J^*$. Furthermore if there exists a function $\bar{J} \in B(S)$ such that $\lim_{k \to \infty} (T_{\mu_0} \cdots T_{\mu_k})(\bar{J})(x)$ exists and is a real number for all $x \in S$ and all sequences $\{\mu_k\}$, $\mu_k \in M$, $\forall k$, then

$$\lim_{k \to \infty} (T_{\mu_0} \cdots T_{\mu_k})(J)(x) = \lim_{k \to \infty} (T_{\mu_0} \cdots T_{\mu_k})(\bar{J})(x) \qquad \forall J \in B(S), \quad x \in S,$$
$$\mu_k \in M, \quad k = 0, 1, \ldots,$$

$$J^*(x) = \inf_{\substack{\mu_i \in M \\ i = 0, 1, \ldots}} \lim_{k \to \infty} (T_{\mu_0} \cdots T_{\mu_k})(J)(x) \qquad \forall J \in B(S), \quad x \in S.$$

(c) Prove (a) and (b) when the contraction assumption (C) is replaced by the assumption that there exists a scalar $\alpha$ with $0 < \alpha < 1$, a scalar $\lambda > 0$, and an integer $m \geqslant 1$ such that

$$\begin{aligned} \|(T_{\mu_0} T_{\mu_1} \cdots T_{\mu_{m-1}})(J) - (T_{\mu_0} T_{\mu_1} \cdots T_{\mu_{m-1}})(J')\| &\leqslant \alpha \|J - J'\|, \\ \forall J, J' \in B(S), \qquad \mu_0, \ldots, \mu_{m-1} &\in M, \\ \|T_\mu(J) - T_\mu(J')\| &\leqslant \lambda \|J - J'\|, \\ \forall J, J' \in B(S), \quad \mu &\in M. \end{aligned} \tag{C'}$$

(d) Assume that $S$ is a finite set, $S = \{1, 2, \ldots, n\}$. Under Eqs. (M) and (C′) show that $\{J^*(1), \ldots, J^*(n)\}$ is a solution of each of the following problems:

$$\min_{\substack{T(J)(i) \leqslant J(i) \\ i = 1, \ldots, n}} \sum_{i=1}^{n} J(i), \qquad \max_{\substack{T(J)(i) \geqslant J(i) \\ i = 1, \ldots, n}} \sum_{i=1}^{n} J(i).$$

(e) Assume that $S$ and $C$ are finite sets and Eqs. (M) and (C′) hold. Consider the following generalized policy iteration algorithm:

(1) Start with an initial $\mu^0 \in M$.

(2) Given $\mu^i$ calculate $J_{\mu^i}$ and $\mu^{i+1} \in M$ such that $T_{\mu^{i+1}}(J_{\mu^i}) = T(J_{\mu^i})$.

Repeat the process until $J_{\mu^{\bar{k}+1}} = J_{\mu^{\bar{k}}}$ for some index $\bar{k}$.

Show that the algorithm will yield $J^*$ after a finite number of iterations.

(f) Assume that Eqs. (M) and (C) hold. Let $J \in B(S)$ and let $\mu^k$ be such that $T_{\mu^k}[T^{k-1}(J)] = T^k(J)$ and $\mu^k \in M$. Show that

$$\|J^* - J_{\mu^k}\| \leqslant [2\alpha^k/(1 - \alpha)] \|T(J) - J\|.$$

Show also that

$$\|J^* - T^k(J)\| \leqslant b_k,$$

where $b_k$ is defined recursively by

$$b_1 = \frac{\alpha \|T(J) - J\|}{1 - \alpha}, \qquad b_k = \min\left\{\alpha b_{k-1}, \frac{\alpha \|T^k(J) - T^{k-1}(J)\|}{1 - \alpha}\right\}.$$

*Hints*: (a) Show that

$$\|T_\mu(J) - T_\mu(J')\| \leqslant \alpha \|J - J'\| \qquad \forall J, J' \in B(S), \quad \mu \in M,$$

$$\|T(J) - T(J')\| \leqslant \alpha \|J - J'\| \qquad \forall J, J' \in B(S),$$

and use the contraction mapping fixed point theorem.

(b) Use the fact that $J_\mu \geqslant T(J_\mu) \geqslant \cdots \geqslant T^k(J_\mu) \geqslant \cdots$ to show that $\inf_\mu J_\mu(x) \geqslant J^*(x)$. To prove the reverse inequality, for any $\varepsilon > 0$, let $\mu \in M$ be such that

$$J^*(x) \geqslant T_\mu(J^*)(x) - \varepsilon \qquad \forall x \in S.$$

Use the contraction assumption and the above inequality to show that for all $k = 1, 2, \ldots,$

$$T_\mu^k(J^*)(x) \geqslant T_\mu^{k+1}(J^*)(x) - \alpha^k \varepsilon \qquad \forall x \in S,$$

and conclude that

$$J^*(x) \geqslant J_\mu(x) - [\varepsilon/(1 - \alpha)] \qquad \forall x \in S.$$

Use the above argument to show that

$$J^*(x) = \inf_{\mu_i,\ i=0,1,\ldots} \lim_{k\to\infty} (T_{\mu_0} \cdots T_{\mu_k})(J^*)(x) \qquad \forall x \in S.$$

(c) Show that

$$\| T_\mu^m(J) - T_\mu^m(J') \| \leqslant \alpha \| J - J' \| \qquad \forall J, J' \in B(S), \quad \mu \in M,$$

$$\| T^m(J) - T^m(J') \| \leqslant \alpha \| J - J' \| \qquad \forall J, J' \in B(S).$$

(f) Show that the following relations hold:

$$\| J^* - T^k(J) \| \leqslant \frac{\alpha}{1 - \alpha} \| T^k(J) - T^{k-1}(J) \| \leqslant \frac{\alpha^k}{1 - \alpha} \| T(J) - J \|,$$

$$\begin{aligned} \| J_{\mu^k} - T^k(J) \| &= \| J_{\mu^k} - T_{\mu^k}[T^{k-1}(J)] \| \\ &\leqslant \frac{\alpha}{1 - \alpha} \| T^k(J) - T^{k-1}(J) \| \leqslant \frac{\alpha^k}{1 - \alpha} \| T(J) - J \|. \end{aligned}$$

**5.** Consider a problem similar to that of Section 6.1 except for the fact that when we are at state $x_k$ there is a probability $\beta$, where $0 < \beta < 1$, that the next state $x_{k+1}$ will be determined according to $x_{k+1} = f(x_k, u_k, w_k)$ and a probability $(1 - \beta)$ that the system will move to a termination state where it stays permanently thereafter at no cost. Show that even if $\alpha = 1$ (no discounting) the problem can be put into the discounted cost framework. Use the results of Problem 4 to analyze the case where $\beta$ depends on $u$ and

$$0 < \inf_{u \in C} \beta(u) \leqslant \sup_{u \in C} \beta(u) < 1.$$

**6.** Consider a problem similar to Problem (D) under Assumption B except for the fact that the discount factor $\alpha$ depends on the current state $x_k$, the control $u_k$, and the disturbance $w_k$, i.e., the cost functional has the form

$$J_\pi(x_0) = \lim_{N\to\infty} \mathop{E}_{\substack{w_k \\ k=0,1,\ldots}} \left\{ \sum_{k=0}^{N-1} \alpha_{\pi,k}\, g[x_k, \mu_k(x_k), w_k] \right\},$$

where

$$\alpha_{\pi,k} = \alpha[x_0, \mu_0(x_0), w_0] \alpha[x_1, \mu_1(x_1), w_1] \cdots \alpha[x_k, \mu_k(x_k), w_k],$$

with $\alpha(x, u, w)$ a given function satisfying

$$0 \leqslant \inf\{\alpha(x, u, w) | x \in S, u \in C, w \in D\}$$
$$\leqslant \sup\{\alpha(x, u, w) | x \in S, u \in C, w \in D\} < 1.$$

Show that the results and algorithms of Sections 6.1 and 6.2 have direct counterparts for such problems.

**7.** Let $\bar{J}: S \to R$ be any bounded function on $S$ and consider the successive approximation method of Section 6.2 with a starting function $J: S \to R$ of the form

$$J(x) = \bar{J}(x) + r \qquad \forall x \in S,$$

where $r$ is some scalar. Show that the bounds $T^k(J)(x) + c_k$, $T^k(J)(x) + \bar{c}_k$ on $J^*(x)$ of Proposition 4 are independent of the scalar $r$ for all $x \in S$. Show also that if $S$ consists of a single element $\tilde{x}$ (i.e., $S = \{\tilde{x}\}$), then

$$T(J)(\tilde{x}) + c_1 = T(J)(\tilde{x}) + \bar{c}_1 = J^*(\tilde{x}).$$

**8.** Consider Problem (D) under Assumption P or N. Show that for any probability distribution $\{p_1, p_2, \ldots\}$ defined on $\{x^1, x^2, \ldots\}$, a countable subset of $S$, we have

$$\inf_{\pi \in \Pi} \sum_{i=1}^{\infty} p_i J_\pi(x^i) = \sum_{i=1}^{\infty} p_i J^*(x^i).$$

*Hint*: If $J^*(x^i)$ is equal to $+\infty$ (under Assumption P) or $-\infty$ (under Assumption N) for some $i$ for which $p_i > 0$ the result is evident. So assume $J^*(x^i) \neq \pm\infty$ for all $i$. We have $\sum_{i=1}^{\infty} p_i J^*(x^i) \leqslant \inf_\pi \sum_{i=1}^{\infty} p_i J_\pi(x^i)$ so it remains to prove the reverse inequality. For a given $\varepsilon > 0$ consider for each $x^i$ an admissible policy of the form

$$\pi_0(x^i) = \{\tilde{\mu}_0(\cdot), \mu_1^0(x^i, \cdot), \mu_2^0(x^i, \cdot), \ldots\},$$

such that

$$\sum_{i=1}^{\infty} p_i J_{\pi_0(x^i)}(x^i) \leqslant \sum_{i=1}^{\infty} p_i J^*(x^i) + \varepsilon.$$

Notice that $\pi_0(x^i)$ represents a different policy for each $x^i$ and $\mu_j^0(x^i, x_j)$ represents the control applied when the initial state is $x^i$ and the state at time $j$ is $x_j$.

Let Assumption P hold and consider a sequence $\{\varepsilon_k\}$ with $\varepsilon_k > 0$, $\forall k$, and $\sum_{k=1}^{\infty} \varepsilon_k = \varepsilon$. Write

$$\sum_{i=1}^{\infty} p_i J_{\pi_0(x^i)}(x^i) = \sum_{i=1}^{\infty} p_i \mathop{E}_{w_0} \{g[x^i, \tilde{\mu}_0(x^i), w_0]\}$$
$$+ \sum_{i=1}^{\infty} p_i \mathop{E}_{x_1} \{J^1_{\pi_0(x^i)}(x_1) | x_0 = x^i, u_0 = \tilde{\mu}_0(x^i)\},$$

where

$$J^1_{\pi_0(x^i)}(x_1) = \lim_{N\to\infty} \underset{\substack{w_k \\ k=1,2,\dots}}{E} \left\{ \sum_{k=1}^{N-1} \alpha^k g[x_k, \mu_k^0(x^i, x_k), w_k] \right\}.$$

Show that one may find a policy of the form

$$\pi_1 = \{\tilde{\mu}_0(\cdot), \tilde{\mu}_1(\cdot), \mu_2^1(x_1, \cdot), \mu_3^1(x_1, \cdot), \dots\},$$

such that

$$\sum_{i=1}^{\infty} p_i J_{\pi_1}(x^i) \leqslant \sum_{i=1}^{\infty} p_i J_{\pi_0(x^i)}(x^i) + \varepsilon_1,$$

and with $\tilde{\mu}_1, \mu_j^1$ satisfying $\tilde{\mu}_1(x_1) \in U(x_1)$ $\forall x_1 \in S$, and $\mu_j^1(x_1, x_j) \in U(x_j)$, $\forall x_1, x_j \in S$, $j = 2, 3, \dots$. Note that the values of $\mu_j^1$ depend on $x_1$ and $x_j$ but not on $x^i$. Notice also that the policy $\pi_1$ is not an admissible policy according to our definition since the control $\mu_j^1(x_1, x_j)$ applied at time $j \geqslant 2$ depends on the state $x_1$ (which has occurred at time 1) as well as the current state $x_j$.

Similarly proceeding, show that one may find a policy of the form

$$\pi_k = \{\tilde{\mu}_0(\cdot), \tilde{\mu}_1(\cdot), \dots, \tilde{\mu}_k(\cdot), \mu_{k+1}^k(x_k, \cdot), \mu_{k+2}^k(x_k, \cdot), \dots\},$$

such that

$$\sum_{i=1}^{\infty} p_i J_{\pi_k}(x^i) \leqslant \sum_{i=1}^{\infty} p_i J_{\pi_{k-1}}(x^i) + \varepsilon_k,$$

where

$$\pi_{k-1} = \{\tilde{\mu}_0(\cdot), \tilde{\mu}_1(\cdot), \dots, \tilde{\mu}_{k-1}(\cdot), \mu_k^{k-1}(x_{k-1}, \cdot), \mu_{k+1}^{k-1}(x_{k-1}, \cdot), \dots\}.$$

Show that we have for all $k$ and for all $N \leqslant k$

$$\sum_{i=1}^{\infty} p_i \underset{\substack{w_j \\ j=0,1,\dots,N}}{E} \left\{ g[x^i, \tilde{\mu}_0(x^i), w_0] + \sum_{j=1}^{N} \alpha^j g[x_j, \tilde{\mu}_j(x_j), w_j] \right\} - \sum_{j=1}^{k} \varepsilon_j$$

$$\leqslant \sum_{i=1}^{\infty} p_i J_{\pi_k}(x^i) - \sum_{j=1}^{k} \varepsilon_j \leqslant \sum_{i=1}^{\infty} p_i J_{\pi_{k-1}}(x^i) - \sum_{j=1}^{k-1} \varepsilon_j$$

$$\leqslant \cdots \leqslant \sum_{i=1}^{\infty} p_i J_{\pi_0(x^i)}(x^i) \leqslant \sum_{i=1}^{\infty} p_i J^*(x^i) + \varepsilon.$$

Consider the policy $\tilde{\pi} = \{\tilde{\mu}_0, \tilde{\mu}_1, \dots\}$ and use the above inequalities to show that

$$\sum_{i=1}^{\infty} p_i J_{\tilde{\pi}}(x^i) \leqslant \sum_{i=1}^{\infty} p_i J^*(x^i) + 2\varepsilon.$$

Under Assumption N show that for each $N$ there exists an admissible policy of the form $\pi_N = \{\tilde{\mu}_0, \tilde{\mu}_1^N, \ldots, \tilde{\mu}_N^N, \mu, \mu, \ldots\}$ such that

$$\sum_{i=1}^{\infty} p_i J_{\pi_N}^N(x^i) = \sum_{i=1}^{\infty} p_i \underset{\substack{w_k \\ k=0,\ldots,N}}{E} \left\{ g[x^i, \tilde{\mu}_0(x^i), w_0] + \sum_{k=1}^{N} \alpha^k g[x_k, \tilde{\mu}_k^N(x_k), w_k] \right\}$$

$$\leqslant \sum_{i=1}^{\infty} p_i \underset{\substack{w_k \\ k=0,\ldots,N}}{E} \left\{ g[x^i, \tilde{\mu}_0(x^i), w_0] + \sum_{k=1}^{N} \alpha^k g[x_k, \mu_k^0(x^i, x_k), w_k] \right\} + \varepsilon.$$

Show that

$$\liminf_{N\to\infty} \sum_{i=1}^{\infty} p_i J_{\pi_N}(x^i) \leqslant \liminf_{N\to\infty} \sum_{i=1}^{\infty} p_i J_{\pi_N}^N(x^i) \leqslant \sum_{i=1}^{\infty} p_i J^*(x^i) + 2\varepsilon.$$

Hence there exists an $N$ such that

$$\sum_{i=1}^{\infty} p_i J_{\pi_N}(x^i) \leqslant \sum_{i=1}^{\infty} p_i J^*(x^i) + 3\varepsilon.$$

For detailed proofs see Theorems 4.3 and 4.4 of Strauch [S17] or Bertsekas and Shreve [B16].

**9.** Let $S = [0, \infty)$, $C = U(x) = (0, \infty)$, be the state and control spaces, respectively, let the system equation be

$$x_{k+1} = (2/\alpha)x_k + u_k, \qquad k = 0, 1, \ldots,$$

where $\alpha$ is the discount factor, and let

$$g(x_k, u_k) = x_k + u_k$$

be the cost per stage. Show that for this deterministic problem Assumption P is satisfied and that $J^*(x) = \infty \ \forall x \in S$, but $T^k(J_0)(0) = 0$ for all $k$ [$J_0$ is the zero function, $J_0(x) = 0 \ \forall x \in S$].

**10.** Verify that all the results of Section 6.4 proved under Assumption P hold if the cost per stage $g$ satisfies

$$0 \leqslant g(x, u, w) \leqslant +\infty \qquad \forall (x, u, w) \in S \times C \times D.$$

**11.** Consider Problem (D) for the case of a deterministic problem involving a linear system

$$x_{k+1} = Ax_k + Bu_k, \qquad k = 0, 1, \ldots,$$

where the pair $(A, B)$ is controllable and $x_k \in R^n$, $u_k \in R^m$. Assume no constraints on the control and a cost per stage $g$ satisfying

$$0 \leqslant g(x, u) \qquad \forall (x, u) \in R^n \times R^m.$$

Assume furthermore that $g$ is continuous in $x$ and $u$, and that $g(x_n, u_n) \to +\infty$ if $\{x_n\}$ is bounded and $|u_n| \to +\infty$.

(a) Show that for a discount factor $\alpha \in (0, 1)$ the optimal cost satisfies $0 \leqslant J^*(x) < +\infty$, $\forall x \in R^n$. Furthermore there exists an optimal stationary policy and

$$\lim_{k \to \infty} T^k(J_0)(x) = J^*(x) \qquad \forall x \in R^n.$$

(b) Show that the same holds true except perhaps for $J^*(x) < +\infty$ when the system is not linear but rather is of the form $x_{k+1} = f(x_k, u_k)$, with $f: R^n \times R^m \to R^n$ being a continuous function.

(c) Prove the same results assuming that the control is constrained to lie in a compact set $U \subset R^m (U(x) = U, \forall x \in R^n)$ in place of the assumption $g(x_n, u_n) \to +\infty$ if $\{x_n\}$ is bounded and $|u_n| \to +\infty$.

*Hint*: Show that $T^k(J_0)$ is real valued and continuous for every $k$ and use Proposition 13.

**12.** Under Assumption B let $\mu: S \to C$ be such that $\mu(x) \in U(x)$ $\forall x \in S$, and

$$\begin{aligned} T_\mu(J^*)(x) &= \mathop{E}_{w} \{g[x, \mu(x), w] + \alpha J^*[f(x, \mu(x), w)]\} \\ &\leqslant \inf_{u \in U(x)} \{g(x, u, w) + \alpha J^*[f(x, u, w)]\} + \varepsilon \\ &= J^*(x) + \varepsilon \qquad \forall x \in S. \end{aligned}$$

Show that

$$J^*(x) \leqslant J_\mu(x) \leqslant J^*(x) + [\varepsilon/(1 - \alpha)] \qquad \forall x \in S.$$

*Hint*: Show that $T_\mu^k(J^*)(x) \leqslant J^*(x) + \sum_{i=0}^{k-1} \alpha^i \varepsilon$.

**13.** *Generalized Policy Iteration Algorithm* The purpose of this problem is to provide a policy iteration algorithm for the case where the state space and the control space are not necessarily finite sets. Under Assumption B let $\{\mu, \mu, \ldots\}$ be an admissible stationary policy and let $\tilde{J}_\mu: S \to R$ be such that

$$\sup_{x \in S} |\tilde{J}_\mu(x) - J_\mu(x)| \leqslant \gamma.$$

Let $J': S \to R$ be such that

$$\sup_{x \in S} |J'(x) - T(\tilde{J}_\mu)(x)| \leqslant \delta$$

and assume that

$$\sup_{x \in S} |J'(x) - \tilde{J}_\mu(x)| \leqslant \varepsilon.$$

Show that for all $x \in S$ there holds

$$J^*(x) \leqslant J_\mu(x) \leqslant J^*(x) + [(\delta + \varepsilon)/(1 - \alpha)] + \gamma. \tag{102}$$

Consider the following policy iteration algorithm, for fixed $\gamma$, $\delta$, $\varepsilon > 0$.

(1) Start with an admissible stationary policy $\pi^0 = \{\mu^0, \mu^0, \ldots\}$.

(2) Given $\{\mu^i, \mu^i, \ldots\}$ find $\tilde{J}_{\mu^i}: S \to R$ such that $|\tilde{J}_{\mu^i}(x) - J_{\mu^i}(x)| \leqslant \gamma\alpha^i$ for all $x \in S$.

(3) Find $\mu^{i+1}: S \to C$ with $\mu^{i+1}(x) \in U(x)$ for all $x \in S$ such that

$$T_{\mu^{i+1}}(\tilde{J}_{\mu^i})(x) \leqslant T(\tilde{J}_{\mu^i})(x) + \delta\alpha^i \qquad \forall x \in S.$$

If $\sup_{x \in S} |T_{\mu^{i+1}}(\tilde{J}_{\mu^i})(x) - \tilde{J}_{\mu^i}(x)| \leqslant \varepsilon$ stop. Otherwise replace $\mu^i$ by $\mu^{i+1}$ and go to Step 2.

Show that the algorithm will terminate after a finite number of iterations (say $k$) and that

$$J^*(x) \leqslant J_{\mu^k}(x) \leqslant J^*(x) + \frac{\alpha^{k-1}\delta + \varepsilon}{1 - \alpha} + \gamma\alpha^{k-1} \qquad \forall x \in S.$$

*Hint*: To show inequality (102) use the fact that for any $\beta > 0$ there exists a $k$ such that for all $x \in S$

$$|T^{k+1}(\tilde{J}_\mu)(x) - J^*(x)| \leqslant \beta.$$

Then use the inequalities

$$\begin{aligned} |\tilde{J}_\mu(x) - J^*(x)| &\leqslant |\tilde{J}_\mu(x) - T(\tilde{J}_\mu)(x)| + |T(\tilde{J}_\mu)(x) - T^2(\tilde{J}_\mu)(x)| \\ &\quad + \cdots + |T^{k+1}(\tilde{J}_\mu)(x) - J^*(x)| \\ &\leqslant \sup_{x \in S} |\tilde{J}_\mu(x) - T(\tilde{J}_\mu)(x)|(1 + \alpha + \cdots + \alpha^k) + \beta \\ &\leqslant \frac{1}{1 - \alpha} \sup_{x \in S} |\tilde{J}_\mu(x) - T(\tilde{J}_\mu)(x)| + \beta, \end{aligned}$$

to show that

$$|\tilde{J}_\mu(x) - J^*(x)| \leqslant \frac{1}{1 - \alpha} \sup_{x \in S} |\tilde{J}_\mu(x) - T(\tilde{J}_\mu)(x)| \qquad \forall x \in S.$$

To show that the policy iteration algorithm will terminate in a finite number of iterations, assume the contrary, i.e., we have

$$\sup_{x \in S} |T_{\mu^{i+1}}(\tilde{J}_{\mu^i})(x) - \tilde{J}_{\mu^i}(x)| > \varepsilon$$

for all $i$ and the algorithm generates an infinite sequence of policies $\{\pi^i\}$. Show first that for all $x \in S$ and $i = 0, 1, \ldots,$ we have

$$T_{\mu^{i+1}}(J_{\mu^i})(x) \leqslant T(J_{\mu^i})(x) + (\delta + 2\gamma\alpha)\alpha^i \leqslant J_{\mu^i}(x) + (\delta + 2\gamma\alpha)\alpha^i.$$

Use this inequality to show that for all $x \in S$ and $i$, $k$,

$$T^k_{\mu^{i+1}}(J_{\mu^i})(x) \leqslant T(J_{\mu^i})(x) + (\delta + 2\gamma\alpha)\alpha^i \sum_{j=0}^{k-1} \alpha^j,$$

and conclude that for all $x \in S$ and $i = 0, 1, \ldots,$

$$J^*(x) \leqslant J_{\mu^{i+1}}(x) \leqslant T(J_{\mu^i})(x) + \lambda\alpha^i,$$

where

$$\lambda = (\delta + 2\gamma\alpha)/(1 - \alpha).$$

Show that for all $x \in S$ and $i = 1, 2, \ldots,$

$$J^*(x) \leqslant J_{\mu^i}(x) \leqslant T^i(J_{\mu^0})(x) + i\alpha^{i-1}\lambda,$$

and conclude that

$$\lim_{i \to \infty} \sup_{x \in S} |J_{\mu^i}(x) - J^*(x)| = 0.$$

Use this equality to reach a contradiction.

**14.** The purpose of this problem is to show that the successive approximation method of Section 6.2 will yield an optimal policy after a finite number of iterations when $S$, $C$, and $D$ are finite sets. Under Assumption B let $J: S \to R$ be a function such that for some $\varepsilon > 0$ and all $x \in S$ we have

$$|J(x) - J^*(x)| \leqslant \varepsilon.$$

Let $\mu(x)$ be such that for all $x \in S$ we have $\mu(x) \in U(x)$ and

$$\begin{aligned} T_\mu(J)(x) &= \underset{w}{E}\{g[x, \mu(x), w] + \alpha J[f(x, \mu(x), w)]\} \\ &= \min_{u \in U(x)} \underset{w}{E}\{g(x, u, w) + \alpha J[f(x, u, w)]\} = T(J)(x). \end{aligned}$$

(a) Show that for all $x \in S$

$$|T_\mu(J)(x) - J(x)| \leqslant (1 + \alpha)\varepsilon.$$

(b) Using the above inequality show that for all $x \in S$,

$$|T_\mu(J)(x) - J_\mu(x)| \leqslant [\alpha(1 + \alpha)/(1 - \alpha)]\varepsilon.$$

(c) Show that for all $x \in S$,

$$J^*(x) \leqslant J_\mu(x) \leqslant J^*(x) + [2\varepsilon/(1 - \alpha)].$$

(d) Assume that the state, control, and disturbance spaces are finite sets. Show that the successive approximation method after some index will yield an optimal policy at every iteration, i.e., for any starting function $J : S \to R$ there exists an index $\bar{k}$ such that if $\mu^*$ is such that

$$T_{\mu^*}[T^k(J)] = T^{k+1}(J) \qquad \text{and} \qquad k \geqslant \bar{k},$$

then $\{\mu^*, \mu^*, \ldots\}$ is optimal.

**15.** Under Assumption P or N show that if $\tilde{J}: S \to R$ is a bounded function satisfying $\tilde{J} = T(\tilde{J})$, then $\tilde{J} = J^*$.

**16.** Prove the following strengthened version of part (b) of Proposition 9: Under Assumption N if $\tilde{J}: S \to [-\infty, \infty)$ is bounded above and satisfies $\tilde{J} \leqslant T(\tilde{J})$, then $\tilde{J} \leqslant J^*$.

**17.** *Linear–Quadratic Problems with Nonstationary Disturbances* Consider the linear–quadratic problem of Section 6.5 with the only difference that the disturbances $w_k$ have zero mean but their covariance matrices are nonstationary and uniformly bounded over $k$. Show that the optimal control law remains unchanged.

**18.** *Periodic Linear–Quadratic Problems* Consider the linear system

$$x_{k+1} = A_k x_k + B_k u_k + w_k, \qquad k = 0, 1, \ldots,$$

and the quadratic cost functional

$$J_\pi(x_0) = \lim_{N \to \infty} \underset{\substack{w_k \\ k=0,\ldots,N-1}}{E} \left\{ \sum_{k=0}^{N-1} \alpha^k [x_k' Q_k x_k + \mu_k(x_k)' R_k \mu_k(x_k)] \right\},$$

where the matrices above have appropriate dimensions, $Q_k$ and $R_k$ are positive semidefinite and positive definite, respectively, for all $k$, and $0 < \alpha < 1$. Assume that the system and cost functional are periodic with period $p$ (cf. Section 6.7), that the controls are unconstrained, and that the disturbances are independent, have zero mean, and finite covariance matrices. Assume further that the following (controllability) condition is in effect.

Given any initial state $\bar{x}_0$ there exists a finite sequence of controls $\{\bar{u}_0, \bar{u}_1, \ldots, \bar{u}_r\}$ such that $\bar{x}_{r+1} = 0$ where $\bar{x}_{r+1}$ is generated by

$$\bar{x}_{k+1} = A_k \bar{x}_k + B_k \bar{u}_k, \qquad k = 0, 1, \ldots, r.$$

Show that there is an optimal periodic policy $\pi^*$ of the form

$$\pi^* = \{\mu_0^*, \mu_1^*, \ldots, \mu_{p-1}^*, \mu_0^*, \mu_1^*, \ldots, \mu_{p-1}^*, \ldots\},$$

where $\mu_0^*, \ldots, \mu_{p-1}^*$ are given by

$$\mu_i^*(x) = -\alpha(\alpha B_i' K_{i+1} B_i + R_i)^{-1} B_i' K_{i+1} A_i x, \qquad i = 0, \ldots, p-2,$$

$$\mu_{p-1}^*(x) = -\alpha(\alpha B_{p-1}' K_0 B_{p-1} + R_{p-1})^{-1} B_{p-1}' K_0 A_{p-1} x$$

and the matrices $K_0, K_1, \ldots, K_{p-1}$ satisfy the coupled set of $p$ algebraic Riccati equations given by

$$K_i = A_i'[\alpha K_{i+1} - \alpha^2 K_{i+1} B_i(\alpha B_i' K_{i+1} B_i + R_i)^{-1} B_i' K_{i+1}] A_i + Q_i,$$
$$i = 0, 1, \ldots, p-2,$$
$$K_{p-1} = A_{p-1}'[\alpha K_0 - \alpha^2 K_0 B_{p-1}(\alpha B_{p-1}' K_0 B_{p-1} + R_{p-1})^{-1} B_{p-1}' K_0] A_{p-1} + Q_{p-1}.$$

*Hint*: Use Eqs. (99) and the compactness condition (101).

**19.** *Discounted-Linear-Quadratic Problems with Imperfect State Information* Consider the linear-quadratic problem of Section 6.5 with the difference that the controller, instead of having perfect state information, has access to measurements of the form

$$z_k = Cx_k + v_k, \qquad k = 0, 1, \ldots.$$

Similarly, as in Section 4.3, the disturbances $v_k$ are independent, have identical statistics, zero mean, and finite covariance matrix. Assume that for every admissible policy $\pi$ the matrices

$$E\{[x_k - E\{x_k | I_k\}][x_k - E\{x_k | I_k\}]' | \pi\}$$

are uniformly bounded over $k$, where $I_k$ is the information vector defined in Section 4.3. Show that the optimal policy is $\pi^* = \{\mu^*, \mu^*, \ldots\}$, where $\mu^*$ is given by

$$\mu^*(I_k) = -\alpha(\alpha B'KB + R)^{-1} B'KA\, E\{x_k | I_k\} \qquad \forall I_k, \quad k = 0, 1, \ldots.$$

Show also that the same is true if $w_k, v_k$ are nonstationary with zero mean and covariance matrices that are uniformly bounded over $k$.

*Hint*: Combine the theory of Sections 4.3 and 6.7. Use the compactness condition of Proposition 13.

**20.** *Periodic Inventory Control Problems* In the inventory control problem of Section 6.6 consider the case where the statistics of the demands $w_k$, the prices $c_k$, and the holding and the shortage costs are periodic with period $p$. Show that there exists an optimal periodic policy of the form $\pi^* = \{\mu_0^*, \ldots, \mu_{p-1}^*, \mu_0^*, \ldots, \mu_{p-1}^*, \ldots\}$,

$$\mu_i^*(x) = \begin{cases} S_i^* - x & \text{if } x \leq S_i^*, \\ 0 & \text{otherwise,} \end{cases} \qquad i = 0, 1, \ldots, p-1,$$

where $S_0^*, \ldots, S_{p-1}^*$ are appropriate scalars.

*Hint* Use Eqs. (99) and the compactness condition (101).

**21.** Consider the problem of this chapter under Assumption B for the case where the sets $S$, $C$, and $D$ are finite sets. Using the notation of Section 6.2 consider the controlled system

$$p_{k+1} = p_k P_{\mu_k}, \qquad k = 0, 1, \ldots,$$

where $p_k$ is a probability distribution over $S$ viewed as a row vector, and $P_{\mu_k}$ is the transition probability matrix corresponding to a function $\mu_k: C \to S$ with $\mu_k(i) \in U(i)$ for all $i \in S$. The state is $p_k$ and the control is $\mu_k$. Consider also the cost functional

$$\lim_{N \to \infty} \sum_{k=0}^{N-1} \alpha^k p_k g_{\mu_k}.$$

Show that the optimal value function and an optimal policy for the deterministic problem involving the system and the cost functional above yield the optimal value and an optimal policy for the problem of this chapter.

**22.** Let Assumption P hold and assume that $\pi^* = \{\mu_0^*, \mu_1^*, \ldots\} \in \Pi$ satisfies $J^* = T_{\mu_k^*}(J^*)$ for all $k$. Show that $\pi^*$ is optimal, i.e., $J_{\pi^*} = J^*$.

**23.** Under Assumption P show that given $\varepsilon > 0$ there exists a policy $\pi_\varepsilon \in \Pi$ such that $J_{\pi\varepsilon}(x) \leqslant J^*(x) + \varepsilon$ for all $x \in S$, and that for $\alpha < 1$ the policy $\pi_\varepsilon$ can be taken stationary.

*Hint*: Let $\{\varepsilon_k\}$ be a sequence such that $\varepsilon_k > 0$ for all $k$ and $\sum_{k=0}^{\infty} \alpha^k \varepsilon_k = \varepsilon$. For each $k$ let $\mu_k: C \to S$ be such that $\bar{\mu}_k(x) \in U(x)$ and $T_{\bar{\mu}_k}(J^*)(x) \leqslant T(J^*)(x) + \varepsilon_k$ for all $x \in S$. Let $\pi_\varepsilon = \{\bar{\mu}_0, \bar{\mu}_1, \ldots\}$.

**24.** Under Assumption P show that if there exists an optimal policy, i.e., a policy $\pi^* \in \Pi$ such that $J_{\pi^*} = J^*$, then there exists an optimal stationary policy.

**25.** Use the following counterexample to show that the result of Problem 24 may fail to hold under Assumption N if $J^*(x) = -\infty$ for some $x \in S$. Let $S = D = \{0, 1\}$, $f(x, u, w) = w$, $g(x, u, w) = u$, $U(0) = (-\infty, 0]$, $U(1) = \{0\}$, $p(w = 0 \mid x = 0, u) = \frac{1}{2}$, $p(w = 1 \mid x = 1, u) = 1$. Show that $J^*(0) = -\infty$, $J^*(1) = 0$, and that the admissible nonstationary policy $\{\mu_0^*, \mu_1^*, \ldots\}$ with $\mu_k^*(0) = -(2/\alpha)^k$ is optimal. Show that any admissible stationary policy $\{\mu, \mu, \ldots\}$ satisfies $J_\mu(0) = [2/(2-\alpha)]\mu(0)$, $J_\mu(1) = 0$ (see [B22], [D8], [O2] for related analysis).

**26.** Show that the result of Problem 23 holds under Assumption $N$ if $S$ is finite set, $\alpha = 1$ and $J^*(x) > -\infty$ for all $x \in S$. Construct a counterexample to show that the result can fail to hold if $S$ is countable and $\alpha < 1$ (even if $J^*(x) > -\infty$ for all $x \in S$).

*Chapter 7*

# Minimization of Total Expected Value — Undiscounted Cost

In this chapter we consider infinite horizon problems with undiscounted cost functionals. The basic problem we consider is identical with Problem (D) of the previous chapter except for the absence of the discount factor. Again we assume stationarity of the system and the cost per stage. However, nonstationary or periodic problems may be treated by reduction to the stationary case similarly as in Section 6.7.

PROBLEM (U) Consider the stationary discrete-time dynamic system

$$x_{k+1} = f(x_k, u_k, w_k), \qquad k = 0, 1, \ldots, \tag{1}$$

where the state $x_k$, $k = 0, 1, \ldots$, is an element of a space $S$, the control $u_k$, $k = 0, 1, \ldots$, is an element of a space $C$, and the random disturbance $w_k$, $k = 0, 1, \ldots$, is an element of a space $D$. It is assumed that $D$ is a countable set. The control $u_k$ is constrained to take values in a given subset $U(x_k)$ of $C$ that depends on the current state $x_k$ [$u_k \in U(x_k)$, $\forall x_k \in S$, $k = 0, 1, \ldots$]. The random disturbances $w_k$, $k = 0, 1, \ldots$, have identical statistics and are characterized by probabilities $P(\cdot \mid x_k, u_k)$ defined on $D$, where $P(w_k \mid x_k, u_k)$ denotes the probability of $w_k$ occurring when the current state and control are $x_k$ and $u_k$, respectively. The probability of $w_k$ may depend explicitly on $x_k$ and $u_k$ but not on values of prior disturbances $w_{k-1}, \ldots, w_0$.

Given an initial state $x_0$, the problem is to find a control law, or policy, $\pi = \{\mu_0, \mu_1, \ldots\}$, where $\mu_k: S \to C$, $\mu_k(x_k) \in U(x_k)$ for all $x_k \in S$, $k = 0, 1, \ldots,$ that minimizes the cost functional

$$J_\pi(x_0) = \lim_{N \to \infty} \underset{\substack{w_k \\ k=0,1,\ldots}}{E} \left\{ \sum_{k=0}^{N-1} g[x_k, \mu_k(x_k), w_k] \right\} \tag{2}$$

subject to the system equation constraint (1). The real-valued function $g: S \times C \times D \to R$ is given.

One of the main difficulties introduced by the absence of the discount factor is that now the value of the cost functional (2) may be infinite for some (possibly all) initial states even when costs per stage are bounded. For this reason Problem (U) makes sense in the case where $g \geqslant 0$ only when the optimal value function

$$J^*(x) = \inf_\pi J_\pi(x) \qquad \forall x \in S$$

takes finite values for at least some initial states, for otherwise every admissible policy is optimal. This type of difficulty is of the same nature as the one we encountered in connection with discounted problems with unbounded costs per stage (Section 6.4). As in Section 6.4 we shall allow the possibility that the functions $J_\pi$ and $J^*$ may take the value $+\infty$ or $-\infty$, i.e., they may be extended real valued.

As in the previous chapter we need to impose assumptions on the cost per stage function $g$ that guarantee that the limit in (2) exists in the sense that it is a real number or $\pm\infty$. We shall be considering the following two assumptions, which closely parallel the corresponding assumptions P and N of Section 6.4.

**Assumption P** The function $g$ in the cost functional (2) satisfies

$$0 \leqslant g(x, u, w) \qquad \forall (x, u, w) \in S \times C \times D.$$

**Assumption N** The function $g$ in the cost functional (2) satisfies

$$g(x, u, w) \leqslant 0 \qquad \forall (x, u, w) \in S \times C \times D.$$

There is, however, an important difference between the assumptions above and those of Section 6.4. While in Section 6.4 the assumptions made could be replaced by the assumption that $g(x, u, w)$ is bounded below or above without affecting the results obtained, this is not true anymore for Problem (U)—a complication due again to the absence of the discount factor.

Nonetheless, the results to be obtained in the next section for Problem (U) under Assumptions P and N are very similar to those of Section 6.4. Once we develop these results we shall examine some special cases in Sections 7.2–7.4.

## 7.1 Convergence and Existence Results

As in the previous chapter we shall consider mappings $T$, $T_\mu$ operating on functions $J$ that are defined on $S$ and take values in $[0, +\infty]$, when working under Assumption P, and in $[-\infty, 0]$ when working under Assumption N. The mapping $T$ is defined by

$$T(J)(x) = \inf_{u \in U(x)} \mathop{E}_{w} \{g(x, u, w) + J[f(x, u, w)]\}, \tag{3}$$

and the mapping $T_\mu$ is defined for every admissible stationary policy $\pi = \{\mu, \mu, \ldots\}$ by

$$T_\mu(J)(x) = \mathop{E}_{w} \{g[x, \mu(x), w] + J[f(x, \mu(x), w)]\}. \tag{4}$$

Again we write for any two functions $J$, $J'$ mapping $S$ into $[0, +\infty]$ or $[-\infty, 0]$,

$$J \leqslant J' \qquad \text{if} \quad J(x) \leqslant J'(x) \quad \forall x \in S.$$

Clearly we have the monotonicity relations

$$J \leqslant J' \Rightarrow T(J) \leqslant T(J'), \qquad T_\mu(J) \leqslant T_\mu(J'). \tag{5}$$

We again denote by $J^*(x)$ the optimal value of Problem (U) when the initial state is $x$, and by $J_\mu(x)$ the value of the cost functional (2) corresponding to an admissible stationary policy $\{\mu, \mu, \ldots\}$ and an initial state $x$.

The following result can be proved by an almost verbatim repetition of the proof of Proposition 8 in Section 6.4.

**Proposition 1** (Optimality Equation) Under either Assumption P or N the optimal value function $J^*$ of Problem (U) satisfies

$$J^*(x) = \inf_{u \in U(x)} \mathop{E}_{w} \{g(x, u, w) + J^*[f(x, u, w)]\} \quad \forall x \in S, \tag{6}$$

or in terms of the mapping $T$ of (3),

$$J^* = T(J^*). \tag{7}$$

**Corollary 1.1** Let $\{\mu, \mu, \ldots\}$ be an admissible stationary policy. Then under Assumption P or N we have

$$J_\mu(x) = \mathop{E}_{w} \{g[x, \mu(x), w] + J_\mu[f(x, \mu(x), w)]\}, \tag{8}$$

or equivalently in terms of the mapping $T_\mu$ of (4)

$$J_\mu = T_\mu(J_\mu). \tag{9}$$

The following propositions and corollaries may be proved by essentially repeating the proofs of the corresponding results of Section 6.4.

**Proposition 2** (a) Under Assumption P if $\tilde{J}: S \to [0, +\infty]$ is a function that satisfies $\tilde{J} = T(\tilde{J})$, then $J^* \leqslant \tilde{J}$.

(b) Under Assumption N if $\tilde{J}: S \to [-\infty, 0]$ is a function that satisfies $\tilde{J} = T(\tilde{J})$, then $\tilde{J} \leqslant J^*$.

**Corollary 2.1** Let $\pi = \{\mu, \mu, \ldots\}$ be an admissible stationary policy.

(a) Under Assumption P if $\tilde{J}: S \to [0, +\infty]$ is a function that satisfies $\tilde{J} = T_\mu(\tilde{J})$, then $J_\mu \leqslant \tilde{J}$.

(b) Under Assumption N if $\tilde{J}: S \to [-\infty, 0]$ is a function that satisfies $\tilde{J} = T_\mu(\tilde{J})$, then $\tilde{J} \leqslant J_\mu$.

**Corollary 2.2** (Necessary and Sufficient Condition for Optimality under Assumption N) In order for an admissible policy $\pi^* = \{\mu^*, \mu^*, \ldots\}$ to be optimal under Assumption N it is necessary and sufficient that

$$J_{\mu^*} = T_{\mu^*}(J_{\mu^*}) = T(J_{\mu^*}),$$

or equivalently

$$\begin{aligned} J_{\mu^*}(x) &= \underset{w}{E}\,\{g[x, \mu^*(x), w] + J_{\mu^*}[f(x, \mu^*(x), w)]\} \\ &\leqslant \underset{w}{E}\,\{g(x, u, w) + J_{\mu^*}[f(x, u, w)]\} \qquad \forall x \in S, \quad u \in U(x). \end{aligned} \tag{10}$$

**Proposition 3** (Necessary and Sufficient Condition for Optimality under Assumption P) In order for an admissible policy $\pi^* = \{\mu^*, \mu^*, \ldots\}$ to be optimal under Assumption P, it is necessary and sufficient that

$$J^* = T_{\mu^*}(J^*) = T(J^*),$$

or equivalently

$$\begin{aligned} J^*(x) &= \underset{w}{E}\,\{g[x, \mu^*(x), w] + J^*[f(x, \mu^*(x), w)]\} \\ &\leqslant \underset{w}{E}\,\{g(x, u, w) + J^*[f(x, u, w)]\} \qquad \forall x \in S, \quad u \in U(x). \end{aligned} \tag{11}$$

### *Successive Approximation—Policy Iteration—Linear Programming*

We turn now to the question of convergence of the DP algorithm to the optimal value function $J^*$. Similar results as those of Section 6.4 may be obtained. Let $J_0$ be the zero function on $S$, i.e.,

$$J_0(x) = 0 \qquad \forall x \in S.$$

Consider the DP algorithm that generates $T(J_0)$, $T^2(J_0), \ldots, T^k(J_0), \ldots$. Each of the functions $T^k(J_0)$ represents the optimal value function for a corresponding finite horizon problem with $k$ stages. We have under Assumption P that

$$J_0 \leqslant T(J_0) \leqslant \cdots \leqslant T^k(J_0) \leqslant \cdots,$$

while under Assumption N

$$J_0 \geqslant T(J_0) \geqslant \cdots \geqslant T^k(J_0) \geqslant \cdots.$$

In either case the pointwise limit function $J_\infty$, where

$$J_\infty(x) = \lim_{k\to\infty} T^k(J_0)(x) \qquad \forall x \in S, \tag{12}$$

is well defined as a function from $S$ into $[0, +\infty]$ under Assumption P or into $[-\infty, 0]$ under Assumption N. As in Section 6.4, we have

**Proposition 4** (a) Let Assumption P hold and assume that $J_\infty = T(J_\infty)$. Then

$$J_\infty = J^*.$$

(b) Let Assumption N hold. Then

$$J_\infty = J^*.$$

As in the corresponding discounted case it need not be true that $J_\infty = J^*$ under Assumption P (a counterexample is provided by Problem 9 of Chapter 6 by setting $\alpha = 1$). As in Section 6.4 we have the following sufficient condition for $J_\infty = J^*$.

**Proposition 5** Let Assumption P hold and assume that the sets

$$U_k(x, \lambda) = \left\{ u \in U(x) \,\middle|\, \underset{w}{E}\, \{g(x, u, w) + T^k(J_0)[f(x, u, w)]\} \leqslant \lambda \right\} \tag{13}$$

are compact subsets of a Euclidean space for every $x \in S$, $\lambda \in R$, and for all $k$ greater than some integer $\bar{k}$. Then

$$J_\infty = T(J_\infty) = J^*.$$

Furthermore, there exists an optimal stationary policy.

Note that when $U(x)$ is a finite set for every $x \in S$ the assumption of the above proposition is satisfied.

Propositions 4 and 5 form the basis for a successive approximation method for computing in the limit the optimal value function $J^*$. An alternative method for computing $J^*$ under Assumption N is based on the following proposition. A related result under P is given in Problem 16.

**Proposition 6** Under Assumption N if $\tilde{J}: S \to [-\infty, 0]$ is a function such that $\tilde{J} \leqslant T(\tilde{J})$, then $\tilde{J} \leqslant J^*$.

*Proof* Since $\tilde{J} \leqslant J_0$ we have

$$T^k(\tilde{J}) \leqslant T^k(J_0),$$

from which, using the fact that $\lim_{k\to\infty} T^k(J_0)(x) = J_\infty(x) = J^*(x)$, we obtain

$$\limsup_{k\to\infty} T^k(\tilde{J})(x) \leqslant J^*(x) \qquad \forall x \in S.$$

On the other hand, we have $\tilde{J} \leqslant T(\tilde{J})$ implying $\tilde{J}(x) \leqslant \limsup_{k\to\infty} T^k(\tilde{J})(x)$ for all $x \in S$, and the result follows. Q.E.D.

Based on the preceding proposition we have that if $S$ is a finite set, $S = \{1, 2, \ldots, n\}$, and $C$ is also a finite set, $C = \{u^1, \ldots, u^m\}$, then $J^*(1), \ldots, J^*(n)$ solve the linear programming problem (cf. end of Section 6.2)

$$\max_{\lambda_i \leqslant 0} \sum_{i=1}^{n} \lambda_i$$

subject to

$$\lambda_i \leqslant \bar{g}(i, u^k) + \sum_{j=1}^{n} p_{ij}(u^k)\lambda_j \qquad i = 1, 2, \ldots, n, \quad u^k \in U(i),$$

where the notation corresponds to that of Section 6.2.

Under Assumption P it is possible to use a policy iteration algorithm in an effort to find $J^*$ and an optimal stationary policy. The algorithm is motivated by the following proposition, which parallels Proposition 7 of Section 6.2. A similar proposition cannot be proved under Assumption N in the absence of additional assumptions.

**Proposition 7** Let Assumption P hold and let $\{\mu, \mu, \ldots\}$ be an admissible stationary policy. If $\{\bar{\mu}, \bar{\mu}, \ldots\}$ is another admissible policy such that

$$\begin{aligned} &\mathop{E}_{w} \{g[x, \bar{\mu}(x), w] + J_\mu[f(x, \bar{\mu}(x), w)]\} \\ &\qquad = \min_{u \in U(x)} \mathop{E}_{w} \{g(x, u, w) + J_\mu[f(x, u, w)]\} \qquad \forall x \in S, \end{aligned}$$

then

$$J_{\bar{\mu}}(x) \leqslant J_\mu(x) \qquad \forall x \in S.$$

*Proof* Using the hypothesis and the facts $J_\mu = T_\mu(J_\mu)$, $J_\mu \geqslant 0$ (by Assumption P) we have, for any initial state $x_0$,

$$\begin{aligned}
J_{\bar{\mu}}(x_0) &= \lim_{N\to\infty} \underset{w_k}{E}\left\{\sum_{k=0}^{N-1} g[x_k, \bar{\mu}(x_k), w_k]\right\} \\
&\leqslant \liminf_{N\to\infty} \underset{w_k}{E}\left\{J_\mu(x_N) + g[x_{N-1}, \bar{\mu}(x_{N-1}), w_{N-1}] + \sum_{k=0}^{N-2} g[x_k, \bar{\mu}(x_k), w_k]\right\} \\
&\leqslant \liminf_{N\to\infty} \underset{w_k}{E}\left\{J_\mu(x_N) + g[x_{N-1}, \mu(x_{N-1}), w_{N-1}] + \sum_{k=0}^{N-2} g[x_k, \bar{\mu}(x_k), w_k]\right\} \\
&= \liminf_{N\to\infty} \underset{w_k}{E}\left\{J_\mu(x_{N-1}) + \sum_{k=0}^{N-2} g[x_k, \bar{\mu}(x_k), w_k]\right\} \\
&\ \ \vdots \\
&\leqslant J_\mu(x_0)
\end{aligned}$$

and the result is proved. Q.E.D.

Notice that the proposition states that the policy $\{\bar{\mu}, \bar{\mu}, \ldots\}$ is no worse than $\{\mu, \mu, \ldots\}$ but does not guarantee a strict improvement. When $J_{\bar{\mu}} = J_\mu$ the most that one can obtain is that

$$J_\mu = T(J_\mu).$$

In the discounted case with bounded costs per stage, the above equality implied optimality of $\{\mu, \mu, \ldots\}$, but this is not necessarily true under our present assumptions. Nonetheless, we shall be able to use Proposition 7 to construct a policy iteration algorithm in Section 7.4.

## 7.2 Optimal Stopping

Consider a situation where at each state $x$ of the state space there are two possible actions available. We may either stop (control $u^1$) and pay a terminal cost $t(x)$, or pay a cost $c(x)$ and continue the process (control $u^2$) according to the system equation

$$x_{k+1} = f_c(x_k, w_k), \qquad k = 0, 1, \ldots. \tag{14}$$

The objective is to find the optimal stopping policy that minimizes the total expected costs over an infinite number of stages. It is assumed that the input disturbances $w_k$ in (14) have the same probability distribution for all $k$, which depends only on the current state $x_k$.

In order to put this problem within the framework of Problem (U) we introduce an additional state $s$ (termination state) and we complete the system equation (14) as in Section 3.4 by letting

$$x_{k+1} = s \qquad \text{if} \quad u_k = u^1 \quad \text{or} \quad x_k = s.$$

No further cost is incurred once the system reaches the termination state. In other words if the termination action is taken, the system is driven to the termination state $s$, and if the system is already in the termination state, it remains there permanently (i.e., $s$ is an absorbing state).

We shall assume in this section that

$$t(x) \geqslant 0, \qquad c(x) \geqslant 0 \qquad \forall x \in S. \tag{15}$$

The case where $t(x) \leqslant 0$ and $c(x) \leqslant 0$ for all $x \in S$ is treated in Problem 10. Actually whenever there exists an $\varepsilon > 0$ such that $c(x) \geqslant \varepsilon$ for all $x \in S$, the results to be obtained apply also to the case where

$$\inf_{x \in S} t(x) > -\infty,$$

i.e., when $t(x)$ is bounded below by some scalar rather than bounded by zero. The reason is that if $c(x)$ is assumed to be greater than $\varepsilon > 0$ for all $x \in S$, any policy that will not stop within a finite expected number of stages results in infinite cost and can be excluded from consideration. As a result if we reformulate the problem and add a constant $r$ to $t(x)$ so that $t(x) + r \geqslant 0$ for all $x \in S$, the optimal cost $J^*(x)$ will be merely increased by $r$, while optimal policies will remain unaffected.

Now under our assumptions the problem clearly falls within the framework of Problem (U) provided the disturbance space $D$ is a countable set. Furthermore, Assumption P is satisfied by virtue of (15). The mapping $T$ of (3) takes the form

$$T(J)(x) = \min[t(x), c(x) + \underset{w}{E}\{J[f_c(x, w)]\}] \qquad \forall x \in S, \tag{16}$$

where $t(x)$ is the cost of the termination action $u^1$ and $c(x) + E_w\{J[f_c(x, w)]\}$ is the cost of the continuation action $u^2$. To be precise we should also define $T(J)(s) = 0$ where $s$ is the termination state. However, in what follows the value of various functions at $s$ is immaterial and will not be explicitly considered.

By Proposition 1 the optimal value function $J^*$ satisfies

$$J^* = T(J^*).$$

Since the control space has only two elements, by Proposition 5 we have

$$\lim_{k \to \infty} T^k(J_0)(x) = J^*(x) \qquad \forall x \in S, \tag{17}$$

where $J_0$ is the zero function ($J_0(x) = 0$, $\forall x \in S$). By Proposition 3, there exists a stationary optimal policy $\{\mu^*, \mu^*, \ldots\}$ described as follows:

$$\text{Stop} \quad \text{if} \quad t(x) < c(x) + \underset{w}{E}\{J^*[f_c(x, w)]\}.$$

$$\text{Continue} \quad \text{if} \quad t(x) \geqslant c(x) + \underset{w}{E}\{J^*[f_c(x, w)]\}.$$

Let us denote by $S^*$ the optimal stopping set (which may be empty)

$$S^* = \left\{x \in S \,|\, t(x) < c(x) + \underset{w}{E}\{J^*[f_c(x, w)]\}\right\},$$

and by $\bar{S}^*$ its complement in $S$:

$$\bar{S}^* = \{x \in S \,|\, x \notin S^*\} = \left\{x \in S \,|\, t(x) \geqslant c(x) + \underset{w}{E}\{J^*[f_c(x, w)]\}\right\}.$$

The set $S^*$ is the set of states where stopping is optimal. Consider also the sets

$$S_k = \left\{x \in S \,|\, t(x) < c(x) + \underset{w}{E}\{T^k(J_0)[f_c(x, w)]\}\right\},$$

and their complements in $S$,

$$\bar{S}_k = \left\{x \in S \,|\, t(x) \geqslant c(x) + \underset{w}{E}\{T^k(J_0)[f_c(x, w)]\}\right\}.$$

The stopping sets $S_1, S_2, \ldots, S_k, \ldots$ determine the optimal policy for finite horizon versions of the stopping problem and are determined via the successive approximation method. Since we have

$$J_0 \leqslant T(J_0) \leqslant \cdots \leqslant T^k(J_0) \leqslant \cdots \leqslant J^*,$$

it follows that

$$S_1 \subset \cdots \subset S_k \subset \cdots \subset S^* \qquad \text{and} \qquad \bar{S}_1 \supset \cdots \supset \bar{S}_k \supset \cdots \supset \bar{S}^*.$$

Now for any state $\tilde{x} \in \bigcap_{k=1}^{\infty} \bar{S}_k$ we have for all $k$,

$$t(\tilde{x}) \geqslant c(\tilde{x}) + \underset{w}{E}\{T^k(J_0)[f_c(\tilde{x}, w)]\}, \qquad k = 0, 1, \ldots,$$

and by taking limits and using (17) we obtain

$$t(\tilde{x}) \geqslant c(\tilde{x}) + \underset{w}{E}\{J^*[f_c(\tilde{x}, w)]\},$$

from which $\tilde{x} \in \bar{S}^*$. Hence

$$\bar{S}^* \supset \bigcap_{k=1}^{\infty} \bar{S}_k,$$

which, in conjunction with the fact, that $\bar{S}_k \supset \bar{S}^*$ for all $k$, yields

$$\bar{S}^* = \bigcap_{k=1}^{\infty} \bar{S}_k,$$

or equivalently

$$S^* = \bigcup_{k=1}^{\infty} S_k. \tag{18}$$

In other words, the *optimal stopping set $S^*$ for the infinite horizon problem is equal to the union of all the finite horizon stopping sets $S_k$.*

Notice that in the case where the state space is a finite set, (18) shows that the successive approximation method will determine the optimal stopping set in a finite number of iterations. Also when the state space is a finite set, some additional results of analytical and computational importance may be proved for the optimal stopping problem. These results will be developed in the context of a more general problem in Section 7.4. The remainder of this section is devoted to examples.

ASSET SELLING EXAMPLE Consider the asset selling example of Sections 3.4 and 6.1 (see also Problem 3.10) where the rate of interest $r$ is zero and there is instead a maintenance cost $c > 0$ per period for which the house remains unsold. We have the following equation for the optimal cost:

$$J^*(x) = \max\left[x, \; -c + \underset{w}{E}\{J^*(w)\}\right]. \tag{19}$$

In this equation $x$ takes values in a bounded interval of the form $[0, M]$, where $M$ is some positive scalar that bounds the possible offers from above. Notice that in this particular case we consider maximization of total expected reward and the termination reward $x$ is positive. Hence assumption (15) is not satisfied. Since, however, the termination cost is bounded, our analysis of this section is still applicable [cf. the discussion following (15)].

Now from (19) we obtain an optimal stationary policy of the form

If the current offer exceeds $-c + E_w\{J^*(w)\}$, sell the asset; otherwise do not sell.

The threshold level $-c + E_w\{J^*(w)\}$ may be obtained in a similar manner as in Section 3.4.

HYPOTHESIS TESTING EXAMPLE—SEQUENTIAL PROBABILITY RATIO TEST Consider the hypothesis testing problem of Section 4.5 for the case where the number of possible observations is unlimited. The state space is $S = [0, 1]$, i.e., the space of the sufficient statistic

$$p_k = P(x_k = x^0 \mid z_0, z_1, \ldots, z_k)$$

augmented with a termination state $s$, which will not be explicitly considered in the expressions below. To each state $p \in [0, 1]$ we may assign the termination cost

$$t(p) = \min[(1 - p)L_0, pL_1],$$

i.e., the cost associated with optimal choice between the distributions $f_0$ and $f_1$. The mapping $T$ of (16) takes the form

$$T(J)(p) = \min\left[(1 - p)L_0, pL_1, c + \underset{z}{E}\left\{J\left[\frac{pf_0(z)}{pf_0(z) + (1 - p)f_1(z)}\right]\right\}\right],$$

where the expectation over $z$ above is taken with respect to the probability distribution

$$P(z) = pf_0(z) + (1 - p)f_1(z) \qquad \forall z \in Z.$$

The optimal value function $J^*$ satisfies

$$J^*(p) = \min\left[(1 - p)L_0, pL_1, c + \underset{z}{E}\left\{J^*\left[\frac{pf_0(z)}{pf_0(z) + (1 - p)f_1(z)}\right]\right\}\right], \qquad (20)$$

and is obtained in the limit through the equation

$$J^*(p) = \lim_{k \to \infty} T^k(J_0)(p) \qquad \forall p \in [0, 1],$$

where $J_0$ is the zero function on $[0, 1]$.

Now consider the functions $T^k(J_0)$, $k = 0, 1, \ldots$. It is clear that

$$J_0 \leqslant T(J_0) \leqslant \cdots \leqslant T^k(J_0) \leqslant \cdots \leqslant \min[(1 - p)L_0, pL_1].$$

Furthermore, in view of the analysis of Section 4.5 we have that the function $T^k(J_0)$ is concave on $[0, 1]$ for all $k$. Hence the pointwise limit function $J^*$ is also concave on $[0, 1]$. In addition, we have clearly from (20) that

$$J^*(0) = J^*(1) = 0 \qquad \text{and} \qquad J^*(p) \leqslant \min[(1 - p)L_0, pL_1].$$

It follows from (20) and Fig. 7.1 that [provided $c < L_0 L_1/(L_0 + L_1)$] there exist two scalars $\bar{\alpha}, \bar{\beta}$ with $0 < \bar{\beta} \leqslant \bar{\alpha} < 1$ that determine an optimal stationary policy of the form

| | |
|---|---|
| Accept $f_0$ | if $p \geqslant \bar{\alpha}$. |
| Accept $f_1$ | if $p \leqslant \bar{\beta}$. |
| Continue the observations | if $\bar{\beta} < p < \bar{\alpha}$. |

In view of the optimality of the stationary policy above, the employment of the sequential probability ratio test described in Section 4.5 is justified when the number of possible observations is large and tends to infinity.

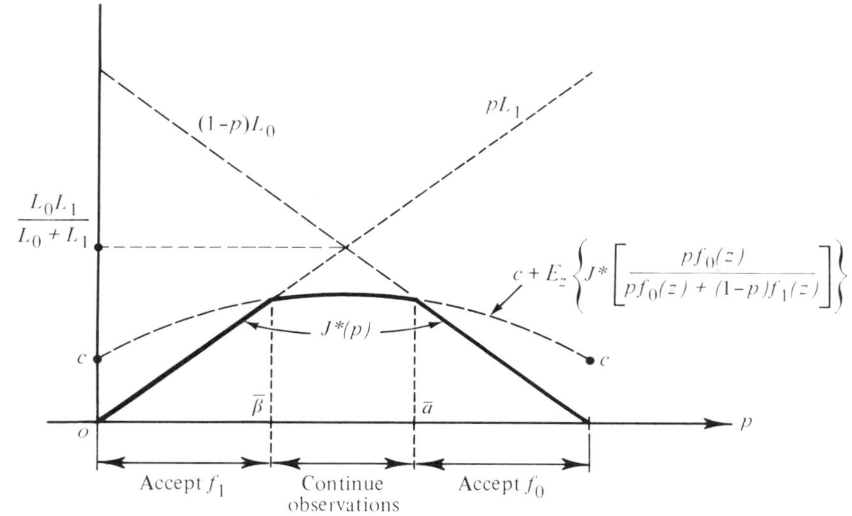


FIGURE 7.1

## 7.3 Optimal Gambling Strategies

A gambler enters a certain game played as follows. The gambler may stake at any time $k$ any amount $u_k \geqslant 0$ that does not exceed his current fortune $x_k$ (defined to be his initial capital plus his gain or minus his loss thus far). He wins his stake back and as much more with probability $p$ and he loses his stake with probability $(1 - p)$. Thus the gambler's fortune evolves according to the equation

$$x_{k+1} = x_k + w_k u_k, \qquad k = 0, 1, \ldots, \tag{21}$$

where $w_k = 1$ with probability $p$ and $w_k = -1$ with probability $(1 - p)$. Several games, such as playing red and black in roulette, fit the description given above.

The gambler enters the game with an initial capital $x_0$ and his goal is to increase his fortune up to a level $X$. He continues gambling until he either reaches his goal or loses his entire initial capital, at which point he leaves the game. The problem is to determine the optimal gambling strategy for maximizing the probability of reaching his goal. By a gambling strategy we mean a rule that specifies what the stake should be at time $k$ when the gambler's fortune is $x_k$ for every $x_k$ with $0 < x_k < X$.

The problem may be cast within the framework of Problem (U) where we consider maximization in place of minimization. Let us assume for convenience that fortunes are normalized so that $X = 1$. The state space is the set $[0, 1] \cup \{s\}$, where $s$ is a termination state to which the system moves with certainty from both states 0 and 1 with corresponding rewards 0 and 1. When $x_k \neq 0$, $x_k \neq 1$, the system evolves according to Eq. (21). The control constraint set is specified by

$$0 \leqslant u_k \leqslant x_k, \qquad 0 \leqslant u_k \leqslant 1 - x_k.$$

The reward per stage when $x_k \neq 0, x_k \neq 1$, is zero. Under these circumstances the probability of reaching the goal is equal to the total expected reward. Assumption N holds since the problem under consideration is equivalent to a problem of minimizing expected total cost with nonpositive costs per stage.

The mapping $T$ of (3) takes the form

$$T(J)(x) = \sup_{\substack{0 \leqslant u \leqslant x \\ 0 \leqslant u \leqslant 1-x}} [pJ(x+u) + (1-p)J(x-u)] \qquad \forall x \in (0, 1), \tag{22}$$
$$T(J)(0) = 0, \qquad T(J)(1) = 1$$

for any function $J: [0, 1] \to [0, +\infty]$. Actually for this problem one may restrict attention to functions $J$ taking values in the interval $[0, 1]$ with $J(0) = 0$, $J(1) = 1$.

Consider now the case where

$$0 < p < \tfrac{1}{2},$$

i.e., the game is unfair to the gambler. A discretized version of the case where $\frac{1}{2} \leqslant p < 1$ is considered in Problem 15. When $0 < p < \frac{1}{2}$ it is intuitively clear that if the gambler follows a very conservative strategy and stakes a very small amount at each time, he is all but certain to lose his capital. For example, if the gambler adopts a strategy of betting $1/n$ at each time, then it may be shown [A8, p. 182] that his probability of attaining the target fortune of unity starting with an initial capital $i/n$, $0 < i < n$, is given by

$$\left[\left(\frac{1-p}{p}\right)^i - 1\right]\left[\left(\frac{1-p}{p}\right)^n - 1\right]^{-1}.$$

If $0 < p < \frac{1}{2}$, $n$ tends to infinity, and $i/n$ tends to a constant, the probability above tends to zero, thus indicating that placing consistently small bets is a bad strategy.

From the preceding discussion one is led to consider a policy of placing large bets and in particular the so-called *bold strategy* whereby the gambler stakes at each time $k$ his entire fortune $x_k$ or just enough to reach his goal, whichever is least. In other words the bold strategy is the stationary policy $\pi^* = \{\mu^*, \mu^*, \ldots\}$ with $\mu^*$ given by

$$\mu^*(x) = \begin{cases} x & \text{if } 0 < x \leqslant \frac{1}{2}, \\ 1 - x & \text{if } \frac{1}{2} \leqslant x < 1. \end{cases}$$

We shall prove that the bold strategy is indeed an optimal policy. To this end it is sufficient, by Corollary 2.2, to show that for every initial fortune $x \in [0, 1]$ the value of the reward fortune $J_{\mu^*}(x)$ corresponding to the bold strategy $\{\mu^*, \mu^*, \ldots\}$ satisfies the sufficiency condition

$$T(J_{\mu^*}) = J_{\mu^*},$$

or equivalently

$$J_{\mu^*}(0) = 0, \qquad J_{\mu^*}(1) = 1,$$
$$J_{\mu^*}(x) \geqslant pJ_{\mu^*}(x + u) + (1 - p)J_{\mu^*}(x - u)$$
$$\forall x \in (0, 1), \quad u \in [0, x] \cap [0, 1 - x]. \tag{23}$$

Now by using the definition of the bold strategy and the fact that

$$J_{\mu^*} = T_{\mu^*}(J_{\mu^*})$$

(cf. Corollary 1.1) we obtain that the function $J_{\mu^*}$ must satisfy

$$J_{\mu^*}(0) = 0, \qquad J_{\mu^*}(1) = 1, \tag{24}$$

$$J_{\mu^*}(x) = \begin{cases} pJ_{\mu^*}(2x) & \text{if } 0 < x \leqslant \frac{1}{2}, \\ p + (1 - p)J_{\mu^*}(2x - 1) & \text{if } \frac{1}{2} \leqslant x < 1. \end{cases} \tag{25}$$

We prove the following lemma showing that $J_{\mu^*}$ is uniquely defined from the above relations.

**Lemma** For every $p$, with $0 < p \leqslant \frac{1}{2}$, there is only one bounded function on $[0, 1]$ satisfying (24), and (25), the function $J_{\mu^*}$. Furthermore, $J_{\mu^*}$ is continuous and strictly increasing on $[0, 1]$.

*Proof* Suppose that there existed two bounded functions $J_1 : [0, 1] \to R$, $J_2 : [0, 1] \to R$ such that $J_i(0) = 0$, $J_i(1) = 1$, $i = 1, 2$, and

$$J_i(x) = \begin{cases} pJ_i(2x) & \text{if } 0 < x \leqslant \frac{1}{2}, \\ p + (1 - p)J_i(2x - 1) & \text{if } \frac{1}{2} \leqslant x < 1, \end{cases} \qquad i = 1, 2.$$

Then we have

$$J_1(2x) - J_2(2x) = (J_1(x) - J_2(x))/p \qquad \text{if } 0 \leqslant x \leqslant \tfrac{1}{2}, \tag{26}$$

$$J_1(2x - 1) - J_2(2x - 1) = (J_1(x) - J_2(x))/1 - p \qquad \text{if } \tfrac{1}{2} \leqslant x \leqslant 1. \tag{27}$$

Let $z$ be any real number with $0 \leqslant z \leqslant 1$. Define

$$z_1 = \begin{cases} 2z & \text{if } 0 \leqslant z \leqslant \frac{1}{2}, \\ 2z - 1 & \text{if } \frac{1}{2} < z \leqslant 1, \end{cases}$$

$$\vdots$$

$$z_k = \begin{cases} 2z_{k-1} & \text{if } 0 \leqslant z_{k-1} \leqslant \frac{1}{2}, \\ 2z_{k-1} - 1 & \text{if } \frac{1}{2} < z_{k-1} \leqslant 1, \end{cases}$$

for $k = 1, 2, \ldots$. Then from (26) and (27) it follows (using $p \leqslant \frac{1}{2}$) that

$$|J_1(z_k) - J_2(z_k)| \geqslant |J_1(z) - J_2(z)|/(1 - p)^k \qquad k = 1, 2, \ldots.$$

Since $J_1(z_k) - J_2(z_k)$ is bounded it follows that $J_1(z) - J_2(z) = 0$, for otherwise the right-hand side of the inequality would tend to $+\infty$. Since $z \in [0, 1]$ is arbitrary we obtain $J_1 = J_2$. Hence $J_{\mu^*}$ is the unique bounded function on $[0, 1]$ satisfying (24) and (25).

To show that $J_{\mu^*}$ is strictly increasing and continuous we consider the mapping $T_{\mu^*}$, which operates on functions $J : [0, 1] \to [0, 1]$ and is defined by

$$T_{\mu^*}(J)(x) = \begin{cases} pJ(2x) + (1 - p)J(0) & \text{if } 0 < x \leqslant \frac{1}{2}, \\ pJ(1) + (1 - p)J(2x - 1) & \text{if } \frac{1}{2} \leqslant x < 1, \end{cases} \tag{28}$$

$$T_{\mu^*}(J)(0) = 0, \qquad T_{\mu^*}(J)(1) = 1.$$

Consider the functions $J_0, T_{\mu^*}(J_0), \ldots, T_{\mu^*}^k(J_0), \ldots$, where $J_0$ is the zero function ($J_0(x) = 0$ for all $x \in [0, 1]$). We have

$$J_{\mu^*}(x) = \lim_{k \to \infty} T_{\mu^*}^k(J_0)(x) \qquad \forall x \in [0, 1]. \tag{29}$$

Furthermore, the functions $T_{\mu^*}^k(J_0)$ can be shown to be monotonically nondecreasing in the interval $[0, 1]$. Hence, by (29), $J_{\mu^*}$ is also monotonically nondecreasing.

Consider now for $n = 0, 1, \ldots$ the sets

$$S_n = \{x \in [0, 1] \mid x = k2^{-n},\ k = \text{nonnegative integer}\}.$$

The following fact may be verified in a straightforward manner concerning the functions $T_{\mu^*}^k(J_0)$, $k = 0, 1, \ldots$ :

$$T_{\mu^*}^m(J_0)(x) = T_{\mu^*}^n(J_0)(x) \qquad \forall x \in S_{n-1}, \quad m \geqslant n \geqslant 1.$$

As a result of the above equality and (29),

$$J_{\mu^*}(x) = T_{\mu^*}^n(J_0)(x) \qquad \forall x \in S_{n-1}, \quad n \geqslant 1. \tag{30}$$

A further fact that may be verified by using induction and (28) and (30) is that for any nonnegative integers $k$, $n$ for which $0 \leqslant k2^{-n} < (k+1)2^{-n} \leqslant 1$, we have

$$p^n \leqslant J_{\mu^*}[(k+1)2^{-n}] - J_{\mu^*}(k2^{-n}) \leqslant (1-p)^n. \tag{31}$$

Since any number in $[0, 1]$ can be approximated arbitrarily closely from above and below by numbers of the form $k2^{-n}$ and since $J_{\mu^*}$ has been shown to be monotonically nondecreasing it follows immediately from (31) that $J_{\mu^*}$ is continuous and strictly increasing. Q.E.D.

We are now in a position to prove the following proposition.

**Proposition 8** The bold strategy is an optimal stationary gambling policy.

*Proof* We shall prove the sufficiency condition

$$J_{\mu^*}(x) \geqslant pJ_{\mu^*}(x+u) + (1-p)J_{\mu^*}(x-u)$$
$$\forall x \in [0, 1], \quad u \in [0, x] \cap [0, 1-x]. \tag{23}$$

In view of the continuity of $J_{\mu^*}$ established in the previous lemma it is sufficient to establish (23) for all $x \in [0, 1]$ and $u \in [0, x] \cap [0, 1-x]$ that belong to the union $\bigcup_{n=0}^{\infty} S_n$ of the sets $S_n$ defined by

$$S_n = \{z \in [0, 1] \mid z = k2^{-n}, k = \text{nonnegative integer}\}. \tag{32}$$

We shall use induction. By using the fact that $J_{\mu^*}(0) = 0, J_{\mu^*}(\frac{1}{2}) = p, J_{\mu^*}(1) = 1$, we can show that (23) holds for all $x$ and $u$ in $S_0$ and $S_1$. Assume that (23) holds for $x, u \in S_n$ and $n \geqslant 1$. We will show that it holds for all $x, u \in S_{n+1}$.

For any $x$, $u \in S_{n+1}$ with $u \in [0, x] \cap [0, 1-x]$ there are four possibilities:

(1) $x + u \leqslant \frac{1}{2}$,
(2) $x - u \geqslant \frac{1}{2}$,
(3) $x - u \leqslant x \leqslant \frac{1}{2} \leqslant x + u$,
(4) $x - u \leqslant \frac{1}{2} \leqslant x \leqslant x + u$.

We shall prove (23) for each of the cases above.

*Case 1* If $x$, $u \in S_{n+1}$, then $2x \in S_n$, $2u \in S_n$, and by the induction hypothesis

$$J_{\mu^*}(2x) - pJ_{\mu^*}(2x+2u) - (1-p)J_{\mu^*}(2x-2u) \geqslant 0. \tag{33}$$

If $x + u \leqslant \frac{1}{2}$, then by (25)

$$J_{\mu^*}(x) - pJ_{\mu^*}(x+u) - (1-p)J_{\mu^*}(x-u) = p[J_{\mu^*}(2x) - pJ_{\mu^*}(2x+2u) - (1-p)J_{\mu^*}(2x-2u)]$$

and using (33) the desired relation (23) is proved for the case under consideration.

*Case 2* If $x, u \in S_{n+1}$, then $(2x - 1) \in S_n$, $2u \in S_n$, and by the induction hypothesis

$$J_{\mu^*}(2x - 1) - pJ_{\mu^*}(2x + 2u - 1) - (1 - p)J_{\mu^*}(2x - 2u - 1) \geqslant 0.$$

If $x - u \geqslant \frac{1}{2}$, then by (25)

$$\begin{aligned} J_{\mu^*}(x) &- pJ_{\mu^*}(x + u) - (1 - p)J_{\mu^*}(x - u) \\ &= p + (1 - p)J_{\mu^*}(2x - 1) - p[p + (1 - p)J_{\mu^*}(2x + 2u - 1)] \\ &\quad -(1 - p)[p + (1 - p)J_{\mu^*}(2x - 2u - 1)] \\ &= (1 - p)[J_{\mu^*}(2x - 1) - pJ_{\mu^*}(2x + 2u - 1) \\ &\quad -(1 - p)J_{\mu^*}(2x - 2u - 1)] \geqslant 0, \end{aligned}$$

and (23) follows from the above relations.

*Case 3* Using (25) we have

$$\begin{aligned} &J_{\mu^*}(x) - pJ_{\mu^*}(x + u) - (1 - p)J_{\mu^*}(x - u) \\ &\quad = pJ_{\mu^*}(2x) - p[p + (1 - p)J_{\mu^*}(2x + 2u - 1)] - p(1 - p)J_{\mu^*}(2x - 2u) \\ &\quad = p[J_{\mu^*}(2x) - p - (1 - p)J_{\mu^*}(2x + 2u - 1) - (1 - p)J_{\mu^*}(2x - 2u)]. \end{aligned}$$

Now we must have $x \geqslant \frac{1}{4}$, for otherwise $u < \frac{1}{4}$ and $x + u < \frac{1}{2}$. Hence $2x \geqslant \frac{1}{2}$ and the sequence of equalities above can be continued as follows:

$$\begin{aligned} J_{\mu^*}(x) &- pJ_{\mu^*}(x + u) - (1 - p)J_{\mu^*}(x - u) \\ &= p[p + (1 - p)J_{\mu^*}(4x - 1) - p - (1 - p)J_{\mu^*}(2x + 2u - 1) \\ &\quad -(1 - p)J_{\mu^*}(2x - 2u)] = p(1 - p)[J_{\mu^*}(4x - 1) \\ &\quad -J_{\mu^*}(2x + 2u - 1) - J_{\mu^*}(2x - 2u)] = (1 - p)[J_{\mu^*}(2x - \tfrac{1}{2}) \\ &\quad -pJ_{\mu^*}(2x + 2u - 1) - pJ_{\mu^*}(2x - 2u)]. \end{aligned}$$

Since $p \leqslant (1 - p)$, the last expression is greater than or equal to both

$$(1 - p)[J_{\mu^*}(2x - \tfrac{1}{2}) - pJ_{\mu^*}(2x + 2u - 1) - (1 - p)J_{\mu^*}(2x - 2u)]$$

and

$$(1 - p)[J_{\mu^*}(2x - \tfrac{1}{2}) - (1 - p)J_{\mu^*}(2x + 2u - 1) - pJ_{\mu^*}(2x - 2u)].$$

Now for $x$, $u \in S_{n+1}$ and $n \geqslant 1$, we have $(2x - \frac{1}{2}) \in S_n$, $(2u - \frac{1}{2}) \in S_n$ if $(2u - \frac{1}{2}) \in [0, 1]$, and $(\frac{1}{2} - 2u) \in S_n$ if $(\frac{1}{2} - 2u) \in [0, 1]$. By the induction hypothesis the first or the second of the above expressions is nonnegative, depending on whether $2x + 2u - 1 \geqslant 2x - \frac{1}{2}$ or $2x - 2u \geqslant 2x - \frac{1}{2}$, i.e., $u \geqslant \frac{1}{4}$ or $u \leqslant \frac{1}{4}$. Hence (23) is proved for Case 3.

*Case 4* The proof resembles the one for Case 3. Using (25) we have

$$\begin{aligned} &J_{\mu^*}(x) - pJ_{\mu^*}(x + u) - (1 - p)J_{\mu^*}(x - u) \\ &\quad = p + (1 - p)J_{\mu^*}(2x - 1) - p[p + (1 - p)J_{\mu^*}(2x + 2u - 1)] \\ &\qquad - (1 - p)pJ_{\mu^*}(2x - 2u) = p(1 - p) + (1 - p)[J_{\mu^*}(2x - 1) \\ &\qquad - pJ_{\mu^*}(2x + 2u - 1) - pJ_{\mu^*}(2x - 2u)]. \end{aligned}$$

We must have $x \leqslant \frac{3}{4}$ for otherwise $u < \frac{1}{4}$ and $x - u > \frac{1}{2}$. Hence $0 \leqslant 2x - 1 \leqslant \frac{1}{2} \leqslant 2x - \frac{1}{2} \leqslant 1$ and using (25) we have

$$(1 - p)J_{\mu^*}(2x - 1) = (1 - p)pJ_{\mu^*}(4x - 2) = p[J_{\mu^*}(2x - \tfrac{1}{2}) - p].$$

Using the relations above we obtain

$$\begin{aligned} &J_{\mu^*}(x) - pJ_{\mu^*}(x + u) - (1 - p)J_{\mu^*}(x - u) \\ &\quad = p(1 - p) + p[J_{\mu^*}(2x - \tfrac{1}{2}) - p] - p(1 - p)J_{\mu^*}(2x + 2u - 1) \\ &\qquad - p(1 - p)J_{\mu^*}(2x - 2u) \\ &\quad = p[(1 - 2p) + J_{\mu^*}(2x - \tfrac{1}{2}) - (1 - p)J_{\mu^*}(2x + 2u - 1) \\ &\qquad - (1 - p)J_{\mu^*}(2x - 2u)]. \end{aligned}$$

The relations above are equal to both

$$\begin{aligned} &p[(1 - 2p)[1 - J_{\mu^*}(2x + 2u - 1)] + J_{\mu^*}(2x - \tfrac{1}{2}) \\ &\qquad - pJ_{\mu^*}(2x + 2u - 1) - (1 - p)J_{\mu^*}(2x - 2u)] \end{aligned}$$

and

$$\begin{aligned} &p[(1 - 2p)[1 - J_{\mu^*}(2x - 2u)] + J_{\mu^*}(2x - \tfrac{1}{2}) \\ &\qquad - (1 - p)J_{\mu^*}(2x + 2u - 1) - pJ_{\mu^*}(2x - 2u)]. \end{aligned}$$

Since $0 \leqslant J_{\mu^*}(2x + 2u - 1) \leqslant 1$ and $0 \leqslant J_{\mu^*}(2x - 2u) \leqslant 1$, the expressions above are greater than or equal to both

$$p[J_{\mu^*}(2x - \tfrac{1}{2}) - pJ_{\mu^*}(2x + 2u - 1) - (1 - p)J_{\mu^*}(2x - 2u)]$$

and

$$p[J_{\mu^*}(2x - \tfrac{1}{2}) - (1 - p)J_{\mu^*}(2x + 2u - 1) - pJ_{\mu^*}(2x - 2u)]$$

and the result follows as in Case 3. Q.E.D.

We note that the bold strategy is not the unique optimal stationary gambling strategy. For a characterization of all such optimal strategies see

Dubins and Savage [D8, p. 90]. Several other gambling situations where strategies of the bold type are optimal are described in reference [D8, Chapters 5 and 6].

## 7.4 The First Passage Problem†

In this section we shall make the following assumptions within the framework of the problem of this chapter:

**C.1** The state space $S$, the control space $C$, and the disturbance space $D$ are finite sets. The state space is denoted by

$$S = \{0, 1, \ldots, n\}.$$

**C.2** The state 0 is an absorbing state, i.e.,

$$x_k = 0 \quad \Rightarrow \quad x_{k+1} = 0 \qquad \forall u_k \in C, \quad w_k \in D.$$

Furthermore there is no cost incurred while the system is in the absorbing state, i.e.,

$$E\{g(0, u, w) \,|\, x = 0, u\} = 0 \qquad \forall u \in U(0).$$

**C.3** There exists a positive integer $m$ such that for every admissible policy $\pi = \{\mu_0, \mu_1, \ldots\}$ there holds

$$P(x_m = 0 \,|\, x_0 = i, \pi) > 0 \qquad \text{for} \quad \text{all } i = 1, 2, \ldots, n,$$

where $P(x_m = 0 \,|\, x_0 = i, \pi)$ denotes the probability that at time $m$ the system will be in the absorbing state given that the policy $\pi$ is used and the initial state is $i$.

In the problem above control continues until the system passes through the absorbing state for the first time. At this point the system operation essentially terminates. Thus the objective in the problem is to reach the absorbing state with the least possible cost. One may easily show using C.3 that termination will occur with probability one for every policy employed. It is also possible to show that if $P(x_m = 0 \,|\, x_0 = i, \pi) > 0$, for some integer $m$, then we have $P(x_{m'} = 0 \,|\, x_0 = i, \pi) > 0$ for some integer $m'$ with $m' \leqslant n$, i.e., it is possible to reach the absorbing state within $n$ steps. As a result we may take $m = n$ in C.3, and furthermore C.3 is equivalent to the seemingly weaker condition that for each $\pi$ there is an $m$ such that $P(x_m = 0 \,|\, x_0 = i, \pi) > 0$, $i = 1, 2, \ldots, n$. While Assumption C.3 is somewhat restrictive we shall be able to relax it somewhat later in this section.

† This section requires some familiarity with the basic notions associated with finite state Markov chains. A summary together with references is given in Appendix D.

Under C.1–C.3 one may prove a number of important results that are not available under either Assumption P or N. In fact, it turns out that *it is not necessary to assume Assumption P or N*, i.e., the costs per stage $g$ need not be either nonnegative or nonpositive. The basic reason is that under C.1–C.3 the mapping $T$ of (3) is an $m$-stage contraction mapping over the set of all functions $J: S \to R$ with $J(0) = 0$, where $m$ is the positive integer in C.3. In other words (see Section 6.3) we have for some $\rho < 1$ and for any two functions $J, J': S \to R$ with $J(0) = J'(0) = 0$,

$$\|T^m(J) - T^m(J')\| \leqslant \rho \|J - J'\|,$$

or equivalently

$$\max_{i=0,1,\ldots,n} |T^m(J)(i) - T^m(J')(i)| \leqslant \rho \max_{i=0,1,\ldots,n} |J(i) - J'(i)|.$$

Furthermore, for any admissible stationary policy $\pi = \{\mu, \mu, \ldots\}$ the mapping $T_\mu$ of (4) is also an $m$-stage contraction mapping, i.e.,

$$\|T_\mu^m(J) - T_\mu^m(J')\| \leqslant \rho_\mu \|J - J'\|,$$

for some $\rho_\mu < 1$ and every two functions $J, J': S \to R$ with $J(0) = J'(0) = 0$. These facts, which are proved in the proposition below, have important analytical and computational consequences.

**Proposition 9** Under C.1–C.3 there exists a scalar $\rho < 1$ such that for all $J, J': S \to R$ with $J(0) = J'(0) = 0$ we have

$$\max_{i=0,1,\ldots,n} |T^m(J)(i) - T^m(J')(i)| \leqslant \rho \max_{i=0,1,\ldots,n} |J(i) - J'(i)|, \tag{34}$$

$$T^k(J)(0) = T^k(J')(0) = 0, \qquad k = 0, 1, \ldots.$$

*Proof* For any $u \in U(i)$, $i = 0, 1, \ldots, n$, let us denote by $p_{ij}(u)$ the transition probability

$$p_{ij}(u) = P[W_{ij}(u) | x = i, u] \tag{35}$$

where $W_{ij}(u) \subset D$ is the set

$$W_{ij}(u) = \{w \in D \mid f(i, u, w) = j\}. \tag{36}$$

In other words $p_{ij}(u)$ is the probability that the next state will be $j$ given that the present state is $i$ and the control $u$ is applied as explained in the beginning of Section 6.2. Denote also

$$\bar{g}(i, u) = \underset{w}{E}\,\{g(i, u, w) | i, u\}, \qquad i = 0, 1, \ldots, n. \tag{37}$$

We have by C.2,

$$p_{0j}(u) = 0 \quad \forall j = 1, \ldots, n, \quad u \in U(0), \qquad p_{00}(u) = 1 \quad \forall u \in U(0), \tag{38}$$

$$\bar{g}(0, u) = 0 \qquad \forall u \in U(0). \tag{39}$$

By (3), C.1, C.2, and (37–39) we have for any two functions $J, J': S \to R$ with $J(0) = J'(0) = 0$,

$$T(J)(i) = \min_{u \in U(i)} \left[ \bar{g}(i, u) + \sum_{j=1}^{n} p_{ij}(u) J(j) \right], \qquad i = 1, \ldots, n, \tag{40}$$

$$\begin{aligned} T(J')(i) &= \min_{u \in U(i)} \left[ \bar{g}(i, u) + \sum_{j=1}^{n} p_{ij}(j) J'(j) \right], \qquad i = 1, \ldots, n, \\ T(J)(0) &= T(J')(0) = 0. \end{aligned} \tag{41}$$

Let $\mu(i)$, $\mu'(i)$ attain the minimum in (40) and (41), respectively. We have

$$T(J')(i) - T(J)(i) \leqslant \sum_{j=1}^{n} p_{ij}[\mu(i)] \, |J(j) - J'(j)|,$$

$$T(J)(i) - T(J')(i) \leqslant \sum_{j=1}^{n} p_{ij}[\mu'(i)] \, |J(j) - J'(j)|.$$

Combining the above two inequalities we obtain

$$|T(J)(i) - T(J')(i)| \leqslant \sum_{j=1}^{n} p_{ij}[\bar{\mu}_0(i)] \, |J(j) - J'(j)|,$$

where for each $i$, $\bar{\mu}_0(i)$ is equal to the value $\mu(i)$ or $\mu'(i)$ that yields the larger value of $\sum_{j=1}^{n} p_{ij} |J(j) - J'(j)|$.

By repeating the argument above we obtain that for every $k$ and some $\bar{\mu}_{k-1}: S \to C$ with $\bar{\mu}_{k-1}(i) \in U(i)$, $i = 1, \ldots, n$, we have

$$|T^k(J)(i) - T^k(J')(i)| \leqslant \sum_{j=1}^{n} p_{ij}[\bar{\mu}_{k-1}(i)] \, |T^{k-1}(J)(j) - T^{k-1}(J')(j)|. \tag{42}$$

By using the vector notation

$$E_k = \begin{bmatrix} |T^k(J)(1) - T^k(J')(1)| \\ \vdots \\ |T^k(J)(n) - T^k(J')(n)| \end{bmatrix}, \qquad k = 0, 1, \ldots,$$

and the matrix notation

$$P_{k-1} = \begin{bmatrix} p_{11}[\bar{\mu}_{k-1}(1)] & \cdots & p_{1n}[\bar{\mu}_{k-1}(1)] \\ \vdots & & \vdots \\ p_{n1}[\bar{\mu}_{k-1}(n)] & \cdots & p_{nn}[\bar{\mu}_{k-1}(n)] \end{bmatrix}, \qquad k = 1, 2, \ldots,$$

we may write (42) as

$$E_k \leqslant P_{k-1} E_{k-1}, \qquad k = 1, 2, \ldots,$$

where the inequality is coordinatewise. Using the fact that $p_{ij}$ is nonnegative we have $E_k \leqslant P_{k-1} P_{k-2} \cdots P_0 E_0$, and for $k = m$ we obtain

$$E_m \leqslant P_{m-1} P_{m-2} \cdots P_0 E_0.$$

Now the matrix $P_{m-1} P_{m-2} \cdots P_0$ is the $m$-step transition probability matrix corresponding to a policy of the form $\bar{\pi} = \{\bar{\mu}_{m-1}, \bar{\mu}_{m-2}, \ldots, \bar{\mu}_0, \mu, \mu, \ldots\}$, where $\bar{\mu}_0, \ldots, \bar{\mu}_{m-1}$ have been defined above and $\mu$ is any function satisfying $\mu(i) \in U(i)$ for all $i$ (see Appendix D). Note that $\bar{\pi}$ is admissible.

Thus the inequality above is equivalent to the following inequality, which holds for every $i = 1, \ldots, n$:

$$|T^m(J)(i) - T^m(J')(i)| \leqslant \sum_{j=1}^{n} P(x_m = j \mid x_0 = i, \bar{\pi}) |J(j) - J'(j)|.$$

Hence

$$\max_{i=1,\ldots,n} |T^m(J)(i) - T^m(J)(i)|$$

$$\leqslant \left[ \max_{i=1,\ldots,n} \sum_{j=1}^{n} P(x_m = j \mid x_0 = i, \bar{\pi}) \right] \max_{i=1,\ldots,n} |J(i) - J'(i)|.$$

By C.3 we have

$$\rho = \max_{i=1,\ldots,n} \sum_{j=1}^{n} P(x_m = j \mid x_0 = i, \bar{\pi}) < 1,$$

and (34) follows. Q.E.D.

By employing our usual device of restricting the control constraint set so that there is only one admissible policy we obtain for every admissible stationary policy $\{\mu, \mu, \ldots\}$ and any two functions $J, J' : S \to R$ with $J(0) = J'(0)$, the following relation:

$$\max_{i=0,1,\ldots,n} |T_\mu^m(J)(i) - T_\mu^m(J')(i)| \leqslant \rho_\mu \max_{i=0,1,\ldots,n} |J(i) - J'(i)|,$$

where $\rho_\mu$ is a constant depending on $\mu$ and satisfying $\rho_\mu < 1$. However, it is possible to obtain a stronger result for the problem of this section, which we state below. The proof is obtained by essentially repeating the proof of Proposition 9 and is left to the reader.

**Corollary 9.1** Let $\pi = \{\mu_0, \mu_1, \ldots\}$ be an admissible policy. Then under C.1–C.3, we have for all $J, J' : S \to R$ with $J(0) = J'(0) = 0$,

$$\max_{i=0,1,\ldots,n} |(T_{\mu_0} T_{\mu_1} \cdots T_{\mu_{m-1}})(J)(i) - (T_{\mu_0} T_{\mu_1} \cdots T_{\mu_{m-1}})(J')(i)|$$

$$\leqslant \rho_\pi \max_{i=0,1,\ldots,n} |J(i) - J'(i)|, \tag{43}$$

where

$$\rho_\pi = \max_{i=1,\ldots,n} \sum_{j=1}^{n} P(x_m = j \mid x_0 = i, \pi) < 1.$$

Having established the $m$-stage contraction properties of Proposition 9 and Corollary 9.1 we are able to state a number of important analytical and computational results for the first passage problem under C.1–C.3. The following proposition guarantees the convergence of the successive approximation method to the optimal value function $J^*$ starting from an arbitrary function $J: S \to R$ with $J(0) = 0$. Also $J^*$ and $J_\mu$ can be obtained as unique fixed points of the equations $J = T(J)$ and $J = T_\mu(J)$, respectively.

**Proposition 10** Let C.1–C.3 hold. Then for any function $J: S \to R$ with $J(0) = 0$,

$$J^*(x) = \lim_{k \to \infty} T^k(J)(x) \qquad \forall x \in S,$$

and for every admissible stationary policy $\{\mu, \mu, \ldots\}$,

$$J_\mu(x) = \lim_{k \to \infty} T_\mu^k(J)(x) \qquad \forall x \in S.$$

Furthermore, $J^*$ and $J_\mu$ are unique solutions of the equations $J = T(J)$ and $J = T_\mu(J)$, respectively, within the class of functions $J: S \to R$ with $J(0) = 0$. In addition if $\mu^*(i)$ attains for $i = 1, \ldots, n$ the minimum in the right-hand side of the equation

$$J^*(i) = \min_{u \in U(i)} \left[ \bar{g}(i, u) + \sum_{j=1}^{n} p_{ij}(u) J^*(j) \right], \qquad i = 1, \ldots, n,$$

then $\pi^* = \{\mu^*, \mu^*, \ldots\}$ is an optimal stationary policy.

The proof of Proposition 10 may be obtained through arguments similar to those used in Section 6.1 or by simple modifications of proofs of corresponding results of Problem 4 in Chapter 6 and is left to the reader. As indicated also by Problem 4 in Chapter 6, it is possible to compute optimal stationary policies under C.1–C.3 by using the method of policy iteration or linear programming (cf. Section 6.2). The development and proof of validity of these algorithms is left again as an exercise for the reader.

We now consider the first passage problem under a different set of assumptions.

**C.1′** Same as C.1.

**C.2′** Same as C.2.

**C.3′** There exists a stationary admissible policy $\bar{\pi} = \{\bar{\mu}, \bar{\mu}, \ldots\}$ and a positive integer $m$ such that

$$P(x_m = 0 \mid x_0 = i, \bar{\pi}) > 0 \qquad \forall i = 1, 2, \ldots, n, \tag{44}$$

where $P(x_m = 0 \mid x_0 = i, \bar{\pi})$ denotes the probability that at time $m$ the system will be in state $x = 0$ given that the policy $\bar{\pi}$ is used and the initial state is $i$.

**C.4′** There holds

$$g(i, u, w) > 0 \qquad \forall u \in U(i), \quad i = 1, \ldots, n, \quad w \in D. \tag{45}$$

Notice that C.3′ is a much less restrictive assumption than C.3 since now we require that (44) hold for a single stationary policy $\bar{\pi}$ rather than for every policy as in C.3. For example, C.3 will not hold in a finite state version of the stopping problem of Section 7.2 since by using the policy that continues operation at every stage regardless of the current state, the absorbing (termination) state will never be reached. On the other hand C.3′ is satisfied for the stopping problem of Section 7.2 since (44) holds for the policy $\bar{\pi}$ that terminates at every state $i = 1, 2, \ldots, n$. For this policy (44) is satisfied with $m = 1$. Notice, however, that by (45) we require now that expected costs per stage corresponding to the states $i = 1, 2, \ldots, n$ are strictly positive.

We now introduce the following definition.

**Definition** An admissible stationary policy $\pi = \{\mu, \mu, \ldots\}$ will be called a *proper* policy if there exists a positive integer $m$ such that

$$P(x_m = 0 \mid x_0 = i, \pi) > 0 \qquad \forall i = 1, 2, \ldots, n, \tag{46}$$

where $P(x_m = 0 \mid x_0 = i, \pi)$ denotes the probability that at time $m$ the system will be in the absorbing state $x = 0$ given that the policy $\pi$ is used and the initial state is $i$.

By C.3′ there exists at least one proper policy. Let us denote for any $\mu: S \to C$ by $P_\mu$ the transition probability matrix

$$P_\mu = \begin{bmatrix} 1 & 0 & \cdots & 0 \\ p_{10}[\mu(1)] & p_{11}[\mu(1)] & \cdots & p_{1n}[\mu(1)] \\ \vdots & \vdots & \vdots & \vdots \\ p_{n0}[\mu(n)] & p_{n1}[\mu(n)] & \cdots & p_{nn}[\mu(n)] \end{bmatrix} \tag{47}$$

where $p_{ij}$ is the transition probability defined by (35) and (36). The first row of $P_\mu$ has the form $(1, 0, \ldots, 0)$ in view of the fact that $x = 0$ is an absorbing state [cf. (38)]. Note that by Corollary 9.1 if $\pi = \{\mu, \mu, \ldots\}$ is proper, then by restricting the control constraint set so that there is only one admissible policy, namely $\pi$, we obtain a problem for which C.1–C.3 applies. Then application of Proposition 10 yields:

**Corollary 10.1** If $\pi = \{\mu, \mu, \ldots\}$ is a proper policy, then

$$J_\mu = \{J_\mu(0), J_\mu(1), \ldots, J_\mu(n)\}$$

satisfies $J_\mu = T_\mu(J_\mu)$ or equivalently

$$J_\mu = g_\mu + P_\mu J_\mu, \tag{48}$$

where $P_\mu$ is matrix (47) and $g_\mu$ is the vector

$$g_\mu = \begin{bmatrix} g_\mu(0) \\ g_\mu(1) \\ \vdots \\ g_\mu(n) \end{bmatrix} = \begin{bmatrix} \underset{w}{E}\{g[0, \mu(0), w] \,|\, 0, \mu(0)\} \\ \underset{w}{E}\{g[1, \mu(1), w] \,|\, 1, \mu(1)\} \\ \vdots \\ \underset{w}{E}\{g[n, \mu(n), w] \,|\, n, \mu(n)\} \end{bmatrix} = \begin{bmatrix} \bar{g}[0, \mu(0)] \\ \bar{g}[1, \mu(1)] \\ \vdots \\ \bar{g}[n, \mu(n)] \end{bmatrix}$$

Furthermore, $J_\mu$ is the unique function $J : S \to R$ satisfying $J = T_\mu(J)$ and $J(0) = 0$. In addition, for every function $J : S \to R$ with $J(0) = 0$, we have

$$\lim_{k \to \infty} T_\mu^k(J)(i) = J_\mu(i), \qquad i = 0, 1, \ldots, n. \tag{49}$$

It is important to realize that in view of Assumption C.4′ we are essentially interested in proper policies only, for if $\pi$ is not a proper policy, then there exists a state $i$ from which the absorbing state will never be reached using $\pi$. Since the expected cost incurred per stage will be bounded below by a positive number [by (45) and finiteness of the state space] we will have $J_\pi(i) = +\infty$, while with every proper policy the corresponding cost will be finite. This observation is the key to understanding the results to be obtained. In fact this observation yields immediately the following result.

**Proposition 11** Under C.1′–C.4′ there exist optimal stationary policies. Each one of these policies must be a proper policy.

*Proof* Under C.1′–C.4′ Assumption P is satisfied and since the control space is finite there exist optimal stationary policies by Proposition 5. Also by C.3′ there exists a proper policy $\bar{\pi} = \{\bar{\mu}, \bar{\mu}, \ldots\}$ and by Corollary 10.1 we have $0 \leqslant J_{\bar{\mu}}(i) < +\infty$, $\forall i$. Hence $0 \leqslant J^*(i) < +\infty$, $\forall i$. Since for every stationary policy that is not proper the associated cost is infinite for some initial states, such a policy cannot be optimal. Q.E.D.

We now arrive at our main result.

**Proposition 12** Under C.1′–C.4′ in order for a proper policy $\pi^* = \{\mu^*, \mu^*, \ldots\}$ to be optimal it is necessary and sufficient that

$$J_{\mu^*} = T_{\mu^*}(J_{\mu^*}) = T(J_{\mu^*}), \tag{50}$$

or equivalently

$$\begin{aligned} J_{\mu^*}(i) &= g_{\mu^*}(i) + \sum_{j=0}^{n} p_{ij}[\mu^*(i)]J_{\mu^*}(j) \\ &= \min_{u \in U(i)} \mathop{E}_{w} \{g(i, u, w) + J_{\mu^*}[f(i, u, w)]\}, \qquad i = 0, 1, \ldots, n. \end{aligned}$$

*Proof* Necessity of (50) is clear from the theory of Section 7.1 (Proposition 1 and Corollary 1.1). To prove sufficiency we have that for any optimal proper policy $\bar{\pi} = \{\bar{\mu}, \bar{\mu}, \ldots\}$ the condition $J_{\mu^*} = T(J_{\mu^*})$ implies

$$J_{\mu^*} \leqslant T_{\bar{\mu}}(J_{\mu^*}) \leqslant T_{\bar{\mu}}^2(J_{\mu^*}) \leqslant \cdots \leqslant \lim_{k \to \infty} T_{\bar{\mu}}^k(J_{\mu^*}).$$

By using (49) and the optimality of the proper policy $\bar{\pi}$ we obtain

$$J_{\mu^*}(i) \leqslant \lim_{k \to \infty} T_{\bar{\mu}}^k(J_{\mu^*})(i) = J_{\bar{\mu}}(i) = J^*(i) \qquad \forall i \in S.$$

Hence $\pi^*$ is optimal. Q.E.D.

A related result is the following.

**Proposition 13** Under C.1′–C.4′ the optimal value function $J^*$ is the only real-valued function with $J(0) = 0$, $J(i) \geqslant 0$, $i = 1, \ldots, n$, that satisfies the equation $J = T(J)$.

*Proof* Let $J : S \to R$ be a real-valued function with $J(0) = 0$, $J(i) \geqslant 0$, such that $J = T(J)$. Let $\mu$ be such that $J = T_\mu(J)$. By Corollary 2.1 we have $J_\mu \leqslant J$ and hence $\{\mu, \mu, \ldots\}$ is a proper policy. Hence by Corollary 10.1 we have $J_\mu = J$. Thus

$$J_\mu = T_\mu(J_\mu) = T(J_\mu)$$

and by Proposition 12, $\{\mu, \mu, \ldots\}$ is optimal or equivalently $J = J_\mu = J^*$. Q.E.D.

It is possible to combine Proposition 12 with Proposition 7 to show the validity of the policy iteration algorithm for the first passage problem under Assumption C.1′–C.4′. Indeed if $\pi = \{\mu, \mu, \ldots\}$ is a proper policy and $\bar{\mu}$ is such that $\bar{\mu}(i) \in U(i)$, $i = 0, 1, \ldots, n$, and

$$T_{\bar{\mu}}(J_\mu) = T(J_\mu),$$

by Proposition 7, we have

$$J_{\bar{\mu}}(i) \leqslant J_\mu(i), \qquad i = 0, 1, \ldots, n,$$

and $\{\bar{\mu}, \bar{\mu}, \ldots\}$ is proper. If we had $J_{\bar{\mu}} = J_\mu$, then in view of the relations

$$J_{\bar{\mu}} = T_{\bar{\mu}}(J_{\bar{\mu}}) \leqslant T_{\bar{\mu}}(J_\mu) = T(J_\mu) \leqslant T_\mu(J_\mu) = J_\mu,$$

we would obtain $T(J_\mu) = J_\mu$ and by Proposition 12, $\pi = \{\mu, \mu, \ldots\}$ would be optimal. Hence either $\pi = \{\mu, \mu, \ldots\}$ is optimal or $\bar{\pi} = \{\bar{\mu}, \bar{\mu}, \ldots\}$ is a strictly better proper policy. It follows that by policy iteration we can obtain an optimal proper policy in a finite number of steps provided that we start with a proper policy.

## 7.5 Notes

The material of Section 7.1 is very similar to that of Section 6.4. For further discussion see the references cited in Chapter 6. The gambling problem and its solution are taken from the fascinating work of Dubins and Savage [D8]. The first passage problem was first formulated by Eaton and Zadeh [E1]. The presentation given here differs somewhat from presentations in other sources [D4, K10, P1] in that it makes direct use of the general results of Section 7.1. A policy of the type described in Problem 7 is called a *one-stage lookahead policy* [R4, p. 138]. The general framework and the results of Problem 9 are due to the author [B13, B16]. The results of Problems 13 and 14 are also due to the author [B6, B9].

## Problems

**1.** Do Problems 8, 22–25, and 11 [except for part (a)] of Chapter 6 for the case of Problem (U) of this chapter (i.e., $\alpha = 1$).

**2.** Let Assumption P hold and consider the case $S = D = \{1, 2, \ldots, n\}$, $x_{k+1} = w_k$. The mapping $T$ is represented as

$$T(J)(i) = \inf_{u \in U(i)} \sum_{j=1}^{n} p_{ij}(u)[g(i, u, j) + J(j)], \qquad i = 1, \ldots, n,$$

where $p_{ij}(u)$ denotes the transition probability that the next state will be $j$ when the current state is $i$ and control $u$ is applied. Assume that $p_{ij}(u)$ and

$g(i, u, j)$ are continuous on $U(i)$ for all $i, j$ and that the sets $U(i)$ are compact subsets of $R^m$ for all $i$. Show that we have $\lim_{k\to\infty} T^k(J_0)(i) = J^*(i)$, where $J_0(i) = 0$, $i = 1, \ldots, n$. Show also that there exists an optimal stationary policy.

**3.** Consider the problem of finding a scalar sequence $\{u_0, u_1, \ldots\}$ satisfying $\sum_{k=0}^{\infty} u_k \leqslant c$, $u_k \geqslant 0$, $\forall k$, and maximizing $\sum_{k=0}^{\infty} g(u_k)$, where $c > 0$ is given and $g(u) \geqslant 0$ for all $u \geqslant 0$, $g(0) = 0$. Assume that $g$ is monotonically nondecreasing on $[0, \infty)$. Show that the optimal value of the problem is $J^*(c)$, where $J^*$ is a monotonically nondecreasing function on $[0, \infty)$ satisfying $J^*(0) = 0$ and

$$J^*(x) = \sup_{0 \leqslant u \leqslant x} \{g(u) + J^*(x - u)\} \qquad \forall x \in [0, \infty).$$

**4.** *Deterministic Linear–Quadratic Problems* Consider the deterministic linear–quadratic problem involving the system

$$x_{k+1} = Ax_k + Bu_k,$$

and the cost functional

$$J_\pi(x_0) = \sum_{k=0}^{\infty} x_k' Q x_k + \mu_k(x_k)' R \mu_k(x_k).$$

It is assumed that $R$ is positive definite symmetric, $Q$ is of the form $C'C$, and that the pairs $(A, B)$, $(A, C)$ are controllable and observable, respectively. Use the theory of Sections 3.1 and 7.1 to show that the stationary policy $\pi^* = \{\mu^*, \mu^*, \ldots\}$ with

$$\mu^*(x) = -(B'KB + R)^{-1} B'KAx$$

is optimal, where $K$ is the unique positive semidefinite symmetric solution of the algebraic Riccati equation (cf. Section 3.1)

$$K = A'[K - KB(B'KB + R)^{-1} B'K]A + Q.$$

Provide a similar result under an appropriate controllability assumption for the case of a periodic deterministic linear system and a periodic quadratic cost functional (cf. Section 6.7).

**5.** Prove Proposition 10 in Section 7.4. Also devise a policy iteration algorithm for the first passage problem under conditions C.1–C.3 and show that it will yield an optimal stationary policy in a finite number of iterations.

**6.** Consider the first passage problem under Assumptions C1′–C.4′. Show that if $J: S \to R$ is a function with $J(i) \geqslant 0$, $i = 0, 1, \ldots, n$, and $J(0) = 0$, then $\lim_{k\to\infty} T^k(J)(i) = J^*(i)$, $i = 0, \ldots, n$.

**7.** Consider the stopping problem under C.1′–C.4′ and let $B$ be the set of states

$$B = \left\{ i = 1, 2, \ldots, n \,\middle|\, t(i) \leqslant c(i) + \sum_{j=1}^{n} p_{ij} t(j) \right\}.$$

Assume that $p_{ij} = 0$ if $i \in B$ and $j \neq B$. Show that an optimal policy is to stop if and only if the current state is in $B$.

**8.** Consider the first passage problem under C.1′–C.4′. Let $\pi = \{\mu, \mu, \ldots\}$, $\pi' = \{\mu', \mu', \ldots\}$ be two proper policies. Define $\bar{\mu}$ as

$$\bar{\mu}(i) = \begin{cases} \mu(i) & \text{if} \quad J_\mu(i) \leqslant J_{\mu'}(i), \\ \mu'(i) & \text{if} \quad J_\mu(i) > J_{\mu'}(i). \end{cases}$$

Show that $\{\bar{\mu}, \bar{\mu}, \ldots\}$ is proper and that

$$J_{\bar{\mu}}(i) \leqslant \min\{J_\mu(i), J_{\mu'}(i)\}, \qquad i = 0, 1, \ldots, n.$$

**9.** Let $S$ and $C$ be two sets, and let $F(S)$ be the set of all functions $J : S \to [-\infty, +\infty]$. Consider a mapping $H : S \times C \times F(S) \to [-\infty, +\infty]$ having the following properties:

(a) $J \leqslant J' \Rightarrow H(x, u, J) \leqslant H(x, u, J')$, $\forall (x, u) \in S \times C$, $J, J' \in F(S)$.

(b) If $\{J^k\}$ is any sequence with $J^k \in F(S)$, $J^k \leqslant J^{k+1}$ for all $k$ and such that $\bar{J}(x) = \lim_{k \to \infty} J^k(x)$ for all $x \in S$, then

$$\lim_{k \to \infty} H(x, u, J^k) = H(x, u, \bar{J}) \qquad \forall (x, u) \in S \times C.$$

(c) There exists a scalar $\alpha > 0$ such that for all scalars $r > 0$ and functions $J \in F(S)$ there holds

$$H(x, u, J) \leqslant H(x + u, J + re) \leqslant H(x, u, J) + \alpha r \qquad \forall (x, u) \in S \times C,$$

where $e(x) = 1$ for all $x \in S$.

For each $x \in S$ let $U(x)$ be a given nonempty subset of $C$. Let $M$ be the set of all functions $\mu : S \to C$ with $\mu(x) \in U(x)$, $\forall x \in S$. Define the mappings $T : F(S) \to F(S)$, $\tilde{T} : F(S) \to F(S)$ by

$$T(J)(x) = \inf_{\mu \in M} H[x, \mu(x), J], \qquad \tilde{T}(J)(x) = \sup_{\mu \in M} H[x, \mu(x), J] \qquad \forall x \in S,$$

and for every $\mu \in M$ the mapping $T_\mu : F(S) \to F(S)$,

$$T_\mu(J)(x) = H[x, \mu(x), J] \qquad \forall x \in S.$$

Let $J_0 \in F(S)$ be a function such that $J_0(x) > -\infty$ for all $x \in S$ and

$$J_0(x) \leqslant H(x, u, J_0) \qquad \forall (x, u) \in S \times C.$$

Let $\Pi$ be the set of all sequences of functions $\pi = \{\mu_0, \mu_1, \ldots\}$, where $\mu_k \in M$, $k = 0, 1, \ldots,$ Define for all $x \in S$

$$J_\pi(x) = \lim_{k \to \infty} (T_{\mu_0} T_{\mu_1} \cdots T_{\mu_k})(J_0)(x),$$

$$J^*(x) = \inf_{\pi \in \Pi} J_\pi(x), \qquad \tilde{J}^*(x) = \sup_{\pi \in \Pi} J_\pi(x),$$

$$J_\infty(x) = \lim_{k \to \infty} T^k(J_0)(x), \qquad \tilde{J}_\infty(x) = \lim_{k \to \infty} \tilde{T}^k(J_0)(x).$$

Show that

(a) $T^k(J_0)(x) = \inf_\pi (T_{\mu_0} T_{\mu_1} \cdots T_{\mu_{k-1}})(J_0)(x)$, $\forall x \in S$.
(b) $\tilde{T}^k(J_0)(x) = \sup_\pi (T_{\mu_0} T_{\mu_1} \cdots T_{\mu_{k-1}})(J_0)(x)$, $\forall x \in S$.
(c) $J^*$ and $\tilde{J}^*$ are fixed points of $T$ and $\tilde{T}$, respectively, i.e.,

$$J^* = T(J^*), \qquad \tilde{J}^* = \tilde{T}(\tilde{J}^*).$$

(d) $J' \in F(S),\ J' \geqslant J_0,\ J' \geqslant T(J') \Rightarrow J' \geqslant J^*$,
$J' \in F(S),\ J' \geqslant J_0,\ J' \geqslant \tilde{T}(J') \Rightarrow J' \geqslant \tilde{J}^*$.

(e) Define for each $\mu \in M$, the function $J_\mu \in F$ by

$$J_\mu(x) = \lim_{k \to \infty} T_\mu^k(J_0)(x) \qquad \forall x \in S.$$

Show that

$$J_\mu = T_\mu(J_\mu),$$
$$J_\mu = J^* \Leftrightarrow T_\mu(J^*) = T(J^*),$$
$$J_\mu = \tilde{J}^* \Leftrightarrow T_\mu(J_\mu) = \tilde{T}(J_\mu).$$

(f) Show that if there exists a $\pi^* \in \Pi$ such that $J_{\pi^*} = J^*$ then there exists a $\mu^* \in M$ such that $J_{\mu^*} = J^*$.

(g) $\tilde{J}^* = J_\infty$.

(h) Assume that there exists a positive integer $\bar{k}$ such that for all $k \geqslant \bar{k}$, $x \in S$, $\lambda \in R$, the set

$$U_k(x, \lambda) = \{u \in U(x) \mid H[x, u, T^k(J_0)] \leqslant \lambda\}$$

is a compact subset of a Euclidean space. Then

$$J^* = T(J^*) = J_\infty = T(J_\infty),$$

and there exists an optimal stationary policy $\pi^* = \{\mu^*, \mu^*, \ldots\}$.

*Hints*: (a) For any fixed $k$ and any $\varepsilon > 0$ let $\bar{\mu}_i \in M$, $i = 0, \ldots, k-1$, be such that

$$T_{\bar{\mu}_i}[T^{k-i-1}(J_0)] \leqslant T^{k-i}(J_0) + \varepsilon e.$$

Use property (c) to show that

$$\inf_{\pi}(T_{\mu_0} \cdots T_{\mu_{k-1}})(J_0) \leqslant T^k(J_0) + \sum_{i=0}^{k-1} \alpha^i \varepsilon e.$$

(b) For each $i = 0, 1, \ldots, k-1$ consider a sequence $\{\mu_i^n\} \subset M$ such that

$$\lim_{n \to \infty} T_{\mu_i^n}[\tilde{T}^{k-i-1}(J_0)] = \tilde{T}^{k-i}(J_0),$$

$$T_{\mu_i^n}[\tilde{T}^{k-i-1}(J_0)] \leqslant T_{\mu_i^{n+1}}[\tilde{T}^{k-i-1}(J_0)], \qquad n = 0, 1, \ldots.$$

Show that

$$\sup_\pi (T_{\mu_0} T_{\mu_1} \cdots T_{\mu_{k-1}})(J_0) \geqslant \lim_{\substack{n_i \to \infty \\ i=0,\ldots\ldots,k-1}} (T_{\mu_0^{n_0}} T_{\mu_1^{n_1}} \cdots T_{\mu_{k-1}^{n_{k-1}}})(J_0) = \tilde{T}^k(J_0).$$

(c)–(h) Adapt the proofs of corresponding results in Section 6.4. See references [B13, B16] for more detailed analysis and proofs.

**10.** Consider the stopping problem of Section 7.2 under the assumption that

$$t(x) \leqslant 0, \qquad c(x) \leqslant 0 \qquad \forall x \in S.$$

Consider the mapping $T$ defined by

$$T(J)(x) = \min\left[t(x),\ c(x) + \underset{w}{E}\{J[f_c(x, w)]\}\right].$$

(a) Show that the optimal value function $J^*$ satisfies

$$J^* = T(J^*), \qquad J^* = \lim_{k \to \infty} T^k(J_0),$$

where $J_0(x) = 0$, $\forall x \in S$. Verify also that if $S$ is a finite set, then $J^*$ may be obtained by linear programming.

(b) Consider the case where $c(x) = 0$, $\forall x \in S$. Show that

$$J^* = \lim_{k \to \infty} T^k(t),$$

where $T^k(t)$ denotes the function obtained after $k$ applications of the mapping $T$ on the function $t(\cdot)$.

(c) Let $S = \{1, 2, \ldots\}$, $f_c(i, w) = i + 1$, $c(i) = 0$ for all $i \in S$, $w \in D$, and $t(i) = -1 + (1/i)$ for all $i \in S$. Show that $J^*(i) = -1$ for all $i$ and that there does not exist an optimal policy for this problem (even though the control space is a finite set).

**11.** Let $z_0, z_1, \ldots$ be a sequence of independent and identically distributed random variables taking values on a countable set $Z$. We know that the probability distribution of the $z_k$'s is one out of $n$ distributions $f_1, f_2, \ldots, f_n$, and we are trying to decide which distribution is the correct one. At each time $k$ after observing $z_1, \ldots, z_k$ we may either stop the observations and accept one of the $n$ distributions as correct, or take another observation at a cost $c > 0$. The cost for accepting $f_i$ given that $f_j$ is correct is $L_{ij}, i, j = 1, \ldots, n$. We assume $L_{ij} > 0$ for $i \neq j$, $L_{ii} = 0$, $i = 1, \ldots, n$. The a priori distribution of $f_1, \ldots, f_n$ is denoted

$$P_0 = \{p_0^1, p_0^2, \ldots, p_0^n\}, \quad p_0^i \geqslant 0, \qquad \sum_{i=1}^{n} p_0^i = 1.$$

Show that the optimal cost $J^*(P_0)$ is a concave function of $P_0$. Characterize the optimal acceptance regions and show how they can be obtained in the limit by means of a successive approximation method.

**12.** Show that a finite horizon problem with $N$ stages that falls within the framework of the basic problem of Chapter 2 can be viewed as a (stationary) first passage problem (not necessarily with finite state, control, and disturbance space) for which assumptions similar to C.2 and C.3 of Section 7.4 are satisfied. Show also that a contraction condition such as (34) holds for this problem.

*Hint*: If $S_0, S_1, \ldots, S_N$ are the state spaces for the stages $0, 1, \ldots, N$, define a new state space $S$ by $S = \{(x, k) | x \in S_k,\ k = 0, 1, \ldots, N\} \cup \{T\}$, where $T$ is a termination (absorbing) state to which the system is driven with certainty from every state in $\{(x, N) | x \in S_N\}$ similar to the constructions of Section 6.7.

**13.** *Infinite Time Reachability* Consider the stationary system

$$x_{k+1} = f(x_k, u_k, w_k), \qquad k = 0, 1, \ldots,$$

of the problem of this chapter, where the disturbance space $D$ is an arbitrary (not necessarily countable) set. The disturbances $w_k$ can take values in a subset $W(x_k, u_k)$ of $D$ that may depend on $x_k$ and $u_k$. This problem deals with the following question: Given a nonempty subset $X$ of the state space $S$, under what conditions does there exist an admissible policy $\{\mu_0, \mu_1, \ldots\}$ with $\mu_k(x_k) \in U(x_k)$ for all $x_k \in S$ and $k = 0, 1, \ldots$, such that the state of the (closed-loop) system

$$x_{k+1} = f[x_k, \mu_k(x_k), w_k] \tag{51}$$

belongs to the set $X$ for all $k$ and all possible values $w_k \in W[x_k, \mu_k(x_k)]$, i.e.,

$$x_k \in X \qquad \forall w_k \in W[x_k, \mu_k(x_k)], \qquad k = 0, 1, \ldots. \tag{52}$$

The set $X$ is said to be *infinitely reachable* if there exists an admissible policy $\{\mu_0, \mu_1, \ldots\}$ and *some* initial state $x_0 \in X$ for which relations (51) and (52) are satisfied. It is said to be *strongly reachable* if there exists an admissible policy $\{\mu_0, \mu_1 \ldots\}$ such that for *all* initial states $x_0 \in X$ relations (51) and (52) are satisfied.

Consider the function $R$ mapping any subset $Z$ of the state space $S$ into a subset $R(Z)$ of $S$ defined by

$$R(Z) = \{x \mid \text{there exists } u \in U(x) \text{ such that } f(x, u, w) \in Z, \forall w \in W(x, u)\} \cap Z.$$

(a) Show that the set $X$ is strongly reachable if and only if $R(X) = X$.

Given $X$ consider the set $X^*$ defined as follows: $x_0 \in X^*$ if and only if $x_0 \in X$ and there exists an admissible policy $\{\mu_0, \mu_1, \ldots\}$ such that (51) and (52) are satisfied when $x_0$ is taken as the initial state of the system.

(b) Show that a set $X$ is infinitely reachable if and only if it contains a nonempty strongly reachable set. Furthermore, the largest such set is $X^*$ in the sense that $X^*$ is strongly reachable whenever nonempty and if $\tilde{X} \subset X$ is another strongly reachable set, then $\tilde{X} \subset X^*$.

(c) Show that if $X$ is infinitely reachable, there exists an admissible stationary policy $\{\mu, \mu, \ldots\}$ such that if the initial state $x_0$ belongs to $X^*$, then all subsequent states of the closed-loop system $x_{k+1} = f[x_k, \mu(x_k), w_k]$ are guaranteed to belong to $X^*$.

(d) Given $X$ consider the sets $R(X), \ldots, R^k(X), \ldots$, where $R^k(X)$ denotes the set obtained after $k$ applications of the mapping $R$ on $X$. Show that

$$X^* \subset \bigcap_{k=1}^{\infty} R^k(X).$$

(e) Given $X$, consider for each $x \in X$ and $k = 1, 2, \ldots$ the set

$$U_k(x) = \{u \mid f(x, u, w) \in R^k(X), \forall w \in W(x, u)\}.$$

Show that if there exists an index $\bar{k}$ such that for all $x \in X$ and $k \geq \bar{k}$ the set $U_k(x)$ is a compact subset of a Euclidean space, then $X^* = \bigcap_{k=1}^{\infty} R^k(X)$.

*Hint*: Use the results of Problem 9. See Bertsekas [B9] for detailed proofs.

**14.** *Infinite Time Reachability for Linear Systems* Consider the linear stationary system

$$x_{k+1} = Ax_k + Bu_k + Gw_k,$$

where $x_k \in R^n$, $u_k \in R^m$, $w_k \in R^r$, and the matrices $A, B, G$ are known and have appropriate dimensions. The matrix $A$ is assumed invertible. The controls $u_k$ and the disturbances $w_k$ are restricted to take values in the ellipsoids $U = \{u \mid u'Ru \leq 1\}$ and $W = \{w \mid w'Qw \leq 1\}$, respectively, where $R$ and $Q$ are positive definite symmetric matrices of appropriate dimensions. Show that in order for the ellipsoid $X = \{x \mid x'Kx \leqslant 1\}$, where $K$ is a positive definite symmetric matrix, to be strongly reachable (in the terminology of Problem 13), it is sufficient that for some positive definite matrix $M$ and for some scalar $\beta \in (0, 1)$ we have

$$K = A'\left[(1-\beta)K^{-1} - \frac{1-\beta}{\beta} GQ^{-1}G' + BR^{-1}B'\right]^{-1} A + M, \tag{53}$$

$$K^{-1} - \frac{1}{\beta} GQ^{-1}G': \quad \text{positive definite.} \tag{54}$$

Show also that if (53) and (54) are satisfied, the stationary policy $\{\mu^*, \mu^*, \ldots\}$, where

$$\mu^*(x) = -(R + B'FB)^{-1}B'FAx = Lx,$$

$$F = \left[(1-\beta)K^{-1} - \frac{1-\beta}{\beta} GQ^{-1}G'\right]^{-1}$$

achieves reachability of the ellipsoid $X = \{x \mid x'Kx \leqslant 1\}$. Furthermore, the matrix $(A + BL)$ is a stable matrix. (For a proof together with a computational procedure for finding matrices $K$ satisfying (53), (54) see Bertsekas [B6, B9].)

**15.** *Gambling Strategies for Favorable Games* A gambler plays a game such as the one of Section 7.3 but where the probability of winning $p$ satisfies $\frac{1}{2} \leqslant p < 1$. His objective is to reach a final fortune $n$, where $n$ is a positive integer with $n \geqslant 2$. His initial fortune is a positive integer $i$ with $0 < i < n$ and his stake at time $k$ can take only integer values $u_k$ satisfying $0 \leqslant u_k \leqslant x_k$, $0 \leqslant u_k \leqslant n - x_k$, where $x_k$ is his fortune at time $k$. Show that the strategy that always stakes one unit is optimal [i.e., $\mu^*(x) = 1$ for all integers $x$ with $0 < x < n$ is optimal].

*Hint*: Use Proposition 10 to show that

$$J_{\mu^*}(i) = [((1-p)/p)^i - 1][((1-p)/p)^n - 1]^{-1}, \qquad 0 \leqslant i \leqslant n, \quad \tfrac{1}{2} < p < 1,$$

$$J_{\mu^*}(i) = i/n, \qquad 0 \leqslant i \leqslant n, \quad p = \tfrac{1}{2}$$

(or see Ash [A8, p. 182] for a proof). Then use the sufficiency condition of Corollary 2.2.

**16.** Under Assumption P show that if $\tilde{J}: S \to [0, +\infty]$ is a function such that $T(\tilde{J}) \leqslant \tilde{J}$, then $J^* \leqslant \tilde{J}$. Devise a mathematical programming procedure for solving the problem when $S$, $C$, and $D$ are finite sets.

*Chapter 8*

# Minimization of Average Expected Value

The results of the previous chapter are applicable to problems where the infimum of the total expected value of the cost functional may be either finite or infinite for any given initial state. While for several classes of problems this infimum is finite for at least some initial states, in many problems under the positivity assumption $P$ the total expected value of the cost functional is infinite for every initial state and every admissible policy. Under these circumstances the framework adopted in the previous chapter is clearly inadequate since within this framework every policy is optimal. On the other hand, in many situations it turns out that while the total expected value

$$\lim_{N \to \infty} E\left\{\sum_{k=0}^{N-1} g[x_k, \mu_k(x_k), w_k]\right\} \tag{1}$$

corresponding to every admissible policy $\{\mu_0, \mu_1, \ldots\}$ and initial state $x_0$ is infinite, the limit

$$\lim_{N \to \infty} (1/N)\, E\left\{\sum_{k=0}^{N-1} g[x_k, \mu_k(x_k), w_k]\right\} \tag{2}$$

exists and is finite for every initial state and admissible policy. This is true in particular if the state space $S$, the control space $C$, and the disturbance space $D$ are finite sets, i.e., the system controlled is a finite state Markov chain. Expression (2) may be viewed as *expected cost per stage* and is a reasonably meaningful criterion for optimization. This chapter will deal with a problem similar to the problem that was the subject of the previous chapter except for the fact that the expected cost per stage (2) is minimized in place of the total expected cost of (1). Furthermore, *in the first three sections we shall restrict ourselves to the case of finite state space, control space, and disturbance space.* For this reason it is perhaps convenient to switch at the outset to a notation that is better suited to finite state systems and is described in the problem formulated below.

Let $S = \{1, \ldots, n\}$ denote the state space.† To each state $i \in S$ and each control $u$ in the finite control space $C$ there corresponds a set of transition probabilities $p_{ij}(u), j = 1, \ldots, n$, where $p_{ij}(u)$ denotes the probability that the next state will be $j$ given that the present state is $i$ and control $u$ is applied. These transition probabilities specify completely the system together with the statistical description of the uncertainty as discussed in Sections 1.4 and 6.2. Each time the system is in state $i \in S$ and control $u \in C$ is applied, we incur an expected cost denoted $g(i, u)$. The objective is to minimize the average cost per stage

$$J_\pi(x_0) = \lim_{N \to \infty} (1/N)\, E\left\{\sum_{k=0}^{N-1} \bar{g}[x_k, \mu_k(x_k)]\right\}, \tag{3}$$

over all admissible policies $\pi = \{\mu_0, \mu_1, \ldots\}$ with $\mu_k : S \to C$, $\mu_k(i) \in U(i)$, $\forall i \in S$ for which $J_\pi(x_0)$ is well defined as a real number for every initial state $x_0 \in S$.

Let us now provide a preliminary discussion of the problem that motivates some of the results to be obtained in the next section. Given any stationary admissible policy $\pi = \{\mu, \mu, \ldots\}$ let us denote by $P_\mu$ the *transition probability matrix* having elements $p_{ij}[\mu(i)]$:

$$P_\mu = \begin{bmatrix} p_{11}[\mu(1)] & \cdots & p_{1n}[\mu(1)] \\ \vdots & & \vdots \\ p_{n1}[\mu(n)] & \cdots & p_{nn}[\mu(n)] \end{bmatrix} \tag{4}$$

By the definition of $p_{ij}$ we have

$$p_{ij}[\mu(i)] \geqslant 0 \quad \forall i, j, \qquad \sum_{j=1}^{n} p_{ij}[\mu(i)] = 1 \quad \forall i.$$

† In the first three sections of this chapter we shall make use of some of the notions and results associated with finite state Markov chains. A summary of these results together with references is provided in Appendix D.

The matrix $P_\mu$ may be used to express the $m$-step transition probabilities corresponding to a stationary policy $\pi = \{\mu, \mu, \ldots\}$:

$$P(x_{k+m} = j \mid x_k = i, \pi),$$

i.e., the probability that the state will be $j$ at time $(k + m)$ given that the state is $i$ at time $k$ and the policy $\pi$ is used. We have

$$p_{ij}[\mu(i)] = P(x_{k+1} = j \mid x_k = i, \pi),$$

and it is an elementary matter to show (see Appendix D) that

$$[P_\mu^m]_{ij} = P(x_{k+m} = j \mid x_k = i, \pi),$$

where $[P_\mu^m]_{ij}$ is the element of the $i$th row and $j$th column of the matrix $P_\mu^m$ (i.e., $P_\mu$ raised to the $m$th power).

Let us now consider the value of the cost functional $J_\pi(x_0)$ of (3). As before we use the notation

$$J_\pi(i) = J_\mu(i), \qquad i = 1, \ldots, n,$$

for stationary policies $\pi = \{\mu, \mu, \ldots\}$. Denote

$$J_\mu = \begin{bmatrix} J_\mu(1) \\ J_\mu(2) \\ \vdots \\ J_\mu(n) \end{bmatrix}, \qquad g_\mu = \begin{bmatrix} \bar{g}[1, \mu(1)] \\ \bar{g}[2, \mu(2)] \\ \vdots \\ \bar{g}[n, \mu(n)] \end{bmatrix} \tag{5}$$

With this notation it is easy to see that

$$J_\mu = \lim_{N\to\infty} (1/N) \sum_{k=0}^{N-1} P_\mu^k g_\mu. \tag{6}$$

The following result shows that $J_\mu$ is well defined. It is a standard result on transition probability matrices and it is provided here only for the sake of the following discussion, rather than for obtaining any concrete results. Its proof may be found in [K7].

**Lemma 1** For any $n \times n$ stochastic matrix $P$, i.e., a matrix with elements $p_{ij}$ satisfying

$$p_{ij} \geqslant 0, \quad i, j = 1, \ldots, n, \qquad \sum_{j=1}^{n} p_{ij} = 1, \quad i = 1, \ldots, n,$$

we have

$$\lim_{N\to\infty} (1/N) \sum_{k=0}^{N-1} P^k = P^*, \tag{7}$$

where $P^*$ is a stochastic matrix with the following properties:

(a) $P^* = PP^* = P^*P = P^*P^*$.

(b) $(I - P + P^*)$ is an invertible matrix, where $I$ denotes the $n \times n$ identity matrix.

Denoting now

$$P_\mu^* = \lim_{N\to\infty} (1/N) \sum_{k=0}^{N-1} P_\mu^k, \tag{8}$$

and using Lemma 1, we have from (6) that

$$J_\mu = P_\mu^* g_\mu. \tag{9}$$

Thus for every admissible stationary policy the corresponding average cost per stage is well defined and conveniently characterized by the above equation. Consider also the vector

$$h_\mu = (I - P_\mu + P_\mu^*)^{-1}(I - P_\mu^*)g_\mu, \tag{10}$$

where the inverse above exists by part (b) of Lemma 1. We have

$$(I - P_\mu + P_\mu^*)h_\mu = (I - P_\mu^*)g_\mu, \tag{11}$$

and multiplying both sides by $P_\mu^*$ and using part (a) of Lemma 1 we obtain

$$P_\mu^* h_\mu = 0. \tag{12}$$

Using (12) and (9) we may write (11) as

$$J_\mu + h_\mu = g_\mu + P_\mu h_\mu. \tag{13}$$

This equation is satisfied by every admissible stationary policy $\pi = \{\mu, \mu, \ldots\}$ and corresponds to the familiar functional equations satisfied by the cost corresponding to stationary policies in the discounted and undiscounted total cost cases of the previous two chapters. In those cases the average cost per stage $J_\mu$ was zero whenever the total cost was finite and the corresponding functional equation had the form

$$h_\mu = g_\mu + \alpha P_\mu h_\mu,$$

where $0 < \alpha < 1$ in the discounted case and $\alpha = 1$ in the undiscounted case.

As a result of the preceding discussion we have that with every admissible stationary policy $\pi = \{\mu, \mu, \ldots\}$ there is associated a vector of average costs per stage $J_\mu$ defined by (8) and (9) and satisfying the functional equation (13). An interesting question is whether there exists a stationary policy $\{\mu^*, \mu^*, \ldots\}$ that is optimal in the sense

$$J_{\mu^*} \leqslant J_\mu, \qquad \forall \mu,$$

where this inequality is considered to be componentwise [i.e., we write $J_{\mu^*} \leqslant J_\mu$ if $J_{\mu^*}(i) \leqslant J_\mu(i)$ for each initial state $i \in S$]. The answer to this question is affirmative (see [B19, D4]) but we shall not prove this fact. Instead we shall concentrate on the important and frequently encountered case where the optimal cost vector $J_{\mu^*}$ is of the form

$$J_{\mu^*} = \lambda e, \tag{14}$$

where $\lambda$ is a scalar and $e$ is the unit vector on $R^n$, i.e.,

$$e = [1, 1, \ldots, 1]'.$$

Equation (14) will hold, for example, if the matrix $P_{\mu^*}^*$ of (8) corresponding to $\{\mu^*, \mu^*, \ldots\}$ has identical rows. When (14) holds the optimal cost per stage is the same for every initial state. In the next section we provide conditions that ensure that (14) holds. Furthermore, under these conditions we show that optimal stationary policies $\{\mu^*, \mu^*, \ldots\}$ can be obtained from the optimality equation

$$J_{\mu^*} + h_{\mu^*} = g_{\mu^*} + P_{\mu^*}h_{\mu^*} = \min_{\mu \in M}(g_\mu + P_\mu h_{\mu^*}), \tag{15}$$

where $M$ is the (finite) set of all functions $\mu: S \to C$ with $\mu(i) \in U(i)$, $i = 1, \ldots, n$, and the minimization is considered to be componentwise. Some connections with the discounted cost problem are also established in the next section. Subsequent sections provide computational algorithms for obtaining an optimal policy. The last section treats a problem involving a linear system and a quadratic cost functional.

## 8.1 Existence Results

Our first result provides a condition for existence of a stationary optimal policy satisfying a certain optimality equation. This condition is not readily verifiable but is implied by other more natural conditions, which we shall provide subsequently.

**Proposition 1** Assume that there exists a function $h: S \to R$ and a constant $\lambda$ such that

$$\lambda + h(i) = \min_{u \in U(i)} \left[ \bar{g}(i, u) + \sum_{j=1}^{n} p_{ij}(u)h(j) \right] \qquad \forall i = 1, \ldots, n. \tag{16}$$

Then if $\mu^*(i)$ attains the minimum in (16) for every $i = 1, \ldots, n$, the stationary policy $\pi^* = \{\mu^*, \mu^*, \ldots\}$ is optimal. Furthermore, the optimal value of the cost functional $J_\pi(i)$ of (3) is equal to $\lambda$ for every $i = 1, \ldots, n$, i.e.,

$$\lambda = J_{\mu^*}(i) = \inf_{\pi} J_\mu(i) \qquad \forall i = 1, 2, \ldots, n. \tag{17}$$

*Proof* For any admissible policy $\pi = \{\mu_0, \mu_1, \ldots\}$, and any initial state $x_0 \in S$ let $E\{\cdot \mid x_0, \pi\}$, $E\{\cdot \mid x_{k-1}, u_{k-1}, \pi\}$ denote conditional expectation given the policy $\pi$ is used and $x_0$ or $x_{k-1}$ and $u_{k-1} = \mu_{k-1}(x_{k-1})$ have occurred, respectively. We have for any $k \geq 1$,

$$\underset{x_k}{E}\{h(x_k) \mid x_0, \pi\} = \underset{x_{k-1}, u_{k-1}}{E}\left\{\underset{x_k}{E}\{h(x_k) \mid x_{k-1}, u_{k-1}, \pi\} \mid x_0, \pi\right\}.$$

It follows that for every $N \geq 1$,

$$E\left\{\sum_{k=1}^{N}[h(x_k) - E\{h(x_k) \mid x_{k-1}, u_{k-1}, \pi\}] \mid x_0, \pi\right\} = 0. \tag{18}$$

We have

$$\begin{aligned}
&\underset{x_k}{E}\{h(x_k) \mid x_{k-1}, u_{k-1}, \pi\} \\
&= \sum_{j=1}^{n} p_{x_{k-1}, j}(u_{k-1}) h(j) \\
&= \bar{g}(x_{k-1}, u_{k-1}) + \sum_{j=1}^{n} p_{x_{k-1}, j}(u_{k-1}) h(j) - \bar{g}(x_{k-1}, u_{k-1}) \\
&\geq \min_{u \in U(x_{k-1})}\left[\bar{g}(x_{k-1}, u) + \sum_{j=1}^{n} p_{x_{k-1}, j}(u) h(j)\right] - \bar{g}(x_{k-1}, u_{k-1}) \\
&= \lambda + h(x_{k-1}) - \bar{g}(x_{k-1}, u_{k-1}),
\end{aligned} \tag{19}$$

with equality above when $\pi = \{\mu^*, \mu^*, \ldots\}$ (by the definition of $\mu^*$). Hence from (18) and (19) we have for every $N \geq 1$,

$$0 \leq E\left\{\sum_{k=0}^{N-1}[h(x_{k+1}) - \lambda - h(x_k) + \bar{g}(x_k, u_k)] \mid x_0, \pi\right\},$$

or equivalently

$$\lambda \leq (1/N)\, E\{h(x_N) \mid x_0, \pi\} - (1/N) h(x_0) + (1/N)\, E\left\{\sum_{k=0}^{N-1} \bar{g}(x_k, u_k) \mid x_0, \pi\right\},$$

with equality if $\pi = \{\mu^*, \mu^*, \ldots\}$. Taking the limit as $N \to \infty$, we obtain

$$\lambda \leq \lim_{N \to \infty} (1/N)\, E\left\{\sum_{k=0}^{N-1} \bar{g}(x_k, u_k) \mid x_0, \pi\right\} = J_\pi(x_0),$$

for every $x_0 \in S$ and every admissible policy $\pi$. Furthermore equality holds above when $\pi = \{\mu^*, \mu^*, \ldots\}$ and the result follows. Q.E.D.

We note that the result of the previous proposition depends on the finiteness of the state space only to the extent that $h$ is a bounded function, and indeed an extension of it can be shown to hold for an arbitrary state space [R3] assuming that $h$ is bounded.

Now given a stationary policy $\pi = \{\mu, \mu, \ldots\}$ we may consider as in the past two chapters a problem where the constraint set $U(i)$ is replaced by the set $\tilde{U}(i) = \{\mu(i)\}$, i.e., $\tilde{U}(i)$ contains a single element, the control $\mu(i)$. Since for the resulting problem there is only one admissible policy, the policy $\{\mu, \mu, \ldots\}$, application of Proposition 1 yields the following corollary.

**Corollary 1.1** Let $\pi = \{\mu, \mu, \ldots\}$ be an admissible stationary policy. Assume that there exists a function $h_\mu : S \to R$ and a constant $\lambda_\mu$ such that

$$\lambda_\mu + h_\mu(i) = \bar{g}[i, \mu(i)] + \sum_{j=1}^{n} p_{ij}[\mu(i)] h_\mu(j) \qquad \forall i = 1, \ldots, n.$$

Then the value of the cost functional (3) corresponding to $\pi$ is the same for every initial state and is given by

$$J_\mu(i) = \lambda_\mu \qquad \forall i = 1, 2, \ldots, n.$$

We now turn to obtaining conditions that guarantee the existence of $\lambda$ and $h$ satisfying (16). At the same time we shall be able to establish a connection with the discounted cost problem of Chapter 6.

Consider the discounted cost functional

$$\lim_{N \to \infty} E\left\{\sum_{k=0}^{N-1} \alpha^k \bar{g}[x_k, \mu_k(x_k)]\right\}, \qquad 0 < \alpha < 1.$$

Let us denote by $J_\alpha(i)$ the optimal value of this cost functional corresponding to $\alpha$ and the initial state $i \in S$. We have from the results of Chapter 6 that $J_\alpha(\cdot)$ is the unique solution of the optimality equation

$$J_\alpha(i) = \min_{u \in U(i)} \left[\bar{g}(i, u) + \alpha \sum_{j=1}^{n} p_{ij}(u) J_\alpha(j)\right], \qquad i = 1, \ldots, n. \tag{20}$$

Let $s$ be an arbitrary state in $S$ and let us define

$$h_\alpha(i) = J_\alpha(i) - J_\alpha(s), \qquad i = 1, \ldots, n. \tag{21}$$

We have, by using (21) to eliminate $J_\alpha(i)$ from (20),

$$\begin{aligned} h_\alpha(i) + J_\alpha(s) &= \min_{u \in U(i)} \left[\bar{g}(i, u) + \alpha \sum_{j=1}^{n} p_{ij}(u)[h_\alpha(j) + J_\alpha(s)]\right] \\ &= \alpha J_\alpha(s) + \min_{u \in U(i)} \left[\bar{g}(i, u) + \alpha \sum_{j=1}^{n} p_{ij}(u) h_\alpha(j)\right] \end{aligned}$$

from which

$$(1 - \alpha) J_\alpha(s) + h_\alpha(i) = \min_{u \in U(i)} \left[\bar{g}(i, u) + \alpha \sum_{j=1}^{n} p_{ij}(u) h_\alpha(j)\right]. \tag{22}$$

It follows from (22) and the finiteness of $S$ and $C$ that if for some sequence $\{\alpha_m\}$ with $0 < \alpha_m < 1$ and $\alpha_m \to 1$ we have

$$\lim_{m\to\infty} (1 - \alpha_m)J_{\alpha_m}(s) = \lambda, \tag{23}$$

$$\lim_{m\to\infty} h_{\alpha_m}(i) = h(i), \qquad i = 1, \ldots, n \tag{24}$$

(i.e., the limits above exist), then the constant $\lambda$ and the function $h: S \to R$ defined by (23) and (24) satisfy

$$\lambda + h(i) = \min_{u\in U(i)} \left[ \bar{g}(i, u) + \sum_{j=1}^{n} p_{ij}(u)h(j) \right], \qquad i = 1, \ldots, n,$$

and condition (16) is satisfied. The following proposition states that a sufficient condition for existence of a sequence $\{\alpha_m\}$ such that the limits in (23) and (24) exist is that the differences $[J_\alpha(i) - J_\alpha(s)]$ are uniformly bounded over $\alpha$.

**Proposition 2** Assume that there exists a constant $L$ such that for some state $s \in S$ we have

$$|J_\alpha(i) - J_\alpha(s)| \leq L \qquad \forall i \in S, \quad \alpha \in (0, 1). \tag{25}$$

Then:

(a) There exists a constant $\lambda$ and a function $h: S \to R$ satisfying (16).
(b) For some sequence $\alpha_m \to 1$ we have

$$h(i) = \lim_{m\to\infty} [J_{\alpha_m}(i) - J_{\alpha_m}(s)], \qquad i = 1, \ldots, n.$$

(c) $\lim_{\alpha\to 1} (1 - \alpha)J_\alpha(i) = \lambda$, $\forall i = 1, \ldots, n$.

*Proof* Let $\{\alpha_k\}$ be any sequence such that $\alpha_k \to 1$. By (25) the sequences $\{J_{\alpha_k}(i) - J_{\alpha_k}(s)\}$ are bounded. Hence there exists a subsequence of $\{\alpha_k\}$, say $\{\alpha_m\}$, such that $\{J_{\alpha_m}(i) - J_{\alpha_m}(s)\}$ converges to a limit $h(i)$ for each $i \in S$, and part (b) is proved. Now by Proposition 6.1 and finiteness of the state space and control space we have

$$|J\alpha_m(s)| \leq M(1 - \alpha_m)^{-1},$$

where $M$ is some constant. Hence the sequence $\{(1 - \alpha_m)|J_{\alpha_m}(s)|\}$ is bounded. Thus there exists a subsequence of $\{\alpha_m\}$, say $\{\alpha_{m'}\}$ such that

$$(1 - \alpha_{m'})J_{\alpha_{m'}}(s) \to \lambda.$$

From (22) we have

$$(1 - \alpha_{m'})J_{\alpha_{m'}}(s) + h_{\alpha_{m'}}(i) = \min_{u\in U(i)} \left[ \bar{g}(i, u) + \alpha_{m'} \sum_{j=1}^{n} p_{ij}(u)h_{\alpha_{m'}}(j) \right].$$

Taking the limit above and interchanging limit and minimization [using the finiteness of $U(i)$] we obtain

$$\lambda + h(i) = \min_{u \in U(i)} \left[ \bar{g}(i, u) + \sum_{j=1}^{n} p_{ij}(u) h(j) \right],$$

and part (a) is proved.

To prove (c) note that by the proof of (a) and (b) for any sequence $\alpha_k \to 1$ and every subsequence of $\{\alpha_k\}$, say $\{\alpha_m\}$, such that $\lim_{m\to\infty} (1 - \alpha_m) J_{\alpha_m}(s)$ exists, we have that $\{(\Delta - \alpha_k) J_{\alpha_k}(s)\}$ is bounded and

$$\lim_{m\to\infty} (1 - \alpha_m) J_{\alpha_m}(s) = \lambda,$$

where $\lambda$ is the optimal value of the problem [by part (a) and Proposition 1]. It follows that $\lim_{k\to\infty} (1 - \alpha_k) J_{\alpha_k}(s) = \lambda$ and hence $\lim_{\alpha\to 1} (1 - \alpha) J_\alpha(s) = \lambda$. Now if $|J_\alpha(i) - J_\alpha(s)|$ is uniformly bounded for some $s$ it is also uniformly bounded for every $s \in S$ and hence by repeating the proof for every $s \in S$ we obtain $\lim_{\alpha\to 1} (1 - \alpha) J_\alpha(i) = \lambda$ for all $i \in S$. Q.E.D.

We are now ready to state and prove the following proposition, which combined with Proposition 1 provides one of the main results of this section.

Recall that to every admissible stationary policy $\{\mu, \mu, \ldots\}$ there corresponds a transition probability matrix $P_\mu$ defined by (4). For any positive integer $m$ we shall denote by $p_{ij}^m(\mu)$ the element in the $i$th row and $j$th column of the matrix $P_\mu^m$ ($P_\mu$ raised to the $m$th power):

$$P_\mu^m = [p_{ij}^m(\mu)].$$

The scalar $p_{ij}^m(\mu)$ is the probability that the state will be $j$ after $m$ stages when the initial state is $i$ and the stationary policy $\pi = \{\mu, \mu, \ldots\}$ is used:

$$p_{ij}^m(\mu) = P(x_m = j \mid x_0 = i, \pi).$$

For any two states $i, s \in S$ let us denote by $K_{is}(\mu)$ the smallest index $k$ for which $x_k = s$ when $x_0 = i$ and the stationary policy $\pi = \{\mu, \mu, \ldots\}$ is used:

$$K_{is}(\mu) = \inf\{k \mid x_k = s, x_0 = i, x_j \neq s \text{ for } 1 \leqslant j < k\}.$$

We call $K_{is}(\mu)$ the *first passage time* from $i$ to $s$ associated with $\mu$. For each $i$, $s$, and $\mu$, $K_{is}(\mu)$ may be viewed as a random variable taking positive integer values or the value $+\infty$ with probabilities

$$q_{is}^k = P(K_{is}(\mu) = k) = P(x_k = s, x_j \neq s, 1 \leqslant j < k \mid x_0 = i, \pi),$$

$$P(K_{is}(\mu) = \infty) = 1 - \sum_{k=1}^{\infty} q_{is}^k.$$

We define the *mean first passage time* $E\{K_{is}(\mu)\}$ associated with $\mu$ by

$$E\{K_{is}(\mu)\} = \begin{cases} \sum_{k=1}^{\infty} k q_{is}^k & \text{if } \sum_{k=1}^{\infty} q_{is}^k = 1, \\ +\infty & \text{otherwise.} \end{cases}$$

One may show (see Appendix D) that for any given state $s \in S$,

$$E\{K_{is}(\mu)\} < \infty \quad \forall i \in S \Rightarrow p_{is}^{m_i}(\mu) > 0 \quad \text{for some } m_i \geqslant 1 \quad \forall i \in S, \tag{26}$$

where $p_{is}^{m_i}$ was defined above. Furthermore, if the stochastic matrix $P_\mu$ corresponds to an irreducible Markov chain (see Appendix D), then we have $E\{K_{ij}(\mu)\} < \infty$ for all $i, j \in S$.

**Proposition 3** Suppose that there exists a state $s \in S$ such that for every admissible stationary policy $\pi = \{\mu, \mu, \ldots\}$ and every state $i \in S$ we have

$$E\{K_{is}(\mu)\} < \infty. \tag{27}$$

Then there exists a constant $\lambda$ and a function $h : S \to R$ satisfying (16), i.e.,

$$\lambda + h(i) = \min_{u \in U(i)} \left[ \bar{g}(i, u) + \sum_{j=1}^{n} p_{ij}(u) h(j) \right], \qquad i = 1, \ldots, n.$$

*Proof* We assume without loss of generality that

$$0 \leqslant \bar{g}(i, u) \leqslant M \qquad \forall i \in S, \quad u \in C, \tag{28}$$

where $M$ is a constant. This is true since, by the finiteness of $S$ and $C$, $\bar{g}(i, u)$ is bounded, and furthermore the addition of a constant to $\bar{g}(i, u)$ merely adds the same constant to the cost functional (3) for every admissible policy. Let $\alpha$ be any discount factor, $0 < \alpha < 1$, and $\{\mu_\alpha, \mu_\alpha, \ldots\}$ a policy that minimizes the corresponding discounted cost. We have, for every $i \in S$,

$$J_\alpha(i) = E\left\{ \sum_{k=0}^{K_{is}(\mu_\alpha)-1} \alpha^k \bar{g}[x_k, \mu_\alpha(x_k)] + \sum_{k=K_{is}(\mu_\alpha)}^{\infty} \alpha^k \bar{g}[x_k, \mu_\alpha(x_k)] \,|\, x_0 = i \right\}. \tag{29}$$

By (28), the first term on the right is less than or equal to $M\ E\{K_{is}(\mu_\alpha)\}$. The second term is equal to

$$E\{\alpha^{K_{is}(\mu_\alpha)}\}\, E\left\{ \sum_{k=K_{is}(\mu_\alpha)}^{\infty} \alpha^{k-K_{is}(\mu_\alpha)} \bar{g}[x_k, \mu_\alpha(x_k)] \,|\, x_{K_{is}(\mu_\alpha)} = s \right\} = E\{\alpha^{K_{is}(\mu_\alpha)}\} J_\alpha(s),$$

which is in turn less than or equal to $J_\alpha(s)$. Hence if $Q$ is an integer such that $E\{K_{is}(\mu)\} \leqslant Q$ for all $i$ and $\mu$, we obtain

$$J_\alpha(i) \leqslant M\ E\{K_{is}(\mu_\alpha)\} + J_\alpha(s) \leqslant MQ + J_\alpha(s),$$

or

$$J_\alpha(i) - J_\alpha(s) \leqslant MQ. \tag{30}$$

Also by (29),

$$J_\alpha(i) \geqslant J_\alpha(s)\, E\{\alpha^{K_{is}(\mu_\alpha)}\},$$

or equivalently

$$J_\alpha(s) - J_\alpha(i) \leqslant [1 - E\{\alpha^{K_{is}(\mu_\alpha)}\}]J_\alpha(s). \tag{31}$$

By Proposition 1 in Chapter 6 we have

$$J_\alpha(s) \leqslant M/(1-\alpha), \tag{32}$$

while using the fact that $0 < \alpha < 1$, we have

$$E\{\alpha^{K_{is}(\mu_\alpha)}\} \geqslant \alpha^{E\{K_{is}(\mu_\alpha)\}} \geqslant \alpha^Q, \tag{33}$$

where $Q$ is the integer for which $E\{K_{is}(\mu)\} \leqslant Q$. From (31)–(33) we obtain

$$J_\alpha(s) - J_\alpha(i) \leqslant [(1-\alpha^Q)/(1-\alpha)]M = \sum_{i=0}^{Q-1} \alpha^i M \leqslant QM. \tag{34}$$

Combining (30) and (34),

$$|J_\alpha(i) - J_\alpha(s)| \leqslant MQ \qquad \forall i \in S, \quad \alpha \in (0, 1)$$

and the result follows by Proposition 2. Q.E.D.

Condition (27) is satisfied in particular *if every stationary policy gives rise to an irreducible Markov chain* (see Appendix D). It is to be noted that from the preceding proof it is evident that it is sufficient that (27) holds only for stationary policies that minimize the expected discounted cost for some discount factor. It is possible to obtain other conditions that guarantee the existence of $\lambda$ and $h$ such that (16) holds. One such condition is the following.

**Weak Accessibility Condition** For any two states $i, j \in S$ there exists an admissible stationary policy $\pi = \{\mu, \mu, \ldots\}$ and an integer $m$ such that

$$p_{ij}^m(\mu) = P(x_m = j \mid x_0 = i, \pi) > 0. \tag{35}$$

**Proposition 4** Suppose that the weak accessibility condition holds. Then there exists a constant $\lambda$ and a function $h: S \to R$ satisfying (16), i.e.,

$$\lambda + h(i) = \min_{u \in U(i)} \left[ \bar{g}(i, u) + \sum_{j=1}^{n} p_{ij}(u) h(j) \right], \qquad i = 1, \ldots, n.$$

*Proof* See the Appendix to this chapter.

Combination of Propositions 1 and 3 or Propositions 1 and 4 yields:

**Theorem** Under the assumption of Proposition 3 or under the weak accessibility condition there exists a constant $\lambda$ and a function $h: S \to R$ such that

$$\lambda + h(i) = \min_{u \in U(i)} \left[ \bar{g}(i, u) + \sum_{j=1}^{n} p_{ij}(u) h(j) \right], \qquad i = 1, \ldots, n. \qquad (16')$$

Furthermore, the optimal value of the cost functional $J_\pi(i)$ of (3) is equal to $\lambda$ for every $i = 1, \ldots, n$, i.e.,

$$\lambda = \inf_\pi J_\pi(i), \qquad i = 1, \ldots, n.$$

In addition if $\mu^*(i)$ attains the minimum in the right-hand side of (16′) for every $i = 1, \ldots, n$, then the stationary policy $\pi^* = \{\mu^*, \mu^*, \ldots\}$ is optimal.

The conditions listed above are probably the weakest known that guarantee that the optimal average cost per stage is independent of the initial state. It is clear, of course, that some sort of accessibility condition must be satisfied by the transition probability matrices corresponding to stationary policies or at least to optimal stationary policies. For if there existed two states none of which could be reached from the other no matter which policy we use, then it can be only by accident that the same optimal cost per stage will correspond to each one. An extreme example of this type of situation is to consider a problem where the state is forced to stay the same regardless of the control applied, i.e., each state is absorbing. Then the optimal average cost per stage for each state $i$ will be $\min_{u \in U(i)} \bar{g}(i, u)$ and this cost may be different for different states.

Finally, we state the following corollary of Proposition 3, which is obtained in the same way as Corollary 1.1.

**Corollary 3.1** Let $\pi = \{\mu, \mu, \ldots\}$ be an admissible stationary policy and assume that there exists a state $s \in S$ such that $E\{K_{is}(\mu)\} < \infty$ for all $i \in S$. Then there exists a constant $\lambda_\mu$ and a function $h_\mu: S \to R$ such that

$$J_\mu(i) = \lambda_\mu, \qquad i = 1, \ldots, n, \qquad (36)$$

and furthermore

$$\lambda_\mu + h_\mu(i) = \bar{g}[i, \mu(i)] + \sum_{j=1}^{n} p_{ij}[\mu(i)] h_\mu(j), \qquad i = 1, 2, \ldots, n. \qquad (37)$$

Equation (37) represents a system of $n$ linear equations with $(n + 1)$ unknowns—the scalars $\lambda_\mu, h_\mu(1), h_\mu(2), \ldots, h_\mu(n)$. We may add one additional equation to this system by requiring that

$$h_\mu(s) = 0. \qquad (38)$$

This can be done since if $\{\lambda_\mu, h_\mu(1), \ldots, h_\mu(n)\}$ is a solution of (37) so is $\{\lambda_\mu, h_\mu(1) + r, \ldots, h_\mu(n) + r\}$, where $r$ is any scalar. Corollary 3.1 states that under the assumption $E\{K_{is}(\mu)\} < \infty$, system (37) and (38) has at least one solution. We now show that the same assumption guarantees that the system of equations (37) and (38) has a unique solution.

**Proposition 5** For any admissible stationary policy $\pi = \{\mu, \mu, \ldots\}$ for which there exists an $s \in S$ such that $E\{K_{is}(\mu)\} < \infty$ for all $i \in S$, the system of equations (37) and (38) has a unique solution.

*Proof* Let $\{\lambda, h(1), \ldots, h(n)\}$ and $\{\lambda', h'(1), \ldots, h'(n)\}$ be two solutions. We have $\lambda = \lambda' = \lambda_\mu$ by Corollary 1.1. Hence from (37) we obtain for every $m \geqslant 1$,

$$h - h' = P_\mu(h - h') = P_\mu^m(h - h'),$$

or equivalently

$$h(i) - h'(i) = \sum_{j=1}^{n} p_{ij}^m(\mu)[h(j) - h'(j)] \qquad \forall i = 1, \ldots, n.$$

From (26) and for a fixed $i$ we obtain for some $m_i \geqslant 1$ and $\varepsilon_i > 0$,

$$p_{is}^{m_i}(\mu) \geqslant \varepsilon_i > 0$$

and from (38), $h(s) - h'(s) = 0$. Hence

$$\begin{aligned} |h(i) - h'(i)| &\leqslant \sum_{j=1}^{n} p_{ij}^{m_i}(\mu)|h(j) - h'(j)| \\ &= \sum_{j \neq s} p_{ij}^{m_i}(\mu)|h(j) - h'(j)| \\ &\leqslant (1 - \varepsilon_i) \max_j |h(j) - h'(j)|. \end{aligned}$$

Thus we obtain

$$\max_j |h(j) - h'(j)| \leqslant (1 - \varepsilon) \max_j |h(j) - h'(j)|,$$

where

$$\varepsilon = \min[\varepsilon_1, \ldots, \varepsilon_n] > 0.$$

Hence $h(j) = h'(j)$ for all $j$. Q.E.D.

We close this section with an example.

Machine replacement example Consider a machine that can be in any one of $n$ states, $S = \{1, 2, \ldots, n\}$. The implication here is that state $i$ is better than state $i + 1$, $i = 1, 2, \ldots, n - 1$, and state 1 corresponds to a machine in perfect condition. The operating cost per unit time for which the machine starts in state $i$ is denoted $g_i$, and we assume

$$0 \leqslant g_1 \leqslant g_2 \leqslant \cdots \leqslant g_n. \tag{39}$$

During a time period of operation the transition probabilities satisfy

$$p_{ij} = 0 \qquad \text{if} \quad j < i, \tag{40}$$

$$p_{ii} < 1, \qquad i = 1, \ldots, n, \tag{41}$$

i.e., the machine cannot go to a better state with usage. We also assume that

$$i \leqslant i' \Rightarrow \sum_{j=k}^{n} p_{ij} \leqslant \sum_{j=k}^{n} p_{i'j} \qquad \forall k = 1, 2, \ldots, n. \tag{42}$$

At the beginning of each period the state of the machine is determined and a decision is made whether to replace the machine at a cost $R > 0$ with a new machine that is in state 1 or to continue operation. Thus there are two possible controls—replace and do not replace. The problem is to find a policy that minimizes the average cost per period.

It is to be noted that the hypothesis of Proposition 3 is not satisfied for this problem. Indeed, consider the policy that never replaces. Then assumptions (40) and (41) imply that for this policy the only state that can be reached from every other state is the state $n$. Consider also the policy that replaces the machine at every state. Then, assuming $p_{1n} = 0$, state $n$ cannot be reached from any state $i \neq n$. Notice also that one cannot guarantee in the absence of further assumptions that the weak accessibility condition stated after Proposition 3 is satisfied. We shall be able, however, to argue in terms of Proposition 2.

Consider the corresponding discounted problem with a discount factor $\alpha < 1$. Then we have

$$J_\alpha(i) = \min\left[ R + g_1 + \alpha \sum_{j=1}^{n} p_{1j} J_\alpha(j),\ g_i + \alpha \sum_{j=1}^{n} p_{ij} J_\alpha(j) \right], \qquad i = 1, 2, \ldots, n.$$

It follows that

$$J_\alpha(i) - J_\alpha(1) = \min\left[ R,\ (g_i - g_1) + \alpha \sum_{j=1}^{n} (p_{ij} - p_{1j}) J_\alpha(j) \right] \leqslant R. \tag{43}$$

It is possible to show (as in the second example of Section 6.1) that in view of (39)–(42), we have for all $\alpha \in (0, 1)$,

$$0 \leqslant J_\alpha(i) - J_\alpha(1), \qquad i = 1, 2, \ldots, n.$$

Furthermore $J_\alpha(i) - J_\alpha(1)$ is nondecreasing in $i$. Hence by Proposition 2 there exists a scalar $\lambda$ and a nondecreasing function $h(i)$, $i = 1, \ldots, n$, such that

$$\lambda + h(i) = \min\left[R + g_1 + \sum_{j=1}^{n} p_{1j}h(j),\ g_i + \sum_{j=1}^{n} p_{ij}h(j)\right], \qquad i = 1, 2, \ldots, n,$$

and the policy that chooses the minimizing action above is average cost optimal. Let

$$i^* = \max\left\{i \mid g_i + \sum_{j=1}^{n} p_{ij}h(j) \leqslant R + g_1 + \sum_{j=1}^{n} p_{1j}h(j)\right\}.$$

Then the policy that replaces if the current state is greater than $i^*$ and does not replace otherwise is optimal.

## 8.2 Successive Approximation

Since, as seen in Chapter 6, the method of successive approximation may be used for computing the optimal discounted cost function $J_\alpha$ of (20) and furthermore under the assumption of Proposition 2 we have

$$\lim_{\alpha \to 1}(1 - \alpha)J_\alpha(i) = \lambda, \qquad i = 1, 2, \ldots, n,$$

one expects that a limiting form (as $\alpha \to 1$) of the successive approximation method may be used for the average cost problem. Indeed, this is the case under an assumption that we shall introduce shortly. Prior to proceeding with precise formulations and results let us provide a heuristic discussion that indicates the appropriate limiting form of the successive approximation method.

Let $J: S \to R$ be any function on $S$, $\alpha \in (0, 1)$ be a discount factor, and consider the following mapping $T_\alpha$, which is familiar from Chapter 6:

$$T_\alpha(J)(i) = \min_{u \in U(i)}\left[\bar{g}(i, u) + \alpha \sum_{j=1}^{n} p_{ij}(u)J(j)\right], \qquad i = 1, \ldots, n.$$

As discussed in Section 6.2, the successive approximation method for the $\alpha$-discounted problem consists of the iteration

$$T_\alpha^{k+1}(J)(i) = \min_{u \in U(i)}\left[\bar{g}(i, u) + \alpha \sum_{j=1}^{n} p_{ij}(u)T_\alpha^k(J)(j)\right], \qquad i = 1, \ldots, n. \quad (44)$$

We have

$$\lim_{k \to \infty} T_\alpha^k(J)(i) = J_\alpha(i), \qquad i = 1, \ldots, m,$$

for an arbitrary function $J: S \to R$, where $J_\alpha$ is the $\alpha$-discounted optimal value function. Since we may have $T_\alpha^k(J)(i) \to \infty$ as $k \to \infty$ and $\alpha \to 1$ we shall rewrite (44) in terms of quantities that have finite limits as $k \to \infty$ and $\alpha \to 1$.

For any fixed state $s \in S$ let us denote, for all $i$ and $k$,

$$h_{\alpha,k}(i) = T_\alpha^k(J)(i) - T_\alpha^k(J)(s). \tag{45}$$

Using the notation above we may write (44) as

$$[T_\alpha^{k+1}(J)(s) - \alpha T_\alpha^k(J)(s)] + h_{\alpha,k+1}(i) = \min_{u \in U(i)} \left[ \bar{g}(i,u) + \alpha \sum_{j=1}^{n} p_{ij}(u) h_{\alpha,k}(j) \right]. \tag{46}$$

Let us denote, for all $i = 1, 2, \ldots, n$ and $k$,

$$H_{\alpha,k}(i) = h_{\alpha,k}(i) + [T_\alpha^k(J)(s) - \alpha T_\alpha^{k-1}(J)(s)], \tag{47}$$

From (45) and (47) we have

$$h_{\alpha,k}(i) = H_{\alpha,k}(i) - H_{\alpha,k}(s), \qquad i = 1, 2, \ldots, n. \tag{48}$$

Now we may write (46) [or, equivalently, (44)] as

$$H_{\alpha,k+1}(i) = \min_{u \in U(i)} \left[ \bar{g}(i,u) + \alpha \sum_{j=1}^{n} p_{ij}(u) h_{\alpha,k}(j) \right], \tag{49}$$

with $h_{\alpha,k}(i)$ defined by

$$h_{\alpha,k}(i) = H_{\alpha,k}(i) - H_{\alpha,k}(s). \tag{50}$$

Algorithm (49) and (50) with a starting function $h_{\alpha,0}: S \to R$ satisfying $h_{\alpha,0}(s) = 0$ may be viewed as an alternative implementation of the successive approximation algorithm (44) and may equally well be used for solution of the $\alpha$-discounted problem. In the limit algorithm (49) and (50) will yield functions $h_\alpha$ and $H_\alpha$ via the relations

$$h_\alpha(i) = \lim_{k \to \infty} h_{\alpha,k}(i), \qquad i = 1, 2, \ldots, n, \tag{51}$$

$$H_\alpha(i) = \lim_{k \to \infty} H_{\alpha,k}(i), \qquad i = 1, 2, \ldots, n. \tag{52}$$

In addition we will have [cf. (45) and (47)]

$$h_\alpha(i) = J_\alpha(i) - J_\alpha(s), \qquad i = 1, 2, \ldots, n, \tag{53}$$

$$H_\alpha(s) = (1 - \alpha) J_\alpha(s), \tag{54}$$

from which the optimal values $J_\alpha(1), J_\alpha(2), \ldots, J_\alpha(n)$ may be recovered.

Now if we formally take the limit as $\alpha \to 1$ in algorithm (49) and (50) and denote

$$h_k(i) = \lim_{\alpha \to 1} h_{\alpha, k}(i), \tag{55}$$

$$H_k(i) = \lim_{\alpha \to 1} H_{\alpha, k}(i), \tag{56}$$

we obtain the algorithm

$$H_{k+1}(i) = \min_{u \in U(i)} \left[ \bar{g}(i, u) + \sum_{j=1}^{n} p_{ij}(u) h_k(j) \right], \tag{57}$$

$$h_k(i) = H_k(i) - H_k(s), \tag{58}$$

where $h_0 : S \to R$ is a function with $h_0(s) = 0$. Let us suppose for a moment that the algorithm yields functions $H$ and $h$ via

$$h(i) = \lim_{k \to \infty} h_k(i), \tag{59}$$

$$H(i) = \lim_{k \to \infty} H_k(i). \tag{60}$$

Then assuming that the limit with respect to $\alpha$ and $k$ may be interchanged,

$$\lim_{\alpha \to 1} h_\alpha(i) = \lim_{\alpha \to 1} \lim_{k \to \infty} h_{\alpha, k}(i) = \lim_{k \to \infty} \lim_{\alpha \to 1} h_{\alpha, k}(i) = \lim_{k \to \infty} h_k(i),$$

$$\lim_{\alpha \to 1} H_\alpha(i) = \lim_{\alpha \to 1} \lim_{k \to \infty} H_{\alpha, k}(i) = \lim_{k \to \infty} \lim_{\alpha \to 1} H_{\alpha, k}(i) = \lim_{k \to \infty} H_k(i),$$

we obtain from (53), (54), (59), and (60):

$$h(i) = \lim_{\alpha \to 1} [J_\alpha(i) - J_\alpha(s)], \qquad i = 1, 2, \ldots, n,$$

$$H(s) = \lim_{\alpha \to 1} (1 - \alpha) J_\alpha(s), \qquad i = 1, 2, \ldots, n.$$

However, by Proposition 2 these imply, under the corresponding assumptions, that the quantity $H(s)$ obtained from algorithm (57) and (58) is equal to the optimal average cost per stage of the problem and the function $h$ enters in the optimality equation (16).

The conclusion from this informal discussion is that algorithm (57) and (58) is the natural candidate as a successive approximation method for the problem of this chapter and, under appropriate assumptions, should yield in the limit the optimal average cost per stage.

The validity of this conjecture is established in the following proposition. In fact, we need not require that $h_0(s) = 0$ since by (58) we will have $h_k(s) = 0$ for all $k \geqslant 1$ even if $h_0(s) \neq 0$.

**Proposition 6** Assume that there exists an $\varepsilon > 0$ and a positive integer $m$ such that for some state $s \in S$ and all admissible policies $\pi = \{\mu_0, \mu_1, \ldots\}$ we have

$$[P_{\mu_m} P_{\mu_{m-1}} \cdots P_{\mu_0}]_{is} \geqslant \varepsilon > 0 \qquad \forall i = 1, 2, \ldots, n, \tag{61}$$

where $P_{\mu_k}$, $k = 0, \ldots, m$, denotes the transition probability matrix corresponding to $\mu_k$ as in (4) and $[P_{\mu_m} P_{\mu_{m-1}} \cdots P_{\mu_0}]_{is}$ denotes the element in the $i$th row and $s$th column of the matrix $P_{\mu_m} \cdots P_{\mu_0}$. Consider the algorithm

$$H_{k+1}(i) = \min_{u \in U(i)} \left[ \bar{g}(i, u) + \sum_{j=1}^{n} p_{ij}(u) h_k(j) \right], \qquad i = 1, 2, \ldots, n, \quad k = 0, 1, \ldots, \tag{62}$$

$$h_{k+1}(i) = H_{k+1}(i) - H_{k+1}(s), \qquad i = 1, 2, \ldots, n, \quad k = 0, 1, \ldots, \tag{63}$$

where $h_0 : S \to R$ is an arbitrary function. Then the limits

$$h(i) = \lim_{k \to \infty} h_k(i), \qquad i = 1, 2, \ldots, n, \tag{64}$$

$$H(i) = \lim_{k \to \infty} H_k(i), \qquad i = 1, 2, \ldots, n, \tag{65}$$

exist and we have

$$H(s) = \lambda = \inf_{\pi} J_\pi(i), \qquad i = 1, 2, \ldots, n, \tag{66}$$

i.e., $\lambda$ is the optimal average cost per stage of the problem. In addition, $\lambda$ and $h$ satisfy

$$\lambda + h(i) = \min_{u \in U(i)} \left[ \bar{g}(i, u) + \sum_{j=1}^{n} p_{ij}(u) h(j) \right]. \tag{67}$$

*Proof* Let $\mu_k(i) \in U(i)$ attain the minimum in (62) for every $k$ and $i$. Denote for all $k$

$$H_k = \begin{bmatrix} H_k(1) \\ \vdots \\ H_k(n) \end{bmatrix} \qquad h_k = \begin{bmatrix} h_k(1) \\ \vdots \\ h_k(n) \end{bmatrix} \qquad g_{\mu_k} = \begin{bmatrix} \bar{g}[1, \mu_k(1)] \\ \vdots \\ \bar{g}[n, \mu_k(n)] \end{bmatrix}$$

We have

$$H_{k+1} = g_{\mu_k} + P_{\mu_k} h_k \leqslant g_{\mu_{k-1}} + P_{\mu_{k-1}} h_k,$$
$$H_k = g_{\mu_{k-1}} + P_{\mu_{k-1}} h_{k-1} \leqslant g_{\mu_k} + P_{\mu_k} h_{k-1},$$

and hence

$$h_{k+1} = g_{\mu_k} + P_{\mu_k} h_k - H_{k+1}(s)e \leqslant g_{\mu_{k-1}} + P_{\mu_{k-1}} h_k - H_{k+1}(s)e,$$
$$h_k = g_{\mu_{k-1}} + P_{\mu_{k-1}} h_{k-1} - H_k(s)e \leqslant g_{\mu_k} + P_{\mu_k} h_{k-1} - H_k(s)e,$$

where $e$ is the unit vector $e = [1, 1, \ldots, 1]'$. From the relations above we obtain

$$P_{\mu_k}(h_k - h_{k-1}) - [H_{k+1}(s) - H_k(s)]e \leqslant h_{k+1} - h_k$$
$$\leqslant P_{\mu_{k-1}}(h_k - h_{k-1}) - [H_{k+1}(s) - H_k(s)]e.$$

Since this relation holds for every $k \geqslant 1$, by iterating we obtain

$$\begin{aligned} &P_{\mu_k} P_{\mu_{k-1}} \cdots P_{\mu_{k-m}}(h_{k-m} - h_{k-m-1}) - [H_{k+1}(s) - H_{k-m}(s)]e \\ &\quad \leqslant h_{k+1} - h_k \\ &\quad \leqslant P_{\mu_{k-1}} P_{\mu_{k-2}} \cdots P_{\mu_{k-m-1}}(h_{k-m} - h_{k-m-1}) - [H_{k+1}(s) - H_{k-m}(s)]e \end{aligned} \tag{68}$$

Write

$$q_k = h_{k+1} - h_k.$$

Then (68) yields

$$q_k(i) \leqslant \sum_{j=1}^{n} [P_{\mu_{k-1}} P_{\mu_{k-2}} \cdots P_{\mu_{k-m-1}}]_{ij} q_{k-m-1}(j) - [H_{k+1}(s) - H_{k-m}(s)]. \tag{69}$$

Now we have for all $k$ by hypothesis $[P_{\mu_{k-1}} \cdots P_{\mu_{k-m-1}}]_{is} \geqslant \varepsilon > 0$, and by (63), $q_k(s) = 0$. Hence from (69)

$$\max_j q_k(j) \leqslant (1 - \varepsilon) \max_j q_{k-m-1}(j) - [H_{k+1}(s) - H_{k-m}(s)]. \tag{70}$$

Using a similar argument, from (68) we also obtain

$$\min_j q_k(j) \geqslant (1 - \varepsilon) \min_j q_{k-m-1}(j) - [H_{k+1}(s) - H_{k-m}(s)]. \tag{71}$$

From (70) and (71) we have

$$\max_j q_k(j) - \min_j q_k(j) \leqslant (1 - \varepsilon)[\max_j q_{k-m-1}(j) - \min_j q_{k-m-1}(j)],$$

which implies that for all $k$ greater than some index and some $B > 0$ we have

$$\max_j q_k(j) - \min_j q_k(j) \leqslant B(1 - \varepsilon)^{k/(m+1)}.$$

Since $q_k(s) = 0$, it follows that

$$|h_{k+1}(i) - h_k(i)| = |q_k(i)| \leqslant \max_j q_k(j) - \min_j q_k(j) \leqslant B(1 - \varepsilon)^{k/(m+1)}.$$

This in turn implies that $\{h_k(i)\}$ converges to some real number $h(i)$:

$$h(i) = \lim_{k\to\infty} h_k(i), \qquad i = 1, 2, \ldots, n.$$

From (62) we see that the sequence $\{H_k\}$ also converges to a vector $H \in R^n$, and we have

$$H(i) = \min_{u\in U(i)}\left[\bar{g}(i, u) + \sum_{j=1}^{n} p_{ij}(u)h(j)\right], \tag{72}$$

as well as

$$h(i) = H(i) - H(s). \tag{73}$$

From (72) and (73) we have

$$H(s) + h(i) = \min_{u\in U(i)}\left[\bar{g}(i, u) + \sum_{j=1}^{n} p_{ij}(u)h(j)\right],$$

and by Proposition 1, $H(s) = \lambda = \inf_\pi J_\pi(i)$. Q.E.D.

As for discounted problems, one may obtain upper and lower bounds on the optimal average cost per stage $\lambda$ via the successive approximation algorithm.

**Proposition 7** Under the assumption of Proposition 6 for algorithm (62) and (63) there holds

$$H_k(s) + c_k \leqslant H_{k+1}(s) + c_{k+1} \leqslant \lambda \leqslant H_{k+1}(s) + \bar{c}_{k+1} \leqslant H_k(s) + \bar{c}_k, \tag{74}$$

where for all $k \geqslant 1$,

$$c_k = \min_i[H_{k+1}(i) - H_k(i)], \tag{75}$$

$$\bar{c}_k = \max_i[H_{k+1}(i) - H_k(i)]. \tag{76}$$

*Proof* Let $\mu_k(i)$ attain the minimum in (62) for each $k$ and $i$. We have

$$\begin{aligned}
H_{k+1}(i) - H_k(i) &= \bar{g}[i, \mu_k(i)] + \sum_{j=1}^{n} p_{ij}[\mu_k(i)][H_k(j) - H_k(s)] - H_k(i)\\
&\geqslant \bar{g}[i, \mu_k(i)] + \sum_{j=1}^{n} p_{ij}[\mu_k(i)][H_k(j) - H_k(s)]\\
&\quad - \bar{g}[i, \mu_k(i)] - \sum_{j=1}^{n} p_{ij}[\mu_k(i)][H_{k-1}(j) - H_{k-1}(s)]\\
&\geqslant \min_j[H_k(j) - H_{k-1}(j)] + H_{k-1}(s) - H_k(s).
\end{aligned}$$

Taking the minimum over $i$ we obtain

$$H_{k-1}(s) + c_{k-1} \leqslant H_k(s) + c_k.$$

A similar argument shows that for all $k$

$$H_k(s) + \bar{c}_k \leqslant H_{k-1}(s) + \bar{c}_{k-1}.$$

Since $c_k$ tends to zero by Proposition 6 we have $\{H_k(s) + c_k\} \to \lambda$ and since $\{H_k(s) + c_k\}$ is a nondecreasing sequence it follows that $H_k(s) + c_k \leqslant \lambda$. Similarly $H_k(s) + \bar{c}_k \geqslant \lambda$ and the result is proved. Q.E.D.

We now demonstrate the successive approximation algorithm and the error bounds (74) by means of an example.

EXAMPLE 1 Consider an undiscounted version of the example of Section 6.2. We have

$$S = \{1, 2\}, \qquad C = \{u^1, u^2\},$$

$$P(u^1) = \begin{bmatrix} p_{11}(u^1) & p_{12}(u^1) \\ p_{21}(u^1) & p_{22}(u^1) \end{bmatrix} = \begin{bmatrix} \frac{3}{4} & \frac{1}{4} \\ \frac{3}{4} & \frac{1}{4} \end{bmatrix}$$

$$P(u^2) = \begin{bmatrix} p_{11}(u^2) & p_{12}(u^2) \\ p_{21}(u^2) & p_{22}(u^2) \end{bmatrix} = \begin{bmatrix} \frac{1}{4} & \frac{3}{4} \\ \frac{1}{4} & \frac{3}{4} \end{bmatrix}$$

and

$$\bar{g}(1, u^1) = 2, \qquad \bar{g}(1, u^2) = 0.5, \qquad \bar{g}(2, u^1) = 1, \qquad \bar{g}(2, u^2) = 3.$$

Letting $s = 1$ be the reference state, algorithm (62) and (63) takes the form

$$H_{k+1}(i) = \min\left\{\bar{g}(i, u^1) + \sum_{j=1}^{2} p_{ij}(u^1)h_k(j),\right.$$
$$\left.\bar{g}(i, u^2) + \sum_{j=1}^{2} p_{ij}(u^2)h_k(j)\right\}, \qquad i = 1, 2,$$

$$h_{k+1}(1) = 0,$$
$$h_{k+1}(2) = H_{k+1}(2) - H_{k+1}(1),$$

and the constants $c_k$, $\bar{c}_k$ of (75) and (76) are given by

$$c_k = \min\{H_{k+1}(1) - H_k(1), H_{k+1}(2) - H_k(2)\},$$
$$\bar{c}_k = \max\{H_{k+1}(1) - H_k(1), H_{k+1}(2) - H_k(2)\}.$$

The results of the computation starting with $h_0(1) = h_0(2) = 0$ are shown in Table 8.1.

TABLE 8.1

| $k$ | $h_k(1)$ | $h_k(2)$ | $H_k(1)$ | $H_{k-1}(1) + c_{k-1}$ | $H_{k-1}(1) + \bar{c}_{k-1}$ |
|---|---|---|---|---|---|
| 0 | 0.00000 | 0.00000 | | | |
| 1 | 0.00000 | 0.50000 | 0.50000 | | |
| 2 | 0.00000 | 0.25000 | 0.87500 | 0.62500 | 0.87500 |
| 3 | 0.00000 | 0.37500 | 0.68750 | 0.68750 | 0.81250 |
| 4 | 0.00000 | 0.31250 | 0.78125 | 0.71875 | 0.78125 |
| 5 | 0.00000 | 0.34375 | 0.73438 | 0.73438 | 0.76563 |
| 6 | 0.00000 | 0.32813 | 0.75781 | 0.74219 | 0.75781 |
| 7 | 0.00000 | 0.33594 | 0.74609 | 0.74609 | 0.75391 |
| 8 | 0.00000 | 0.33203 | 0.75195 | 0.74805 | 0.75195 |
| 9 | 0.00000 | 0.33398 | 0.74902 | 0.74902 | 0.75098 |
| 10 | 0.00000 | 0.33301 | 0.75049 | 0.74951 | 0.75049 |
| 11 | 0.00000 | 0.33350 | 0.74976 | 0.74976 | 0.75024 |
| 12 | 0.00000 | 0.33325 | 0.75012 | 0.74988 | 0.75012 |
| 13 | 0.00000 | 0.33337 | 0.74994 | 0.74994 | 0.75006 |
| 14 | 0.00000 | 0.33331 | 0.75003 | 0.74997 | 0.75003 |
| 15 | 0.00000 | 0.33334 | 0.74998 | 0.74998 | 0.75002 |
| 16 | 0.00000 | 0.33333 | 0.75001 | 0.74999 | 0.75001 |
| 17 | | | 0.75000 | 0.75000 | 0.75000 |

## 8.3 Policy Iteration

The policy iteration algorithm for solving the average cost problem is similar to those described in the past two chapters. Given a stationary policy one obtains an improved policy by means of a minimization process until no further improvement is possible. We shall assume throughout that there exists a state $s \in S$ such that for every admissible stationary policy $\pi = \{\mu, \mu, \ldots\}$ we have

$$E\{K_{is}(\mu)\} < \infty \qquad \forall i = 1, \ldots, n,$$

as in Proposition 3.

Let $\pi^k = \{\mu^k, \mu^k, \ldots\}$ be an admissible stationary policy obtained at the $k$th iteration of the algorithm. We determine the average cost per stage $\lambda_{\mu^k}$ corresponding to $\pi^k$ by solving the system of $(n+1)$ equations

$$\lambda_{\mu^k} + h_{\mu^k}(i) = \bar{g}[i, \mu^k(i)] + \sum_{j=1}^{n} p_{ij}[\mu^k(i)] h_{\mu^k}(j), \qquad i = 1, \ldots, n, \tag{77}$$

$$h_{\mu^k}(s) = 0. \tag{78}$$

This system has a unique solution by Proposition 5. Subsequently we find a policy $\pi^{k+1} = \{\mu^{k+1}, \mu^{k+1}, \ldots\}$ where $\mu^{k+1}(i)$ is such that

$$\bar{g}[i, \mu^{k+1}(i)] + \sum_{j=1}^{n} p_{ij}[\mu^{k+1}(i)]h_{\mu^k}(j)$$
$$= \min_{u \in U(i)}\left[\bar{g}(i, u) + \sum_{j=1}^{n} p_{ij}(u)h_{\mu^k}(j)\right], \qquad i = 1, \ldots, n, \tag{79}$$

where we set $\mu^{k+1}(i) = \mu^k(i)$ if $\mu^k(i)$ attains the minimum above. Let $\lambda_{\mu^{k+1}}$ and $h_{\mu^{k+1}}(i)$, $i = 1, \ldots, n$, be the unique solution of the system of equations

$$\lambda_{\mu^{k+1}} + h_{\mu^{k+1}}(i) = \bar{g}[i, \mu^{k+1}(i)] + \sum_{j=1}^{n} p_{ij}[\mu^{k+1}(i)]h_{\mu^{k+1}}(j), \qquad i = 1, \ldots, n,$$
$$h_{\mu^{k+1}}(s) = 0.$$

We claim that

$$\lambda_{\mu^k} \geqslant \lambda_{\mu^{k+1}}. \tag{80}$$

Indeed, by switching to vector–matrix notation and using (77)–(80), we can write

$$\lambda_{\mu^k} e + h_{\mu^k} = g_{\mu^k} + P_{\mu^k} h_{\mu^k} \geqslant g_{\mu^{k+1}} + P_{\mu^{k+1}} h_{\mu^k}, \tag{81}$$

and

$$\lambda_{\mu^k} e \geqslant g_{\mu^{k+1}} + (P_{\mu^{k+1}} - I)h_{\mu^k},$$

where $e = [1, 1, \ldots, 1]'$ and $I$ is the $n \times n$ identity matrix. By multiplying both sides of the vector inequality by the matrix $P_{\mu^{k+1}}$ the inequality is preserved since $P_{\mu^{k+1}}$ has nonnegative elements, and we have

$$\lambda_{\mu^k} P_{\mu^{k+1}} e \geqslant P_{\mu^{k+1}} g_{\mu^{k+1}} + (P_{\mu^{k+1}}^2 - P_{\mu^{k+1}})h_{\mu^k}.$$

Since $P_{\mu^{k+1}} e = e$, we obtain

$$\lambda_{\mu^k} e \geqslant P_{\mu^{k+1}} g_{\mu^{k+1}} + (P_{\mu^{k+1}}^2 - P_{\mu^{k+1}})h_{\mu^k}.$$

Similarly we have for every $i \geqslant 0$,

$$\lambda_{\mu^k} e \geqslant P_{\mu^{k+1}}^i g_{\mu^{k+1}} + (P_{\mu^{k+1}}^{i+1} - P_{\mu^{k+1}}^i)h_{\mu^k}.$$

Hence by summing over $i$ we obtain for every $N \geqslant 1$,

$$\lambda_{\mu^k} e \geqslant (1/N)\left(\sum_{i=0}^{N-1} P_{\mu^{k+1}}^i g_{\mu^{k+1}}\right) + (1/N)(P_{\mu^{k+1}}^N - I)h_{\mu^k}. \tag{82}$$

Since $(P_{\mu^{k+1}}^N - I)h_{\mu^k}$ is bounded we have $\lim_{N\to\infty}(1/N)(P_{\mu^{k+1}}^N - I)h_{\mu^k} = 0$ and furthermore

$$\lim_{N\to\infty} (1/N)\left(\sum_{i=0}^{N-1} P_{\mu^{k+1}}^i g_{\mu^{k+1}}\right) = \lambda_{\mu^{k+1}} e,$$

by (6). Hence from (82) we obtain

$$\lambda_{\mu^k} e \geqslant \lim_{N\to\infty} (1/N)\left(\sum_{i=0}^{N-1} P^i_{\mu^{k+1}} g_{\mu^{k+1}}\right) = \lambda_{\mu^{k+1}} e,$$

and (80) is proved.

Thus the policy $\{\mu^{k+1}, \mu^{k+1}, \ldots\}$ obtained via (79) is as good or better than the policy $\{\mu^k, \mu^k, \ldots\}$. Now consider the generated sequence $\{\lambda_{\mu^k}\}$. Since it is a nonincreasing sequence and furthermore the set of all stationary policies is finite, we must have for some $\lambda$ and some index $\bar{k}$,

$$\lambda_{\mu^k} = \lambda \qquad \forall k \geqslant \bar{k}. \tag{83}$$

If, in addition, we have for some policy $\pi^* = \{\mu^*, \mu^*, \ldots\}$,

$$\mu^k = \mu^* \qquad \forall k \geqslant \bar{k}, \tag{84}$$

then $\pi^*$ must be optimal in view of construction (77) and (79) and Proposition 1. Relation (84) can be guaranteed *if different policies have different average costs per stage associated with them.* Another assumption that, as we prove below, guarantees (84) is when for all $k \geqslant \bar{k}$ the matrices

$$P^*_{\mu^k} = \lim_{N\to\infty} (1/N) \sum_{i=0}^{N-1} P^i_{\mu^k}$$

of (7) have identical rows each element of which is positive. This occurs if under each transition probability matrix $P_{\mu^k}$ the resulting Markov chain is irreducible (see Appendix D). Indeed under these circumstances if $\beta_k$ is the (row) vector consisting of the row elements of $P^*_{\mu^k}$,

$$P^*_{\mu^k} = \begin{bmatrix} \beta_k \\ \vdots \\ \beta_k \end{bmatrix}$$

then from (81) we have

$$\lambda_{\mu^k} \beta_{k+1} e + \beta_{k+1} h_{\mu^k} \geqslant \beta_{k+1} g_{\mu^{k+1}} + \beta_{k+1} P_{\mu^{k+1}} h_{\mu^k}, \tag{85}$$

and by Lemma 1 we obtain $\beta_{k+1} g_{\mu^{k+1}} = \lambda_{\mu^{k+1}}$ and $\beta_{k+1} P_{\mu^{k+1}} = \beta_{k+1}$. Thus (85) is equivalent to

$$\lambda_{\mu^k} \geqslant \lambda_{\mu^{k+1}}.$$

Since the elements of $\beta_{k+1}$ are positive we have equality above if and only if equality holds in (81) or equivalently $\mu^k(i)$ attains the minimum in (79). Thus relation (83) implies that $\mu^k(i)$ attains the minimum in (79). Hence (84) holds and $\mu^k$ is optimal.

We state the last conclusion from the preceding discussion as a proposition.

**Proposition 8** Assume that for every admissible stationary policy $\{\mu, \mu, \ldots\}$ the transition probability matrix $P_\mu$ gives rise to an irreducible Markov chain. Then the policy iteration algorithm will yield an optimal policy in a finite number of iterations.

We now demonstrate the policy iteration algorithm by means of the example of the previous section.

EXAMPLE 1 (CONTINUED) Let

$$\mu^0(1) = u^1, \qquad \mu^0(2) = u^2.$$

We take $s = 1$ as the reference state and we obtain $\lambda_{\mu^0}$, $h_{\mu^0}(1)$, $h_{\mu^0}(2)$ from the system of equations

$$\lambda_{\mu^0} + h_{\mu^0}(1) = \bar{g}(1, u^1) + p_{11}(u^1)h_{\mu^0}(1) + p_{12}(u^1)h_{\mu^0}(2),$$
$$\lambda_{\mu^0} + h_{\mu^0}(2) = \bar{g}(2, u^2) + p_{21}(u^2)h_{\mu^0}(1) + p_{22}(u^2)h_{\mu^0}(2),$$
$$h_{\mu^0}(1) = 0.$$

Substituting the data of the problem

$$\lambda_{\mu^0} = 2 + \tfrac{1}{4}h_{\mu^0}(2), \qquad \lambda_{\mu^0} + h_{\mu^0}(2) = 3 + \tfrac{3}{4}h_{\mu^0}(2),$$

from which

$$\rhd \qquad \lambda_{\mu^0} = 2.5, \qquad h_{\mu^0}(1) = 0, \qquad h_{\mu^0}(2) = 2. \qquad \lhd$$

We now find $\mu^1(1)$, $\mu^1(2)$ by the minimization indicated in (79). We determine

$$\min[\bar{g}(1, u^1) + p_{11}(u^1)h_{\mu^0}(1) + p_{12}(u^1)h_{\mu^0}(2),$$
$$\bar{g}(1, u^2) + p_{11}(u^2)h_{\mu^0}(1) + p_{12}(u^2)h_{\mu^0}(2)]$$
$$= \min[2 + \tfrac{1}{4} \times 2, 0.5 + \tfrac{3}{4} \times 2] = \min[2.5, 2],$$

$$\min[\bar{g}(2, u^1) + p_{21}(u^1)h_{\mu^0}(1) + p_{22}(u^1)h_{\mu^0}(2),$$
$$\bar{g}(2, u^2) + p_{21}(u^2)h_{\mu^0}(1) + p_{22}(u^2)h_{\mu^0}(2)]$$
$$= \min[1 + \tfrac{1}{4} \times 2, 3 + \tfrac{3}{4} \times 2] = \min[1.5, 4.5].$$

The minimization yields

$$\rhd \qquad \mu^1(1) = u^2, \qquad \mu^1(2) = u^1. \qquad \lhd$$

We obtain $\lambda_{\mu^1}$, $h_{\mu^1}(1)$, $h_{\mu^1}(2)$ from the system of equations

$$\lambda_{\mu^1} + h_{\mu^1}(1) = \bar{g}(1, u^2) + p_{11}(u^2)h_{\mu^1}(1) + p_{12}(u^2)h_{\mu^1}(2),$$
$$\lambda_{\mu^1} + h_{\mu^1}(2) = \bar{g}(2, u^1) + p_{21}(u^1)h_{\mu^1}(1) + p_{22}(u^1)h_{\mu^1}(2),$$
$$h_{\mu^1}(1) = 0.$$

By substitution of the data of the problem, we obtain

▷ $$\lambda_{\mu^1} = 0.75, \qquad h_{\mu^1}(1) = 0, \qquad h_{\mu^1}(2) = \tfrac{1}{3}.$$ ◁

We find $\mu^2(1)$, $\mu^2(2)$ by determining the minimum in

$$\begin{aligned}\min[\bar{g}(1, u^1) &+ p_{11}(u^1)h_{\mu^1}(1) + p_{12}(u^1)h_{\mu^1}(2),\\ &\bar{g}(1, u^2) + p_{11}(u^2)h_{\mu^1}(1) + p_{12}(u^2)h_{\mu^1}(2)]\\ &= \min[2 + \tfrac{1}{4} \times \tfrac{1}{3}, 0.5 + \tfrac{3}{4} \times \tfrac{1}{3}] = \min[2.08, 0.75],\end{aligned}$$

$$\begin{aligned}\min[\bar{g}(2, u^1) &+ p_{21}(u^1)h_{\mu^1}(1) + p_{22}(u^1)h_{\mu^1}(2),\\ &\bar{g}(2, u^2) + p_{21}(u^2)h_{\mu^1}(1) + p_{22}(u^2)h_{\mu^1}(2)]\\ &= \min[1 + \tfrac{1}{4} \times \tfrac{1}{3}, 3 + \tfrac{3}{4} \times \tfrac{1}{3}] = \min[1.08, 3.25].\end{aligned}$$

The minimization yields

$$\mu^2(1) = \mu^1(1) = u^2, \qquad \mu^2(2) = \mu^1(2) = u^1,$$

and hence the policy above is optimal and the optimal average cost per stage is $\lambda_{\mu^1} = 0.75$.

## 8.4 Infinite State Space—Linear Systems with Quadratic Cost Functionals

The standing assumption in the preceding sections has been that the state space is finite and thus the underlying system is a controlled finite state Markov chain. Once one removes the finiteness assumption on the state space many of the results presented in the past three sections no longer hold. For example, while one could restrict attention to stationary policies for finite state systems this is not true anymore where the state space is infinite. For instance, the following example (due to Ross [R4]) shows that for a countable state space the optimal policy may be nonstationary. (This fact is true even if one expands the class of admissible policies to admit randomized policies [R4].)

EXAMPLE Let the state space be $S = \{1, 2, 3, \ldots\}$ and let there be two control actions $C = \{u^1, u^2\}$. The transition probabilities under $u^1$ and $u^2$ are specified by

$$p_{i(i+1)}(u^1) = p_{ii}(u^2) = 1.$$

In other words, the system is deterministic and application of $u^1$ moves the state from $i$ to $(i + 1)$, while application of $u^2$ leaves the state unchanged. The costs per stage are

$$\bar{g}(i, u^1) = 1, \qquad \bar{g}(i, u^2) = 1/i, \qquad i = 1, 2, 3, \ldots.$$

In other words, at state $i$ we may either move to state $(i + 1)$ at the cost of one unit or stay at $i$ at a cost $1/i$.

Now for any stationary policy $\pi = \{\mu, \mu, \ldots\}$ other than the policy for which $\mu(i) = u^1$ for all $i$, let $n(\pi)$ be the smallest integer for which

$$\mu[n(\pi)] = u^2.$$

Then concerning the average cost per stage corresponding to this policy we clearly have

$$J_\pi(i) = 1/n(\pi) > 0 \qquad \forall i \leqslant n(\pi).$$

For the policy where $\mu(i) = u^1$ for all $i$ we have $J_\pi(i) = 1$ for all $i$. Since the optimal cost per stage cannot be less than zero, it is clear that

$$\inf_\pi J_\pi(i) = 0, \qquad i = 1, 2, \ldots.$$

However, the optimal cost is not attained by any stationary policy, so that no stationary policy is optimal. On the other hand consider the nonstationary policy $\pi^*$ that on entering state $i$ chooses $u^2$ for $i$ consecutive times and then chooses $u^1$. If the starting state is $i$, the sequence of costs incurred is

$$\underbrace{\frac{1}{i}, \frac{1}{i}, \ldots, \frac{1}{i}}_{i \text{ times}}, \quad 1, \quad \underbrace{\frac{1}{i+1}, \frac{1}{i+1}, \ldots, \frac{1}{i+1}}_{(i+1) \text{ times}}, \quad 1, \quad \frac{1}{i+2}, \frac{1}{i+2}, \ldots.$$

The average cost corresponding to this policy is

$$J_{\pi^*}(i) = \lim_{m \to \infty} \frac{2m}{\sum_{k=1}^m (i + k)} = 0, \qquad i = 1, 2, 3, \ldots.$$

Hence the nonstationary policy $\pi^*$ is optimal while, as shown above, no stationary policy is optimal.

Generally speaking the analysis of average cost optimization problems involving an infinite state space presents considerable difficulties and as yet there exists little in the way of a complete and powerful theory. However, certain particular special cases can be satisfactorily analyzed and one such case is the average cost version of the linear–quadratic problem examined in Chapters 3, 4, and 6.

Consider an undiscounted version ($\alpha = 1$) for the linear–quadratic problem of Section 6.5 involving the system

$$x_{k+1} = Ax_k + Bu_k + w_k, \qquad k = 0, 1, \ldots, \tag{86}$$

and the cost functional

$$J_\pi(x_0) = \lim_{N\to\infty} (1/N) \underset{\substack{w_k \\ k=0,1,\ldots}}{E} \left\{ \sum_{k=0}^{N-1} [x_k' Q x_k + \mu_k(x_k)' R \mu_k(x_k)] \right\}. \tag{87}$$

We make the same assumptions as in Section 6.5, i.e., that $w_k$ are independent and have zero mean and finite second moments. We also assume that the pair $(A, B)$ is controllable and that the pair $(A, C)$, where $Q = C'C$, is observable. Under these assumptions it was shown in Section 3.1 that the Riccati equation

$$K_0 = 0, \tag{88}$$

$$K_{k+1} = A'[K_k - K_k B(B' K_k B + R)^{-1} B' K_k] A + Q, \tag{89}$$

yields in the limit a matrix $K$,

$$K = \lim_{k\to\infty} K_k, \tag{90}$$

which is the unique solution of the algebraic Riccati equation

$$K = A'[K - KB(B'KB + R)^{-1} B'K] A + Q \tag{91}$$

within the class of positive semidefinite symmetric matrices.

Now the optimal value of the $N$-stage costs

$$(1/N) \underset{\substack{w_k \\ k=0,1,\ldots,N-1}}{E} \left\{ \sum_{k=0}^{N-1} (x_k' Q x_k + u_k' R u_k) \right\} \tag{92}$$

has been derived earlier and was seen to be equal to

$$(1/N)\left[ x_0' K_N x_0 + \sum_{k=0}^{N-1} E\{w' K_k w\} \right].$$

Thus using (90) and the fact that

$$\lim_{N\to\infty} (1/N) \sum_{k=0}^{N-1} E\{w' K_k w\} = E\{w' K w\},$$

the optimal finite horizon costs tend in the limit as $N \to \infty$ to

$$\lambda = E\{w'Kw\}. \tag{93}$$

In addition, the $N$-stage optimal policy in its initial stages tends to the stationary policy

$$\mu^*(x) = -(B'KB + R)^{-1}B'KAx. \tag{94}$$

Furthermore, a simple calculation shows that, by the definition of $\lambda$, $K$, and $\mu^*(x)$, we have

$$\lambda + x'Kx = \min_u E\{x'Qx + u'Ru + (Ax + Bu + w)'K(Ax + Bu + w)\},$$

while the minimum in the right-hand side of the above equation is attained at $u^* = \mu^*(x)$ as given by (94).

Now by repeating the proof of Proposition 1 of this chapter, we obtain

$$\lambda \leqslant (1/N)\, E\{x_N' K x_N | x_0, \pi\} - (1/N)x_0' K x_0 + (1/N)\, E\left\{\sum_{k=0}^{N-1} (x_k' Q x_k + u_k' R u_k) | x_0, \pi\right\},$$

with equality if $\pi = \{\mu^*, \mu^*, \ldots\}$. Hence if $\pi$ is such that $E\{x_N' K x_N | x_0, \pi\}$ is uniformly bounded over $N$, we have by taking the limit above,

$$\lambda \leqslant J_\pi(x) \qquad \forall x \in R^n,$$

with equality if $\pi = \{\mu^*, \mu^*, \ldots\}$. Thus the stationary policy $\{\mu^*, \mu^*, \ldots\}$ as given by (94) is optimal over all policies $\pi$ with $E\{x_N' K x_N | x_0, \pi\}$ bounded uniformly over $N$.

## 8.5 Notes

The average cost problem was first formulated and analyzed by Howard [H15]. Several authors have contributed subsequently to the problem [B2, B25, L1, R4, S7, V2, V4], most notably Blackwell [B19].

In our approach to the results of Section 8.1. we follow Ross [R4]. This approach is generalizable to situations where the state space is infinite. For alternative expositions see references [D4], [K10], and [P1]. The result of the appendix was shown by Bather [B2]. The successive approximation method of Section 8.2 was devised by White [W2]. The error bounds of Proposition 6 are due to Odoni [O1]. Related results for more general situations have been given recently in references [H8] and [H12]. The policy iteration algorithm can be generalized for problems where the optimal average cost per stage is not the same for every initial state (see [B19], [V2], and [D4]). For a computational approach based on linear programming see [M3] and [D4]. For an analysis of average cost Markovian decision problems involving exponential risk-sensitive cost functionals, see the paper by Howard and Matheson [H17]. For analysis of infinite horizon versions of inventory control problems such as the one considered in Section 3.2, see references [13], [H13], [H14], and [V5].

## Problems

**1.** Assume that for some state $s$ we have that $J_\alpha(i) - J_\alpha(s)$ is uniformly bounded for $\alpha \in (0, 1)$. Show that for a sequence $\{\alpha_k\}$ with $\alpha_k \to 1$, $\alpha_k \in (0, 1)$, and a sequence of $\alpha_k$-optimal policies $\{\mu_{\alpha_k}\}$, we have $\mu_{\alpha_k}(i) = \mu^*(i)$ for all $i$ and all $k$ sufficiently large, where $\mu^*$ is average cost optimal.

**2.** Show that if for some sequence $\{\alpha_k\}$ with $\alpha_k \in (0, 1)$, $\alpha_k \to 1$, and a sequence of $\alpha_k$-optimal policies $\{\mu_{\alpha_k}\}$ we have $\mu_{\alpha_k}(i) = \mu^*(i)$ for all $i$ and all $k$ sufficiently large, then $\mu^*$ is average cost optimal.

**3.** *Optimal Control of Deterministic Finite State Systems* Consider a stationary deterministic control system

$$x_{k+1} = f(x_k, u_k), \qquad k = 0, 1, \ldots,$$

where the state $x_k$ belongs to a finite state space $S = \{1, 2, \ldots, n\}$ and the control $u_k$ is constrained in a subset $U(x_k)$ of a finite control space $C$. We say that the system is *completely controllable* if given any two states $i, j \in S$ there exists a sequence of admissible controls that drives the state of the system from the state $i$ to the state $j$ within at most $(n - 1)$ steps. For a completely controllable system and a given initial state $x_0 = i$ consider the problem of finding an admissible control sequence $\{u_0, u_1, \ldots\}$ that minimizes

$$J_\pi(i) = \lim_{N \to \infty} (1/N) \sum_{k=0}^{N-1} g(x_k, u_k),$$

where $g: S \times C \to R$ is given. Show that an optimal control sequence exists and that the optimal cost is the same for every initial state. Show also that there exist optimal control sequences that after a certain time index are periodic.

**4.** Consider a stationary inventory control problem of the type considered in Section 3.2 but with the difference that the stock $x_k$ can only take integer values from 0 to some integer $M$. The amount of the order $u_k$ can take integer values with $0 \leqslant u_k \leqslant M - x_k$ and the random demand $w_k$ can only take nonnegative integer values with $P(w_k = 0) > 0$ and $P(w_k = 1) > 0$. Unsatisfied demand is lost so that stock evolves according to the equation

$x_{k+1} = \max(0, x_k + u_k - w_k)$. The problem is to find an inventory policy that minimizes the average cost per stage. Show that there exists an optimal stationary policy and that the optimal cost is independent of the initial stock $x_0$.

## Appendix Existence Analysis under the Weak Accessibility Condition

In this appendix we provide a proof of Proposition 4 of Section 8.1. In fact, we shall prove a more general result that contains Proposition 4 as a special case. The analysis requires a high degree of mathematical sophistication and is directed toward the advanced reader.

Consider a controlled process in discrete time with a finite state space $S = \{1, 2, \ldots, n\}$. We assume that for any state $i \in S$, the next transition is controlled by choosing a probability vector $p_i = (p_{i1}, p_{i2}, \ldots, p_{in})$ from a closed convex set $D_i \in R^n$. Any selection $p_i \in D_i$, $i = 1, 2, \ldots, n$, defines the rows of a stochastic matrix $P \in D = D_1 \times D_2 \times \cdots \times D_n$. The cost of each transition is prescribed by functions $c_i(p_i)$, $i = 1, \ldots, n$, each assumed convex and continuous on the corresponding set $D_i$. Thus when the current state is $i$ and probability vector $p_i \in D_i$ is selected, the cost incurred is $c_i(p_i)$. If $P \in D$ is used at each time, the corresponding average expected cost is $P^*c$, where

$$P^* = \lim_{N \to \infty} (1/N)\{I + P + P^2 + \cdots + P^{N-1}\},$$

and $c$ is the vector with coordinates $c_1(p_1), \ldots, c_n(p_n)$, where $p_1, \ldots, p_n$ are the rows of $P$. Since the components of $c$ depend continuously on the rows of $P$, the cost $c(P)$ is bounded over $D$ and we have $|J^*(i)| < \infty$ with

$$J^* = \inf_{P \in D} P^*c,$$

where minimization above is considered separately for each coordinate of $P^*c$.

We shall rely on the following accessibility assumption:

*For any pair of states $i, j \in S$, there exists a matrix $P \in D$ and a positive integer $r$ such that $p_{ij}^r > 0$, where $p_{ij}^r$ is the element in the ith row and jth column of the matrix $P^r$ ($P$ to the rth power).*

Based on this assumption we shall prove that there exists a vector $h \in R^n$ and a constant $\lambda$ such that

$$\lambda e + h = \min_{P \in D}\{c + Ph\}, \tag{96}$$

where the minimization above is assumed to be componentwise. By this we mean that the scalar $\lambda$ and the column vector $h = (h_1, h_2, \ldots, h_n)' \in R^n$ satisfy for each $i = 1, 2, \ldots, n$,

$$\lambda + h_i = \min_{p_i \in D_i} \{c_i(p_i) + p_i h\},$$

where $p_i h$ denotes the inner product of the row vector $p_i$ ($i$th row of the stochastic matrix $P$) and the column vector $h$, i.e., $p_i h$ is the $i$th coordinate of the vector $Ph$.

The problem described above is similar in nature to the one considered in the first three sections of this chapter. A control $u$ is identified with its corresponding transition probability vector $p_i(u) = [p_{i1}(u), p_{i2}(u), \ldots, p_{in}(u)]$. There is an important difference, however, in that while here the set of admissible probability vectors $D_i$ is assumed to be a closed convex set, in the problem considered earlier in the chapter the set of admissible controls $U(i)$ and hence also the set of corresponding probability vectors were assumed to be finite. In order to utilize the result of this appendix in proving Proposition 4 we shall need to construct a version of the problem of Sections 8.1–8.3. where *randomization* on the set of admissible controls $U(i)$ is allowed.

Within the framework of the problem of Sections 8.1–8.3 let $i$ be any state ($i = 1, 2, \ldots, n$) and let $u^1, u^2, \ldots, u^{m_i}$ denote the elements of the admissible control set $U(i)$. Consider the case where at each state $i$ it is possible to select, instead of a control $u \in U(i)$, a probability distribution $q = (q^1, q^2, \ldots, q^{m_i})$ over the set $U(i)$. Such a probability distribution will be referred to as a *randomized control*. The set of all randomized controls is denoted $Q_i$. If the current state is $i$ and the randomized control $q$ is selected, the probability that the next state will be $j$ is

$$\sum_{r=1}^{m_i} q^r p_{ij}(u^r), \qquad j = 1, 2, \ldots, n,$$

and the corresponding transition probability vector is

$$p_i(q) = \left[\sum_{r=1}^{m_i} q^r p_{i1}(u^r), \sum_{r=1}^{m_i} q^r p_{i2}(u^r), \ldots, \sum_{r=1}^{m_i} q^r p_{in}(u^r)\right].$$

The set of all possible transition probability vectors as $q$ ranges over $Q_i$ is denoted $D_i$:

$$D_i = \{p_i(q) \mid q \in Q_i\}.$$

The set $D_i$ is clearly the convex hull of the finite set of probability vectors

$$p_i(u^r) = [p_{i1}(u^r), \ldots, p_{in}(u^r)], \qquad r = 1, 2, \ldots, m_i,$$

and hence it is a closed convex set. With each state $i$ and each probability vector $p_i \in D_i$, we associate a transition cost

$$c_i(p_i) = \min\left\{ \sum_{r=1}^{m_i} q^r \bar{g}(i, u^r) \,\middle|\, \sum_{r=1}^{m_i} q^r p_i(u^r) = p_i,\ q \in Q_i \right\}.$$

The cost $c_i(p_i)$ is the least possible expected transition cost associated with randomized controls $q \in Q_i$ that result in a transition probability vector equal to $p_i$. From known facts of the theory of convex functions [R2] it follows that $c_i(\cdot)$ as defined above is a polyhedral convex continuous function over the polyhedron $D_i$. In fact, $c_i$ is the convex hull [R2, p. 36] of (the extended real-valued function) $\tilde{c}_i$ defined by

$$\tilde{c}_i(p_i) = \begin{cases} \bar{g}(i, u^r) & \text{if} \quad p_i = p_i(u^r), \quad r = 1, \ldots, m_i, \\ +\infty & \text{otherwise.} \end{cases}$$

The extreme points of the epigraph [R2] of $c_i$ correspond to a subset of the finite set of points $\{p_i(u^1), \ldots, p_i(u^{m_i})\}$ and the same is true for the extreme points of the epigraph of any function of the form $c_i(p_i) + p_i h$, where $h$ is any vector in $R^n$ and $p_i h$ denotes the inner product of $p_i$ and $h$.

Consider now the problem of this appendix with $D_i$ and $c_i(p_i)$ defined as above and suppose that we are able to prove that (96) holds, i.e.,

$$\lambda + h_i = \min_{p_i \in D_i} \{c_i(p_i) + p_i h\}, \qquad i = 1, \ldots, n.$$

Then in view of the construction of $c_i$ and $D_i$ the minimum on the right-hand side above will be attained at one (or possibly more) of the generating points $p_i(u^1), p_i(u^2), \ldots, p_i(u^{m_i})$ of the set $D_i$, which correspond to nonrandomized controls. As a result, in view of the definition of $c_i$ and $D_i$ we will have

$$\begin{aligned} \lambda + h_i &= \min_{r=1,\ldots,m_i} \left\{ c_i[p_i(u^r)] + \sum_{j=1}^{n} p_{ij}(u^r) h_j \right\} \\ &= \min_{u \in U(i)} \left\{ \bar{g}(i, u) + \sum_{j=1}^{n} p_{ij}(u) h_j \right\}, \qquad i = 1, \ldots, n. \end{aligned}$$

Thus the condition of Proposition 1 of Section 8.1 will be satisfied and the result of Proposition 4 will follow. Thus in order to prove Proposition 4 it is sufficient to prove Eq. (96) within the generalized framework of this appendix.

Consider now the nonlinear operator $M: R^n \to R^n$ defined by

$$Mx = \min_{P \in D} \{c + Px\}, \qquad x \in R^n. \tag{97}$$

Define for all $x = (x_1, \ldots, x_n) \in R^n$,

$$\|x\| = \max_{i \in S} x_i - \min_{j \in S} x_j. \tag{98}$$

Since $\|x - y\| = 0$ if and only if $x$ and $y$ differ by a multiple of the unit vector $e = [1, 1, \ldots, 1]$, (98) defines a norm on the collection of all subsets of $R^n$ elements of which differ by a multiple of the unit vector (i.e., $\|\cdot\|$ is a seminorm on $R^n$ and a norm on the corresponding quotient space of equivalence classes). Now suppose that some vector $h \in R^n$ is a *fixed point* of $M$ in the sense that

$$\|Mh - h\| = 0. \tag{99}$$

Then it follows that

$$\lambda e + h = Mh = \min_{P \in D}\{c + Ph\},$$

for some scalar $\lambda$. Hence proving (96) is equivalent to proving that there exists at least one fixed point of $M$ in the sense of (99).

We begin by deriving some useful properties of the operator $M$ defined by (97). For any $x \in R^n$ we define $H(x) = \max\{x_i | i \in S\}$. Similarly we define $L(x) = \min\{x_i | i \in S\}$. Then $\|x\| = H(x) - L(x)$. We have the following lemma.

**Lemma A.1** For any $x, y \in R^n$ and $\alpha, \beta \in R$:

(1) $H(Mx - My) \leqslant H(x - y)$.
(2) $\|Mx - My\| \leqslant \|x - y\|$.
(3) $\|M(\alpha x) - M(\beta y)\| \leqslant |\alpha| \, \|x - y\| + |\alpha - \beta| \, \|y\|$.

*Proof* (1) Let $P, Q \in D$ be such that

$$Mx = c(P) + Px \qquad \text{and} \qquad My = c(Q) + Qy.$$

Then $Mx \leqslant c(Q) + Qx$ and $Mx - My \leqslant Q(x - y)$. Since $Q$ is a stochastic matrix and $x - y \leqslant H(x - y)e$, we obtain $Mx - My \leqslant H(x - y)e$ from which $H(Mx - My) \leqslant H(x - y)$.

(2) We have

$$\begin{aligned}\|Mx - My\| &= H(Mx - My) + H(My - Mx) \\ &\leqslant H(x - y) + H(y - x) = \|x - y\|.\end{aligned}$$

(3) From part (2), $\|M(\alpha x) - M(\beta y)\| \leqslant \|\alpha x - \beta y\|$, and we have

$$\|\alpha x - \beta y\| = \|\alpha(x - y) + (\alpha - \beta)y\| \leqslant |\alpha| \, \|x - y\| + |\alpha - \beta| \, \|y\|,$$

from which the result follows. Q.E.D.

**Lemma A.2** Let $\{\theta_k\}$ be a nondecreasing sequence with $0 \leqslant \theta_k \leqslant 1$. Consider the sequence $\{z(k)\}$ defined by

$$z(k) = \min_{P \in D}\{c + \theta_k Pz(k-1)\}, \tag{100}$$

with $z(0) = 0$. Then $\{\|z(k)\|\}$ is a bounded sequence.

*Proof* We recall that $c_i(p_i)$ is bounded for $p_i \in D_i$, $i = 1, \ldots, n$. Since $\|z(k)\|$ cannot be affected by adding the same constant to every transition cost $c_i(p_i)$, we may assume that for some $\beta \in R$ we have $0 \leqslant c_i(p_i) \leqslant \beta$ for all $p_i \in D_i$, $i = 1, \ldots, n$. Then it is easy to verify that $\{z(k)\}$ is a nondecreasing sequence, i.e.,

$$z(k+1) \geqslant z(k), \qquad k = 0, 1, \ldots, . \tag{101}$$

Now under the accessibility assumption there exists a stochastic matrix $Q \in D$ with elements $q_{ij}$, $i, j = 1, \ldots, n$ that defines an irreducible Markov chain, i.e., a chain for which every state communicates with every other state. This is a consequence of the accessibility assumption and the convexity of $D$, since we can arrange that

$$q_{ij} > 0 \qquad \text{if} \quad p_{ij} > 0 \quad \text{for some} \quad P \in D, \tag{102}$$

by allowing $P$ to participate in a convex combination forming $Q$. Thus if $P(i, j) \in D$ is a stochastic matrix under which $j$ is accessible from $i$, the matrix $Q = (1/n^2) \sum_{i=1}^{n} \sum_{j=1}^{n} P(i, j)$ defines an irreducible Markov chain. We now associate with $Q$ a set of mean transition times. For each pair of distinct states $i, j \in S$ we denote by $\tau_{ij}$ the expected number of steps required to reach $j$ from $i$ when $Q$ is used as a stationary policy. Then, by considering the first step, we have

$$\tau_{ij} = 1 + \sum_{l \neq j} q_{il}\tau_{lj}, \qquad i, j = 1, \ldots, n, \qquad i \neq j. \tag{103}$$

Finally, prior to establishing the result of the lemma, we prove by induction that

$$z_i(k) \leqslant \beta\tau_{ij} + z_j(k) \qquad \forall i \neq j, \quad k \geqslant 0. \tag{104}$$

Indeed (104) holds for $k = 0$. Assume (104) holds for $k = k$. Since $c(Q) \leqslant \beta e$ and $\theta_{k+1} \leqslant 1$, we have

$$z(k+1) \leqslant c(Q) + \theta_{k+1}Qz(k) \leqslant \beta e + Qz(k).$$

Thus

$$z_i(k+1) \leqslant \beta + \sum_{l \neq j} q_{il}z_l(k) + q_{ij}z_j(k).$$

Using (104) and then (103) we obtain

$$z_i(k+1) \leqslant \beta\left\{1 + \sum_{l \neq j} q_{il}\tau_{lj}\right\} + z_j(k) = \beta\tau_{ij} + z_j(k).$$

Using (101) it follows that

$$z_i(k+1) \leqslant \beta\tau_{ij} + z_j(k+1),$$

i.e., (104) is proved for $k = k + 1$ and the induction proof of (104) is complete.

Now we easily obtain that the sequence $\{\|z(k)\|\}$ is bounded since (104) implies

$$\|z(k)\| \leqslant \max\{\beta\tau_{ij} \mid i, j \in S, i \neq j\}. \quad \text{Q.E.D.}$$

We are now in a position to prove the existence of a fixed point of $M$ in the sense of (99) and hence that (96) holds for some $\lambda \in R$ and $h \in R^n$. Consider the sequence

$$\begin{aligned} y(k) &= \min_{P \in D}\{c + (1 - (1/k))Py(k-1)\}, \qquad k = 1, 2, \ldots, \\ y(0) &= 0. \end{aligned} \tag{105}$$

Let also

$$\tilde{y}(k) = y(k) - L[y(k)]e, \qquad k = 0, 1, \ldots, \tag{106}$$

where $L[y(k)] = \min\{y_i(k) \mid i \in S\}$. Then $L[\tilde{y}(k)] = 0$, $\|\tilde{y}(k)\| = \max_{i \in S} \tilde{y}_i(k)$, and

$$0 \leqslant \tilde{y}_i(k) \leqslant \|\tilde{y}(k)\| = \|y(k)\|. \tag{107}$$

The sequence $\{\|y(k)\|\}$ is bounded by Lemma A.2, and by (107) the sequences $\{\tilde{y}_i(k)\}$, $i = 1, \ldots, n$, are also bounded.

**Proposition** The sequence $\{\tilde{y}(k)\}$ has a limit point $h \in R^n$ such that $\|Mh - h\| = 0$. Hence there exists a scalar $\lambda$ such that

$$\lambda e + h = Mh = \min_{P \in D}\{c + Ph\}.$$

*Proof* As explained above, each sequence $\{\tilde{y}_i(k)\}$, $i = 1, \ldots, n$, is bounded and hence there exists a convergent subsequence of $\{\tilde{y}(k)\}$. Let

$$h = \lim_{r \to \infty} \tilde{y}(k_r), \tag{108}$$

where $\{\tilde{y}(k_r)\}$ is the convergent subsequence. We also have, in view of (106), that

$$\lim_{r \to \infty} \|h - \tilde{y}(k_r)\| = 0.$$

Now by using part (3) of Lemma A.1 we have

$$\|y(k+1) - y(k)\| = \left\| M\left(\frac{k}{k+1}\, y(k)\right) - M\left(\frac{k-1}{k}\, y(k-1)\right)\right\|$$

$$\leqslant \frac{k}{k+1}\, \|y(k) - y(k-1)\| + \frac{\|y(k-1)\|}{k(k+1)}. \tag{109}$$

Let $B$ be a bound for $\|y(k)\|$, i.e., $\|y(k)\| \leqslant B$ for all $k$. Then (109) implies that

$$\|y(k+1) - y(k)\| \leqslant \frac{B}{k+1}\left(1 + \frac{1}{2} + \cdots + \frac{1}{k}\right),$$

and hence $\|y(k+1) - y(k)\| \to 0$ as $k \to \infty$. Hence

$$\|\tilde{y}(k_r + 1) - \tilde{y}(k_r)\| \to 0 \qquad \text{as} \quad r \to \infty. \tag{110}$$

We have for the vector $h$ of (108) that

$$\|Mh - h\| \leqslant \|Mh - \tilde{y}(k_r + 1)\| + \|\tilde{y}(k_r + 1) - \tilde{y}(k_r)\| + \|\tilde{y}(k_r) - h\|. \tag{111}$$

By using part (3) of Lemma A.1 and (106) we obtain

$$\begin{aligned}\|Mh - \tilde{y}(k_r + 1)\| &= \|Mh - y(k_r + 1)\| \\ &= \left\| Mh - M\left(\frac{k_r}{k_r + 1}\, y(k_r)\right)\right\| \\ &\leqslant \|h - y(k_r)\| + \frac{1}{k_r + 1}\, \|y(k_r)\| \\ &= \|h - \tilde{y}(k_r)\| + \frac{1}{k_r + 1}\, \|\tilde{y}(k_r)\|.\end{aligned} \tag{112}$$

Combining (111) and (112) we obtain

$$\|Mh - h\| \leqslant \|\tilde{y}(k_{r+1}) - \tilde{y}(k_r)\| + 2\|\tilde{y}(k_r) - h\| + \frac{B}{k_r + 1},$$

and taking the limit as $r \to \infty$ and using (110) we obtain $\|Mh - h\| = 0$.

Q.E.D.

# Appendixes

*Appendix A*

# Mathematical Review

The purpose of this and the following appendixes is to provide a list of mathematical and probabilistic definitions, notations, relations, and results that are used frequently in the text. For detailed expositions the reader may consult the references listed in each appendix.

## A.1 Sets

If $x$ is a member of the set $S$, we write $x \in S$. We write $x \notin S$ if $x$ is not a member of $S$. A set $S$ may be specified by listing its elements within braces. For example, by writing $S = \{x_1, x_2, \ldots, x_n\}$ we mean that the set $S$ consists of the elements $x_1, \ldots, x_n$. A set $S$ may also be specified in the form

$$S = \{x \mid P(x)\}$$

as the set of elements satisfying property $P$. For example,

$$S = \{x \mid x\text{: real}, 0 \leqslant x \leqslant 1\}$$

denotes the set of all real numbers $x$ satisfying $0 \leqslant x \leqslant 1$.

The *union* of two sets $S$ and $T$ is denoted by $S \cup T$ and the *intersection* of $S$ and $T$ is denoted by $S \cap T$. The union and intersection of a sequence of sets $S_1, S_2, \ldots, S_k, \ldots$ is denoted by $\bigcup_{k=1}^{\infty} S_k$ and $\bigcap_{k=1}^{\infty} S_k$, respectively. If $S$ is a subset of $T$, i.e., if every element of $S$ is also an element of $T$, we write $S \subset T$ or $T \supset S$.

*Finite and Countable Sets*

A set $S$ is said to be *finite* if it consists of a finite number of elements. It is said to be *countable* if one can associate with each element of $S$ a nonnegative integer in a way that to each pair of distinct elements of $S$ there correspond two distinct integers. Thus according to our definition a finite set is also countable but not conversely. A countable set $S$ that is not finite may be represented by listing its elements $x_0, x_1, x_2, \ldots$, i.e., $S = \{x_0, x_1, x_2, \ldots\}$. If $A = \{a_0, a_1, \ldots\}$ is a countable set and $S_{a_0}, S_{a_1}, \ldots$ are each countable sets, then the union $\bigcup_{k=0}^{\infty} S_{a_k}$ (otherwise denoted $\bigcup_{a \in A} S_a$) is also a countable set.

*Sets of Real Numbers*

If $a$ and $b$ are real numbers or $+\infty$, $-\infty$, we denote by $[a, b]$ the set of numbers $x$ satisfying $a \leqslant x \leqslant b$ (including the possibility $x = +\infty$, or $x = -\infty$). A rounded, instead of square, bracket denotes strict inequality in the definition. Thus $(a, b]$, $[a, b)$, and $(a, b)$ denote the set of all $x$ satisfying $a < x \leqslant b$, $a \leqslant x < b$, and $a < x < b$, respectively.

If $S$ is a set of real numbers bounded above, then there is a smallest real number $y$ such that $x \leqslant y$ for all $x \in S$. This number is called the *least upper bound or supremum* of $S$ and is denoted $\sup\{x \mid x \in S\}$. Similarly the greatest real number $z$ such that $z \leqslant x$ for all $x \in S$ is called the *greatest lower bound or infimum* of $S$ and is denoted $\inf\{x \mid x \in S\}$. If $S$ is unbounded above, we write $\sup\{x \mid x \in S\} = +\infty$ and if it is unbounded below, $\inf\{x \mid x \in S\} = -\infty$. If $S$ is the empty set, then by convention we write $\inf\{x \mid x \in S\} = +\infty$ and $\sup\{x \mid x \in S\} = -\infty$.

## A.2 Euclidean Space

The set of all $n$-tuples $x = (x_1, \ldots, x_n)$ where $x_1, \ldots, x_n$ are real numbers constitutes the *n-dimensional Euclidean space* denoted $R^n$. The elements of $R^n$ are referred to as $n$-dimensional vectors or simply vectors when confusion cannot arise. The one-dimensional Euclidean space $R^1$ consists of all the real numbers and is denoted $R$. Vectors in $R^n$ can be added by adding their corresponding components. They can be multiplied by a scalar by multiplication of each component by the scalar. The *inner product* (or scalar product) of two vectors $x = (x_1, \ldots, x_n)$, $y = (y_1, \ldots, y_n)$ is denoted $x'y$ and is equal to

$\sum_{i=1}^{n} x_i y_i$. The *norm* of a vector $x = (x_1, \ldots, x_n) \in R^n$ is denoted $\|x\|$ and is equal to $(x'x)^{1/2} = (\sum_{i=1}^{n} x_i^2)^{1/2}$.

A set of vectors $a_1, a_2, \ldots, a_k$ is said to be *linearly dependent* if there exist scalars $\lambda_1, \lambda_2, \ldots, \lambda_k$, not all zero, such that $\sum_{i=1}^{k} \lambda_i a_i = 0$. If no such set of scalars exists, the vectors are said to be *linearly independent*.

## A.3 Matrices

An $m \times n$ *matrix* is a rectangular array of numbers, called *elements*, arranged in $m$ rows and $n$ columns. The element in the $i$th row and $j$th column of a matrix $A$ is denoted by a subscript $ij$, such as $a_{ij}$, in which case we write $A = [a_{ij}]$. A square matrix (one with $m = n$) with elements $a_{ij} = 0$ for $i \neq j$ and $a_{ii} = 1$, for $i = 1, \ldots, n$, is said to be an *identity matrix*. The *sum* of two $m \times n$ matrices $A$ and $B$ is written as $A + B$ and is the matrix whose elements are the sum of the corresponding elements in $A$ and $B$. The *product of a matrix A and a scalar* $\lambda$, written as $\lambda A$ or $A\lambda$, is obtained by multiplying each element of $A$ by $\lambda$. The *product AB* of an $m \times n$ matrix $A$ and an $n \times p$ matrix $B$ is the $m \times p$ matrix $C$ with elements $c_{ij} = \sum_{k=1}^{n} a_{ik} b_{kj}$. If $b$ is an $n \times 1$ matrix, i.e., an $n$-dimensional column vector, and $A$ is an $m \times n$ matrix, then $Ab$ is an $m$-dimensional (column) vector.

The *transpose* of an $m \times n$ matrix $A$ is the $n \times m$ matrix $A'$ with elements $a'_{ij} = a_{ji}$. A square matrix $A$ is *symmetric* if $A' = A$. A square $n \times n$ matrix $A$ is *nonsingular* if there is an $n \times n$ matrix called the *inverse* of $A$, denoted by $A^{-1}$ such that $A^{-1}A = I = AA^{-1}$, where $I$ is the $n \times n$ identity matrix. A square $n \times n$ matrix $A$ is nonsingular if and only if the $n$ vectors that constitute its rows are linearly independent or equivalently if the $n$ vectors that constitute its columns are linearly independent.

### *Partitioned Matrices*

It is often convenient to partition a matrix into submatrices by drawing partitioning lines through the matrix. For example, the matrix

$$A = \left[\begin{array}{cc|cc} a_{11} & a_{12} & a_{13} & a_{14} \\ \hline a_{21} & a_{22} & a_{23} & a_{24} \\ a_{31} & a_{32} & a_{33} & a_{34} \end{array}\right]$$

may be partitioned into

$$A = \begin{bmatrix} A_{11} & A_{12} \\ A_{21} & A_{22} \end{bmatrix}$$

where

$$A_{11} = [a_{11} \quad a_{12}], \qquad A_{12} = [a_{13} \quad a_{14}],$$

$$A_{21} = \begin{bmatrix} a_{21} & a_{22} \\ a_{31} & a_{32} \end{bmatrix}, \qquad A_{22} = \begin{bmatrix} a_{23} & a_{24} \\ a_{33} & a_{34} \end{bmatrix}.$$

For a partitioned matrix $A = [B \vdots C]$ we use interchangeably the notation $[B, C]$ or $[BC]$. The transpose of the partitioned matrix $A$ above is

$$A' = \begin{bmatrix} A'_{11} & A'_{21} \\ A'_{12} & A'_{22} \end{bmatrix}.$$

Partitioned matrices may be multiplied just as nonpartitioned matrices provided the dimensions involved in the partitions are compatible. Thus if

$$A = \begin{bmatrix} A_{11} & A_{12} \\ A_{21} & A_{22} \end{bmatrix}, \qquad B = \begin{bmatrix} B_{11} & B_{12} \\ B_{21} & B_{22} \end{bmatrix},$$

then

$$AB = \begin{bmatrix} A_{11}B_{11} + A_{12}B_{21} & A_{11}B_{12} + A_{12}B_{22} \\ A_{21}B_{11} + A_{22}B_{21} & A_{21}B_{12} + A_{22}B_{22} \end{bmatrix},$$

provided the dimensions of the submatrices are such that the products $A_{ij}B_{jk}$, $i, j, k = 1, 2$ above can be formed.

### *Rank of a Matrix*

The *rank* of a matrix $A$ is equal to the maximum number of linearly independent row vectors of $A$. It is also equal to the maximum number of linearly independent column vectors. An $m \times n$ matrix is said to be of *full rank* if the rank of $A$ is equal to the minimum of $m$ and $n$. A square matrix is of full rank if and only if it is invertible (i.e., nonsingular).

### *Positive Definite and Semidefinite Matrices*

A square symmetric $n \times n$ matrix $A$ is said to be *positive semidefinite* if $x'Ax \geqslant 0$ for all $x \in R^n$. It is said to be *positive definite* if $x'Ax > 0$ for all nonzero $x \in R^n$. The matrix $A$ is said to be *negative semidefinite* (*definite*) if $(-A)$ is *positive semidefinite* (*definite*).

A positive (negative) definite matrix is invertible and its inverse is also positive (negative) definite. Conversely, an invertible positive (negative) semidefinite matrix is positive (negative) definite. If $A$ and $B$ are $n \times n$ positive semidefinite (definite) matrices, then the matrix $\lambda A + \mu B$ is also positive semidefinite (definite) for all $\lambda > 0$ and $\mu > 0$. If $A$ is an $n \times n$ positive semidefinite matrix and $C$ is an $m \times n$ matrix, then the matrix $CAC'$

is positive semidefinite. If $A$ is positive definite, $C$ has full rank, and $m \leqslant n$, then $CAC'$ is positive definite.

An $n \times n$ positive definite matrix $A$ can be written as $CC'$ where $C$ is a square invertible matrix. If $A$ is positive semidefinite and its rank is $m$, then it can be written $CC'$, where $C$ is an $n \times m$ matrix of full rank.

### *Matrix Inversion Formulas*

The following formulas expressing the inverses of various matrices are often very useful. Let $A$ and $B$ be square invertible matrices and $C$ be a matrix of appropriate dimension. Then, if all the inverses below exist,

$$(A + CBC')^{-1} = A^{-1} - A^{-1}C(B^{-1} + C'A^{-1}C)^{-1}C'A^{-1}.$$

The equation can be verified by multiplying the right-hand side by $A + CBC'$ and showing that the product is the identity matrix.

Consider a partitioned matrix $M$ of the form

$$M = \begin{bmatrix} A & B \\ C & D \end{bmatrix}.$$

Then we have

$$M^{-1} = \left[\begin{array}{c|c} Q & -QBD^{-1} \\ \hline -D^{-1}CQ & D^{-1} + D^{-1}CQBD^{-1} \end{array}\right]$$

where

$$Q = (A - BD^{-1}C)^{-1},$$

provided all the inverses above exist. The proof is obtained by multiplying $M$ with the expression for $M^{-1}$ given above and verifying that the product yields the identity matrix.

## A.4 Topological Concepts in $R^n$

### *Convergence of Sequences*

A sequence of vectors $x_0, x_1, \ldots, x_k, \ldots$ in $R^n$, denoted $\{x_k\}$, is said to *converge to a limit vector* $x$ if $\|x_k - x\| \to 0$ as $k \to \infty$ (that is, if given $\varepsilon > 0$, there is an $N$ such that for all $k \geqslant N$ we have $\|x_k - x\| < \varepsilon$). If $\{x_k\}$ converges to $x$, we write $x_k \to x$ or $\lim_{k\to\infty} x_k = x$. As can be easily verified we have $Ax_k + By_k \to Ax + By$ if $x_k \to x$, $y_k \to y$, and $A, B$ are matrices of appropriate dimension.

A point $x$ is said to be a *limit point* of a sequence $\{x_k\}$ if there is a subsequence of $\{x_k\}$ that converges to $x$, i.e., if there is an infinite subset $K$ of the nonnegative integers such that $\{x_k\}_{k \in K}$ converges to $x$.

A sequence of real numbers $\{r_k\}$ that is monotonically nondecreasing (nonincreasing), i.e., satisfies $r_k \leqslant r_{k+1}$ ($r_k \geqslant r_{k+1}$) for all $k$, must either converge to a real number or be unbounded above (below), in which case we write $\lim_{k \to \infty} r_k = +\infty$ $(-\infty)$. Given any bounded sequence of real numbers $\{r_k\}$ we may consider the sequence $\{s_k\}$, where $s_k = \sup\{r_i \mid i \geqslant k\}$. Since this sequence is monotonically nonincreasing and bounded it must have a limit called the *limit superior* of $\{r_k\}$ and denoted $\limsup_{k \to \infty} r_k$. We define similarly the *limit inferior* of $\{r_k\}$ and denote it $\liminf_{k \to \infty} r_k$. If $\{r_k\}$ is unbounded above, we write $\limsup_{k \to \infty} r_k = +\infty$ and if it is unbounded below, we write $\liminf_{k \to \infty} r_k = -\infty$. We also use this notation if $r_k \in [-\infty, \infty]$ for all $k$.

### *Open, Closed, and Compact Sets*

A subset $S$ of $R^n$ is said to be *open* if for every point $x \in S$ one can find an $\varepsilon > 0$ such that $\{z \mid \|z - x\| < \varepsilon\} \subset S$. A set $S$ is *closed* if and only if its complement in $R^n$ is open. Equivalently $S$ is closed if and only if every convergent sequence $\{x_k\}$ with elements in $S$ converges to a point that also belongs to $S$. A set $S$ is said to be *compact* if and only if it is both closed and bounded (i.e., it is closed and for some $M > 0$ we have $\|x\| \leqslant M$ for all $x \in S$). A set $S$ is compact if and only if every sequence $\{x_k\}$ with elements in $S$ has at least one limit point that belongs to $S$. Another important fact is that if $S_0, S_1, \ldots, S_k, \ldots$ is a sequence of nonempty compact sets in $R^n$ such that $S_k \supset S_{k+1}$ for all $k$, then the intersection $\bigcap_{k=0}^{\infty} S_k$ is a nonempty and compact set.

### *Continuous Functions*

A function $f$ mapping a set $S_1$ into a set $S_2$ is denoted by $f: S_1 \to S_2$. A function $f: R^n \to R^m$ is said to be *continuous* if $f(x_k) \to f(x)$ whenever $x_k \to x$. Equivalently $f$ is continuous if given $x \in R^n$ and $\varepsilon > 0$, there is a $\delta > 0$ such that whenever $\|y - x\| < \delta$ we have $\|f(y) - f(x)\| < \varepsilon$. The function

$$(a_1 f_1 + a_2 f_2)(\cdot) = a_1 f_1(\cdot) + a_2 f_2(\cdot)$$

is continuous for any two scalars $a_1, a_2$ and any two continuous functions $f_1, f_2: R^n \to R^m$. If $S_1, S_2, S_3$ are any sets and $f_1: S_1 \to S_2, f_2: S_2 \to S_3$ are functions, the function $f_2 \cdot f_1: S_1 \to S_3$ defined by $(f_2 \cdot f_1)(x) = f_2[f_1(x)]$ is called the *composition* of $f_1$ and $f_2$. If $f_1: R^n \to R^m$ and $f_2: R^m \to R^p$ are continuous, then $f_2 \cdot f_1$ is also continuous.

## A.5 Convex Sets and Functions

A subset $C$ of $R^n$ is said to be *convex* if for every $x_1, x_2 \in C$ and every scalar $\alpha$ with $0 \leqslant \alpha \leqslant 1$ we have $\alpha x_1 + (1 - \alpha)x_2 \in C$. In words, $C$ is convex if the line segment connecting any two points in $C$ belongs to $C$. A function $f: C \to R$ defined over a convex subset $C$ of $R^n$ is said to be *convex* if for every $x_1, x_2 \in C$ and every scalar $\alpha$ with $0 \leqslant \alpha \leqslant 1$ we have

$$f[\alpha x_1 + (1 - \alpha)x_2] \leqslant \alpha f(x_1) + (1 - \alpha)f(x_2).$$

The function $f$ is said to be *concave* if $(-f)$ is convex. If $f: C \to R$ is convex, then the sets $\Gamma_\lambda = \{x \mid x \in C, f(x) \leqslant \lambda\}$ are also convex for every scalar $\lambda$. An important property is that a real-valued convex function on $R^n$ is always a continuous function.

If $f_1, f_2, \ldots, f_m$ are convex functions over a convex subset $C$ of $R^n$ and $\alpha_1, \alpha_2, \ldots, \alpha_m$ are nonnegative scalars, then the function $\alpha_1 f_1 + \cdots + \alpha_m f_m$ is also convex over $C$. If $f: R^m \to R$ is convex, $A$ is an $m \times n$ matrix, and $b$ is a vector in $R^m$, the function $g: R^n \to R$ defined by $g(x) = f(Ax + b)$ is also convex. If $f: R^n \to R$ is convex, then the function $g(x) = E_w\{f(x + w)\}$, where $w$ is a random vector in $R^n$, is a convex function provided the expected value is well defined and finite for every $x \in R^n$.

For functions $f: R^n \to R$ that are differentiable there are alternative characterizations of convexity. Thus if $\nabla f(x)$ denotes the gradient of $f$ at $x$, i.e.,

$$\nabla f(x) = [\partial f(x)/\partial x^1, \ldots, \partial f(x)/\partial x^n]',$$

the function $f$ is convex if and only if

$$f(y) \geqslant f(x) + \nabla f(x)'(y - x), \qquad \text{for all} \quad x, y \in R^n.$$

If $\nabla^2 f(x)$ denotes the Hessian matrix of $f$ at $x$, i.e., the matrix

$$\nabla^2 f(x) = [\partial^2 f(x)/\partial x^i \, \partial x^j]$$

the elements of which are the second derivatives of $f$ at $x$, then $f$ is convex if and only if $\nabla^2 f(x)$ is a positive semidefinite matrix for every $x \in R^n$. For detailed expositions see references [H6], [H10], and [R6].

*Appendix B*

# On Optimization Theory

Given a real-valued function $f: S \rightarrow R$ defined on a set $S$ and a subset $X \subset S$, by the optimization problem

$$\begin{aligned} &\text{minimize} \quad f(x) \\ &\text{subject to} \quad x \in X, \end{aligned} \tag{B.1}$$

we mean the problem of finding an element $x^* \in X$ (called a *minimizing element* or an *optimal solution*) such that

$$f(x^*) \leqslant f(x) \qquad \forall x \in X.$$

Such an element need not exist. For example, the scalar functions $f(x) = x$ and $f(x) = e^x$ have no minimizing elements over the set of real numbers. The first function decreases without bound to $-\infty$ as $x$ tends toward $-\infty$ while the second decreases toward 0 as $x$ tends toward $-\infty$ but always takes positive values. Given the range of values that $f(x)$ takes as $x$ ranges over $X$, i.e., the set of real numbers

$$\{f(x) | x \in X\}$$

there are two possibilities:

(a) The set $\{f(x) | x \in X\}$ is unbounded below (i.e., contains arbitrarily small real numbers) in which case we write

$$\inf\{f(x)\,|\,x \in X\} = -\infty \qquad \text{or} \qquad \inf_{x \in X} f(x) = -\infty.$$

(b) The set $\{f(x)\,|\,x \in X\}$ is bounded below, i.e., there exists a scalar $M$ such that $M \leqslant f(x)$ for all $x \in X$. The greatest lower bound of $\{f(x)\,|\,x \in X\}$ we denote by

$$\inf\{f(x)\,|\,x \in X\} \qquad \text{or} \qquad \inf_{x \in X} f(x).$$

In either case we call $\inf_{x \in X} f(x)$ the *optimal value* of problem (B.1). If a minimizing element $x^*$ exists, then

$$f(x^*) = \inf_{x \in X} f(x),$$

in which case we also write

$$f(x^*) = \min_{x \in X} f(x) = \inf_{x \in X} f(x),$$

and use the notations $\min_{x \in X} f(x)$ and $\inf_{x \in X} f(x)$ for the optimal value interchangeably.

A maximization problem of the form

$$\begin{array}{ll} \text{maximize} & f(x) \\ \text{subject to} & x \in X, \end{array}$$

may be converted into the minimization problem

$$\begin{array}{ll} \text{minimize} & -f(x) \\ \text{subject to} & x \in X, \end{array}$$

in the sense that both problems above have the same optimal solutions and the optimal value of one is equal to minus the optimal value of the other. This optimal value for the maximization problem is denoted $\sup_{x \in X} f(x)$. When a maximizing element is known to exist we also write interchangeably $\max_{x \in X} f(x)$.

*Existence of Optimal Solutions*

We are often interested in verifying the existence of at least one minimizing element in problem (B.1). Such an element clearly exists when $X$ is a finite set. When $X$ is not finite the existence of a minimizing point in problem (B.1) is guaranteed if $f: R^n \to R$ is a continuous function and $X$ is a compact subset of $R^n$. This is the *Weierstrass theorem*. By a related result existence of a minimizing point is guaranteed if $f: R^n \to R$ is a continuous function, $X = R^n$ and $f(x) \to +\infty$ if $\|x\| \to +\infty$.

*Necessary and Sufficient Conditions for Optimality*

Such conditions are available when $f$ is a differentiable function on $R^n$ and $X$ is a convex subset of $R^n$ (possibly $X = R^n$). Thus if $x^*$ is a minimizing point in problem (B.1), $f: R^n \to R$ is a continuously differentiable function on $R^n$ and $X$ is convex we have

$$\nabla f(x^*)'(x - x^*) \geqslant 0 \qquad \forall x \in X, \tag{B.2}$$

where $\nabla f(x^*)$ denotes the gradient of $f$ at $x^*$. When $X = R^n$, i.e., the minimization is unconstrained, the necessary condition (B.2) is equivalent to the familiar condition

$$\nabla f(x^*) = 0. \tag{B.3}$$

When $f$ is in addition twice continuously differentiable and $X = R^n$, an additional necessary condition is that the *Hessian matrix* $\nabla^2 f(x^*)$ *is positive semidefinite at* $x^*$. An important fact is that *if* $f: R^n \to R$ *is a convex function and* $X$ *is convex, then* (B.2) *is both a necessary and a sufficient condition for optimality* of a point $x^*$.

*Minimization of Quadratic Forms*

Let $f: R^n \to R$ be a quadratic form

$$f(x) = \tfrac{1}{2}x'Qx + b'x,$$

where $Q$ is a symmetric $n \times n$ matrix and $b \in R^n$. If $Q$ is a positive definite matrix, then $f$ is a convex function. Its gradient is given by

$$\nabla f(x) = Qx + b.$$

By (B.3), a point $x^*$ is a minimizing point of $f$ if and only if

$$\nabla f(x^*) = Qx^* + b = 0,$$

which yields

$$x^* = -Q^{-1}b.$$

For detailed expositions see references [A2], [L10], and [Z1].

*Appendix C*

# On Probability Theory

This appendix lists selectively some of the basic probabilistic notions we shall be using. Its main purpose is to familiarize the reader with some of the terminology we shall adopt. It is not meant to be exhaustive and the reader should consult references [A8], [F1], [P2], and [P3] for detailed treatments particularly regarding operations with random variables, conditional probability, Bayes' rule, etc. For an excellent recent treatment of measure theoretic probability theory see the textbook by R. B. Ash, "Real Analysis and Probability," Academic Press, 1972.

*Probability Space*

*A probability space* consists of

(a) a set $\Omega$,

(b) a collection $\mathscr{F}$ of subsets of $\Omega$, called *events*, which includes $\Omega$ and has the following properties:

(1) If $A$ is an event, then the complement $\bar{A} = \{\omega \in \Omega \,|\, \omega \notin A\}$ is also an event. (The complement of $\Omega$ is the empty set and is considered to be an event.)

(2) If $A_1, A_2$ are events, then $A_1 \cap A_2$, $A_1 \cup A_2$ are also events.

(3) If $A_1, A_2, \ldots, A_k, \ldots$ are events, then $\bigcup_{k=1}^{\infty} A_k$ and $\bigcap_{k=1}^{\infty} A_k$ are also events.

(c) a function $P(\cdot)$ assigning to each event $A$ a real number $P(A)$, called the *probability of the event* $A$, and satisfying

(1) $P(A) \geqslant 0$ for every event $A$.
(2) $P(\Omega) = 1$.
(3) $P(A_1 \cup A_2) = P(A_1) + P(A_2)$ for every pair of disjoint events $A_1, A_2$.
(4) $P(\bigcup_{k=1}^{\infty} A_k) = \sum_{k=1}^{\infty} P(A_k)$ for every sequence of mutually disjoint events $A_1, A_2, \ldots, A_k, \ldots$.

The function $P$ is referred to as a *probability measure.*

*Convention for Finite and Countable Probability Spaces*

The case of a probability space where the set $\Omega$ is a countable (possibly finite) set is encountered frequently in this text. Where we specify that $\Omega$ is finite or countable we implicitly assume that the associated collection of events is the collection of *all* subsets of $\Omega$ (including $\Omega$ and the empty set). Under these circumstances the probability of all events is specified by the probability of the elements of $\Omega$ (i.e., of the events consisting of single elements in $\Omega$). Thus if $\Omega$ is a finite set $\Omega = \{\omega_1, \omega_2, \ldots, \omega_n\}$, the probability space is specified by the probabilities $p_1, p_2, \ldots, p_n$, where $p_i$ denotes the probability of the event consisting of $\omega_i$ above. Similarly if $\Omega = \{\omega_1, \omega_2, \ldots, \omega_k, \ldots\}$, the probability space is specified by the corresponding probabilities $p_1, p_2, \ldots, p_k, \ldots$. In either case we refer to $(p_1, p_2, \ldots, p_n)$ or $(p_1, p_2, \ldots, p_k, \ldots)$ as a *probability distribution over* $\Omega$.

*Random Variables*

Given a probability space $(\Omega, \mathscr{F}, P)$, a *random variable* on the probability space is a function $x: \Omega \to R$ such that for every scalar $\lambda$ the set

$$\{\omega \in \Omega \mid x(\omega) \leqslant \lambda\}$$

is an event, i.e., belongs to the collection $F$.

An $n$-dimensional *random vector* $x = (x_1, \ldots, x_n)$ is an $n$-tuple of random variables $x_1, x_2, \ldots, x_n$ each defined on the same probability space.

The *distribution function* (or *cumulative distribution function*) $F: R \to R$ of a random variable $x$ is defined by

$$F(z) = P(\{\omega \in \Omega \mid x(\omega) \leqslant z\}),$$

i.e., $F(z)$ is equal to the probability that the random variable takes a value less than or equal to $z$.

The distribution function $F : R^n \to R$ of a random vector $x = (x_1, x_2, \ldots, x_n)$ is defined by

$$F(z_1, z_2, \ldots, z_n) = P(\{\omega \in \Omega \mid x_1(\omega) \leqslant z_1, x_2(\omega) \leqslant z_2, \ldots, x_n(\omega) \leqslant z_n\}).$$

Given the distribution function of a random vector $x = (x_1, \ldots, x_n)$ the (marginal) distribution function of each random variable $x_i$ is obtained from

$$F_i(z_i) = \lim_{z_j \to \infty,\, j \neq i} F(z_1, z_2, \ldots, z_n).$$

The random variables $x_1, \ldots, x_n$ are said to be *independent* if

$$F(z_1, \ldots, z_n) = F_1(z_1) \cdot F_2(z_2) \cdots F_n(z_n),$$

for all scalars $z_1, \ldots, z_n$.

The *expected value of a random variable* $x$ with distribution function $F$ is defined as

$$E\{x\} = \int_{-\infty}^{\infty} z \, dF(z)$$

provided the integral above is well defined.

The *expected value of a random vector* $x = (x_1, \ldots, x_n)$ is the vector $E\{x\} = (E\{x_1\}, E\{x_2\}, \ldots, E\{x_n\})$.

The *covariance matrix* of a random vector $x = (x_1, \ldots, x_n)$ with expected value $E\{x\} = (\bar{x}_1, \ldots, \bar{x}_n)$ is defined to be the $n \times n$ symmetric positive semidefinite matrix

$$Q_x = \begin{bmatrix} E\{(x_1 - \bar{x}_1)^2\} & \cdots & E\{(x_1 - \bar{x}_1)(x_n - \bar{x}_n)\} \\ & \vdots & \\ E\{(x_n - \bar{x}_n)(x_1 - \bar{x}_1)\} & \cdots & E\{(x_n - \bar{x}_n)^2\} \end{bmatrix}$$

provided the expectations above are well defined.

Two random vectors $x$ and $y$ are said to be *uncorrelated* if

$$E\{(x - E\{x\})(y - E\{y\})'\} = 0$$

where $(x - E\{x\})$ above is viewed as a column vector and $(y - E\{y\})'$ is viewed as a row vector.

The random vector $x = (x_1, \ldots, x_n)$ is said to be characterized by a piecewise continuous *probability density function* $f : R^n \to R$ if $f$ is piecewise continuous and

$$F(z_1, \ldots, z_n) = \int_{-\infty}^{z_1} \int_{-\infty}^{z_2} \cdots \int_{-\infty}^{z_n} f(y_1, \ldots, y_n) \, dy_1 \cdots dy_n,$$

for every $z_1, \ldots, z_n$.

*Conditional Probability*

We shall restrict ourselves to the case where the underlying probability space $\Omega$ is a countable (possibly finite) set and the set of events is the set of all subsets of $\Omega$.

Given two events $A$ and $B$ we define the *conditional probability of $B$ given $A$* by

$$P(B|A) = \begin{cases} P(A \cap B)/P(A) & \text{if} \quad P(A) > 0, \\ 0 & \text{if} \quad P(A) = 0. \end{cases}$$

If $B_1, B_2, \ldots$ are a countable (possibly finite) collection of mutually exclusive and exhaustive events (i.e., the sets $B_i$ are disjoint and their union is $\Omega$) and $A$ is an event, then we have

$$P(A) = \sum_i P(A \cap B_i).$$

From this one may prove that

$$P(A) = \sum_i P(B_i)P(A|B_i).$$

This is called the theorem of *total probability*. From the expressions above we obtain for every $k$

$$P(B_k|A) = P(A \cap B_k)/P(A) = P(B_k)P(A|B_k)/\textstyle\sum_i P(B_i)P(A|B_i),$$

provided $P(A) > 0$. The relation above is referred to as *Bayes' rule*.

Consider now two random vectors $x$ and $y$ on the (countable) probability space taking values in $R^n$ and $R^m$, respectively [i.e., $x(\omega) \in R^n$, $y(\omega) \in R^m$ for all $\omega \in \Omega$]. Given two subsets $X$ and $Y$ of $R^n$ and $R^m$, respectively, we denote

$$P(X|Y) = P(\{\omega|x(\omega) \in X\}|\{\omega|y(\omega) \in Y\}).$$

For a fixed vector $w \in R^n$ we define the *conditional distribution function* of $x$ given $w$ by

$$F(z|w) = P(\{\omega|x(\omega) \leqslant z\}|\{\omega|y(\omega) = w\}),$$

and the *conditional expectation* of $x$ given $w$ by

$$E\{x|w\} = \int_{R_n} z\, dF(z|w),$$

provided the integral above is well defined. Note that $E\{x|w\}$ is a function mapping $w$ into $R^n$. Similarly one may define the conditional covariance of $x$ given $w$, etc.

If $\omega_1, \omega_2, \ldots$ are the elements of $\Omega$, let us denote

$$z_i = x(\omega_i), \qquad w_i = y(\omega_i), \qquad i = 1, 2, \ldots.$$

Also for any vectors $z \in R^n$, $w \in R^m$, let us denote

$$P(z) = P(\{\omega \mid x(\omega) = z\}), \qquad P(w) = P(\{\omega \mid y(\omega) = w\}).$$

We have $P(z) = 0$ if $z \neq z_i$, $i = 1, 2, \ldots$, and $P(w) = 0$ if $w \neq w_i$, $i = 1, 2, \ldots$. Denote also

$$P(z \mid w) = P(\{\omega \mid x(\omega) = z\} \mid \{\omega \mid y(\omega) = w\}).$$

Then if $P(w) > 0$, Bayes' rule yields

$$P(z_i \mid w) = P(z_i)P(w \mid z_i)/\textstyle\sum_j P(z_j)P(w \mid z_j), \qquad i = 1, 2, \ldots.$$

$$P(z \mid w) = 0 \qquad \text{if} \quad z \neq z_i, \quad i = 1, 2, \ldots,$$

where $P(w \mid z) = P(\{\omega \mid y(\omega) = w\} \mid \{\omega \mid x(\omega) = z\})$.

*Appendix D*

# On Finite State Markov Chains

A square $n \times n$ matrix $[p_{ij}]$ is said to be a *stochastic* matrix if all its elements are nonnegative, i.e., $p_{ij} \geqslant 0, i, j = 1, \ldots, n$, and the sum of the elements of each of its rows equals unity, i.e., $\sum_{j=1}^{n} p_{ij} = 1$ for all $j = 1, \ldots, n$.

*Stationary Finite State Markov Chains*

Suppose we are given a stochastic $n \times n$ matrix $P$ together with a finite set $S = \{s^1, \ldots, s^n\}$ called the *state space* with elements $s^1, \ldots, s^n$ called *states*. The pair $(S, P)$ will be referred to as a *stationary finite state Markov chain*. We associate with $(S, P)$ a process whereby an initial state $x_0 \in S$ is chosen in accordance with some initial probability distribution

$$p_0 = (p_0^1, p_0^2, \ldots, p_0^n).$$

Subsequently a transition is made from state $x_0$ to a new state $x_1 \in S$ in accordance with a probability distribution specified by $P$ as follows. The probability that the new state $x_1$ will be $s^j$ is equal to $p_{ij}$ whenever the initial state $x_0$ is $s^i$, i.e.,

$$P(x_1 = s^j | x_0 = s^i) = p_{ij}, \qquad i, j = 1, \ldots, n.$$

Similarly subsequent transitions produce states $x_2, x_3, \ldots$ in accordance with

$$P(x_{k+1} = s^j | x_k = s^i) = p_{ij}, \qquad i, j = 1, \ldots, n. \tag{D.1}$$

The probability that after the $k$th transition the state $x_k$ will be equal to $s_j$ given that the initial state $x_0$ is equal to $s_i$ is denoted

$$p_{ij}^k = P(x_k = s^j | x_0 = s^i), \qquad i, j = 1, \ldots, n. \tag{D.2}$$

These probabilities may be easily calculated to be equal to the elements of the matrix $P^k$ ($P$ raised to the $k$th power), in the sense that $p_{ij}^k$ is the element in the $i$th row and $j$th column of $P^k$:

$$P^k = [p_{ij}^k]. \tag{D.3}$$

Given the initial probability distribution $p_0$ of the state $x_0$ (viewed as a row vector in $R^n$), the probability distribution of the state $x_k$ after $k$ transitions

$$p_k = (p_k^1, p_k^2, \ldots, p_k^n)$$

(viewed again as a row vector) is given by

$$p_k = p_0 P^k, \qquad k = 1, 2, \ldots. \tag{D.4}$$

This relation follows immediately from (D.2) and (D.3) once we write

$$p_k^j = \sum_{i=1}^n P(x_k = s^j | x_0 = s^i) p_0^i = \sum_{i=1}^n p_{ij}^k p_0^i.$$

*Nonstationary Finite State Markov Chains*

Suppose we are given instead of a single stochastic matrix $P$ a sequence $\{P_k\}$ of stochastic $n \times n$ matrices $P_0, P_1, \ldots$. We refer to the pair $(S, \{P_k\})$ as a *nonstationary finite state Markov chain* and we associate with it the following process. The initial state $x_0 \in S$ is chosen in accordance with an initial distribution $p_0 = (p_0^1, p_0^2, \ldots, p_0^n)$. Subsequently a transition is made to a state $x_1 \in S$ in accordance with

$$P(x_1 = s^j | x_0 = s^i) = p_{ij}(0), \qquad i, j = 1, \ldots, n,$$

where $p_{ij}(0)$ is the element in the $i$th row and $j$th column of the stochastic matrix $P_0$ ($P_0 = [p_{ij}(0)]$). Subsequent transitions produce states in accordance with

$$P(x_{k+1} = s^j | x_k = s^i) = p_{ij}(k), \qquad i, j = 1, \ldots, n,$$

where $p_{ij}(k)$ is the element in the $i$th row and $j$th column of $P_k$. The probabilities

$$p_{ij}^k = P(x_k = s^j | x_0 = s^i), \qquad i, j = 1, \ldots, n,$$

can be calculated to be equal to the elements of the matrix $P_0 P_1 \cdots P_{k-1}$:

$$P_0 P_1 \cdots P_{k-1} = [p_{ij}^k].$$

Given the initial probability distribution $p_0$ of the state $x_0$ the probability distribution $p_k = (p_k^1, \ldots, p_k^n)$ of $x_k$ is given by

$$p_k = p_0 P_0 P_1 \cdots P_{k-1}, \qquad k = 1, 2, \ldots.$$

We subsequently restrict ourselves to stationary Markov chains.

*Classification of States of a Markov Chain*

Given a stationary finite state Markov chain $(S, P)$ we say that two states $x^i$ and $x^j$ *communicate* if there exist two positive integers $k_1$ and $k_2$ such that $p_{ij}^{k_1} > 0$ and $p_{ji}^{k_2} > 0$. This definition does not exclude the possibility of a state communicating with itself.

Let $\tilde{S} \subset S$ be a subset of states such that:

(a) All states in $\tilde{S}$ communicate with each other.
(b) If $s^i \in \tilde{S}$ and $s^j \notin \tilde{S}$, then $p_{ij}^k = 0$ for all $k$.

Then we say that $\tilde{S}$ forms an *ergodic class* of states.

If $S$ forms by itself an ergodic class (i.e., all states communicate with each other), then we say that the Markov chain is *irreducible*. It is possible that there exist several ergodic classes. It is also possible to prove that at least one ergodic class must exist. States that do not belong to any ergodic class are called *transient*. Transient states are characterized by the fact that

$$\lim_{k\to\infty} p_{ii}^k = 0 \qquad \text{if and only if} \qquad s^i \text{ is transient.}$$

In other words, if the process starts at a transient state the probability of returning to the same state after $k$ transitions diminishes to zero as $k$ tends to infinity.

It is easy to see from the definitions given that during the process of transition between states once an ergodic class is entered then the process remains within this ergodic class for every subsequent transition. Thus if the process starts within an ergodic class, it stays within that class. If it starts at a transient state, it eventually (with probability one) enters an ergodic class after a number of transitions and subsequently remains there.

*Limiting Probabilities*

An important property of any stochastic matrix $P$ is that the matrix $P^*$ defined by

$$P^* = \lim_{N\to\infty} (1/N) \sum_{k=0}^{N-1} P^k$$

exists [in the sense that the sequences of the elements of $(1/N)\sum_{k=0}^{N-1} P^k$ converge to the corresponding elements of $P^*$]. The elements $p_{ij}^*$ of $P^*$ satisfy

$$p_{ij}^* \geq 0, \qquad \sum_{j=1}^{n} p_{ij}^* = 1, \qquad i = 1, \ldots, n,$$

i.e., $P^*$ is a stochastic matrix.

If $\tilde{S} \subset S$ is an ergodic class and $s^i$, $s^j \in \tilde{S}$, then it may be proved that for all $k$ such that $s^k \in \tilde{S}$,

$$p_{ik}^* = p_{jk}^* > 0,$$

so that *if a Markov chain is irreducible, the matrix $P^*$ has identical rows*. Also if $s^j$ is a transient state, we have

$$p_{ij}^* = 0 \qquad \forall i \in S,$$

so that the columns of the matrix $P^*$ corresponding to transient states are identically zero.

*First Passage Times*

Let us denote by $q_{ij}^k$ the probability that the state will be $s^j$ for the first time after exactly $k \geq 1$ transitions given that the initial state is $s^i$, i.e.,

$$q_{ij}^k = P(x_k = s^j, x_m \neq s^j, 1 \leq m < k \mid x_0 = s^i).$$

Denote also for fixed $i$ and $j$,

$$K_{ij} = \min\{k \geq 1 \mid x_k = s^j, x_0 = s^i\}.$$

Then $K_{ij}$, called the *first passage time from i to j*, may be viewed as a random variable. We have for every $k = 1, 2, \ldots,$

$$P(K_{ij} = k) = q_{ij}^k,$$

and we write

$$P(K_{ij} = \infty) = P(x_k \neq s^j, \forall k = 1, 2, \ldots \mid x_0 = s^i) = 1 - \sum_{k=1}^{\infty} q_{ij}^k.$$

Of course, it is possible that $\sum_{k=1}^{\infty} q_{ij}^k < 1$. This will occur, for example, if $s^j$ cannot be reached from $s^i$ in which case $q_{ij}^k = 0$ for all $k = 1, 2, \ldots$. The *mean first passage time* from $i$ to $j$ is the expected value of $K_{ij}$:

$$E\{K_{ij}\} = \begin{cases} \sum_{k=1}^{\infty} k q_{ij}^k & \text{if } \sum_{k=1}^{\infty} q_{ij}^k = 1, \\ \infty & \text{if } \sum_{k=1}^{\infty} q_{ij}^k < 1. \end{cases}$$

It may be proved that if $s^i$ and $s^j$ belong to the same ergodic class, then

$$E\{K_{ij}\} < \infty.$$

If $s^i$ and $s^j$ belong to two different ergodic classes, then $E\{K_{ij}\} = E\{K_{ji}\} = \infty$. If $s^i$ belongs to an ergodic class and $s^j$ is transient, we have $E\{K_{ij}\} = \infty$.

*Interpretation of Accessibility Conditions*

In Chapter 8 we utilized the condition that there exists a special state $s^k \in S$ such that

$$E\{K_{ik}\} < \infty \qquad \forall s^i \in S. \tag{D.5}$$

Now in view of the preceding discussion the state $s^k$ cannot be transient and thus it must belong to an ergodic class. Furthermore, there cannot be more than one ergodic class since if some state $s^j$ belonged to a different ergodic class than the one of $s^k$ we would have $E\{K_{jk}\} = \infty$. Thus there must exist a single ergodic class and $s^k$ must belong to it. Conversely, condition (D.5) always holds when a single ergodic class exists and $s^k$ belongs to it. In conclusion assuming existence of a state $s^k$ such that (D.5) holds is equivalent to assuming the existence of a single ergodic class.

Assume now that we have a collection of $n \times n$ transition probability matrices $P(\mu)$ parameterized by the elements $\mu$ of some set $M$. Let us denote by $E\{K_{ij}(\mu)\}$ the mean first passage time for going from $s^i$ to $s^j$ when the transition probability matrix is $P(\mu)$. Then clearly from the discussion given earlier it follows that the condition

$$\text{there exists } s^k \in S \text{ such that } E\{K_{ik}(\mu)\} < 0 \text{ for all } s^i \in S,\ \mu \in M,$$

is equivalent to assuming that, for every $\mu$, $P(\mu)$ gives rise to a Markov chain with a single ergodic class and $s^k$ belongs to that class. In particular, the condition above is satisfied if the Markov chain corresponding to $P(\mu)$ is irreducible for every $\mu \in M$. For detailed expositions see references [C3] and [K7].

# References


[A1] Aoki, M., "Optimization of Stochastic Systems—Topics in Discrete-Time Systems." Academic Press, New York, 1967.

[A2] Aoki, M., "Introduction to Optimization Techniques." Macmillan, New York, 1971.

[A3] Aoki, M., and Li, M. T., Optimal discrete-time control systems with cost for observation, *IEEE Trans. Automatic Control* **AC-14** (1969), 165–175.

[A4] Arrow, K. J., "Aspects of the Theory of Risk Bearing." Yrjo Jahnsson Lecture Ser., Helsinki, Finland, 1965.

[A5] Arrow, K. M., Blackwell, D., and Girshick, M. A., Bayes and minimax solutions of sequential design problems, *Econometrica* **17** (1949), 213–244.

[A6] Arrow, K. J., Harris, T., and Marschack, J., Optimal inventory policy, *Econometrica* **19** (1951), 250–272.

[A7] Arrow, K. H., Karlin, S., and Scarf, H., "Studies in the Mathematical Theory of Inventory and Production." Stanford Univ. Press, Stanford, California, 1958.

[A8] Ash, R. B., "Basic Probability Theory." Wiley, New York, 1970.

[A9] Åström, K. J., and Wittenmark, B., On self-tuning regulators, *Automatica* **9** (1973), 185–199.

[A10] Atkinson, R. C., Bower, G. H., and Crothers, E. J., "An Introduction to Mathematical Learning Theory." Wiley, New York, 1965.

[B1] Bar-Shalom, Y., and Tse, E., Dual effect, certainty equivalence, and separation in stochastic control, *IEEE Trans. Automatic Control* **AC-19** (1974), 494–500.

[B2] Bather, J., Optimal decision procedures for finite Markov chains, *Advances in Appl. Probability* **5** (1973), 328–339, 521–540, 541–553.

[B3] Bellman, R., "Dynamic Programming." Princeton Univ. Press, Princeton, New Jersey, 1957.

[B4] Bellman, R., and Dreyfus, S., "Applied Dynamic Programming." Princeton Univ. Press, Princeton, New Jersey, 1962.

[B5] Bertsekas, D. P., On the separation theorem for linear systems, quadratic criteria, and correlated noise, unpubl. rep., Electronic Systems Lab., MIT, Cambridge, Massachusetts, September 1970.

[B6] Bertsekas, D. P., "Control of Uncertain Systems with a Set-Membership Description of the Uncertainty." Ph.D. Dissertation, MIT, Cambridge, Massachusetts, 1971.

[B7] Bertsekas, D. P., Stochastic optimization problems with nondifferentiable cost functionals with an application in stochastic programming, *Proc. IEEE Decision and Control Conf. New Orleans, Louisiana, December 1972.*

[B8] Bertsekas, D. P., On the solution of some minimax problems, *Proc. IEEE Decision and Control Conf. New Orleans, Louisiana, December 1972.*

[B9] Bertsekas, D. P., Infinite time reachability of state space regions by using feedback control, *IEEE Trans. Automatic Control* **AC-17** (1972), 604–613.

[B10] Bertsekas, D. P., "Stochastic optimization problems with nondifferentiable cost functionals, *J. Optimization Theory Appl.* **12** (1973), 218–231.

[B11] Bertsekas, D. P., Linear convex stochastic control problems over an infinite time horizon, *IEEE Trans. Automatic Control* **AC-18** (1973), 314–315.

[B12] Bertsekas, D. P., Necessary and sufficient conditions for existence of an optimal portfolio, *J. Econom. Theory* **8** (1974), 235–247.

[B13] Bertsekas, D. P., Monotone mappings with application in dynamic programming, Coordinated Sci. Lab. Working Paper, Univ. of Illinois, Urbana, Illinois, June 1976, *SIAM J. Control and Optimization* **15** (1977), 438–464.

[B14] Bertsekas, D. P., On error bounds for successive approximation methods, *IEEE Trans. Automatic Control*, **AC-21** (1976), 394–396.

[B15] Bertsekas, D. P., Convergence of discretization procedures in dynamic programming, *IEEE Trans. Automatic Control* **AC-20** (1975), 415–419.

[B16] Bertsekas, D. P., and Shreve, S. E., "Stochastic Optimal Control: The Discrete Time Case," Academic Press, N.Y., 1978.

[B17] Bertsekas, D. P., and Rhodes, I. B., On the minimax reachability of target sets and target tubes, *Automatica* **7** (1971), 233–247.

[B18] Bertsekas, D. P., and Rhodes, I. B., Sufficiently informative functions and the minimax feedback control of uncertain dynamic systems, *IEEE Trans. Automatic Control* **AC-18** (1973), 117–124.

[B19] Blackwell, D., Discrete dynamic programming, *Ann. Math. Statist.* **33** (1962), 719–726.

[B20] Blackwell, D., Discounted dynamic programming, *Ann. Math. Statist.* **36** (1965), 226–235.

[B21] Blackwell, D., Positive dynamic programming, *Proc. 5th Berkeley Symp. Math., Statist., and Probability*, **1** (1965), 415–418.

[B22] Blackwell, D., On stationary policies, *J. Roy. Statist. Soc. Ser. A.* **133** (1970), 33–38.

[B23] Blackwell, D., and Girshick, M. A., "Theory of Games and Statistical Decisions." Wiley, New York, 1954.

[B23a] Blackwell, D., Freedman, D., and Orkin, M., The optimal reward operator in dynamic programming, *Ann. Prob.* **2** (1974), 926–941.

[B24] Blake, I. F., and Thomas, J. B., On a class of processes arising in linear estimation theory, *IEEE Trans. Information Theory* **IT-14** (1968), 12–16.

[B25] Brown, B. W., On the iterative method of dynamic programming on a finite space discrete Markov process, *Ann. Math. Statist.* **36** (1965), 1279–1286.

[C1] Cashman, W. F., and Wonham, W. M., A computational approach to optimal control of stochastic saturating systems, *Internat. J. Control* **10** (1969), 77–98.

[C2] Chernoff, H., "Sequential Analysis and Optimal Design." Regional Conference Series in Applied Mathematics, SIAM, Philadelphia, Pennsylvania, 1972.

[C3] Chung, K. L., "Markov Chains with Stationary Transition Probabilities." Springer-Verlag, Berlin and New York, 1960.

[C4] Cooper, C. A., and Nahi, N. E., An optimal stochastic control problem with observation cost, *Proc. Joint Automatic Control Conf. Atlanta, Georgia, June 1970.*

[C5] Curry, R. E., A new algorithm for suboptimal stochastic control, *IEEE Trans. Automatic Control* **AC-12** (1967), 533–536.

[D1] DeGroot, M. H., "Optimal Statistical Decisions." McGraw-Hill, New York, 1970.

[D2] Denardo, E. V., Contraction mappings in the theory underlying dynamic programming, *SIAM Rev.* **9** (1967), 165–177.

[D3] D'Epenoux, F., Sur un probleme de production et de stockage dans l'aleatoire, *Rev. Francaise Automat. Informat. Recherche Operationnelle*, **14** (1960), (*English Transl.: Management Sci.* **10** (1963), 98–108).

[D4] Derman, C., "Finite State Markovian Decision Processes." Academic Press, New York, 1970.

[D5] Deshpande, J. G., Upadhyay, T. N., and Lainiotis, D. G., Adaptive control of linear control systems, *Automatica* **9** (1973), 107–115.

[D6] Dreyfus, S. E., "Dynamic Programming and the Calculus of Variations." Academic Press, New York, 1965.

[D7] Dreyfus, S. E., and Stalford, H. L., A computational successive improvement scheme for adaptive optimal control processes, *J. Optimization Theory Appl.* **10** (1972), 78–93.

[D8] Dubins, L., and Savage, L. M., "How To Gamble If You Must." McGraw-Hill, New York, 1965.

[D9] Dynkin, E. B., Controlled random sequences, *Theor. Probability Appl.* **10** (1965), 1–14.

[E1] Eaton, J. H. and Zadeh, L. A., Optimal pursuit strategies in discrete state probabilistic systems, *Trans. ASME Ser. D. J. Basic Eng.* **84** (1962), 23–29.

[E2] Eckles, J. E., Optimum maintenance with incomplete information, *Operations Res.* **16** (1968), 1058–1067.

[F1] Feller, W., "An Introduction to Probability Theory and Its Applications," 3rd ed. Wiley, New York, 1968.

[F2] Fishburn, P. C., Utility theory, *Management Sci.* **14** (1968), 335–378.

[F3] Fishburn, P. C., "Utility Theory for Decision Marking." Wiley, New York, 1970.

[F3a] Freedman, D., The optimal reward operator in special classes of dynamic programming problems, *Ann. Prob.* **2** (1974), 942–949.

[F4] Furukawa, N., A Markov decision process with nonstationary transition laws, *Bull. Math. Statist.* **13** (1968), 41–52.

[G1] Groen, G. J., and Atkinson, R. C., Models for optimizing the learning process, *Psychol. Bull.* **66** (1966), 309–320.

[G2] Gunckel, T. L., and Franklin, G. R., A general solution for linear sampled-data control, *Trans. ASME Ser. D. J. Basic Engrg.*, **85** (1963), 197–201.

[H1] Hakansson, N., Optimal investment and consumption strategies for a class of utility functions, *Econometrica* **38** (1970), 587–607.

[H2] Hakansson, N. H., Optimal investment and consumption strategies under risk, an uncertain lifetime, and insurance, *Internat. Econom. Rev.* **10** (1970), 443–466.

[H3] Hakansson, N., Multiperiod mean-variance analysis: toward a general theory of portfolio choice, *J. Finance* (1971), 857–884.

[H4] Hakansson, N. H., On optimal myopic portfolio policies with and without serial correlation of yields, *J. Business* **44** (1971), 325–344.

[H5] Halmos, P., "Measure Theory." Van Nostrand-Reinhold, Princeton, New Jersey, 1950.

[H6] Halmos, P., "Finite-Dimensional Vector Spaces." Van Nostrand-Reinhold, Princeton, New Jersey, 1958.

[H7] Hastings, N. A. J., Some notes on dynamic programming and replacement, *Operational Res. Quart.* **19** (1968), 453–464.

[H8] Hastings, N. A. J., Bounds on the gain of a Markov decision process, *Operations Res.* **19** (1971), 240–243.

[H9] Hinderer, K., "Foundations of Non-Stationary Dynamic Programming with Discrete Time Parameter." Springer-Verlag, Berlin and New York, 1970.

[H10] Hoffman, K., and Kunze, R., "Linear Algebra." Prentice-Hall, Englewood Cliffs, New Jersey, 1961.

[H11] Holt, C. C., Modigliani, F., and Simon, H. A., A linear decision rule for production and employment scheduling, *Management Sci.* **2** (1955), 1–30.

[H12] Hordijk, A., and Tijms, H., The method of successive approximations and Markovian decision problems, *Operations Res.* **22** (1974), 519–521.

[H13] Hordijk, A., and Tijms, H. C., Convergence results and approximations for optimal $(s, S)$ policies, *Management Sci.* **20** (1974), 1432–1438.

[H14] Hordijk, A., and Tijms, H. C., On a conjecture of Iglehart, *Management Sci.* **21** (1975), 1342–1345.

[H15] Howard, R., "Dynamic Programming and Markov Processes." MIT Press, Cambridge, Massachusetts, 1960.

[H16] Howard, R., "Dynamic Probabilistic Systems," Vols. I and II. Wiley, New York, 1971.

[H17] Howard, R., and Matheson, J., Risk-sensitive Markov decision processes, *Management Sci.* **18** (1972), 356–369.

[I1] *IEEE Trans. Automatic Control*, special issue on Linear–Quadratic Gaussian Problem, **AC-16** (1971).

[I2] Iglehart, D. L., Optimality of $(S, s)$ policies in the infinite horizon dynamic inventory problem, *Management Sci.* **9** (1963), 259–267.

[I3] Iglehart, D. L., Dynamic programming and stationary analysis of inventory problems, *in* Scarf, H., Gilford, D., and Shelly, M. (eds.), "Multistage Inventory Models and Techniques." Stanford Univ. Press, Stanford, California, 1963.

[J1] Jacobson, D. H., Optimal stochastic linear systems with exponential performance criteria and their relation to deterministic differential games, *IEEE Trans. Automatic Control* **AC-18** (1973), 124–131.

[J2] Jacobson, D. H., A general result in stochastic optimal control of nonlinear discrete-time systems with quadratic performance criteria, *J. Math. Anal. Appl.* **47** (1974), 153–161.

[J3] Jazwinski, A. H., "Stochastic Processes and Filtering Theory." Academic Press, New York, 1970.

[J4] Jewell, W., Markov renewal programming I and II, *Operations Res.* **11** (1963), 938–971.

[J5] Joseph, P. D., and Tou, J. T., On linear control theory, *AIEE Trans.* **80** (II), (1961), 193–196.

[K1] Kalman, R. E., A new approach to linear filtering and prediction problems, *Trans. ASME Ser. D. J. Basic Engrg.* **82** (1960), 35–45.

[K2] Kalman, R. E., and Koepcke, R. W., Optimal synthesis of linear sampling control systems using generalized performance indexes, *Trans. ASME* **80** (1958), 1820–1826.

[K3] Kalymon, B. A., Stochastic prices in a single-item inventory purchasing model, *Operations Res.* **19** (1971), 1434–1458.

[K4] Kamin, J. H., Optimal portfolio revision with a proportional transaction cost, *Management Sci.* **21** (1975), 1263–1271.

[K5] Karush, W., and Dear, E. E., Optimal stimulus presentation strategy for a stimulus sampling model of learning, *J. Mathematical Psychology* **3** (1966), 15–47.

[K6] Kaufmann, A., and Cruon, R., "Dynamic Programming." Academic Press, New York, 1967.

[K7] Kemeny, J. G., and Snell, J. L., "Finite Markov Chains." Van Nostrand–Reinhold, Princeton, New Jersey, 1960.

[K8] Kleindorfer, P. R., and Glover, K., Linear convex stochastic optimal control with applications in production planning, *IEEE Trans. Automatic Control* **AC-18** (1973), 56–59

[K9] Kucera, V., A contribution to matrix quadratic equations, *IEEE Trans. Automatic Control* **AC-17** (1972), 344–347.

[K10] Kushner, H. "Introduction to Stochastic Control." Holt, New York, 1971.

[L1] Lanery, E., Etude asymptotic des systemes Markoviens a commande, *Rev. Francaise Automat.* Informat. Recherche Operationnelle **1** (1967), 3–57.

[L2] Larson, R. E., "State Increment Dynamic Programming." Amer. Elsevier, New York, 1968.

[L3] Leland, H., On turnpike portfolios, *in* Szegö, G., and Shell, K. (eds.), "Symposium on Mathematical Methods in Investments and Finance." Amer. Elsevier, New York, 1972; also Tech. Rep. No. 49, IMSSS, Stanford, California, October 1971.

[L4] Levhari, D., and Srinivasan, T., Optimal savings under uncertainty, *Rev. Econom. Studies* **36** (1969), 153–163.

[L5] Liusternik, L., and Sobolev, V., "Elements of Functional Analysis." Ungar, New York, 1961.

[L6] Ljung, L., and Wittenmark, B., Asymptotic properties of self-tuning regulators, Rep. 7404, Div. of Autom. Control, Lund Inst. of Technology, Sweden, Feb. 1974.

[L7] Luce, R. D., and Raiffa, H., "Games and Decisions." Wiley, New York, 1957.

[L8] Luenberger, D. G., "Optimization by Vector Space Methods." Wiley, New York, 1969.

[L9] Luenberger, D. G., Cyclic dynamic programming: a procedure for problems with fixed delay, *Operations Res.* **19** (1971), 1101–1110.

[L10] Luenberger, D. G., "Introduction to Linear and Nonlinear Programming." Addison-Wesley, Reading, Massachusetts, 1973.

[M1] Maitra, A., A note on positive dynamic programming, *Ann. Math. Statist.* **40** (1969) 316–319.

[M2] Malinvaud, E., First order certainty equivalence, *Econometrica* **37** (1969), 706–718.

[M3] Manne, A., Linear programming and sequential decisions, *Management Sci.* **6** (1960), 259–267.

[M4] Markovitz, H., "Portfolio Selection." Wiley, New York, 1959.

[M5] McQueen, J., A modified dynamic programming method for Markovian decision problems, *J. Math. Anal. Appl.* **14** (1966), 38–43.

[M6] Meditch, J. S. "Stochastic Optimal Linear Estimation and Control." McGraw-Hill, New York, 1969.

[M7] Mendel, J. M., "Discrete Techniques of Parameter Estimation—The Equation Error Formulation." Dekker, New York, 1973.

[M8] Mossin, J., Optimal multi-period portfolio policies, *J. Business* **41** (1968), 215–229.

[M9] Murphy, W. J., Optimal stochastic control of discrete linear systems with unknown gain, *IEEE Trans. Automatic Control* **AC-13** (1968), 338–344.

[N1] Nahi, N., "Estimation Theory and Applications." Wiley, New York, 1969.

[N2] Nemhauser, G. L., "Introduction to Dynamic Programming." Wiley, New York, 1966.

[N3] von Neumann, J., and Morgenstern, O., "Theory of Games and Economic Behavior." Princeton Univ. Press, Princeton, New Jersey, 1947.

[O1] Odoni, A. R., On finding the maximal gain for Markov decision processes, *Operations Res.* **17** (1969), 857–860.

[O2] Ornstein, D., On the existence of stationary optimal strategies, *Proc. Amer. Math. Soc.* **20** (1969), 563–569.

[O3] Owen, G., "Game Theory." Saunders, Philadelphia, Pennsylvania, 1969.

[P1] Pallu de la Barriere, R., "Optimal Control Theory." Saunders, Philadelphia, Pennsylvania, 1967.

[P2] Papoulis, A., "Probability, Random Variables and Stochastic Processes." McGraw-Hill, New York, 1965.

[P3] Parzen, E., "Modern Probability Theory and Its Applications." Wiley, New York, 1960.

[P4] Payne, H. J., and Silverman, L. M., On the discrete-time algebraic Riccati equation, *IEEE Trans. Automatic Control* **AC-18** (1973), 226–234.

[P5] Phelps, E.,. The accumulation of risky capital: a sequential utility analysis, *Econometrica* **30** (1962), 729–741.

[P6] Pierce, B. D., and Sworder, D. D., Bayes and minimax controllers for a linear system with stochastic jump parameters, *IEEE Trans. Automatic Control* **AC-16** (1971), 300–307.

[P7] Pratt, J. W., Risk aversion in the small and in the large, *Econometrica* **32** (1964), 122–136.

[R1] Raiffa, H., "Decision Analysis: Introductory Lectures on Choices under Uncertainty." Addison-Wesley, Reading, Massachusetts, 1968.

[R2] Rockafellar, R. T., "Convex Analysis." Princeton Univ. Press, Princeton, New Jersey, 1970.

[R3] Ross, S. M., Arbitrary state Markovian decision processes, *Ann. Math. Statist.* **39** (1968), 2118–2122.

[R4] Ross, S. M., "Applied Probability Models with Optimization Applications." Holden-Day, San Francisco, California, 1970.

[R5] Royden, H. L., "Real Analysis." Macmillan New York, 1968.

[R6] Rudin, D., "Principles of Mathematical Analysis." McGraw-Hill, New York, 1964.

[S1] Saridis, G. N., "Self-Organizing Control of Stochastic Systems." Dekker, New York, 1977.

[S2] Saridis, G. N., and Dao, T. K., A learning approach to the parameter-adaptive self-organizing control problem, *Automatica* **8** (1972), 589–597.

[S3] Savage, L. J., The theory of statistical decision, *J. Amer. Statist. Assoc.* **45** (1950), 238–248.

[S4] Savage, L. J., "The Foundations of Statistics." Wiley, New York, 1954.

[S5] Sawaragi, Y., and Yoshikawa, T., Discrete-time Markovian decision processes with incomplete state observation, *Ann. Math. Statist.* **41** (1970), 78–86.

[S6] Scarf, H., The optimality of $(s, S)$ policies for the dynamic inventory problem, *Proc. 1st Stanford Symp. Mathematical Methods in the Social Sciences.* Stanford University Press, Stanford, California, 1960.

[S7] Schweitzer, P. J., Perturbation theory and finite Markov chains, *J. Appl. Prob.* **5** (1968), 401–413.

[S8] Shapley, L. S., Stochastic games, *Proc. Nat. Acad. Sci. U.S.A.* **39** (1953),

[S9] Sharpe, W., "Portfolio Theory and Capital Markets." McGraw-Hill, New York, 1970.

[S10] Simon, H. A., Dynamic programming under uncertainty with a quadratic criterion function, *Econometrica* **24** (1956), 74–81.

[S11] Sirjaev, A. N., Some new results in the theory of controlled random processes, *Selected Transl. Math. Statist. Probability* **8** (1970), 49–130.

[S12] Smallwood, R. D., The analysis of economic teaching strategies for a simple learning model, *J. Mathematical Psychology* **8** (1971), 285–301.

[S13] Smallwood, R. D., and Sondik, E. J., The optimal control of partially observable Markov processes over a finite horizon, *Operations Res.* **11** (1973), 1071–1088.

[S14] Sondik, E. J., "The Optimal Control of Partially Observable Markov Processes." Ph.D. Dissertation, Department of Engineering–Economic Systems, Stanford Univ. Stanford, California, June 1971.

[S15] Stein, G., and Saridis, G. N., A parameter adaptive control technique, *Automatica* **5** (1969), 731–739.

[S16] Stratonovich, R. L., On the theory of optimal control: sufficient coordinates, *Automat. Remote Control* **23** (1963), 847–854.

[S17] Strauch, R., Negative dynamic programming, *Ann. Math. Statist.* **37** (1966), 871–890.

[S18] Striebel, C. T., Sufficient statistics in the optimal control of stochastic systems, *J. Math. Anal. Appl.* **12** (1965), 576–592.

[S18a] Striebel, C., "Optimal Control of Discrete Time Stochastic Systems." Springer-Verlag, New York, 1975.

[S19] Sussman, R., "Optimal Control of Systems with Stochastic Disturbances." Electronics Research Lab., Univ. California, Berkeley, Rep. No. 63-20, November 1963.

[S20] Sworder, D. D., Bayes' controllers with memory for a linear system with jump parameters, *IEEE Trans. Automatic Control* **AC-17** (1972), 119–121.

[T1] Thau, F. E., and Witsenhausen, H. S., A comparison of closed-loop and open-loop optimum systems, *IEEE Trans. Automatic Control* **AC-11** (1966), 619–621.

[T2] Theil, H., Econometric models and welfare maximization, *Weltwirtsch. Arch.* **72** (1954), 60–83.

[T3] Tse, E., and Athans, M., Adaptive stochastic control for a class of linear systems, *IEEE Trans. Automatic Control* **AC-17** (1972), 38–52.

[T4] Tse, E., and Bar-Shalom, Y., An actively adaptive control for linear systems with random parameters via the dual control approach, *IEEE Trans. Automatic Control* **AC-18** (1973), 109–117.

[T5] Tse, E., Bar-Shalom, Y., and Meier, L., III, Wide-sense adaptive dual control for nonlinear stochastic systems, *IEEE Trans. Automatic Control* **AC-18** (1973), 98–108.

[V1] Vajda, S., Stochastic programming, *in* Abadie, J. (ed.), "Integer and Nonlinear Programming," Chapter 14. North-Holland Publ., Amsterdam, 1970.

[V2] Veinott, A. F., Jr., On finding optimal policies in discrete dynamic programming with no discounting, *Ann. Math. Statist.* **37** (1966), 1284–1294.

[V3] Veinott, A. F., Jr., The status of mathematical inventory theory, *Management Sci.* **12** (1966), 745–777.

[V4] Veinott, A. F., Jr., Discrete dynamic programming with sensitive discount optimality criteria, *Ann. Math. Statist.* **40** (1969), 1635–1660.

[V5] Veinott, A. F., and Wagner, H. M., Computing optimal $(s, S)$ policies, *Management Sci.* **11** (1965), 525–552.

[V6] Vershik, A. M., Some characteristic properties of Gaussian stochastic processes, *Theory Probability Appl.*, **9** (1964), 353–356.

[W1] Wald, A., "Sequential Analysis." Wiley, New York, 1947.

[W2] White, D. J., Dynamic programming, Markov chains, and the method of successive approximations, *J. Math. Anal. Appl.* **6** (1963), 373–376.

[W3] White, D., "Dynamic Programming." Holden-Day, San Francisco, California, 1969.

[W4] Witsenhausen, H. S., "Minimax Control of Uncertain Systems." Ph.D. Dissertation, Department of Electrical Engineering, MIT, Cambridge, Massachusetts, May 1966.

[W5] Witsenhausen, H. S., Inequalities for the performance of suboptimal uncertain systems, *Automatica* **5** (1969), 507–512.

[W6] Witsenhausen, H. S., On performance bounds for uncertain systems, *SIAM J. Control* **8** (1970), 55–89.

[W7] Witsenhausen, H. S., Separation of estimation and control for discrete-time systems, *Proc. IEEE* **59** (1971), 1557–1566.

[W8] Witsenhausen, H. S., A standard form for sequential stochastic control, *Math. Systems Theory* **7** (1973), 5–11.

[W9] Wittenmark, B., Stochastic adaptive control methods: a survey, *Internat. J. Control* **21** (1975), 705–730.

[W10] Wong, P. J., and Luenberger, D. G., Reducing the memory requirements of dynamic programming, *Operations Res.* **16** (1968), 1115–1125.

[W11] Wonham, W. M., On the separation theorem of stochastic control, *SIAM J. Control* **6** (1968), 312–326.

[W12] Wonham, W. M., On a matrix Riccati equation of stochastic control, *SIAM J. Control* **6** (1968), 681–697.

[Z1] Zangwill, W. I., "Nonlinear Programming: A Unified Approach." Prentice Hall, Englewood Cliffs, New Jersey, 1969.

# Index